AF411175

Advances in the Experimentation and Numerical Modeling of Material Joining Processes

Advances in the Experimentation and Numerical Modeling of Material Joining Processes

Editor

Raul D. S. G. Campilho

Basel • Beijing • Wuhan • Barcelona • Belgrade • Novi Sad • Cluj • Manchester

Editor
Raul D. S. G. Campilho
Mechanical Engineering
INEGI
Porto
Portugal

Editorial Office
MDPI
St. Alban-Anlage 66
4052 Basel, Switzerland

This is a reprint of articles from the Special Issue published online in the open access journal *Materials* (ISSN 1996-1944) (available at: www.mdpi.com/journal/materials/special_issues/ modeling_joining).

For citation purposes, cite each article independently as indicated on the article page online and as indicated below:

Lastname, A.A.; Lastname, B.B. Article Title. *Journal Name* **Year**, *Volume Number*, Page Range.

ISBN 978-3-03928-592-1 (Hbk)
ISBN 978-3-03928-591-4 (PDF)
doi.org/10.3390/books978-3-03928-591-4

Contents

About the Editor

Raul D. S. G. Campilho

Raul Duarte Salgueiral Gomes Campilho is an Assistant Professor with Habilitation at ISEP—School of Engineering and a researcher in INEGI. His main research interests are composite materials, material-joining processes, numerical modelling, and industrial process improvement by automation and robotics.

Preface

The field of material joining processes plays a fundamental role in the design of structures with diverse applications. This MDPI reprint, titled "Advances in the Experimentation and Numerical Modelling of Material Joining Processes", addresses the latest developments and insights in this field of research. The reprint includes a wide range of topics, ranging from adhesive-bonded joints to welding, friction stir welding, hybrid joining, and miscellaneous joining techniques.

The opening chapter serves as the editorial to the Special Issue that originated this reprint. Emphasis was given to the main limitations, possible improvements, active research lines, and prospects of material joining processes. Chapters 2 to 10 focus on adhesive-bonded joints and welding techniques. From exploring the tensile behaviour of adhesive single-lap joints to retaining mechanical properties in gas metal arc welded joints, these chapters offer a detailed examination of the experimental and numerical aspects of material joining. Friction stir welding, a cutting-edge technique with diverse applications, is the subject of chapters 11 to 17. These chapters describe the current progress on this topic, with both research and review works, including the microstructure and mechanical property analysis of dissimilar alloy-joining and the effects of tool tilt angles on friction stir welded joints. The exploration of corrosion behaviour in friction stir welded aluminium alloys and probeless friction stir spot welding further enriches the understanding of this evolving field. Chapters 18 and 19 present hybrid joining solutions, namely laser-arc welding, and weld-bonded joints. The concluding chapters, 20 to 22, cover a diverse array of topics. These include infiltration and solidification processes for interpenetrating phase composites, the microstructures and properties of steel-based alloys via additive manufacturing, and concrete blocks with wood chips used in structural walls in seismic areas.

As editor of this reprint, I am grateful to the esteemed contributors who have shared their expertise, insights, and innovative research. The collaborative effort of researchers from around the world has resulted in a rich compilation that not only reflects the current state of the field but also provides guidelines for future exploration and innovation. I hope that this reprint serves as a valuable resource for researchers, engineers, and practitioners, enabling a deeper understanding of material-joining techniques and inspiring further advancements in this crucial field.

Raul D. S. G. Campilho
Editor

Editorial

Advances in the Experimentation and Numerical Modeling of Material Joining Processes

Raul D. S. G. Campilho [1,2]

1 ISEP—School of Engineering, Polytechnic of Porto, R. Dr. António Bernardino de Almeida, 431, 4200-072 Porto, Portugal; rds@isep.ipp.pt; Tel.: +351-939-526-892
2 INEGI—Institute of Science and Innovation in Mechanical and Industrial Engineering, Pólo FEUP, Rua Dr. Roberto Frias, 400, 4200-465 Porto, Portugal

check for
updates

Citation: Campilho, R.D.S.G. Advances in the Experimentation and Numerical Modeling of Material Joining Processes. *Materials* **2024**, *17*, 130. https://doi.org/10.3390/ma17010130

Received: 15 December 2023
Accepted: 25 December 2023
Published: 27 December 2023

Material joining processes are a critical factor in engineering structures since they influence such structures' structural integrity, performance, and longevity [1]. While significant progress has been made in the improvement of diverse joining processes [2], current design limitations require further consideration and innovative solutions [3]. This Special Issue, entitled "Advances in the Experimentation and Numerical Modeling of Material Joining Processes", published by MDPI, aims to explore the current state of these processes, identify design limitations, propose avenues for improvement, and discuss ongoing research and prospects. Both experimental and numerical simulation approaches are examined to provide a complete perspective of the field. This Special Issue addresses a wide scope of joining methods, ranging from traditional techniques such as welding and brazing to modern innovations such as friction stir welding (FSW) and laser welding. Historically, mechanical fastening and welding techniques have been indispensable in material joining. Mechanical fastening stands out as a prevalent approach in the industry [4]. Initially employed for metal joining, this method has gained acceptance for use in the production of composite materials [5]. Screws are a robust and cost-effective non-permanent solution, although they result in negative effects with regard to weight and aesthetics [6]. Permanent fastening mechanisms, such as snap-fits, are incorporated directly into components during molding, increasing durability and eliminating the risk of loosening over time [7]. Variants of traditional welding include extrusion welding and gas–metal–arc welding, which are in prevalent use [8,9]. FSW, a solid-state joining process, excels in joining difficult-to-weld materials through use of traditional methods [10]. Laser welding presents a high-energy-density approach, allowing for precise control and minimal thermal distortion [11]. Hybrid joining techniques, e.g., combining laser and arc welding, are gaining prominence [12]. Adhesive bonding has gained increasing importance in engineering, with it offering weight reduction and improved fatigue performance [13]. Associated with all joining processes, the integration of numerical modeling is crucial for a deeper understanding of complex phenomena. In fact, computational simulations help to predict strength and analyze thermal distributions, stress concentrations, and material flow during joining processes [14,15].

Current design limitations

- Traditional screws and fasteners add weight and may pose size constraints when lightweight and compact designs are crucial, such as in the aerospace and electronics industries [16]. In some cases, mechanical fasteners may be sensitive to temperature variations, leading to thermal expansion or contraction, which can affect joint integrity [17].
- Welding and brazing often face challenges related to heat-affected zones (HAZs), residual stresses, and microstructural alterations [18]. Despite extensive industrial application, strength and material compatibility limitations exist [19].

- FSW has benefits such as reduced heat input and minimized distortion [20] but is limited by tool wear, process stability, and adaptability to different materials, which pose significant design challenges [21].
- Laser welding, as a high-energy-density approach, presents a unique set of design limitations, such as optimal parameter selection to balance efficiency and thermal stress [22]. Hybrid laser welding introduces additional complexities, namely the interactions between different heat sources, equipment synchronization and coordination, and welded material compatibility with different processes [23].
- Adhesive bonding faces challenges related to joint strength, durability, loading rate, and process optimization [24], added to the non-existing standardized adhesive joint characterization protocols. The comparability of results becomes difficult due to variations in testing methods and reporting parameters [25]. Furthermore, the absence of a standardized predictive approach presents challenges in the design process [26].
- In the numerical modeling domain, challenges persist in accurately representing the material joining processes, due to recurring simplifying assumptions that limit the predictive capabilities [27]. Computational simulations require refinement and need to account for all associated phenomena to enhance predictive accuracy [28].

Improving existing design processes

- In mechanical fastening, increasing durability and fatigue strength are essential for applications under repetitive loading or dynamic stress [29]. Industrially, simplifying the assembly process by designing fastening mechanisms that are user-friendly and require minimal manual effort is a crucial aspect to enhance productivity [30].
- Welding and brazing can benefit from studies on the optimization of process parameters, including heat input and post-weld cooling rates, with the aim of mitigating challenges associated with HAZ and respective microstructural alterations, and residual stresses in the welded parts [31]. Integrating advanced materials and alloys can increase joint strength and material compatibility, leading to robust and reliable designs [32].
- In FSW, tool wear and process stability issues are crucial for process improvement [33]. Research efforts are being made to develop wear-resistant tool materials and optimize tool geometries [34]. An improved understanding of material flow during FSW may improve control over the process, ensuring high-quality joints [35].
- Laser welding process improvement involves balancing the process parameters to achieve optimal efficiency and minimal thermal stress [36]. Under this scope, recent advancements in laser technology and real-time monitoring systems have contributed to efficient joining [37]. Hybrid joining between laser welding and other techniques requires understanding process interactions and the exploitation of synergistic effects [38].
- Adhesive bonding can benefit from adhesives with improved strength, durability, and tailoring at the molecular level, offering given requirements for specific applications [39]. Comprehensive design guidelines should be developed [40] together with robust methodologies for adhesive joint characterization, either experimental or numerical [13,41]. Advances in non-destructive evaluation techniques can provide real-time monitoring of joint behavior under operation [42].
- Numerical modeling plays a fundamental role in improving design processes by replicating complex phenomena and reducing the testing requirements [43,44]. Refining computational models and removing any geometrical and material simplifying to accurately simulate material behavior, thermal distributions, and stress concentrations has led to improvements in predictive capabilities [45].

Current lines of research

- In mechanical fastening, recent investigations have used titanium alloys or high-strength stainless steels to enhance fastener properties, such as improving strength, reducing weight, and increasing corrosion resistance [46]. Lightweight materials and

- composites as fasteners are an ongoing topic of research to achieve a high strength-to-weight ratio [47].
- Traditional welding topics include advanced filler materials, gas compositions, and shielding atmospheres, with the aim of enhancing the high robustness and quality of welds [48]. Research works on innovative heat management methods, including pulsed welding and tailored energy inputs, show potential to minimize thermal distortions [49].
- Research on FSW is focused on understanding and mitigating tool wear [50], namely through the proposal of innovative tool geometries and adaptive control strategies, which extend tool life and maintain process stability [51]. New tool materials to improve process behavior under varying process conditions are also addressed [52].
- Laser welding investigations focus on optimizing beam characteristics, such as wavelength and focus, to increase precision and control [53]. Real-time monitoring and feedback systems are being developed to dynamically adjust the process parameters, ensuring consistent and high-quality joints [54]. The integration of robotics into hybrid laser welding systems is gaining attention for use in increasing production rates [55].
- Adhesive bonding research is currently exploring advanced and tailored adhesive formulations with improved mechanical properties, durability, and adaptability to diverse substrates [56]. Novel surface treatments and pre-bonding techniques [57], and adhesive testing procedures under extreme conditions, such as high temperatures or corrosive environments, are other active fields of research [58].
- Computational models are becoming increasingly sophisticated, accurately simulating the complex dynamics of material joining processes [59]. Multiphysics simulations, involving thermal, mechanical, and metallurgical aspects, are innovations in numerical modeling research [60].

Prospects

- Artificial intelligence (AI) and machine learning should bring new insights through integration into numerical modeling to achieve enhanced accuracy and the optimization of the different processes [61]. AI algorithms can learn from vast experimental datasets, enabling the identification of patterns and correlations to improve predictions [62].
- Innovative materials can bring new developments in material joining. The improvement of advanced alloys, composites, and smart materials with tailored properties is ongoing [63], potentially leading to new approaches. Self-healing materials, for example, offer the possibility for joints to self-repair and lifespan extension [64].
- Hybrid joining methods are a promising avenue for the future. The synergistic combination of welding with adhesive bonding [65] or friction stir welding with adhesive bonding [66] presents opportunities to overcome individual method limitations.
- Further investigations of meshless methods and the eXtended Finite Element Method (XFEM) are anticipated to aid in overcoming conventional FEM techniques in evaluation crack initiation and propagation [67]. These methods present alternative approaches to deliver a more precise and efficient representation of joint behavior [68,69].
- Environmental sustainability is a critical aspect in material joining. Efforts are underway toward developing eco-friendly joining processes, which aid in reducing either energy consumption or the environmental impact [70]. Green welding technologies, such as friction stir scribe welding and cold metal transfer welding, aim to reduce the carbon footprint of such processes [71].

In conclusion, this Special Issue, "Advances in the Experimentation and Numerical Modeling of Material Joining Processes", aims to highlight the state of the art of the most diverse material joining processes, including respective challenges and opportunities. In parallel, this editorial paper has also aided in highlighting the main limitations, suggested improvements to be made, active research lines, and prospects. It has thus been shown that, currently, significant limitations exist but that these limitations also find a response in the current topics being addressed worldwide by academicians and industrialists. Moreover, innovative works are identified that address current topics such as AI, smart materials, hybrid joining, numerical methods, and green technology, which may bring significant improvements in joining processes and help to progress the performance of structures even further.

Conflicts of Interest: The author declares no conflicts of interest.

References

1. Messler, R.W. *Joining of Materials and Structures: From Pragmatic Process to Enabling Technology*; Butterworth-Heinemann: Oxford, UK, 2004; pp. 1–790.
2. Çam, G.; İpekoğlu, G. Recent developments in joining of aluminum alloys. *Int. J. Adv. Manuf. Technol.* **2017**, *91*, 1851–1866. [CrossRef]
3. Pramanik, A.; Basak, A.; Dong, Y.; Sarker, P.; Uddin, M.; Littlefair, G.; Dixit, A.; Chattopadhyaya, S. Joining of carbon fibre reinforced polymer (CFRP) composites and aluminium alloys—A review. *Compos. Part A Appl. Sci. Manuf.* **2017**, *101*, 1–29. [CrossRef]
4. Mehmanparast, A.; Lotfian, S.; Vipin, S.P. A review of challenges and opportunities associated with bolted flange connections in the offshore wind industry. *Metals* **2020**, *10*, 732. [CrossRef]
5. Kahraman, A.D.; Kahraman, F. Mechanical fastening methods of polymer-based composites. *Int. J. Adv. Nat. Sci. Eng. Res.* **2023**, *7*, 234–239.
6. Katsivalis, I.; Thomsen, O.T.; Feih, S.; Achintha, M. Strength evaluation and failure prediction of bolted and adhesive glass/steel joints. *Glass Struct. Eng.* **2018**, *3*, 183–196. [CrossRef]
7. Troughton, M.J. (Ed.) Chapter 18—Mechanical fastening. In *Handbook of Plastics Joining*, 2nd ed.; William Andrew Publishing: Boston, MA, USA, 2009; pp. 175–201.
8. Tawfeek, T. Study the influence of gas metal arc welding parameters on the weld metal and heat affected zone microstructures of low carbon steel. *Int. J. Eng. Technol.* **2017**, *9*, 2013–2019. [CrossRef]
9. Lu, X.; Zhang, C.; Zhao, G.; Guan, Y.; Chen, L.; Gao, A. State-of-the-art of extrusion welding and proposal of a method to evaluate quantitatively welding quality during three-dimensional extrusion process. *Mater. Des.* **2016**, *89*, 737–748. [CrossRef]
10. Ahmed, M.M.; Seleman, M.M.E.-S.; Fydrych, D.; Gürel, Ç. Review on friction stir welding of dissimilar magnesium and aluminum alloys: Scientometric analysis and strategies for achieving high-quality joints. *J. Magnes. Alloys*, 2023, *in press*.
11. Sadeghian, A.; Iqbal, N. A review on dissimilar laser welding of steel-copper, steel-aluminum, aluminum-copper, and steel-nickel for electric vehicle battery manufacturing. *Opt. Laser Technol.* **2022**, *146*, 107595. [CrossRef]
12. Liu, Q.; Wu, D.; Wang, Q.; Zhang, P.; Yan, H.; Sun, T.; Li, R. Progress and perspectives of joints defects of laser-arc hybrid welding: A review. *Int. J. Adv. Manuf. Technol.* 2023, *in press*. [CrossRef]
13. Campilho, R.D.S.G. Design of adhesive bonded joints. *Processes* **2023**, *11*, 3369. [CrossRef]
14. Faria, R.V.F.; Campilho, R.D.S.G.; Gonçalves, P.J.A.; Rocha, R.J.B. Advanced numerical methods for the strength prediction of hybrid adhesively bonded T-peel joints. *J. Adhes* **2022**, *98*, 154–179. [CrossRef]
15. Leśniak, D.; Libura, W.; Leszczyńska-Madej, B.; Bogusz, M.; Madura, J.; Płonka, B.; Boczkal, S.; Jurczak, H. FEM numerical and experimental work on extrusion welding of 7021 aluminum alloy. *Materials* **2023**, *16*, 5817. [CrossRef] [PubMed]
16. Borba, N.Z.; Kötter, B.; Fiedler, B.; dos Santos, J.F.; Amancio-Filho, S. Mechanical integrity of friction-riveted joints for aircraft applications. *Compos. Struct.* **2020**, *232*, 111542. [CrossRef]
17. den Otter, C.; Maljaars, J. Preload loss of stainless steel bolts in aluminium plated slip resistant connections. *Thin-Walled Struct.* **2020**, *157*, 106984. [CrossRef]
18. Messler, R.W., Jr. *A Practical Guide to Welding Solutions: Overcoming Technical and Material-Specific Issues*; John Wiley & Sons: Hoboken, NJ, USA, 2019.
19. Verma, R.P.; Pandey, K.; András, K.; Khargotra, R.; Singh, T. Difficulties and redressal in joining of aluminium alloys by GMA and GTA welding: A review. *J. Mater. Res. Technol.* **2023**, *23*, 2576–2586. [CrossRef]
20. Choudhary, A.K.; Jain, R. Fundamentals of friction stir welding, its application, and advancements. In *Welding Technology*; Davim, J.P., Ed.; Springer: Berlin, Germany, 2021; pp. 41–90.
21. Saha, R.; Biswas, P. Current status and development of external energy-assisted friction stir welding processes: A review. *Weld. World* **2022**, *66*, 577–609. [CrossRef]
22. Li, Y.; Xiong, M.; He, Y.; Xiong, J.; Tian, X.; Mativenga, P. Multi-objective optimization of laser welding process parameters: The trade-offs between energy consumption and welding quality. *Opt. Laser Technol.* **2022**, *149*, 107861. [CrossRef]
23. Suder, W.; Ganguly, S.; Williams, S.; Yudodibroto, B.Y. Penetration and mixing of filler wire in hybrid laser welding. *J. Mater. Process. Technol.* **2021**, *291*, 117040. [CrossRef]
24. Viana, G.M.S.O.; Costa, M.; Banea, M.D.; da Silva, L.F.M. A review on the temperature and moisture degradation of adhesive joints. *Proc. Inst. Mech. Eng. L* **2017**, *231*, 488–501. [CrossRef]
25. Tserpes, K.; Barroso-Caro, A.; Carraro, P.A.; Beber, V.C.; Floros, I.; Gamon, W.; Kozłowski, M.; Santandrea, F.; Shahverdi, M.; Skejić, D.; et al. A review on failure theories and simulation models for adhesive joints. *J. Adhes* **2022**, *98*, 1855–1915. [CrossRef]
26. Campilho, R.D.S.G. *Strength Prediction of Adhesively-Bonded Joints*; CRC Press: Boca Raton, FL, USA, 2017.

27. Lambiase, F.; Genna, S.; Kant, R. A procedure for calibration and validation of FE modelling of laser-assisted metal to polymer direct joining. *Opt. Laser Technol.* **2018**, *98*, 363–372. [CrossRef]
28. AbuShanab, W.S.; Abd Elaziz, M.; Ghandourah, E.I.; Moustafa, E.B.; Elsheikh, A.H. A new fine-tuned random vector functional link model using Hunger games search optimizer for modeling friction stir welding process of polymeric materials. *J. Mater. Res. Technol.* **2021**, *14*, 1482–1493. [CrossRef]
29. Kraemer, F.; Stähler, M.; Klein, M.; Oechsner, M. Influence of lubrication systems on the fatigue strength of bolted joints. *Appl. Sci.* **2022**, *12*, 2778. [CrossRef]
30. Zhu, L.; Bouzid, A.-H.; Hong, J. A method to reduce the number of assembly tightening passes in bolted flange joints. *J. Manuf. Sci. Eng.* **2021**, *143*, 121006. [CrossRef]
31. Vajari, K.K.; Saffar, S. Investigation of gas tungsten arc welding parameters on residual stress of heat affected zone in inconel X750 super alloy welding using finite element method. *Int. J. Mech. Ind. Eng.* **2023**, *17*, 347–357.
32. Bayock, F.N.; Kah, P.; Salminen, A.; Belinga, M.; Yang, X. Feasibility study of welding dissimilar advanced and ultra high strength steels. *Rev. Adv. Mater. Sci.* **2020**, *59*, 54–66. [CrossRef]
33. Emamian, S.S.; Awang, M.; Yusof, F.; Sheikholeslam, M.; Mehrpouya, M. Improving the friction stir welding tool life for joining the metal matrix composites. *Int. J. Adv. Manuf. Technol.* **2020**, *106*, 3217–3227. [CrossRef]
34. Maji, P.; Karmakar, R.; Nath, R.K.; Paul, P. An overview on friction stir welding/processing tools. *Mater Today Proc.* **2022**, *58*, 57–64. [CrossRef]
35. Chen, G.; Li, H.; Wang, G.; Guo, Z.; Zhang, S.; Dai, Q.; Wang, X.; Zhang, G.; Shi, Q. Effects of pin thread on the in-process material flow behavior during friction stir welding: A computational fluid dynamics study. *Int. J. Mach. Tools Manuf.* **2018**, *124*, 12–21. [CrossRef]
36. Nisar, S.; Noor, A.; Shah, A.; Siddiqui, U.; Khan, S.Z. Optimization of process parameters for laser welding of A5083 aluminium alloy. *Opt. Laser Technol.* **2023**, *163*, 109435. [CrossRef]
37. Kim, H.; Nam, K.; Oh, S.; Ki, H. Deep-learning-based real-time monitoring of full-penetration laser keyhole welding by using the synchronized coaxial observation method. *J. Manuf. Process.* **2021**, *68*, 1018–1030. [CrossRef]
38. Zhang, C.; Gao, M.; Zeng, X. Influences of synergy effect between laser and arc on laser-arc hybrid welding of aluminum alloys. *Opt. Laser Technol.* **2019**, *120*, 105766. [CrossRef]
39. Bovone, G.; Dudaryeva, O.Y.; Marco-Dufort, B.; Tibbitt, M.W. Engineering hydrogel adhesion for biomedical applications via chemical design of the junction. *ACS Biomater. Sci. Eng.* **2021**, *7*, 4048–4076. [CrossRef] [PubMed]
40. Brunner, A.J. 1—Investigating the performance of adhesively-bonded composite joints: Standards, test protocols, and experimental design. In *Fatigue and Fracture of Adhesively-Bonded Composite Joints*; Vassilopoulos, A.P., Ed.; Woodhead Publishing: Cambridge, UK, 2015; pp. 3–42.
41. Peres, L.M.C.; Arnaud, M.F.T.D.; Silva, A.F.M.V.; Campilho, R.D.S.G.; Machado, J.J.M.; Marques, E.A.S.; dos Reis, M.Q.; Da Silva, L.F.M. Geometry and adhesive optimization of single-lap adhesive joints under impact. *J. Adhes* **2022**, *98*, 677–703. [CrossRef]
42. Broughton, W.R. *Durability Performance of Adhesive Joints*; 1368–6550; National Physical Laboratory: Teddington, UK, 2023.
43. Singh, V.P.; Patel, S.K.; Ranjan, A.; Kuriachen, B. Recent research progress in solid state friction-stir welding of aluminium–magnesium alloys: A critical review. *J. Mater. Res. Technol.* **2020**, *9*, 6217–6256. [CrossRef]
44. Ferreira, J.A.M.; Campilho, R.D.S.G.; Cardoso, M.G.; Silva, F.J.G. Numerical simulation of adhesively-bonded T-stiffeners by cohesive zone models. *Procedia Manuf.* **2020**, *51*, 870–877. [CrossRef]
45. de Sousa, C.C.R.G.; Campilho, R.D.S.G.; Marques, E.A.S.; Costa, M.; da Silva, L.F.M. Overview of different strength prediction techniques for single lap bonded joints. *Proc. Inst. Mech. Eng. L* **2017**, *231*, 210–223. [CrossRef]
46. Li, D.; Uy, B.; Wang, J.; Song, Y. Assessment of titanium alloy bolts for structural applications. *Steel Compos. Struct.* **2022**, *42*, 553–568.
47. Yao, C.; Qi, Z.; Chen, W. Lightweight and high-strength interference-fit composite joint reinforced by thermoplastic composite fastener. *Thin-Walled Struct.* **2022**, *179*, 109471. [CrossRef]
48. Chaturvedi, M.; Vendan, S.A. *Advanced Welding Techniques*; Springer: Berlin, Germany, 2022.
49. Chludzinski, M.; Dos Santos, R.; Churiaque, C.; Fernández-Vidal, S.; Ortega-Iguña, M.; Sánchez-Amaya, J. Pulsed laser butt welding of AISI 1005 steel thin plates. *Opt. Laser Technol.* **2021**, *134*, 106583. [CrossRef]
50. Hasieber, M.; Wenz, F.; Grätzel, M.; Lenard, J.A.; Matthes, S.; Bergmann, J.P. A systematic analysis of maximum tolerable tool wear in friction stir welding. *Weld. World* **2023**, *67*, 325–339. [CrossRef]
51. Meng, X.; Huang, Y.; Cao, J.; Shen, J.; dos Santos, J.F. Recent progress on control strategies for inherent issues in friction stir welding. *Prog. Mater. Sci.* **2021**, *115*, 100706. [CrossRef]
52. Soori, M.; Asmael, M.; Solyalı, D. Recent development in friction stir welding process. *SAE Int. J. Mater. Manuf.* **2021**, *14*, 63–80. [CrossRef] [PubMed]
53. Meng, Y.; Fu, J.; Gong, M.; Zhang, S.; Gao, M.; Chen, H. Laser dissimilar welding of Al/Mg lap-joint with Ti interlayer through optimized 8-shaped beam oscillation. *Opt. Laser Technol.* **2023**, *162*, 109304. [CrossRef]
54. Aminzadeh, A.; Sattarpanah Karganroudi, S.; Meiabadi, M.S.; Mohan, D.G.; Ba, K. A survey of process monitoring using computer-aided inspection in laser-welded blanks of light metals based on the digital twins concept. *Quantum Beam Sci.* **2022**, *6*, 19. [CrossRef]

55. Wu, J.; Zhang, C.; Lian, K.; Cao, H.; Li, C. Carbon emission modeling and mechanical properties of laser, arc and laser–arc hybrid welded aluminum alloy joints. *J. Clean. Prod.* **2022**, *378*, 134437. [CrossRef]
56. Gonçalves, F.A.M.M.; Santos, M.; Cernadas, T.; Alves, P.; Ferreira, P. Influence of fillers on epoxy resins properties: A review. *J. Mater. Sci.* **2022**, *57*, 15183–15212. [CrossRef]
57. Mandolfino, C.; Cassettari, L.; Pizzorni, M.; Saccaro, S.; Lertora, E. A response surface methodology approach to improve adhesive bonding of pulsed laser treated CFRP composites. *Polymers* **2022**, *15*, 121. [CrossRef]
58. Fan, Y.; Zhao, G.; Liu, Z.; Yu, J.; Ma, S.; Zhao, X.; Han, L.; Zhang, M. An experimental study on the mechanical performance of BFRP-Al adhesive joints subjected to salt solutions at elevated temperature. *J. Adhes Sci. Technol.* **2023**, *37*, 1937–1957. [CrossRef]
59. Kupfer, R.; Köhler, D.; Römisch, D.; Wituschek, S.; Ewenz, L.; Kalich, J.; Weiß, D.; Sadeghian, B.; Busch, M.; Krüger, J. Clinching of aluminum materials–methods for the continuous characterization of process, microstructure and properties. *J. Adv. Join. Process.* **2022**, *5*, 100108. [CrossRef]
60. Li, Y.; Yun, Z.; Su, C.; Zhou, X.; Wu, C. Multiphase and multi-physical simulation of open keyhole plasma arc welding. *Case Stud. Therm. Eng.* **2023**, *41*, 102611. [CrossRef]
61. Fernandes, P.H.E.; Silva, G.C.; Pitz, D.B.; Schnelle, M.; Koschek, K.; Nagel, C.; Beber, V.C. Data-driven, physics-based, or both: Fatigue prediction of structural adhesive joints by artificial intelligence. *Appl. Mech.* **2023**, *4*, 334–355. [CrossRef]
62. Samaitis, V.; Yilmaz, B.; Jasiuniene, E. Adhesive bond quality classification using machine learning algorithms based on ultrasonic pulse-echo immersion data. *J. Sound Vib.* **2023**, *546*, 117457. [CrossRef]
63. Rakesh, P.K. Joining processes of advanced materials. In *Joining Processes for Dissimilar and Advanced Materials*; Rakesh, P., Davim, J.P., Eds.; Elsevier: Amsterdam, The Netherlands, 2022; pp. 1–11.
64. Paladugu, S.R.M.; Sreekanth, P.R.; Sahu, S.K.; Naresh, K.; Karthick, S.A.; Venkateshwaran, N.; Ramoni, M.; Mensah, R.A.; Das, O.; Shanmugam, R. A Comprehensive review of self-healing polymer, metal, and ceramic matrix composites and their modeling aspects for aerospace applications. *Materials* **2022**, *15*, 8521. [CrossRef]
65. Wang, Y.; Chai, P.; Ma, H.; Zhang, Y. Characteristics and strength of hybrid friction stir welding and adhesive bonding lap joints for AA2024-T3 aluminium alloy. *J. Adhes* **2022**, *98*, 325–347. [CrossRef]
66. Veligotskyi, N.; Guzanová, A.; Janoško, E. Analysis of the possibilities of joining thin-walled metallic and composite materials. *Mach. Technol. Mater.* **2023**, *17*, 158–163.
67. Gonçalves, D.C.; Sánchez-Arce, I.J.; Ramalho, L.D.C.; Campilho, R.D.S.G.; Belinha, J. Fracture propagation based on meshless method and energy release rate criterion extended to the Double Cantilever Beam adhesive joint test. *Theor. Appl. Fract. Mech.* **2022**, *122*, 103577. [CrossRef]
68. Resende, R.F.P.; Resende, B.F.P.; Sanchez Arce, I.J.; Ramalho, L.D.C.; Campilho, R.D.S.G.; Belinha, J. Elasto-plastic adhesive joint design approach by a radial point interpolation meshless method. *J. Adhes* **2022**, *98*, 2396–2422. [CrossRef]
69. Djebbar, S.C.; Madani, K.; El Ajrami, M.; Houari, A.; Kaddouri, N.; Mokhtari, M.; Feaugas, X.; Campilho, R.D.S.G. Substrate geometry effect on the strength of repaired plates: Combined XFEM and CZM approach. *Int. J. Adhes Adhes* **2022**, *119*, 103252. [CrossRef]
70. Mulcahy, K.R.; Kilpatrick, A.F.; Harper, G.D.; Walton, A.; Abbott, A.P. Debondable adhesives and their use in recycling. *Green Chem.* **2022**, *24*, 36–61. [CrossRef]
71. Narender, M.; Ajay Kumar, V.; Manoj, A. A review on latest trends in derived technologies of friction stir welding. In *Intelligent Manufacturing and Energy Sustainability: Proceedings of ICIMES 2020*; Reddy, A.N.R., Marla, D., Favorskaya, M.N., Satapathy, S.C., Eds.; Springer: Berlin, Germany, 2021; pp. 239–249.

materials

Article

Tensile Behaviour of Double- and Triple-Adhesive Single Lap Joints Made with Spot Epoxy and Double-Sided Adhesive Tape

Przemysław Golewski

Department of Solid Mechanics, Faculty of Civil Engineering and Architecture, Lublin University of Technology, Nadbystrzycka 38, 20-618 Lublin, Poland; p.golewski@pollub.pl

Abstract: Dual adhesives are mainly used to increase the strength of single lap joints (SLJs) by reducing the stress concentration at its ends. However, they can also be used to design the characteristics of the joint so that its operation and failure occur in several stages. This paper presents the results of uniaxial tensile tests for dual-adhesive and triple-adhesive SLJs. The adherends were made of aluminum and glass fiber-reinforced polymer (GFRP) composite. For dual-adhesive SLJs, 10 epoxies and 1.6 mm thick double-sided adhesive tape were used. The stiffest (Epidian 53 (100 g) + "PAC" hardener (80 g)) and most elastic (Scotch-Weld 2216 B/A Translucent) joints were determined, which were then used in a triple-adhesive joint with the same double-sided adhesive tape. Circular holes of different diameters from 8 mm to 20 mm were made in the double-sided adhesive tape, which were filled with liquid epoxy adhesive by injection after the adherends were joined. By using the double-sided adhesive tape, the geometry of the epoxy joints was perfect, free of spews, and had a constant thickness. The effect of the spot epoxy joint diameters and the arrangement of stiff and elastic joints in the SLJs were analyzed using digital image correlation (DIC).

Keywords: dual-adhesive; triple-adhesive; double-sided adhesive tape; DIC; SLJ

Citation: Golewski, P. Tensile Behaviour of Double- and Triple-Adhesive Single Lap Joints Made with Spot Epoxy and Double-Sided Adhesive Tape. *Materials* **2022**, *15*, 7855. https://doi.org/10.3390/ma15217855

Academic Editor: Raul D.S.G. Campilho

Received: 12 October 2022
Accepted: 6 November 2022
Published: 7 November 2022

Publisher's Note: MDPI stays neutral with regard to jurisdictional claims in published maps and institutional affiliations.

1. Introduction

The continuous effort to reduce the weight of structures while increasing their strength has led to the increasing use of polymer–matrix composites (PMC). Despite their many advantages, such as a favorable weight-to-strength ratio, resistance to weathering and the ability to design complex shapes, PMC composites with an epoxy matrix are more difficult to join compared to metals. Metal structures can use techniques such as welding [1], spot welding [2], friction stir welding [3], and clinching [4]. In contrast, epoxy–matrix PMC composites are insulators, do not deform plastically, and suffer thermal degradation when exposed to high temperatures. Therefore, the range of joining techniques available for use with such composites is limited. The universal joining technique is mechanical: removable in the form of bolts [5] and non-removable in the form of rivets [6,7]. Both of these techniques can be used to join composites to other materials, but mechanical joints usually require a hole, and stress concentrations occur, which is dangerous with the brittle linear characteristics of the PMC material [8]. An alternative is adhesive joints [9–11], which do not require interference with the structure of the parts to be joined. Their characteristic feature is that the stress concentrates at the ends of the lap [12], which makes this type of joint sensitive to the technology of their manufacture. In order to increase their strength, a number of techniques are used, such as edge chamfering [13], surface modification [14], modification of the spew geometry [15,16], stepped reduction of the thickness of the parts to be joined to eliminate eccentricity [17,18], and a different length-to-width ratio of the joint [19]. These types of techniques would often be difficult to apply in industrial environments due to the cost of additional machining. Therefore, another technique may be the use of hybrid connections [20–24] in which both a mechanical joint (rivet, bolt, weld) and an adhesive joint are used. Due to differences in stiffness and the occurrence of plastic deformation, the operation of such hybrids can be two-stage, which increases the safety of the structure, for example, if one of the joints is damaged. However, the problem of significant interference with the structure of the materials being joined still remains. Therefore, one of the varieties of hybrid joints is dual-adhesive or mixed adhesive joints, which were first proposed in 1966 by Raphael [25]. This solution is based on the use of layers of different adhesives in one joint. These layers have different stiffnesses. The less rigid layer is placed at the ends of the lap where the greatest deformation of the adhesive occurs. The more rigid layer is inside the overlap, where the deformation is much smaller. Dual adhesive joints are used to increase the strength of the connection. However, they can also be used to design safe, multi-stage joints and to design operating characteristics. Current work in the field of dual-adhesive joints is being carried out towards strengthening tensile lap joints [26] and flexural beams [27]. In the work [28], the authors split lap length in the ratio of 1/3 elastic adhesive, 2/3 rigid adhesive, and the effect of the surface roughness and the lap length were also analyzed. Single adhesive joints were also studied, but the combination of two adhesives produced an increase in strength for all the analysed models. Dual adhesive joints are also subjected to the effects of low and high temperatures [29], and ageing in humid environments [30] and under dynamic loads [31,32], as they have significant potential for application in the automotive, aerospace, and aviation industries, replacing joints with only one adhesive layer. The problem in dual-adhesive joints featuring two liquid adhesives is the method of separating them. Silicone [33,34] and nylon fiber [35] separators are used for this purpose, or alternatively they are allowed to mix with each other at the interface [36]. On the other hand, the authors in the works [37,38] proposed using double-sided adhesive tape as a separator. Such solutions mean that after using only one type of liquid epoxy adhesive, the joint will have two-stage characteristics. In the first stage, the rigid epoxy joint is damaged, while in the next stage, the load is transferred by the double-sided adhesive tape, making the total energy of the damage very high. These types of joints can also be called safety joints. However, until now, no one has used this technique in the configuration: double-sided adhesive tape + 2 types of liquid epoxy adhesives.

The research was divided into two stages. The first stage involved dual-adhesive joints, where an adhesive tape was used in addition to an epoxy joint. The goal of the first stage was to test 10 types of resin + hardener configurations to determine which had the lowest and highest stiffness. This stage was followed by triple-adhesive specimens. In both cases, two materials were combined: 1 mm thick GFRP laminate and 2 mm thick aluminum. The double-sided adhesive tape used in both cases had a thickness of 1.6 mm. Five samples per batch were used, hence the total number of samples was 100.

2. Materials and Methods

In this study, 10 epoxy compositions were used in the form of commercial adhesives and resins along with hardeners.

Two commercial epoxy adhesives were used:

- Distal, produced by Libella, and
- Scotch-Weld 2216 B/A Translucent, produced by 3M.

Other compositions of epoxy adhesives resulted from the use of two resins: Epidian 5 and Epidian 53, as well as two hardeners: PAC and Z1, produced by Ciech, Sarzyna.

PAC is a brown-colored liquid polyamide hardener with relatively low reactivity. The weight ratio of PAC hardener to resin can be varied over a wide range to regulate the reaction rate and properties of the cured material. Compositions richer in PAC hardener are more flexible and resistant to impact, but less hard and less resistant to elevated temperatures. Therefore, this study examined its effect in the range from 60 g to 100 g per 100 g of Epidian 5 resin and from 50 g to 80 g per 100 g of Epidian 53 resin.

Z1 is a thin, colorless liquid. Thanks to its low viscosity, it is used for making epoxy composites by the following processes: hand lamination, vacuum lamination, and repair. The Z1 hardener is very efficient and forms a fairly stiff composite with improved temperature resistance. The addition of Z1 hardener to Epidian 5 and Epidian 53 resins is strictly defined and is, respectively, 10 g and 12 g per 100 g of resin.

Data concerning work life, time to handling strength, and curing time for all epoxy compositions are shown in Table 1.

Table 1. Data for epoxy compositions.

	Work Life (24 °C) [min]	Time to Handling Strength (24 °C) [h]	Curing Time (24 °C) [Days]
Distal Classic	120	24	7
Scotch-Weld 2216 B/A Translucent	120	12–16	30
Epidian 5 + PAC	180	6–8	7–14
Epidian 53 + PAC	180	6–8	7–14
Epidian 5 + Z1	35	6–8	7–14
Epidian 53 + Z1	35	6–8	7–14

The application of the epoxy adhesives was simplified by the use of double-sided adhesive tape. In this case, 3M VHB 5962 tape with a thickness of 1.6 mm was used.

The research was divided into two stages. In the first stage of the study, the dimensions of the lap were 30 mm × 30 mm (Figure 1). The diameter of the epoxy joint was 20 mm. By using double-sided adhesive tape, high repeatability of the geometry could be achieved.

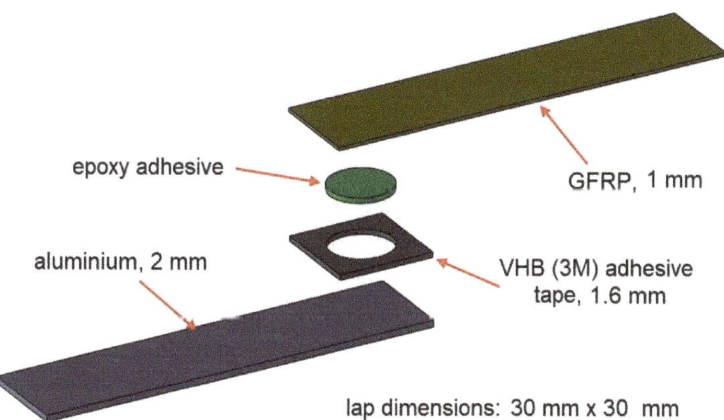

Figure 1. Structure of the dual-adhesive sample for the first stage of the study.

The aim of the first stage was to analyze 10 types of joints differing in the composition of the spot epoxy bond with respect to the maximum strength and stiffness. The following epoxy compositions were used in the samples:

1. Distal Classic (resin 100 g + hardener 74 g),
2. 3M Scotch-Weld 2216 B/A Translucent (resin 100 g + hardener 100 g),
3. Epidian 5 (100 g) + hardener "Z1" (10 g),
4. Epidian 53 (100 g) + hardener "Z1" (12 g),
5. Epidian 5 (100 g) + hardener "PAC" (60 g),
6. Epidian 5 (100 g) + hardener "PAC" (80 g),
7. Epidian 5 (100 g) + hardener "PAC" (100 g),
8. Epidian 53 (100 g) + hardener "PAC" (50 g),
9. Epidian 53 (100 g) + hardener "PAC" (65 g),

10. Epidian 53 (100 g) + hardener "PAC" (80 g).

There were 5 samples in each batch. Hence, the total number of samples in the first stage of testing was 50. As a result of the analysis, the two most extreme adhesive compositions were determined: the stiffest and the most elastic, which were then used in the second stage of testing.

In the second stage of the study, the lap was much more complicated. The length of the lap was 60 mm, with the same width as in the first stage. There were three spot bonds in each lap (Figure 2):

— one central spot bond, whose diameters were: 8 mm, 11 mm, 14 mm, 16 mm, and 20 mm,

— two spot bonds placed at the ends of the lap with the same diameter of 11 mm in each batch.

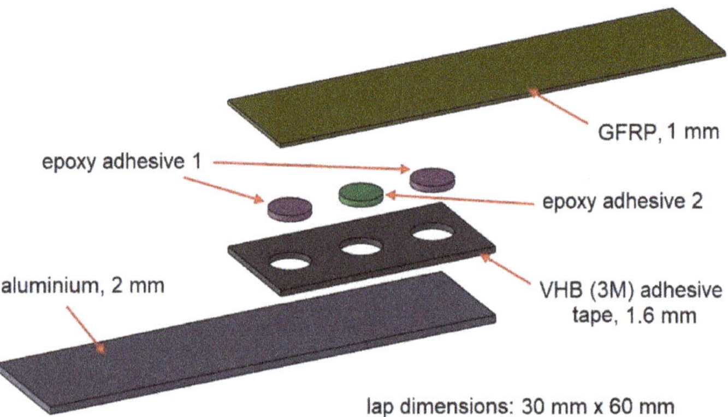

Figure 2. Structure of the triple-adhesive sample in the second stage of study.

Making this type of spot bond was possible through the use of 3M VHB 5962 double-sided adhesive tape with a thickness of 1.6 mm. The samples were divided into two groups: "S" and "E". In the "S" type specimens, the outer joints were stiff, while the central joint was elastic. In the "E" type specimens, the situation was the opposite. A diagram of all types of specimens in the second stage of testing is shown in Figure 3. There were 5 specimens in each batch, so the total number of specimens in the second stage of testing was 50.

The purpose of such changes in geometry and swapping the location of the stiff and elastic bonds was to analyze the effect on the SLJs characteristics, maximum force, and failure energy.

In both the first and second stages of the study, the adherends were made of two different materials: 2 mm-thick AW-6060 aluminum and 1 mm-thick GFRP composite, EPGC 201 from Izo-Erg, Gliwice, Poland. The surfaces were cleaned with Loctite 7061 before bonding.

Manufacturing accurate geometries in both the first and second stages of the study was made possible by the use of double-sided adhesive tape, in which holes were cut with die-cutters before the tape was applied to the substrate. After the holes were made, the tape was then applied to one of the adherends. After removing the protective foil, the second adherend was applied. In this way, the samples were ready for the epoxy adhesive to be applied, which was done by injection into the empty spaces that were formed between the composite and aluminum. Complete samples from the first "S" batches can be seen in Figure 4.

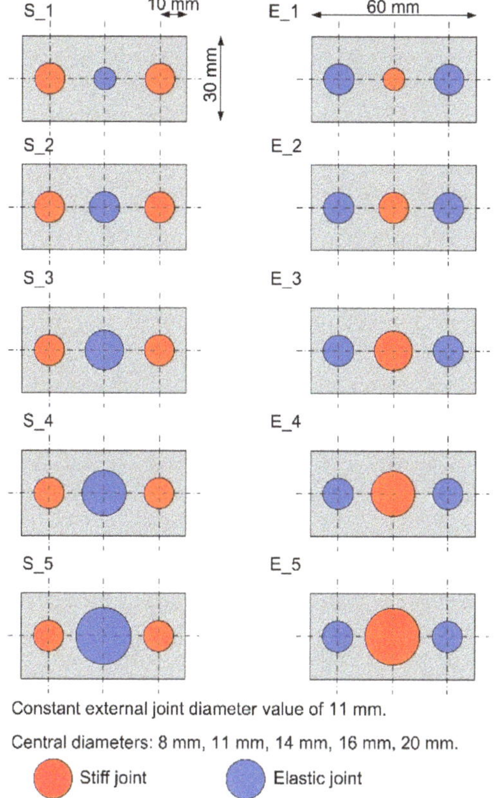

Figure 3. Schemes of lap construction in triple-adhesive samples.

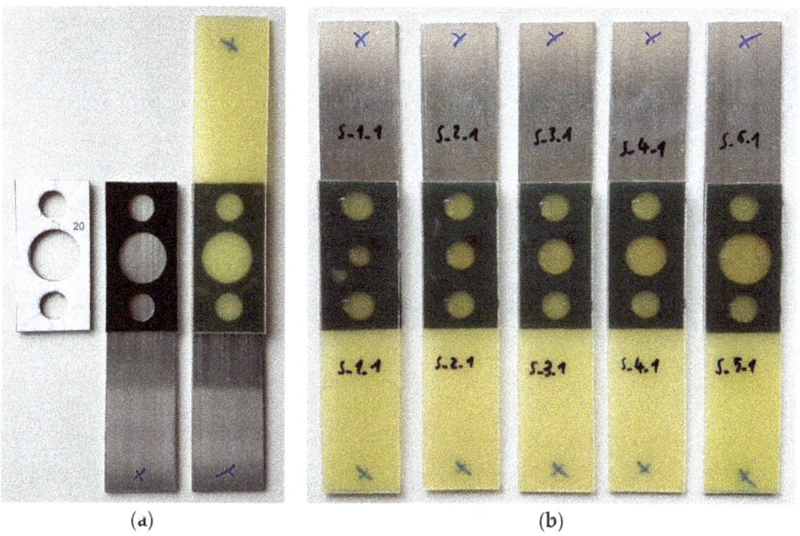

Figure 4. Triple-adhesive samples: (**a**) stages of fabrication; (**b**) completed samples.

The samples were cured at room temperature for one week. After this time, they were subjected to static uniaxial tensile tests on an MTS 100 kN testing machine. According to ASTM D1002-10(2019), the loading rate was 1.27 mm/min. The results from the testing machine were processed in the Diadem software (2019).

An Aramis digital image correlation (DIC) system was used to observe strains on the surfaces of the laps. The results from the Aramis system were processing in the GOM Correlate software (2020).

3. Results and Discussion

The research was divided into two stages, and 50 lap specimens were tested in each stage. In the first stage of testing, the behavior of 10 batches of dual-adhesive joints was determined to identify the two extreme types of joints, taking into account their stiffness.

In the second stage of testing, the behavior of 10 batches of joints differing in both the geometry and location of stiff and elastic joints was determined. The DIC system was also used to observe deformations in this stage.

3.1. Results for the First Stage of the Study

Dual-adhesive specimens with one epoxy bond and a double-sided adhesive tape were used in the first stage of the work. These are materials that differ in both shear strength and stiffness. The failure of this type of joint proceeds in two stages. First, the rigid epoxy joint is damaged. This process occurs quite quickly for displacement in the range of 0.2–1 mm. In the next stage, the double-sided adhesive tape begins to work, which is damaged over a relatively long displacement, as can be seen in the graph in Figure 5b. However, for this part of the study, the first stage of the work and the comparison of 10 epoxy adhesives are the most interesting parts.

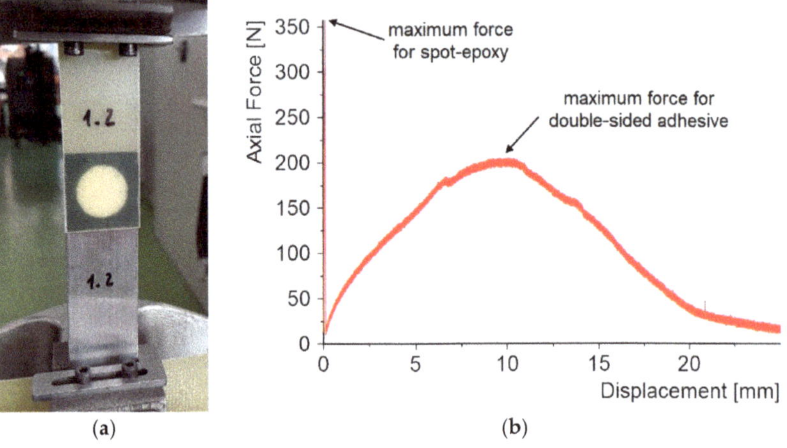

Figure 5. Static tensile test for dual-adhesive type specimens: (**a**) specimen during testing, (**b**) force–displacement diagram (specimen 1_2).

Figure 6 shows the results for the first stage of joint operation.

Figure 6a shows the maximum force values for 10 batches of dual-adhesive specimens. The maximum force value in each case was obtained for the epoxy joint. The highest maximum force (2172 N) was achieved for joints using Scotch-Weld 2216 adhesive (batch 2). The lowest value (588 N) was achieved by joints using Distal Classic adhesive (batch 1) and Epidian 5 (100 g) + "Z1" hardener (10 g) (batch 3).

Figure 6b was created by connecting point (0, 0) with the points corresponding to the maximum force and displacement, i.e., the failure of the epoxy joint, whose characteristics are linear. The type of resin or adhesive used, as well as the proportion of hardener used, have a significant influence on the strength and stiffness. The Scotch-Weld adhesive had the lowest straight angle and the lowest stiffness. The adhesive with Epidian 53 resin (100 g) and PAC hardener (80 g) had the highest stiffness. These two compositions were selected for further testing, for the formation of triple-adhesive joints.

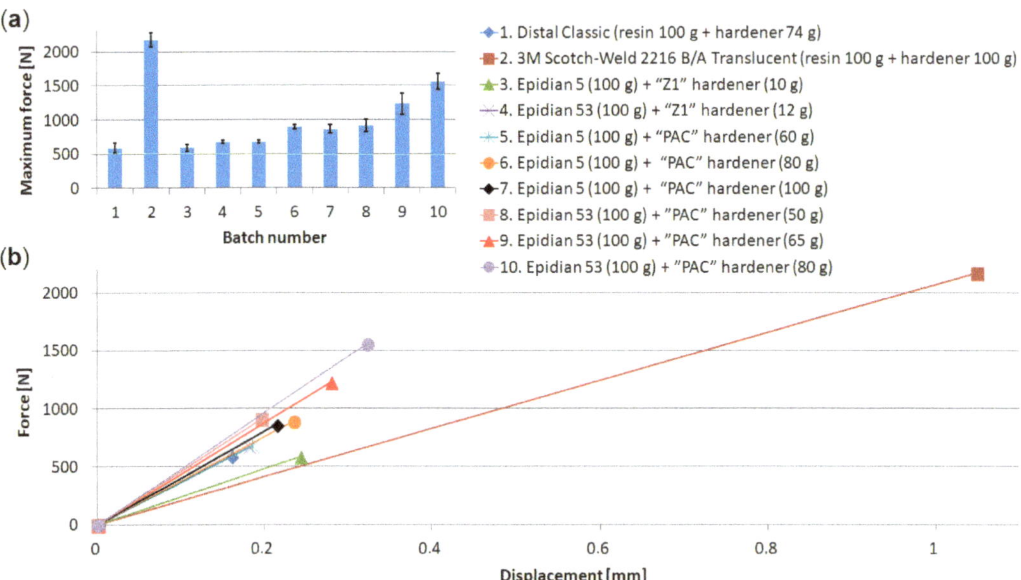

Figure 6. Summary of results of the first stage of joint operation: (**a**) summary of maximum forces, (**b**) comparison of stiffness for epoxy joints.

3.2. Results for the Second Stage of the Study

The adhesive joint is stressed to varying levels during the tensile test of the SLJ. The adhesive material closer to the end of the lap deforms to a greater extent than in the axis of the lap. Therefore, the method of placement of the stiff joint and the elastic joint is of great importance in this case. In classic dual-adhesive joints, the elastic joint is always placed in the end zone of the lap, while the stiff joint is placed closer to the axis of the lap. This arrangement helps increase the strength of the whole connection. However, the question arises: what will the force–displacement diagrams look like when the two adhesives are swapped in place? In addition, the characteristics can be formed by different percentages of the two adhesives: stiff and elastic. Therefore, different diameters of the central joint were considered.

The force–displacement graphs for the "S" and "E" groups for the same geometries are summarized in Figures 7–11. The results for the first batches with the smallest central diameter of 8 mm are shown in Figure 7. With two rigid joints on the outside, we can double the strength compared to the "E" specimens. However, the "E" specimens reach a force of about 1 kN twice, which increases the safety of the joint.

By increasing the diameter of the central joint to 11 mm (Figure 8) in the "E" type specimens there is a decrease in the maximum force, while the force for the second stage is slightly increased. In the "S" type specimens, we encounter the phenomenon that the maximum force is reached in the second stage of the work.

Another increase in the diameter of the central joint to 14 mm (Figure 9) greatly benefits the "S" type specimens. The strength for the first and second stages increases, but the value for the first stage still dominates. In the "E" type specimens, large variations in the results for the second stage were noted, which could be due to possible errors in the preparation of the samples.

The optimal solution in terms of safety was obtained for the "S" type specimens with a central joint diameter of 16 mm (Figure 10). It comes down to the equalization of forces for the first and second stages. In the "E" type specimens, the reverse process occurs. The force value for the first stage increases, while it decreases for the second stage.

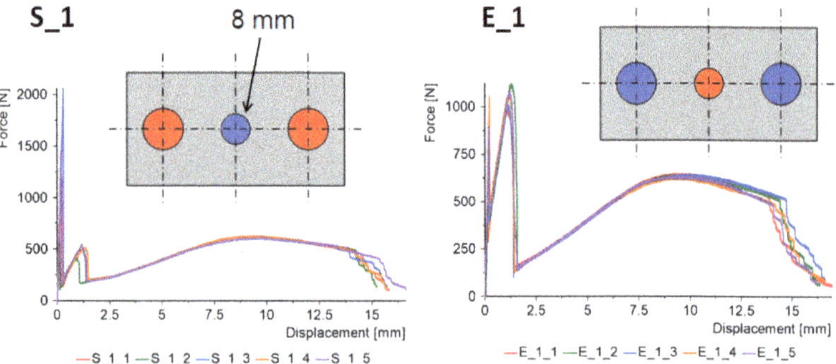

Figure 7. Triple-adhesive samples, batch 1.

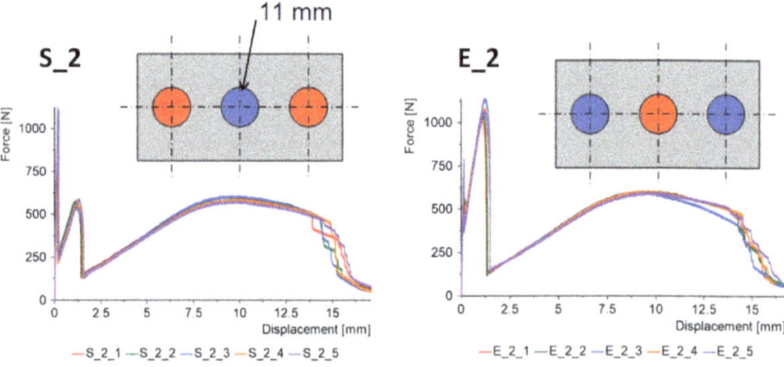

Figure 8. Triple-adhesive samples, batch 2.

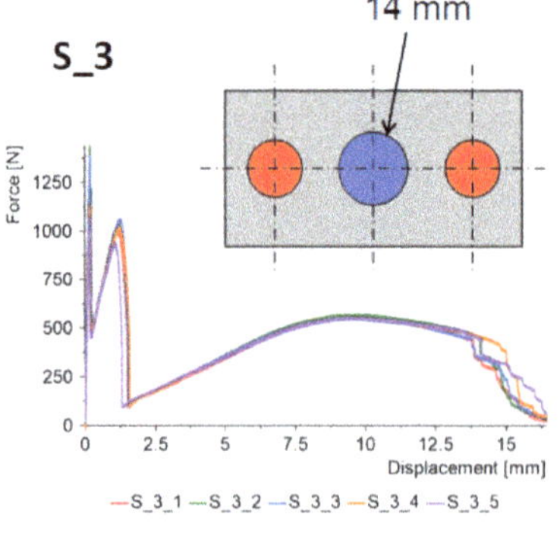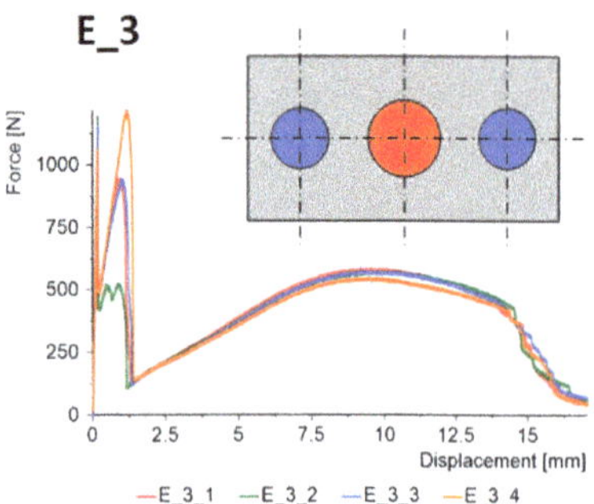

Figure 9. Triple-adhesive samples, batch 3.

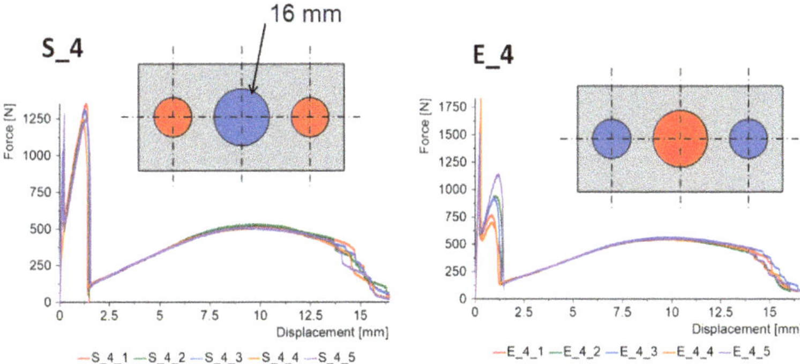

Figure 10. Triple-adhesive samples, batch 4.

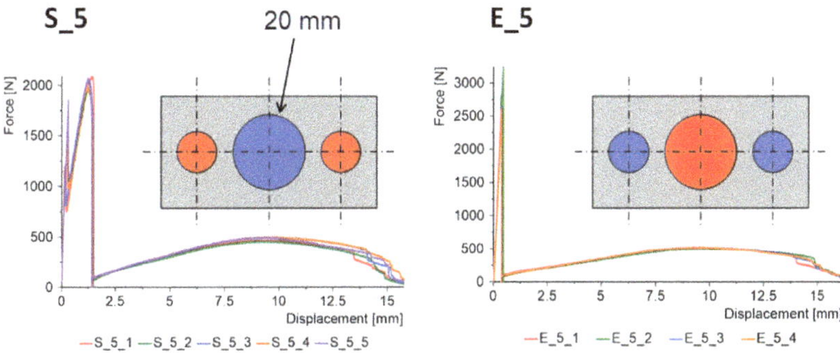

Figure 11. Triple-adhesive samples, batch 5.

The last batches had a central joint diameter of 20 mm (Figure 11). This increase again benefits the "S" type specimens. In this case, the maximum forces for the second stage were obtained at 2 kN. A very interesting phenomenon occurs for the "E" type specimens. One of the working stages of the epoxy joint disappears, which is a negative outcome. Instead, the maximum force is obtained in relation to all the samples tested, with an average for the 5 samples of 2.8 kN.

So far, nothing has been mentioned about the third stage, which concerns the work of the double-sided adhesive tape. In this stage, nothing special happens; similar force values are obtained for all samples.

A common feature of each batch is the occurrence of a third stage, which is related to the work of the double-sided adhesive tape. This stage begins after the damage of all epoxy joints, which occurs for a displacement of about 1.3 mm and continues until a displacement of 15 mm, which corresponds to half the length of the lap. This stage is characterized by a maximum force in the range of 480 N–620 N.

Figures 12 and 13 show images of one specimen from each batch after the uniaxial tensile testing. For both the "S" and "E" group specimens, there is a cohesive damage pattern for double-sided VHB tape. On the other hand, the damage of the spot epoxy bonds is adhesive in nature. The epoxy bonds remain on the composite adherend, which indicates much greater adhesion to the polymer material than to the aluminum. None of the adherends were damaged by exceeding the tensile strength during testing.

The use of two different epoxy bonds and a double-sided adhesive tape caused the SLJ operation to proceed in three stages, as shown in Figure 14.

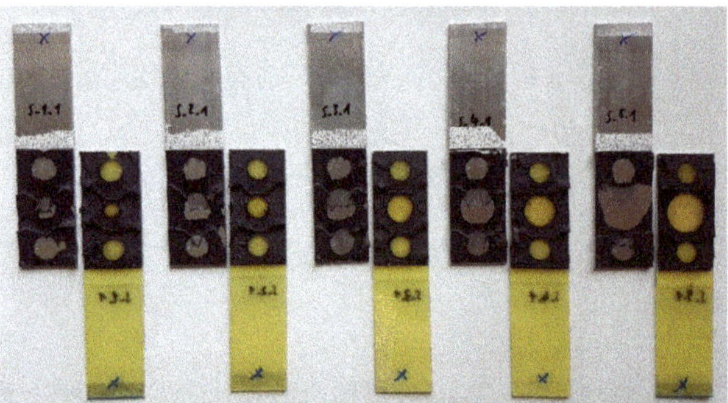

Figure 12. Triple-adhesive group "S" samples after damage.

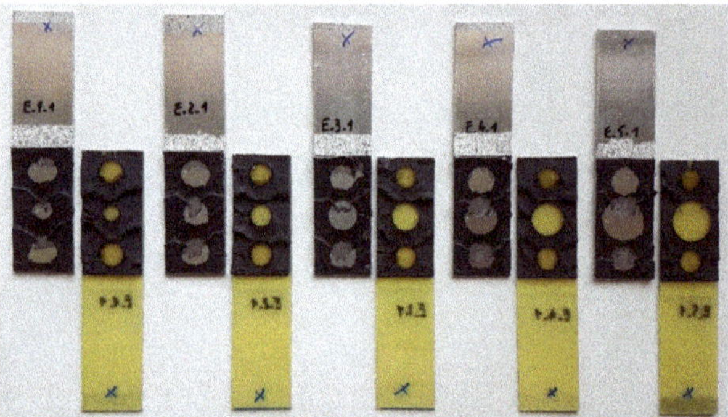

Figure 13. Triple-adhesive group "E" samples after damage.

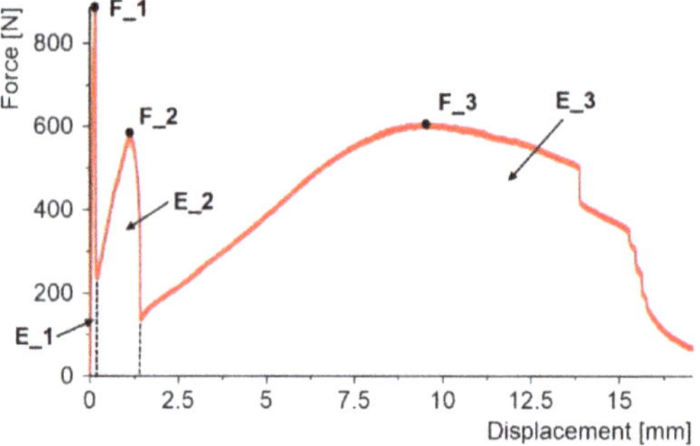

Figure 14. Determination of maximum forces and energies for triple-adhesive specimens.

In the first stage, the load is mainly carried by the stiffer epoxy bond; after its damage, the force does not drop sharply to zero, but the load is taken up by the epoxy bond or bonds with less stiffness. Only after all the epoxy bonds have been damaged is there another drop in force, and the load is carried by the last joint in the form of the VHB tape.

In each of the above-mentioned stages, the maximum value of the force can be distinguished: F_1, F_2, and F_3. The field under the force–displacement diagram corresponds to the energy required to damage each joint and can be divided as indicated in Figure 14: E_1, E_2, and E_3.

Figure 15 collects the average values of the forces for each of the batches, along with a classification of the three stages of operation of the joints. The exception is the "E_5" batch, where the second stage associated with elastic bond work is absent.

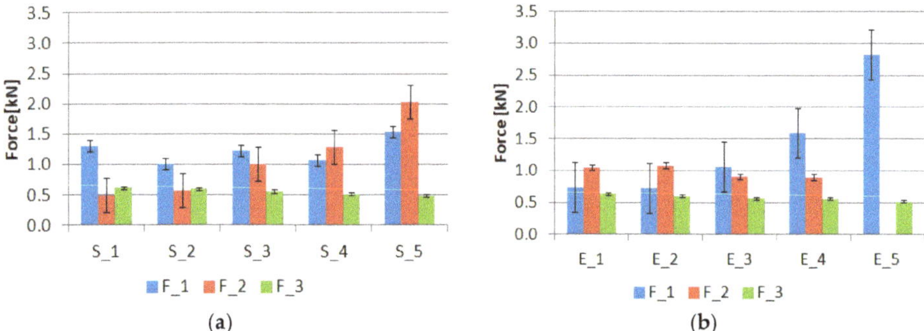

Figure 15. Summary of maximum forces for samples of group: (a) "S" and (b) "E".

Analyzing the samples of the "S" group, which had two external rigid bonds, the following cases can occur depending on the diameter of the internal elastic bond, shown in Table 2.

Table 2. Relation between forces for "S" group.

Case Number	Relation between Forces	Internal Bond Diameter [mm]
1	$F\_1 > F\_2$	8, 11, 14
2	$F\_2 > F\_1$	16, 20
3	$F\_3 > F\_2$	8, 11
4	$F\_2 > F\_3$	14, 16, 20

Thus, there is a strong possibility of shaping the characteristics by making small changes in the value of the elastic bond area. Considering the two extreme cases of S_1 and S_5, when the internal diameter is changed from 8 mm to 20 mm, an increase of 18.5% is obtained for the F_1 force, while an increase of 314% is obtained for the F_2 force.

The value of the F_3 force is reduced by 22.5% due to a decrease in the area of the VHB tape.

Analyzing the samples of the "E" group, which had two external flexible welds, it is also possible to distinguish three stages of operation, except for the E_5 batch in which the second stage is absent. Depending on the internal diameter of the stiff bond, the following cases can occur, shown in Table 3.

Table 3. Relation between forces for "E" group.

Case Number	Relation between Forces	Internal Bond Diameter [mm]
1	$F\_1 > F\_2$	14, 16
2	$F\_2 > F\_1$	8, 11
3	$F\_3 > F\_2$	no such case
4	$F\_2 > F\_3$	8, 11, 14, 16

Thus, using the reverse application of epoxy adhesives, we get different results in almost every batch. The exceptions are the batches S_3 and E_3, for which the F_1, F_2, and F_3 forces contribute, respectively:

- S_3-(F_1 = 1.23 kN, F_2 = 1.02 kN, F_3 = 0.56 kN)
- E_3-(F_1 = 1.06 kN, F_2 = 0.91 kN, F_3 = 0.57 kN)

For these two batches, there is the same order of forces, F_1 > F_2 > F_3, with the values for the forces in the same stages differing by 16%, 12%, and 1.75%. Therefore, it is possible to formulate the hypothesis for the presented triple-adhesive model that there is an arrangement of the lap geometry in which swapping the places of the stiff and elastic bonds does not cause changes in the values of the forces in the work stages. However, this issue requires further research.

The area under the force–displacement diagram indicates the energy required to damage the individual joints. A summary of the average values for the individual "S" and "E" batches is shown in Figure 16.

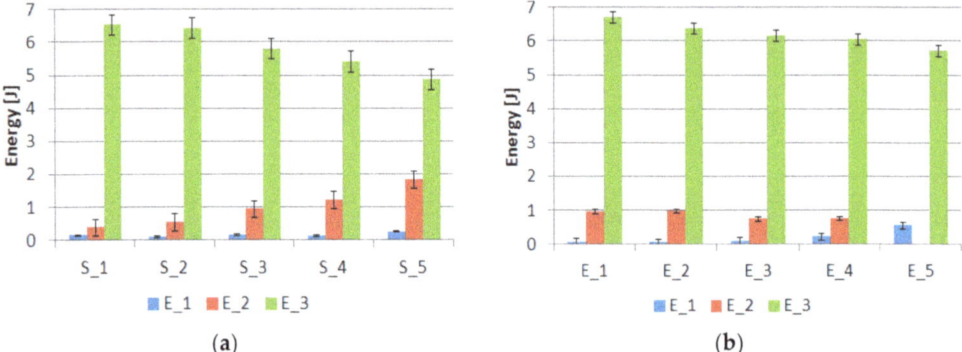

(a)   (b)

Figure 16. Summary of damage energy for samples of group: (a) "S" and (b) "E".

In each case, the energy level associated with the damage of the double-sided adhesive tape dominates. In both groups of "S" and "E" specimens, there is also the effect of decreasing the energy associated with the work of the double-sided tape as the diameter of the internal bond increases. Changing the internal bond from 8 mm to 20 mm results in an energy decrease of 25.5% for the "S" group samples and 14.7% for the "E" group samples. The smaller decrease for the "E" group samples is due to the fact that there is an internal stiff bond, which relieves the double-sided tape much more. The lowest energy, regardless of the type of sample, is characterized by the first stage of joint work. The energy achieved is in the range of 0.08 J–0.57 J. The energy for the second stage is in the range of 0.4 J–1.84 J. The larger effect on the energy in the second stage of the change in diameter can be observed for the samples of the "S" group.

The final stage of the analysis concerns strains on the surface of the composite lap. The purpose of the observation was to determine the strains in the second stages of the joint operation. Figures 17–21 show the results for one sample from each "S" and "E" series and for only the second stage of the epoxy joint operation.

The principal strains are shown for all samples. In addition, the scale range is limited from 0 to 0.2. Hence, the blue fields indicate small or zero strains, while the red fields indicate strains close to 0.2 or greater. Thus, analysis using the Aramis system, which uses digital image correlation, can provide a tool for observing the operation of spot adhesion joints.

In specimens S_1_1 and E_1_1 (Figure 17), both zones of concentrated strain and zones where the values are close to zero are clearly visible. The near-zero values appear as a result of the delamination of the epoxy spot bond from the aluminum adherend. There is a lack of strain symmetry in specimen S_1_1, and the load is mainly transmitted through the upper rigid weld. It is likely that the central flexible bond and the lower stiff bond have partially failed, as can be observed in the force–displacement diagram in the form of upslopes. In specimen E_1_1, there is no doubt about the presence of strain symmetry. This is the stage at which the two elastic bonds are no longer carrying the load, as evidenced by the near circular blue strain fields. The load is carried by the middle part of the specimen, where the stiff bond is working together with the double-sided VHB tape.

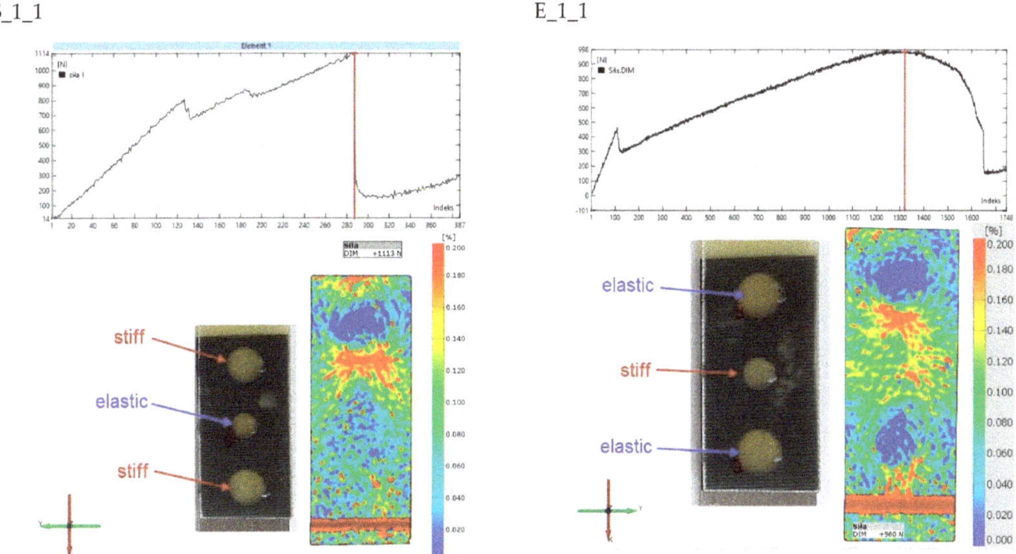

Figure 17. Comparison of principal strain maps for "S" and "E" batch 1 specimens for maximum force in the second stage of operation.

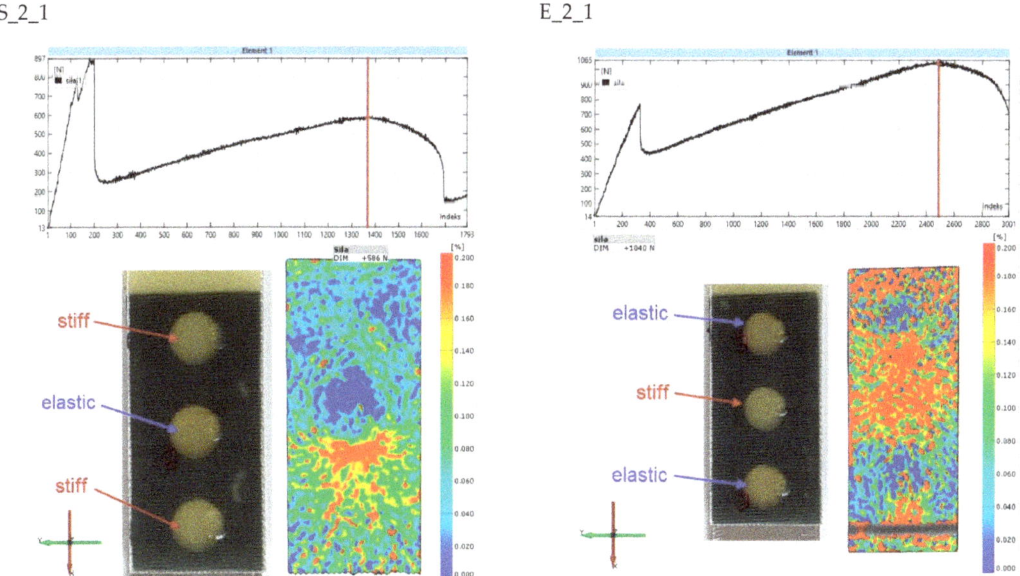

Figure 18. Comparison of principal strain maps for "S" and "E" batch 2 specimens for maximum force in the second stage of operation.

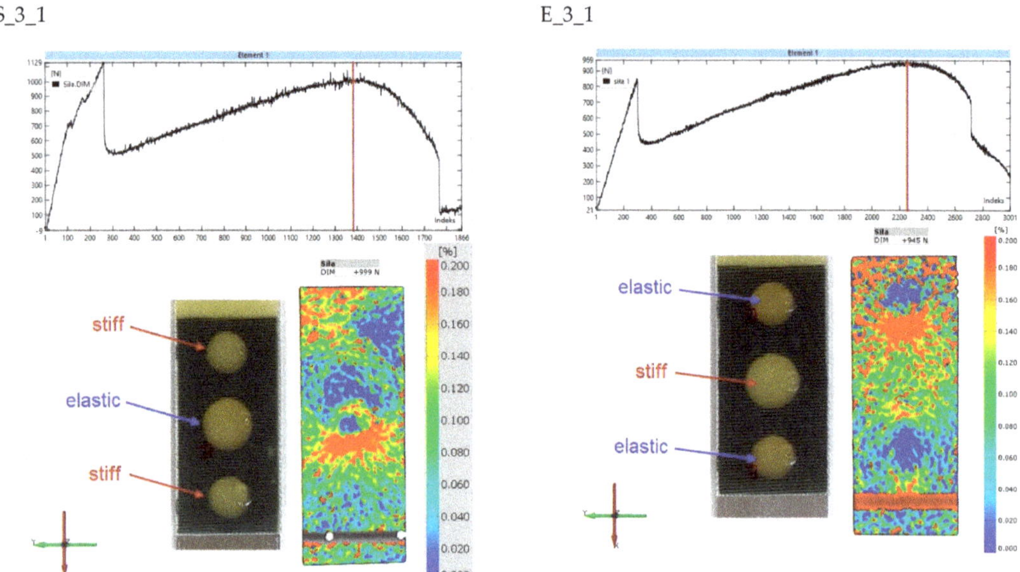

Figure 19. Comparison of principal strain maps for "S" and "E" batch 3 specimens for maximum force in the second stage of operation.

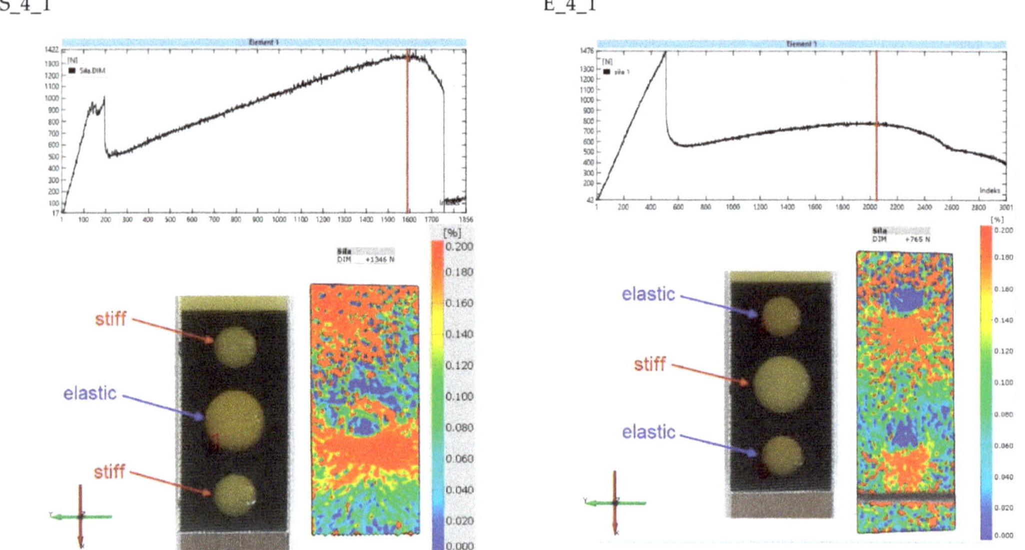

Figure 20. Comparison of principal strain maps for "S" and "E" batch 4 specimens for maximum force in the second stage of operation.

S_5_1   E_5_1

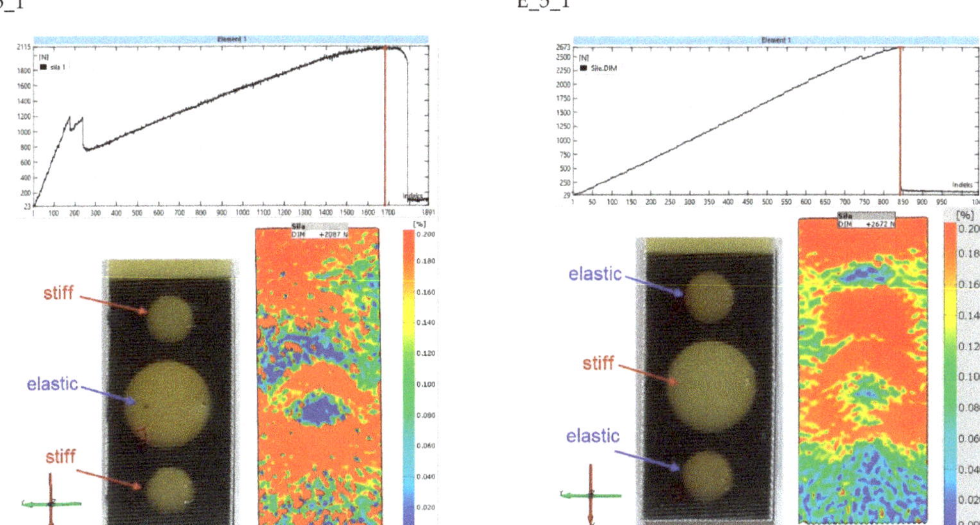

Figure 21. Comparison of principal strain maps for "S" and "E" batch 5 specimens for maximum force in the second stage of operation.

Increasing the internal diameter of the elastic bond to 11 mm (specimen S_2_1, Figure 17), changes the load transfer arrangement. Concentrations occur at the location of the above-mentioned joint, while the upper stiff bond has been damaged. The strain field for specimen E_2_1 is more difficult to analyze, but it can be said with certainty that the load is mainly transferred through the central rigid bond.

The original strain distribution occurs for specimen S_3_1 (Figure 19). In this case, the load in the second stage is carried mainly by the central elastic joint. However, there is a lack of symmetry with respect to the vertical axis, as the upper right part of the lap is unloaded, while there are significant strains on the left side. This may indicate defects in the fabrication of the joint. For the E_3_1 sample, symmetry about the vertical axis is evident. It can also be seen that the lower elastic bond has delaminated, and the load is mainly transferred through the central part of the specimen, where there is a rigid bond.

For the S_4_1 specimen (Figure 20), strain concentrations can be seen in the central joint zone, which in this case is an elastic joint. In this case, it is likely that one of the rigid joints, closer to the edge, is no longer working. The situation is reversed for the E_4_1 specimen, where there are large deformations in the outer joint.

For the S_5_1 specimen (Figure 21), it is difficult to form conclusions, as there is a significant strain concentration almost throughout the lap area. On the other hand, it can be said with certainty that there is little deformation in the end zone of the lap in the E_5_1 specimen, so one of the elastic joints has already stopped working.

In conclusion, although it is not possible to directly observe the work of epoxy bonds, the Aramis DIC system, thanks to its high resolution, is able to record small changes in strains on the lap surface. This is highly advantageous not only for observing the work of individual joints, but also for the quality of their fabrication, which can be verified based on the symmetry of the strain fields with respect to the vertical axis.

4. Conclusions

In this study, uniaxial tensile tests were performed for 50 dual-adhesive samples (the first stage of the work) and 50 triple-adhesive samples (the second stage of the work). The goal of the first stage was to determine the two extremes in terms of the stiffness of the epoxy adhesives. From the 10 compositions analyzed, the most elastic was the 3M Scotch-Weld 2216 B/A Translucent adhesive, while the stiffest was the Epidian 53 (100 g) + "PAC" hardener (80 g). Both adhesives were used in the second stage of the work to makes triple-adhesive joints. In both stages of the work, a 1.6 mm thick VHB double-sided adhesive tape was also used in the lap. The following conclusions were drawn from the work:

1. The use of double-sided adhesive tape makes it possible to ensure a constant thickness of epoxy adhesive, simplifies assembly, and increases the aesthetics of the joint (no excess adhesive on the edges).
2. By changing the amount of PAC hardener added to the resin, an increase in bond strength can be achieved. The change is more evident for Epidian 53 (71% increase) than for Epidian 5 (31.5% increases).
3. By using double-sided adhesive tape and two types of epoxy adhesives with different stiffnesses, a three-stage joint operation was achieved.
4. The triple-adhesive joints provide significant scope for designing the characteristics of joints. Depending on the configuration, maximum force can be achieved in the first or second stage. The multi-stage nature makes these joints safe for use in the automotive industry.
5. By using double-sided adhesive tape, the energy required to damage the joint is several times greater than that for epoxy joints. This also gives the joint a reserve of strength.
6. By using the Aramis system, it was possible to indirectly determine the effort in the individual joints.

Dual and triple adhesive joints may have applications in civil engineering, such as in modular lightweight floor systems made of glass and steel [39]. The joint using double-sided adhesive tape would allow preassembly of the structure and ensure perfect geometry. In the second stage, adhesive joints would be made using epoxy adhesive inserted between the glass and steel surfaces bordered by double-sided adhesive tape. Such a solution will also ensure safety since the failure of the joint is multi-stage and prevents sudden disasters.

Funding: The project/research was financed within the framework of the project of the Lublin University of Technology-Regional Excellence Initiative, funded by the Polish Ministry of Science and Higher Education (contract no. 030/RID/2018/19).

Institutional Review Board Statement: Not applicable.

Informed Consent Statement: Not applicable.

Data Availability Statement: The data presented in this study are available on request from the corresponding author.

Acknowledgments: The author would like to acknowledge to Marcin Kneć and Michał Budka for the help in the samples testing.

Conflicts of Interest: The authors declare no conflict of interest.

References

1. Pinho-da-Cruz, J.A.M.; Ferreira, J.A.M.; Costa, J.D.M.; Borrego, L.F.P. Fatigue analysis of thin AlMgSi welded joints under constant and variable amplitude block loadings. *Thin-Walled Struct.* **2003**, *41*, 389–402. [CrossRef]
2. Sadowski, T.; Kneć, M.; Golewski, P. Fatigue Response of the Hybrid Joints Obtained by Hot Spot Welding and Bonding Techniques. *Key Eng. Mater.* **2014**, *601*, 25–28. [CrossRef]
3. Więckowski, W.; Lacki, P.; Adamus, J. Examinations of steel overlap joints obtained using the friction stir welding technology. *Arch. Metall. Mater.* **2019**, *64*, 393–399. [CrossRef]
4. Balawender, T. The ability to clinching as a function of material hardening behavior. *Acta Metall. Slovaca* **2018**, *24*, 58–64. [CrossRef]
5. Li, X.; Tan, Z.; Wang, L.; Zhang, J.; Xiao, Z.; Luo, H. Experimental investigations of bolted, adhesively bonded and hybrid bolted/bonded single-lap joints in composite laminates. *Mater. Today Commun.* **2020**, *24*, 101244. [CrossRef]
6. Sadowski, T.; Golewski, P. Effect of tolerance in the fitting of rivets in the holes of double lap joints subjected to uniaxial tension. *Key Eng. Mater.* **2014**, *607*, 49–54. [CrossRef]

7. Sadowski, T.; Golewski, P. Modelling and Experimental Testing of Hybrid Joints Made of: Aluminium Adherends, Adhesive Layers and Rivets for Aerospace Applications. *Arch. Metall. Mater.* **2017**, *62*, 1577–1583. [CrossRef]

8. Manzoor, S.; Ud Din, I.; Giasin, K.; Köklü, U.; Khan, K.A.; Panier, S. Three-Dimensional Finite Element Modeling of Drilling-Induced Damage in S2/FM94 Glass-Fiber-Reinforced Polymers (GFRPs). *Materials* **2022**, *15*, 7052. [CrossRef]

9. Negru, R.; Marsavina, L.; Hluscu, M. Experimental and numerical investigations on adhesively bonded joints. *IOP Conf. Ser. Mater. Sci. Eng.* **2016**, *123*, 012012. [CrossRef]

10. Golewski, P.; Sadowski, T.; Pietras, D.; Dudzik, S. The influence of aging in salt chamber on strength of aluminum—CFRP single lap joints. *Mater. Today Proc.* **2021**, *45*, 4264–4267. [CrossRef]

11. Ozel, A.; Yazici, B.; Akpinar, S.; Aydin, M.D.; Temiz, Ş. A study on the strength of adhesively bonded joints with different adherends. *Compos. Part B Eng.* **2014**, *62*, 167–174. [CrossRef]

12. Marsavina, L.; Nurse, A.D.; Braescu, L.; Craciun, E.M. Stress singularity of symmetric free-edge joints with elasto-plastic behaviour. *Comput. Mater. Sci.* **2012**, *52*, 282–286. [CrossRef]

13. Deng, J.; Lee, M.M.K. Effect of plate end and adhesive spew geometries on stresses in retrofitted beams bonded with a CFRP plate. *Compos. Part B Eng.* **2008**, *39*, 731–739. [CrossRef]

14. da Silva, L.F.M.; Carbas, R.J.C.; Critchlow, G.W.; Figueiredo, M.A.V.; Brown, K. Effect of material, geometry, surface treatment and environment on the shear strength of single lap joints. *Int. J. Adhes. Adhes.* **2009**, *29*, 621–632. [CrossRef]

15. Akpinar, S.; Doru, M.O.; Özel, A.; Aydin, M.D.; Jahanpasand, H.G. The effect of the spew fillet on an adhesively bonded single-lap joint subjected to bending moment. *Compos. Part B Eng.* **2013**, *55*, 55–64. [CrossRef]

16. Goglio, L.; Rossetto, M. Stress intensity factor in bonded joints: Influence of the geometry. *Int. J. Adhes. Adhes.* **2010**, *30*, 313–321. [CrossRef]

17. Akpinar, S. The strength of the adhesively bonded step-lap joints for different step numbers. *Compos. Part B Eng.* **2014**, *67*, 170–178. [CrossRef]

18. Brito, R.F.N.; Campilho, R.D.S.G.; Moreira, R.D.F.; Sánchez-Arce, I.J.; Silva, F.J.G. Composite stepped-lap adhesive joint analysis by cohesive zone modelling. *Procedia Struct. Integr.* **2021**, *33*, 665–672. [CrossRef]

19. Gültekin, K.; Akpinar, S.; Özel, A. The effect of the adherend width on the strength of adhesively bonded single-lap joint: Experimental and numerical analysis. *Compos. Part B Eng.* **2014**, *60*, 736–745. [CrossRef]

20. Golewski, P.; Sadowski, T. The influence of single lap geometry in adhesive and hybrid joints on their load carrying capacity. *Materials* **2019**, *12*, 2884. [CrossRef]

21. Sadowski, T.; Nowicki, M.; Golewski, P. The influence of the use of fasteners with different stiffness in hybrid joints subjected to complex mechanical loads. *Arch. Metall. Mater.* **2019**, *64*, 1263–1268. [CrossRef]

22. Sadowski, T.; Golewski, P. Numerical study of the Prestressed connectors and their distribution on the strength of a single lap, a double lap and hybrid joints subjected to uniaxial tensile test. *Arch. Metall. Mater.* **2013**, *58*, 579–585. [CrossRef]

23. El Zaroug, M.; Kadioglu, F.; Demiral, M.; Saad, D. Experimental and numerical investigation into strength of bolted, bonded and hybrid single lap joints: Effects of adherend material type and thickness. *Int. J. Adhes. Adhes.* **2018**, *87*, 130–141. [CrossRef]

24. Li, X.; Cheng, X.; Cheng, Y.; Wang, Z.; Huang, W. Tensile properties of a composite–metal single-lap hybrid bonded/bolted joint. *Chin. J. Aeronaut.* **2021**, *34*, 629–640. [CrossRef]

25. Raphael, C. Variable-adhesive bonded joints. *Appl. Polym. Symp.* **1966**, *3*, 99–108.

26. Jairaja, R.; Naik, G.N. Single and dual adhesive bond strength analysis of single lap joint between dissimilar adherends. *Int. J. Adhes. Adhes.* **2019**, *92*, 142–153. [CrossRef]

27. Bouchikhi, A.S.; Megueni, A.; Gouasmi, S.; Boukoulda, F.B. Effect of mixed adhesive joints and tapered plate on stresses in retrofitted beams bonded with a fiber-reinforced polymer plate. *Mater. Des.* **2013**, *50*, 893–904. [CrossRef]

28. Hirulkar, N.S.; Jaiswal, P.R.; Alessandro, P.; Reis, P. Influence of mechanical surface treatment on the strength of mixed adhesive joint. *Mater. Today Proc.* **2018**, *5*, 18776–18788. [CrossRef]

29. Marques, E.A.S.; Da Silva, L.F.M.; Flaviani, M. Testing and simulation of mixed adhesive joints for aerospace applications. *Compos. Part B Eng.* **2015**, *74*, 123–130. [CrossRef]

30. Zaeri, A.R.; Saeidi Googarchin, H. Experimental investigation on environmental degradation of automotive mixed-adhesive joints. *Int. J. Adhes. Adhes.* **2019**, *89*, 19–29. [CrossRef]

31. Machado, J.J.M.; Marques, E.A.S.; da Silva, L.F.M. Influence of low and high temperature on mixed adhesive joints under quasi-static and impact conditions. *Compos. Struct.* **2018**, *194*, 68–79. [CrossRef]

32. Machado, J.J.M.; Gamarra, P.M.R.; Marques, E.A.S.; da Silva, L.F.M. Numerical study of the behaviour of composite mixed adhesive joints under impact strength for the automotive industry. *Compos. Struct.* **2018**, *185*, 373–380. [CrossRef]

33. da Silva, L.F.M.; Adams, R.D. Adhesive joints at high and low temperatures using similar and dissimilar adherends and dual adhesives. *Int. J. Adhes. Adhes.* **2007**, *27*, 216–226. [CrossRef]

34. da Silva, L.F.M.; Lopes, M.J.C.Q. Joint strength optimization by the mixed-adhesive technique. *Int. J. Adhes. Adhes.* **2009**, *29*, 509–514. [CrossRef]

35. Machado, J.J.M.; Marques, E.A.S.; Silva, M.R.G.; da Silva, L.F.M. Numerical study of impact behaviour of mixed adhesive single lap joints for the automotive industry. *Int. J. Adhes. Adhes.* **2018**, *84*, 92–100. [CrossRef]

36. Chiminelli, A.; Breto, R.; Izquierdo, S.; Bergamasco, L.; Duvivier, E.; Lizaranzu, M. Analysis of mixed adhesive joints considering the compaction process. *Int. J. Adhes. Adhes.* **2017**, *76*, 3–10. [CrossRef]

37. Golewski, P.; Sadowski:, T. The influence of dual adhesive in single lap joints on strength and energy absorption. *Mater. Today Proc.* **2021**, *45*, 4280–4285. [CrossRef]
38. Gajewski, J.; Golewski, P.; Sadowski, T. The use of neural networks in the analysis of dual adhesive single lap joints subjected to uniaxial tensile test. *Materials* **2021**, *14*, 419. [CrossRef]
39. Marchione, F.; Chiappini, G.; Rossi, M.; Scoccia, C.; Munafò, P. Experimental assessment of the static mechanical behaviour of the steel-glass adhesive joint on a 1:2 scale tensegrity floor prototype. *J. Build. Eng.* **2022**, *53*, 104572. [CrossRef]

Article

Retaining Mechanical Properties of GMA-Welded Joints of 9%Ni Steel Using Experimentally Produced Matching Ferritic Filler Metal

Abdel-Monem El-Batahgy [1], Mohamed Raafat Elkousy [2], Ahmed Abd Al-Rahman [2], Andrey Gumenyuk [3], Michael Rethmeier [3,4,5] and Sergej Gook [5,*]

1   Central Metallurgical Research and Development Institute (CMRDI), Cairo 11421, Egypt
2   Metallurgical Engineering Department, Cairo University, Giza 12613, Egypt
3   Federal Institute of Materials Research and Testing (BAM), 12205 Berlin, Germany
4   Institute of Machine Tools and Factory Management, Technische Universität Berlin, 10587 Berlin, Germany
5   Fraunhofer Institute for Production Systems & Design Technology IPK, 10587 Berlin, Germany
*   Correspondence: sergej.gook@ipk.fraunhofer.de; Tel.: +49-(30)39006375

Abstract: Motivated by the loss of tensile strength in 9%Ni steel arc-welded joints performed using commercially available Ni-based austenitic filler metals, the viability of retaining tensile strength using an experimentally produced matching ferritic filler metal was confirmed. Compared to the austenitic Ni-based filler metal (685 MPa), higher tensile strength in gas metal arc (GMA) welded joints was achieved using a ferritic filler metal (749 MPa) due to its microstructure being similar to the base metal (645 MPa). The microstructure of hard martensite resulted in an impact energy of 71 J ($-196\ ^\circ$C), which was two times higher than the specified minimum value of $\geq$34 J. The tensile and impact strength of the welded joint is affected not only by its microstructure, but also by the degree of its mechanical mismatch depending on the type of filler metal. Welds with a harder microstructure and less mechanical mismatch are important for achieving an adequate combination of tensile strength and notched impact strength. This is achievable with the cost-effective ferritic filler metal. A more desirable combination of mechanical properties is guaranteed by applying low preheating temperature (200 $^\circ$C), which is a more practicable and economical solution compared to the high post-weld heat treatment (PWHT) temperature (580 $^\circ$C) suggested by other research.

Keywords: 9%Ni steel; Ni-based austenitic filler metal; matching ferritic filler metal; preheating; post-weld heat treatment; microstructure; mechanical mismatching

Citation: El-Batahgy, A.-M.; Elkousy, M.R.; Al-Rahman, A.A.; Gumenyuk, A.; Rethmeier, M.; Gook, S. Retaining Mechanical Properties of GMA-Welded Joints of 9%Ni Steel Using Experimentally Produced Matching Ferritic Filler Metal. *Materials* 2022, 15, 8538. https://doi.org/10.3390/ma15238538

Academic Editor: Raul D.S.G. Campilho

Received: 25 September 2022
Accepted: 26 November 2022
Published: 30 November 2022

Publisher's Note: MDPI stays neutral with regard to jurisdictional claims in published maps and institutional affiliations.

1. Introduction

The global need to use natural gas as a clean energy source with lower CO_2 emissions than petroleum has been increasing as the world seeks to reduce the problem of environmental contamination. Given the current economic and political situation around the world, many countries are concerned about ensuring their energy stability and are storing large quantities of natural gas. However, the biggest gas fields in the world are located far away from the main areas of consumption, so large quantities of gas have to be transported and stored over long distances. Liquefying the gas is of considerable importance for reducing its volume by about 600-times, which obviously simplifies its transportation and storage processes. Transport ships and storage tanks for liquefied natural gas (LNG) are being built around the world [1,2]. Heat-treated ferritic steel alloyed with 9%Ni is widely accepted as the most important material for the fabrication of large LNG storage tanks and vessels where high strength ($\geq$700 MPa) and pronounced notch impact energy ($\geq$34 J at $-196\ ^\circ$C/$-320\ ^\circ$F) are required [3–5]. Welding is commonly used in the manufacturing of LNG transportation and storage equipment; therefore, understanding the factors that affect the quality of welded joints is crucial to maintaining its mechanical properties. Nowadays, LNG equipment is made of 9%Ni steel using arc-based welding processes, mainly gas metal arc welding (GMAW). In this context, the use of austenitic Ni-based consumables has long been prioritized in the LNG sector due to their high toughness and brittle fracture resistance at $-196\,^{\circ}$C ($-320\,^{\circ}$F) [6–8]. The main limitation of Ni-based fillers is that they do not reach the strength of ferritic steel with 9%Ni. This condition is reflected in design codes for LNG tanks and vessels which base maximum permitted design stresses on the strength of under-matching Ni-based weld joints. Consequently, in order to meet safety requirements and ensure the overall strength of the welded structure, the wall thickness of the LNG units is designed to be excessively thick, which has a negative impact on the economic aspect [9–15]. In other words, retaining a combination of mechanical properties with this steel type is a major challenge. Autogenous laser beam welding has already been used to produce welded joints with 9%Ni steels that have properties similar to those of the base metal (BM) [16–19]. It has been demonstrated in many applications that laser welding is an efficient tool for retaining the properties of various difficult-to-weld materials [20–25]. However, laser welding of 9%Ni steel is still the subject of comprehensive research and will require several more years to clarify and prove its possible actual applications for LNG facilities manufacturing.

Considering the results of laser beam welding, the production of arc-welded joints with a chemical composition and microstructure similar to that of the BM appears to be efficient in maintaining its mechanical properties. In general, ferritic filler metals for joining 9%Ni steel have been under consideration for the last few decades as they would offer an economic benefit in terms of manufacturing costs. Matching filler metals for submerged arc welding (SAW) and gas tungsten arc welding (GTAW) have already been used in the fabrication of 9%Ni steel tubes [26] and in the construction of a laboratory-scale spherical model tank [27]. Recently, research on matching ferritic welding electrodes for shielded metal arc welding (SMAW) have shown attractive results, with an acceptable combination of tensile strength and impact toughness obtained [28]. Nevertheless, these filler metals are not yet extensively used for in-field welding, and more studies are still required in this important area, especially for the commonly used GMA welding process. The objective of the present work was to examine the feasibility of retaining the combination of mechanical properties for GMA-welded joints with 9%Ni steel using an experimentally fabricated matching ferritic filler metal with 11%Ni (ERNi11). For comparison, the commercially available and widely used Ni-based austenitic filler metal ERNiCrMo-3 (AWS-A5.14) was used.

2. Materials and Methods

The BM used was a ferritic 9%Ni steel grade ASTM A553 Type 1 (EN10028-4X8Ni9) with 14.5 mm plate thickness. The chemical composition and the mechanical–technological properties are given in Table 1.

Table 1. Chemical composition (ASTM E-3) and mechanical properties (ASTM A370/ASTM D6110/ASTM E384) of the 9%Ni steel used.

Chemical Composition (wt.%)									
C	Si	Mn	P	S	Al	Ni	Cr	Cu	Fe
0.07	0.20	0.57	0.003	0.001	0.02	9.15	0.03	0.02	balance

Mechanical properties				
Yield strength (MPa)	Tensile strength (MPa)	Elongation (%)	Hardness (HV)	Impact absorbed energy at $-196\,^{\circ}$C ($-320\,^{\circ}$F)
671	745	28	253	178

Specimens with dimensions of $250 \times 100 \times 14.5$ mm^3 and a 60° double V-groove were used for GMA welding tests. This specimen size was defined on the basis of existing laboratory handling techniques. Prior practice with GMA welding tests has shown that a specimen length of at least 250 mm is enough to make a conclusion about weld seam quality. Double-sided butt-welded joints were made using the weld bead deposition sequence shown in Figure 1. Subsequently, preliminary tests were carried out. Based on these tests, the optimum welding parameters (Table 2) were selected. The aim was to obtain well-penetrated weld beads without fusion defects. In addition, undercut-free bead profiles were also a result. In this regard, the combination of 198 A welding current, 29 V arc voltage, and 210 mm/min–230 mm/min welding speed was used, which is in good agreement with well-established welding parameters. Welding was carried out for accurately assembled, aligned, mechanically clamped, and carefully restrained butt joints, as shown in Figure 2. Two types of 1.0 mm diameter filler metals were used, including the commercially available Ni-based austenitic filler metal ERNiCrMo-3 (AWS A5.14) and an experimentally produced matching ferritic filler metal with 11%Ni (ERNi11). Preheating and PWHT were applied to welds made using the matching ferritic filler metal ERNi11. Preheating was conducted at three temperature regimes: 150 °C (302 °F), 200 °C (392 °F), and 250 °C (482 °F). The preheating of the weld specimen was carried out with an oxy-acetylene flame. The preheating and interpass temperatures were monitored with a type K temperature sensor. The interpass temperature was maintained at 150 °C to minimize martensite formation and increase retained austenite in both the WM and HAZ of welded joints made using the ferritic filler metal and the HAZ of welded joints made using the austenitic filler metal. The specimen was welded as soon as the desired temperature was reached. PWHT was performed at a temperature of 580 °C (1076 °F). The holding time was 20 min [28]. The cooling down of samples was conducted in an ambient atmosphere.

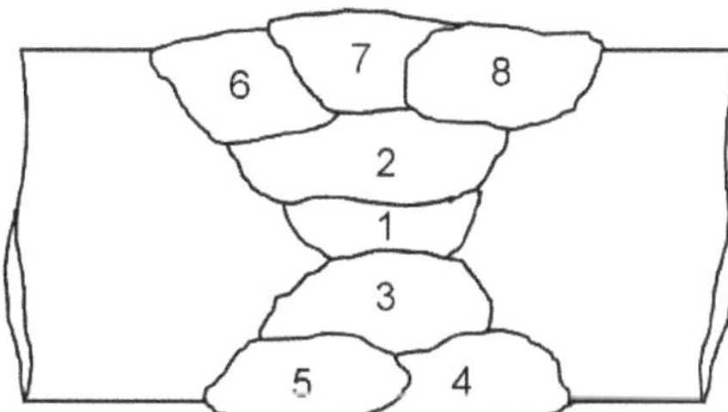

Figure 1. Schematic illustration of the weld bead deposition sequence for two-sided (60° double V-groove) GMA butt-welded joints.

Table 2. Optimum welding parameters used for two-sided GMA butt welded joints.

Pass No.	Current (A)	Voltage (V)	Speed (mm/min)	HI (kJ/mm)
1	198	29	222	1.55
2	198	29	210	1.64
3	198	29	210	1.64
4	198	29	224	1.54
5	198	29	224	1.54
6	198	29	230	1.50
7	198	29	230	1.50
8	198	29	230	1.50

Figure 2. Experimental setup of GMA welding experiments.

After welding, a visual inspection of the GMA welds was performed. The occurrence of external weld defects such as undercuts was evaluated. The selected weld specimens were then subjected to X-ray testing (RT) to identify any internal defects. Samples for metallographic examinations were taken transversally to the welding direction. For metallurgical examinations, picric acid etchant was used for electrolytic etching of welded joints made using the Ni-based austenitic filler metal, while nital etchant was used for chemical etching of joints made using the matching ferritic filler metal. The geometric characteristics of the fusion zone were examined with a low magnification stereoscope, while the microstructures of the weld metal (WM), the material melted and re-solidified by the welding process, the heat-affected zone (HAZ), and the BM were investigated with an optical microscope. Comprehensive microscopic investigations including pattern quality, the phase map, the orientation color map, and grain size distribution were conducted using a scanning electron microscope equipped with an electron backscatter diffraction (EBSD) system. The EBSD, with a 2.53 μm step size, was used for the WM. Grains were detected using a grain detector angle of 3.92° and only grains larger than 100 pixels were considered. Compositional variations across the welded joints were determined using a scanning electron microscope equipped with an energy dispersive X-ray spectroscopy (EDS) unit at an accelerating voltage of 20 kV. Vickers microhardness measurements were performed using a Shimadzu 1000 g machine. The measurements were conducted on polished and etched cross sections at near mid-thickness of its WM, HAZ, and BM while applying a load of 500 g with a dwell time of 20 s. The tensile test was carried out at a constant traverse displacement rate of 2 mm/min using a Shimadzu 1000 kN hydraulic testing machine. Tensile specimens with a gauge length of 40 mm, a thickness of 14 mm, a width of 19 mm (in the range of the gauge length), a clamping range width of 25 mm, and a total length of 200 mm were machined in accordance with the ASME IX standard. The impact test was carried out at −196 °C (−320 °F) using a Roell Amsler 300 J pendulum impact tester. Standard impact test specimens, with the notch location in both the WM and the HAZ, were machined in accordance with the ASME IX standard.

For both tensile and impact tests, three specimens in the as-welded condition were tested for each welded joint. Impact tests were also conducted for both preheated and post-weld heat-treated joints. The average values of each property were considered and compared with those of the BM.

3. Results and Discussion

Visual inspection of GMA-welded joints made with the austenitic Ni base and matching ferritic filler metals revealed no noticeable external weld defects. RT confirmed fully penetrated welds with no unacceptable internal defects. This is mainly due to proper selection of the implemented welding conditions. The welds accepted after RT were sectioned and then metallographically and mechanically examined.

3.1. Metallurgical Examinations

Macro images of cross sections of welded joints made with the austenitic Ni-based filler metal and the matching ferritic filler metal are shown in Figure 3. It can be noticed that comparable fusion zone sizes were obtained due to the similar welding parameters applied to both joints. Weld development is essentially symmetrical about its center. Macroscopic observation confirmed the quality of the welds, which showed no internal defects (lack of fusion, porosity, cracks, etc.).

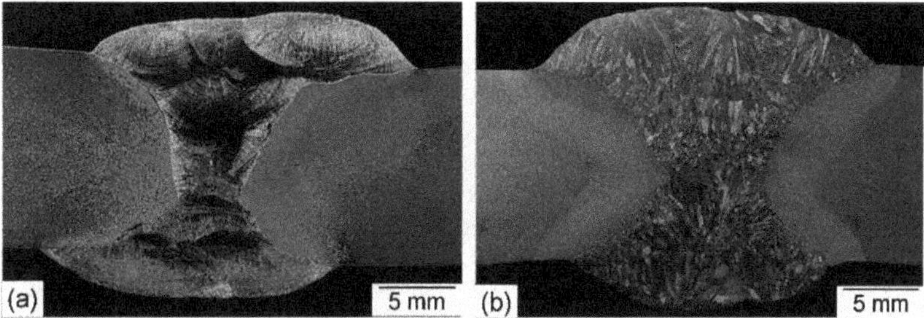

Figure 3. Macro images of cross sections of welded joints produced using the Ni-based austenitic (**a**) and the matching ferritic (**b**) filler metals.

The microstructure of the base metal consists mainly of tempered martensite (Figure 4), and its average hardness value is 253 HV. The microstructure of the WM and the HAZ of a welded joint made using the nickel-based filler metal are shown in Figure 5. The width of the HAZ is 4 mm. Weld soundness was confirmed using a radiographic test and no internal welding defects were detected. The microstructure of the WM is a cast dendritic austenitic structure (Figure 5a), typical to that of the used austenitic filler metal [29]. The microstructure of the HAZ is a coarse-grained martensitic structure and the prior austenite grain boundaries can be seen (Figure 5b). This microstructure was formed due to high-temperature heating during welding followed by rapid cooling.

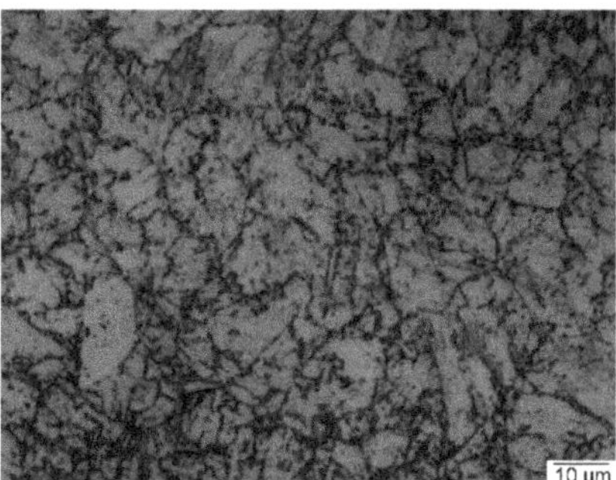

Figure 4. Optical microscopic photograph of the used 9%Ni steel's BM.

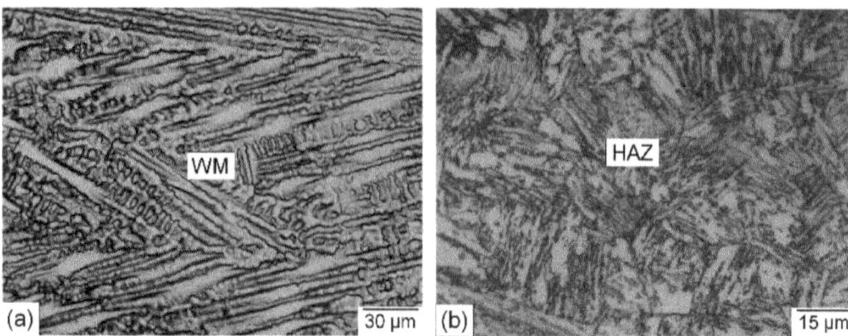

Figure 5. Optical microscopic photographs of a cross section of a welded joint produced using the Ni-based austenitic filler metal: (**a**) WM; (**b**) HAZ.

Optical microscopic photographs of the WM and the HAZ of a welded joint made using the matching ferritic filler metal are shown in Figure 6. Microscopic examination indicated that the width of the WM and HAZ is very close to that of the welded joint produced using the Ni-based austenitic filler metal. It also showed a sound WM where no unacceptable internal defects were found. The main observation is that the WM microstructure is completely different, consisting of a columnar cast dendritic martensitic structure typical for the ferritic filler metal used (Figure 6a). The coarse-grained HAZ microstructure (Figure 6b) is fairly comparable to that of the welded joint produced using the austenitic filler metal (Figure 5b), where a martensitic structure with a small amount of retained austenite was found. This is due to similar heat inputs in both cases. The microstructure of the HAZ bordering the BM exhibited a less martensitic structure due to the lower heating temperature.

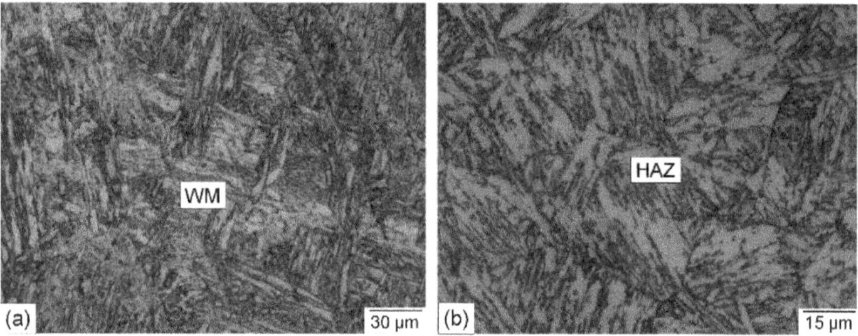

Figure 6. Optical microscopic photographs of a cross section of a welded joint made using the matching ferritic filler metal: (**a**) WM; (**b**) HAZ.

SEM with an EBSD detector was used to study the microstructure in more detail. The pattern quality, phase map, EBSD orientation color map, and grain size distribution of the Ni-based WM are shown in Figure 7. A columnar dendritic structure was obtained for the austenitic weld metal (Figure 7a). The phase map, Figure 7b, confirmed the fully austenitic microstructure. A coarse austenitic structure with high-angle grain boundaries is seen in the orientation color map (Figure 7c). The weld metal is characterized by a significant variation in grain size, as shown by the histogram of grain size distribution. The number of grains detected was 274, the smallest grain size was 10 µm, the largest was 910 µm, and the average grain size was 582 µm (Figure 7d).

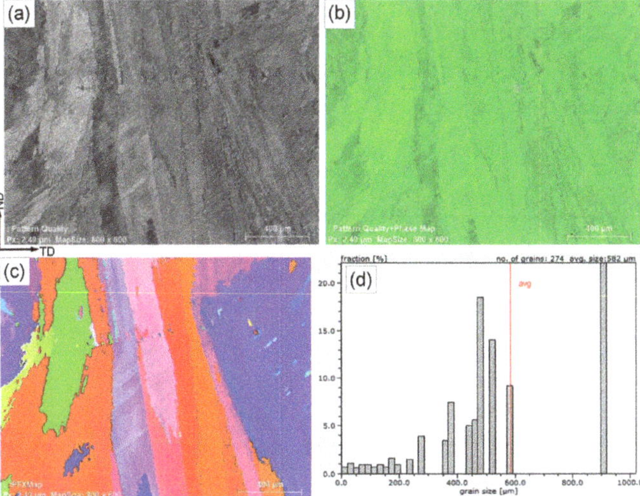

Figure 7. Pattern quality (**a**), phase map (**b**), EBSD orientation color map (**c**), and grain size distribution histogram (**d**) of the weld metal deposited using the Ni-based austenitic filler metal.

For the matching ferritic weld metal, Figure 8 shows the pattern quality, the phase map, the EBSD orientation color map, and the grain size distribution. The coarse columnar dendritic martensitic structure of this ferritic weld metal is confirmed by the pattern quality shown in Figure 8a. The martensitic microstructure and the retained austenite constitute 98.4% and 1.6%, respectively (Figure 8b).

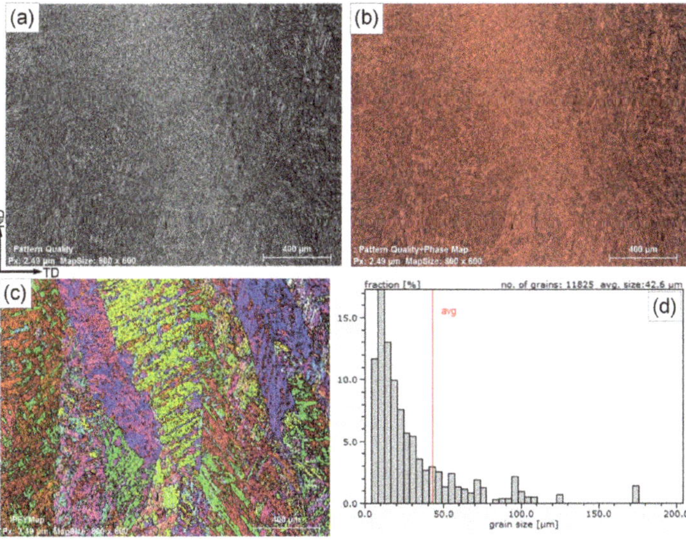

Figure 8. Pattern quality (**a**), phase map (**b**), EBSD orientation color map (**c**), and grain size distribution histogram (**d**) of the weld metal deposited using the matching ferritic filler metal.

Published data [17] on EBSD analyses of a 9%Ni steel's BM showed a microstructure of fine tempered martensite with a 10 μm grain size and a small amount of retained austenite. In contrast to the BM, the orientation color map of the WM provided in Figure 8c shows a coarser martensite structure with small-angle grain boundaries. Significant variation in grain size was detected in the weld by the grain size distribution histogram, with the smallest grain size being 5 μm and the largest 175 μm; the average grain diameter was 42.6 μm, which is much coarser than in the BM, and the total number of examined grains was 11,825 (Figure 8c).

Since the weld metal in GMAW is actually a mixed material of BM and filler wire, the dilution rate (DR) of the BM in WM was estimated for the first pass. Due to the double V-groove geometry, the first pass is expected to have the highest DR of the filler passes. With the welding parameters used (Table 2), the wire feed rates in the welding tests were 10 m/min. Since the welding tests were carried out with a solid wire of 1 mm diameter, the area of the melted welding wire was about 7.85 mm^2 per pass. From the macrographs in Figure 4, it could be determined planimetrically that the area of the first pass was about 13 mm^2. Of this area, 5.15 mm^2 or 40.0% is accounted for by the diluted BM.

3.2. Hardness Measurements

Figure 9 shows representative hardness distributions across welded joints made using the austenitic Ni base and the matching ferritic filler metals, as well as those of the 200 °C preheated joint. The highest variation in the hardness measurements was HV 0.5 ± 3%. Adequate weld zone and HAZ widths of 15.5 mm and 4.0 mm were obtained for all joints. The hardness profiles of the welded joints are significantly influenced by the filler metal type. The Ni-based austenitic filler metal resulted in a weld hardness value of 235 HV, which is close to the hardness of the BM (253 HV). The matching ferritic filler metal resulted in WM with a higher hardness (~345 HV), a result of its martensitic structure. This hard structure is expected to maintain the welded joint's tensile strength while negatively affecting its impact toughness. The hardness values of both the WM and HAZ of the 200 °C preheated joint were significantly decreased (~300 HV) due to its less martensitic structure and an increased fraction of retained austenite. It should be reported that lower preheating temperature (150 °C) resulted in a more martensitic structure due to a higher cooling rate that in turn maintained higher hardness values, which leads to an adverse effect on impact toughness. On the other hand, the 250 °C preheated joint showed a hardness profile similar to that of the 200 °C preheated joint. As a result, 200 °C was considered to be the optimum preheating temperature since it also has a positive effect on the manufacturing cost. It is also important to highlight that PWHT resulted in a hardness profile close to that of the optimum preheated joint and similar to that of previously published research on SMAW with 9%Ni steel [28]. This is due to a tempered martensitic structure with a more stable retained austenite.

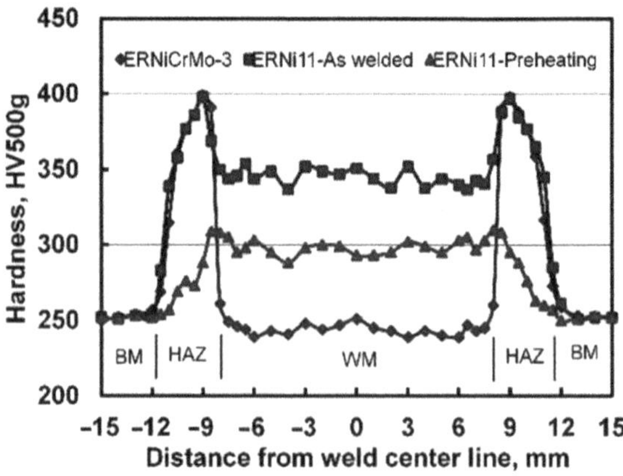

Figure 9. Hardness profiles of as-welded joints produced using the Ni-based austenitic and the matching ferritic filler metals together with that of the 200 °C preheated joint.

The implemented PWHT temperature of 580 °C was decided based on a well-settled and accepted range for 9%Ni steel. PWHT above this temperature can result in a loss of toughness due to unstable temper-austenite, which transforms to martensite at subzero temperatures [30].

3.3. Micro Chemical Analysis

EDS microanalysis was used to study the variations in chemical composition along the HAZ and WM in two specimens, one welded with an austenitic Ni-based filler metal and the other one welded with a matching ferritic filler metal (Figures 10 and 11). For the joint welded with the Ni-based austenitic filler metal, ERNiCrMo-3, the chemical composition of the weld is completely different from that of the BM. Nickel content and chromium content are much higher in the WM compared to the BM, while the iron content in the weld is much lower than that of the 9%Ni BM. On the other hand, the composition of the WM is similar to that of the BM in the case of joints welded with the matching filler metal.

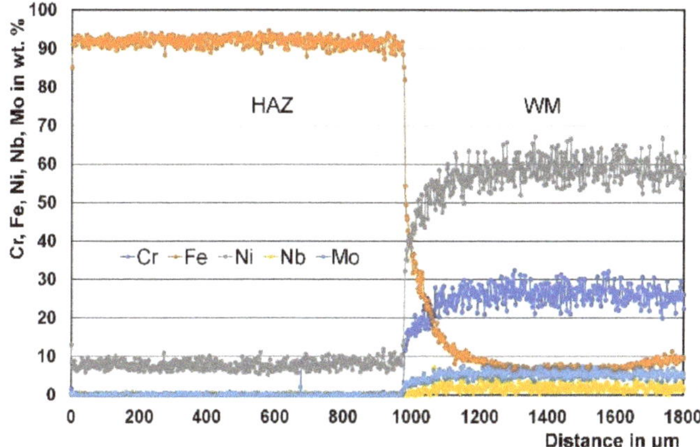

Figure 10. Example of EDS line scan microanalysis through the HAZ and WM deposited using the Ni-based austenitic filler metal.

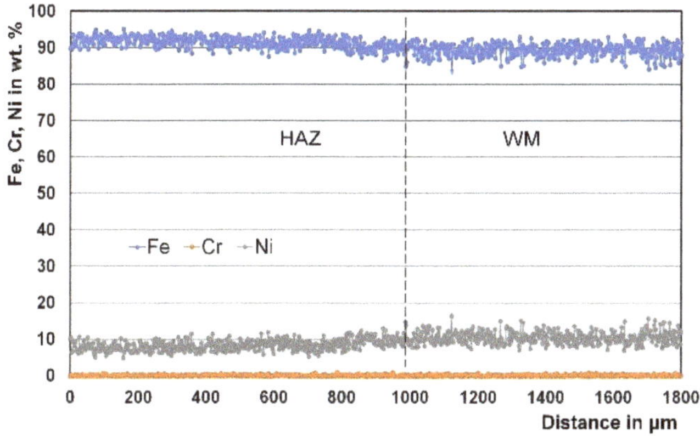

Figure 11. Example of EDS line scan microanalysis through the HAZ and WM deposited using the matching ferritic filler metal.

Regarding welded joints made using the matching ferritic filler metal, no noticeable variation in the chemical composition of its WM or BM was found (Figure 11). In other words, a weld with a chemical composition similar to that of the BM was obtained that in turn resulted in a hard weld metal due to its martensitic structure. Concerning this, higher tensile strength and lower fracture toughness are expected for welded joints made using the matching ferritic filler metal in comparison to those produced using the Ni-based austenitic filler metal.

3.4. Mechanical Properties of Welded Joints

Figure 12 shows pictures of tensile test fracture specimens from welded joints made with austenitic Ni-based and matching ferritic filler metals. The fracture occurred in the deposited weld for the austenitic Ni-based filler metal (Figure 12a), while it occurred in the base metal of the specimen welded with the matching ferritic filler metal (Figure 12b). Concerning this, the ruptures took place in the reduced tensile strength zone of both joints. Unlike expected, welded joints made using the Ni-based austenitic filler metal showed relatively low elongation. This may be explained by a higher tensile strength difference between its weld and the base metal, as well as by a larger difference in the thermal expansion coefficients of its weld and its BM, which led to welding residual stresses and lower overall elongation.

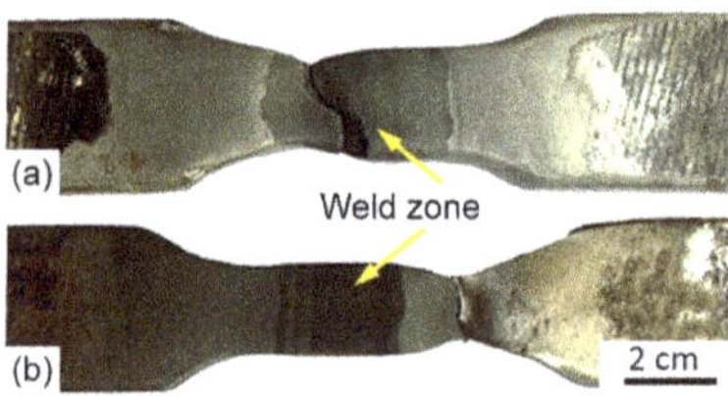

Figure 12. Photographs of tensile fracture specimens from joints produced with the austenitic (**a**) and the ferritic (**b**) filler metals.

The tensile strength of the BM is equal to 745 MPa, the joints welded with the austenitic nickel-based filler metal (ERNiCrMo-3) exhibited tensile strength of 685 MPa, while welding with a similar filler metal (ER11Ni) resulted in tensile strength of 749 MPa (Figure 13). Thus, welding 9%Ni steel with similar ferritic filler metal resulted in higher strength comparable to that of the BM. This can be attributed to the similar microstructures of the WM and BM, which are composed of hard martensite and soft retained austenite.

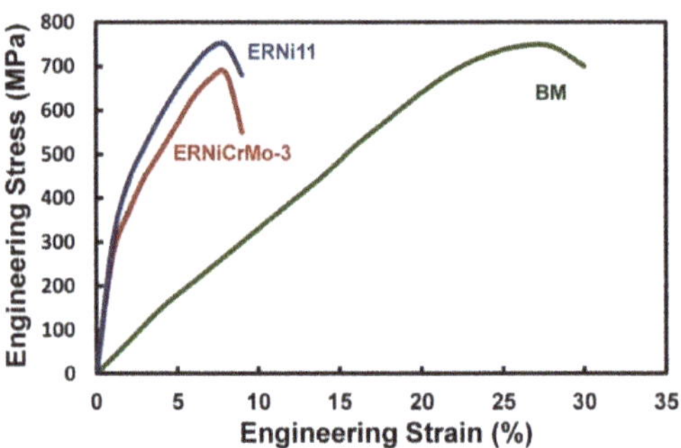

Figure 13. Engineering stress–strain curves of joints produced with the Ni-based austenitic (ERNiCrMo-3) and the matching ferritic (ERNi11) filler metals, as well as the BM.

The weld metal's microstructure is seen as the most important parameter for retaining the tensile strength of the welded joint. Unlike the joints welded with austenitic Ni-based filler metal, the tensile strength of the welded joints with ferritic filler metal complies with the ASME code, which specifies that the minimum tensile strength of the welded joint should be equal to that of the BM. SEM photographs of the fracture surfaces of specimens obtained from welded joints made with the austenitic Ni-base filler metal and the matching ferritic filler metal are shown in Figure 14. Dimples were observed in both specimens, indicating a ductile fracture mode. This is because fractures occurred in the softer Ni-based austenitic WM (Figure 14a) and in the tempered martensitic BM (Figure 14b).

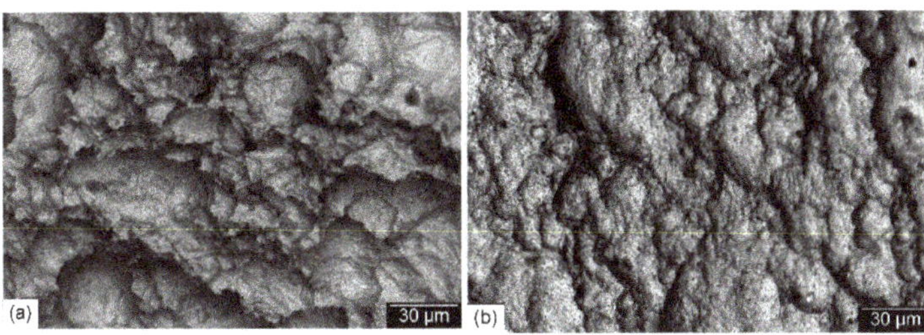

Figure 14. SEM photographs of the tensile fracture surfaces of joints produced with the Ni-based austenitic (**a**) and the matching ferritic (**b**) filler metals.

As for the impact test, a V-notch was cut at the center of the WM and the HAZ using a high precision CNC milling machine to determine the impact toughness of both zones (Figure 15a,b). Figure 16 illustrates the impact test results for the conditions investigated in this study for tests conducted at −196 °C. BM showed the highest impact toughness of 178 J. This can be attributed to its microstructure of tempered martensite and retained austenite. The impact toughness of the HAZ was equal to 66 J and 68 J for joints made using the ferritic filler metal and the Ni-based austenitic filler metal, respectively. This is attributed to a similar microstructure and the heating–cooling cycles in both cases. The impact toughness of the WM produced using a similar ferritic filler metal (ERNi11) is equal to 71 J due to its martensitic structure. This value is still well above the ASME minimum specified value of $\geq$34 J.

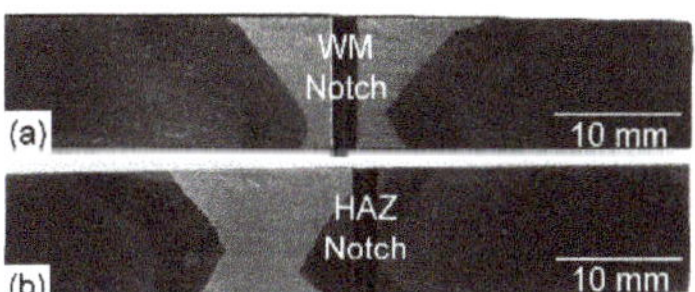

Figure 15. Photographs of V-notch impact test specimens with a notch location in the WM (**a**) and in the HAZ (**b**).

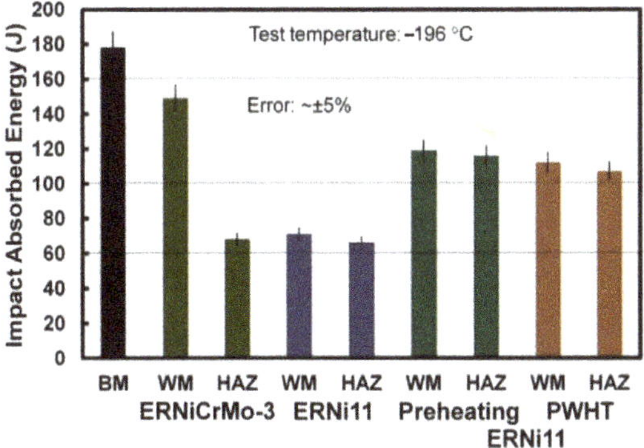

Figure 16. Impact absorbed energy for the WM and HAZ of joints produced with ERNiCrMo-3 and ERNi11 filler metals, together with that of the 200 °C preheated joints, PWHT joints, and the BM.

Applying preheating or post-weld heat treatment on welds produced using ERNi11 decreased the mismatching of toughness between the WM and HAZ; in this case, the absorbed energy was in the range of 107 J to 119 J, as seen in Figure 16.

Figure 17 shows SEM photographs of the impact fracture surface of joints produced with the Ni-based austenitic and the ferritic filler metals. Compared to the WM fracture surface of joints produced using the Ni-based austenitic filler metal (Figure 17a), a less ductile fracture appearance was obtained for joints produced using the matching ferritic filler metal where dimples were not clearly visible (Figure 17b). On the other hand, the impact fracture surface of both preheated and post-weld heat-treated joints showed a ductile fracture surface. SEM photographs of the impact fracture surface of 200 °C preheated and PWHT joints made using the matching ferritic filler metal are shown in Figure 18. The most important thing to note is the clear visible dimples for the fracture surface of the 200 °C preheated joint (Figure 18a), indicating a ductile fracture surface. This is due to the less martensitic structure with an increased fraction of retained austenite. This result is comparable to that of a previous study on laser beam welding with the same material [31]. A ductile impact fracture surface was also obtained for the post-weld heat-treated joint (Figure 18b) due to its WM's tempered martensitic structure with more stable retained austenite. The effect of PWHT is in agreement with a previous study on SMAW with 9%Ni steel [29]. Tensile and impact properties are influenced not only by the microstructure of the fusion zone, but also by the degree of its mechanical mismatching depending on the type of the filler metal. A welded joint with a homogeneous microstructure resulting in lower mechanical mismatching is very important for obtaining an adequate combination of tensile strength and impact toughness. This is achievable with the low-cost matching ferritic filler metal.

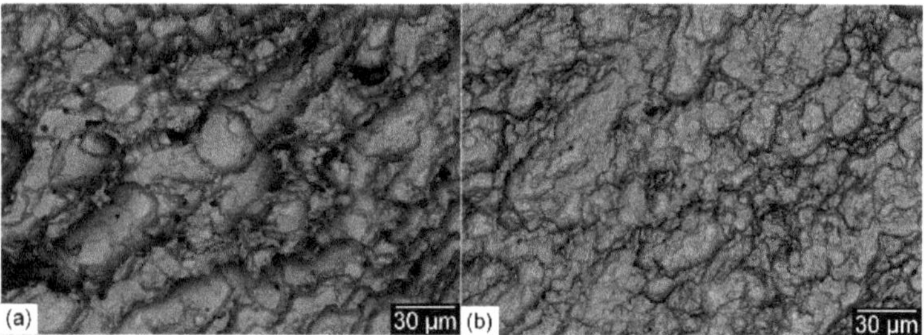

Figure 17. SEM photographs of the impact fracture surface of joints produced with the Ni-based austenitic (**a**) and the ferritic (**b**) filler metals.

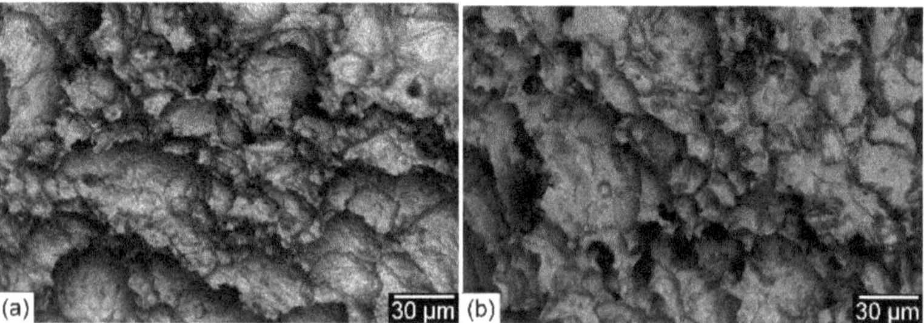

Figure 18. SEM photographs of the impact fracture surface of 200 °C preheated (**a**) and PWHT (**b**) joints produced with the ferritic filler metal.

4. Conclusions

Compared to welded joints made using the austenitic Ni-based filler metal ERNiCrMo-3 (685 MPa), a higher tensile strength (749 MPa) similar to that of the base metal (745 MPa) was obtained using the ferritic filler metal due to its weld metal's martensitic structure. However, this WM's hard martensitic structure resulted in low impact toughness (71 J); however, this is still more than two times higher than the minimum specified value ($\geq$34 J). The most important thing to notice is the lower mismatching for both the tensile strength and impact toughness of the WM and HAZ of this welded joint due to its similar microstructure. The overall tensile and impact properties of the welded joint are affected by its microstructure, as well as its mechanical mismatching as a function of the filler metal. Welded joints with lower mechanical mismatching are important for obtaining an adequate combination of mechanical properties. This is viable using the lower cost matching ferritic filler metal. For further desirable combinations of mechanical properties, either preheating or PWHT should be applied. Preheating results in a less martensitic structure with an increased fraction of retained austenite. PWHT results in a tempered martensitic structure with more stable retained austenite. However, a low preheating temperature of 200 $^\circ$C is an easier, more applicable, and cost-effective solution compared to the high PWHT temperature of 580 $^\circ$C. It should be noted that the preheating of large components such as cryogenic vessels can be very time consuming and expensive. Therefore, the use of mobile inductive preheating systems should be considered, e.g., an inductor that runs along with the welding process and heats the part locally. Such systems are already available on the market.

Author Contributions: Conceptualization, M.R.E., A.-M.E.-B., A.G. and M.R.; Methodology, A.-M.E.-B.; Investigation, A.A.A.-R. and S.G.; Writing—original draft, A.-M.E.-B. and A.A.A.-R.; Writing—review & editing, M.R.E. and A.G.; Visualization, S.G.; Supervision, M.R.E. and M.R. All authors have read and agreed to the published version of the manuscript.

Funding: The current research was funded by both the Science and Technology Development Fund (STDF) of Egypt under Grant GERF ID5111 and the Germany Aerospace Center (DLR) on behalf of the Federal Ministry of Education and Research (BMBF) of Germany under Grant 01DH14012.

Institutional Review Board Statement: Not applicable.

Informed Consent Statement: Not applicable.

Acknowledgments: The authors would like to thank Voestalpine Bohler Welding, Germany GmbH for supplying the experimentally produced matching ferritic filler metal. Appreciation is also extended to PETROJET-Egypt for conducting the welding experiments on site.

Conflicts of Interest: The authors declare no conflict of interest.

References

1. Furuya, H.; Kawabata, T.; Takahashi, Y.; Kamo, T.; Inoue, T.; Okushima, M.; Ando, R.; Onishi, K. Development of low-nickel steel for LNG storage tanks. In Proceedings of the 2013 International Conference and Exhibition on Liquefied Natural Gas, Houston, TX, USA, 16–19 April 2013.
2. Ishimatu, J.; Kawabata, K.; Morita, H.; Ikkai, H.; Suetake, Y. *Building of Advanced Large Sized Membrane Type LNG Carrier*; Mitsubishi Heavy Industries Ltd.: Tokyo, Japan, 2004.
3. Hoshio, M.; Saitoh, N.; Muraoka, H.; Saeki, O. Development of super-9%Ni steel plates with superior low-temperature toughness for LNG storage tanks. *Nippon Steel Tech. Rep.* **2004**, *380*, 17–20.
4. Oh, D.J.; Lee, J.M.; Noh, B.J.; Kim, W.S.; Ando, R.; Matsumoto, T.; Kim, M.H. Investigation of fatigue performance of low temperature alloys for liquefied natural gas storage tanks. *J. Mech. Eng. Sci.* **2015**, *229*, 1300–1314. [CrossRef]
5. Scheid, A.; Félix, L.M.; Martinazzi, D.; Renck, T.; Kwietniewski, C.E.F. The microstructure effect on the fracture toughness of ferritic Ni-alloyed steels. *Mater. Sci. Eng. A* **2016**, *661*, 96–104. [CrossRef]
6. Kobelco. Kobelco's welding consumables for LNG storage tanks made of 9%Ni steel. *Kobelco Weld Today* **2011**, *14*, 3–8.
7. Mathers, G. *Welding of Ferritic Cryogenic Steels*; The Welding Institute: Cambridge, UK, 2012.
8. Gustafsson, M.; Thuvander, M.; Bergqvist, E.L.; Keehan, E.; Karlsson, L. Effect of welding procedure on texture and strength of nickel-based WM. *Sci. Technol. Weld. Jt.* **2007**, *12*, 549–555. [CrossRef]
9. Yoon, Y.K.; Kim, J.H.; Shim, K.T.; Kim, Y.K. Mechanical characteristics of 9%Ni steel welded joint for LNG storage tank at cryogenic. *Int. J. Mod. Phys.* **2012**, *6*, 355–360. [CrossRef]
10. Mu, W.; Li, Y.; Cai, Y.; Wang, M. Cryogenic fracture toughness of 9%Ni steel flux cored arc welds. *J. Mater. Process. Technol.* **2018**, *252*, 804–812. [CrossRef]

11. Jang, J.I.; Ju, J.B.; Lee, B.W.; Kwon, D.; Kim, W.S. Effects of microstructural change on fracture characteristics in coarse-grained heat-affected zones of QLT-processed 9%Ni steel. *Mater. Sci. Eng. A* **2003**, *340*, 68–79. [CrossRef]

12. Mizumoto, M.; Motomatsu, R.; Nagasaki, H.; Iijima, T.; Kobayashi, K.; Mizo, Y. Development of submerged arc welding in vertical up position: A study on welding consumables for 9%Ni steel. *J. Japan Weld. Soc.* **2008**, *82*, 48–49.

13. Park, J.Y.; Lee, J.M.; Kim, M.H. An investigation of the mechanical properties of a weldment of 7% nickel alloy steels. *Metals* **2016**, *6*, 285. [CrossRef]

14. Bourges, P.; Malingraux, M. *Fabrication and Welding of Thick plates in 9%Ni Cryogenic Steel*; Industeel Co.: Le Creusot, France, 2008.

15. Nippes, E.F.; Balaguer, J.P. A study of the weld heat-affected zone toughness of 9%Nickel steel. *Weld. Res.* **1986**, *9*, 237S–243S.

16. Gook, S.; Forquer, M.; El-Batahgy, A.; Gumenyuk, A.; Rethmeier, M. Laser and hybrid laser-arc welding of cryogenic 9%Ni steel for construction of LNG storage tanks. In Proceedings of the 3rd International Conference in Africa and Asia on Welding and Failure Analysis of Engineering Materials, Luxor, Egypt, 2–5 November 2015.

17. El-Batahgy, A.; Gumenyuk, A.; Gook, S.; Rethmeier, M. Comparison between GTA and laser beam welding of 9%Ni steel for critical cryogenic applications. *J. Mater Process. Technol.* **2018**, *261*, 193–201. [CrossRef]

18. Wu, Y.; Cai, Y.; Sun, D.; Zhu, J.; Wu, Y. Microstructure and properties of high-power laser welding of SUS304 to SA553 for cryogenic applications. *J. Mater. Process. Technol.* **2015**, *225*, 56–66. [CrossRef]

19. Huang, Z.; Cai, Y.; Mu, W.; Li, Y.; Hua, X. Effects of laser energy allocation on weld formation of 9%Ni steel made by narrow gap laser welding filled with nickel based alloy. *J. Laser Appl.* **2018**, *30*, 032013. [CrossRef]

20. El-Batahgy, A.; Khourshid, A.; Sharef, T. Effect of laser beam welding parameters on microstructure and properties of duplex stainless steel. *Mater. Sci. Appl.* **2011**, *10*, 1443–1451. [CrossRef]

21. Quiroz, V.; Gumenyuk, A.; Rethmeier, M. Laser beam weldability of high-manganese austenitic and duplex stainless steel sheets. *Weld. World* **2012**, *56*, 9–20. [CrossRef]

22. El-Batahgy, A.; Tsuboi, A. Effect of welding process type on mechanical and corrosion properties of SUS329J4L duplex stainless steel. *J. Strength Fract. Complex.* **2013**, *8*, 189–203. [CrossRef]

23. El-Batahgy, A.; DebRoy, T. Nd-YAG laser beam and GTA welding of Ti-6Al-4V alloy. *Int. J. Eng. Tech. Res.* **2014**, *12*, 43–50.

24. Stavridis, J.; Papacharalampopoulos, A.; Stavropoulos, P. Quality assessment in laser welding: A critical review. *Int. J. Adv. Manuf. Technol.* **2018**, *94*, 1825–1847. [CrossRef]

25. Verma, J.; Taiwade, R.V. Effect of welding processes and conditions on the microstructure, mechanical properties and corrosion resistance of duplex stainless-steel weldments–A review. *J. Manuf. Processes.* **2017**, *25*, 134–152.

26. Agusa, K.; Kosho, M.; Nishiyama, N.; Kitagawa, M.; Nakazawa, M. Production of 9%Ni steel UOE pipe with ferritic filler submerged arc welding. *Trans. Iron Steel Inst. Jpn.* **1986**, *26*, 359–366. [CrossRef]

27. Koshiga, F.; Tanaka, J.; Watanabe, I.; Takamura, T. Matching ferritic consumable welding of 9%Nickel steel to enhance safety and economy. *Weld. J.* **1984**, *4*, 105s–115s.

28. El-Batahgy, A.; Saiyah, A.; Khafagi, S.; Gumenyuk, A.; Gook, S.; Rethmeier, M. Shielded metal arc welding of 9%Ni steel using matching ferritic filler metal. *Sci. Technol. Weld. Join.* **2021**, *26*, 116–122. [CrossRef]

29. Li, X.; Chen, Y.; Hao, B.; Han, Y.; Chu, Y.; Zhang, J. The microstructure and microscopic mechanical performance of welded joint for 9%Ni steel using nickel-based filler metal. *Mat. Res.* **2021**, *5*, 117–125.

30. Zhao, X.-Q.; Pan, T.; Wang, Q.-F.; Su, H.; Yang, C.-F.; Yang, Q.-X. Effect of tempering temperature on microstructure and mechanical properties of steel containing Ni of 9%. *J. Iron Steel Res. Int.* **2011**, *18*, 47–51. [CrossRef]

31. Gook, S.; Krieger, S.; Gumenyuk, A.; El-Batahgy, A.; Rethmeier, M. Notch impact toughness of laser beam welded thick sheets of cryogenic nickel alloyed steel X8Ni9. *Procedia CIRP* **2020**, *94*, 627–631. [CrossRef]

materials

Article

Individual Effects of Alkali Element and Wire Structure on Metal Transfer Process in Argon Metal-Cored Arc Welding

Hanh Van Bui [1], Ngoc Quang Trinh [1,2,*], Shinichi Tashiro [2,*], Tetsuo Suga [2], Tomonori Kakizaki [3], Kei Yamazaki [3], Ackadech Lersvanichkool [4], Anthony B. Murphy [5] and Manabu Tanaka [2]

[1] School of Mechanical Engineering, Hanoi University of Science and Technology, Hanoi 100-000, Vietnam; hanh.buivan@hust.edu.vn

[2] Joining and Welding Research Institute, Osaka University, Osaka 567-0047, Japan; suga@jwri.osaka-u.ac.jp (T.S.); tanaka@jwri.osaka-u.ac.jp (M.T.)

[3] Kobe Steel, Ltd., Fujisawa 251-8551, Japan; kakizaki.tomonori@kobelco.com (T.K.); yamazaki.kei@kobelco.com (K.Y.)

[4] Thai Kobelco Welding Co., Ltd., Muang 10280, Thailand; ackadech.lers@kobelco.com

[5] CSIRO Manufacturing, Lindfield, NSW 2070, Australia; tony.murphy@csiro.au

* Correspondence: n.trinh@jwri.osaka-u.ac.jp (N.Q.T.); tashiro@jwri.osaka-u.ac.jp (S.T.); Tel.: +81-6-6879-6666 (S.T.)

Abstract: This study aimed to clarify the effect of wire structure and alkaline elements in wire composition on metal transfer behavior in metal-cored arc welding (MCAW). A comparison of metal transfer in pure argon gas was carried out using a solid wire (wire 1), a metal-cored wire without an alkaline element (wire 2), and another metal-cored wire with 0.084 mass% of sodium (wire 3). The experiments were conducted under 280 and 320 A welding currents, observed by high-speed imaging techniques equipped with laser assistance and bandpass filters. At 280 A, wire 1 showed a streaming transfer mode, while the others showed a projected one. When the current was 320 A, the metal transfer of wire 2 changed to streaming, while wire 3 remained projected. As sodium has a lower ionization energy than iron, the mixing of sodium vapor into the iron plasma increases its electrical conductivity, raising the proportion of current flowing through metal vapor plasma. As a result, the current flows to the upper region of the molten metal on the wire tip, with the resulting electromagnetic force causing droplet detachment. Consequently, the metal transfer mode in wire 3 remained projected. Furthermore, weld bead formation is the best for wire 3.

Keywords: gas metal arc welding; metal-cored wire; metal transfer; alkali element; current path; metal vapor

Citation: Bui, H.V.; Trinh, N.Q.; Tashiro, S.; Suga, T.; Kakizaki, T.; Yamazaki, K.; Lersvanichkool, A.; Murphy, A.B.; Tanaka, M. Individual Effects of Alkali Element and Wire Structure on Metal Transfer Process in Argon Metal-Cored Arc Welding. *Materials* **2023**, *16*, 3053. https://doi.org/10.3390/ma16083053

Academic Editor: Raul D. S. G. Campilho

Received: 8 March 2023
Revised: 10 April 2023
Accepted: 11 April 2023
Published: 12 April 2023

1. Introduction

Since the gas metal arc welding (GMAW) process was invented and implemented, the transfer behavior of molten metal to the weld pool has been extensively investigated in many experimental and numerical studies due to its importance on the welding performance. For instance, Rhee and Kannatey-Asibu observed the metal transfer phenomenon for several transfer modes [1], and Liu and Siewert measured the transfer rate of molten metal to the weld pool in GMAW [2]. Hu and Tsai measured the arc characteristics and interactive coupling between the arc and molten metal using a unified comprehensive model [3]; furthermore, the melting of the electrode and droplet transfer regime was also predicted [4]. The literature concluded that the metal transfer was strongly influenced by the welding parameters, particularly welding current, shielding gas composition, and type of wire electrode [5]. The welding parameters naturally determine the strength of driving forces acting on the molten metal on the wire. Furthermore, auxiliary forces caused by laser irradiation [6], ultrasonic assistance [7], and wire movement [8] can be applied to improve droplet detachment.

The GMAW process can be carried out using a conventional solid or a tubular wire electrode. When the solid wire is used in pure argon shielding gas, the metal transfer mode is transformed depending on the welding current. In the free-flight transfer mode of the natural metal transfer class reported by Scotti et al. [9], the transfer mode can be classified as a globular, spray, or explosive transfer. The globular transfer was observed at a low welding current, in which the molten metal was transferred in large drops with a diameter greater than the wire diameter. The metal transfer was observed to change from globular to spray mode, with small droplets detaching at a high frequency when the welding current increases to a critical value [10]. In spray transfer mode, the metal transfer can be divided into projected, streaming, and rotating transfers according to increasing welding current [11]. The projected spray transfer is desirable due to a stable arc and metal transfer. A further high welding current can be applied when a welding process with a high deposition rate is required. However, the streaming and rotating transfer with an unstable arc and liquid metal transfer limits the welding current range, as reviewed by Paul et al. [12]. These studies imply that broadening the condition range to achieve a projected spray is advantageous; however, the mechanism to cause the transition of metal transfer mode is dependent on many factors and is not fully understood yet.

On the other hand, the welding process can also be carried out with a tubular wire, known as flux-cored arc welding (FCAW), to extend the application. The tubular wire consists of a metal sheath with flux powder inside. That configuration allows us to adjust the wire's chemical composition or wire structure to improve weldability. For instance, Wang et al. evaluated the metal transfer of flux-cored wire through electrical arc signal, droplet diameter, and high-speed imaging measurement [13]. They reported that several different metal transfer modes could have coexisted simultaneously. Valensi et al. [14] found that the alkaline elements contribute to stabilizing the projected spray arc in an argon-CO_2 gas mixture with 60%vol of CO_2 at a 330 A welding current. Several fluorides, such as CaF_2, KF, or K_2SiF_6, effectively control the hydrogen content contaminated in the weld bead; however, their presence causes an unstable arc with an undesired spatter, as reported in [15]. In addition, a wire with a high flux ratio was observed to reduce the amount of fume formation [16] and increase the metal transfer frequency [17]. The studies above implied that a tubular wire has a more complex metal transfer regime than a conventional solid wire.

Generally, the flux-cored wire can be classified into several groups, such as rutile, metal, basis gas-shielded electrode, and self-shielded electrode, depending on the flux formulation. The flux in the metal-cored wire type principally consists of iron powder to provide a high deposition rate. Starling and Modenesi [18] investigated the droplet transfer regime of three gas-shielded electrode types under 75% argon-25% CO_2 and 100% CO_2 shielding gas. It should be noted that the transfer behavior of metal-cored wire resembles that of solid wire. On the other hand, Trinh et al. [17] compared the weldability of a solid wire to three prototyped metal-cored wires in metal active gas (MAG) consisting of argon +20% CO_2. They stated that the transfer frequency of the solid wire is much lower than that of the metal-cored wires, and the frequency increased with flux ratio and welding current. In addition, the effect of alkaline element proportion on metal transfer under MAG shielding gas was reported by Trinh et al. [19]. The result implied that an alkaline element could improve droplet detachment by forming an additional current path, bypassing the droplet to enhance the electromagnetic force to detach the droplet and reduce the arc pressure beneath the droplet. In the paper, the result was not compared with that for solid wire, so the effect of wire structure was not discussed. A previous study investigated the difference in the metal transfer between a solid and metal-cored wire in pure argon shielding gas [20]. It was reported that the metal-cored wire was not streaming transfer like the solid wire at a high welding current of 280 A. The difference can be explained by the fact that the presence of unmelted flux on the wire tip diminishes the formation of a liquid column necessary for a streaming transfer, which can be referred to as the effect of the wire structure. In this paper, a commercial metal-cored wire was used for MCAW, so the alkali element was not controlled as an experimental parameter.

As described above, the wire structure and the addition of an alkali element are considered to be two dominant factors to govern the metal transfer process in MCAW. However, the previous works [19,20] could not individually evaluate the effect of wire structure and alkaline elements. In this study, for the first time, the individual influences of wire structure and alkaline elements in metal-cored wire on the metal transfer behavior were evaluated. A solid wire, a metal-cored wire without an alkaline element, and a metal-cored wire with sodium as an alkaline element were investigated under a high welding current of 280 and 320 A, especially targeting the transition to streaming transfer in MCAW. The metal transfer behavior of the three wires was measured through a high-speed camera and spectroscopic observation. The comparison of the solid wire and the metal-cored wire without sodium can indicate the effect of wire structure; meanwhile, the comparison of the two metal-cored wires considers the impact of wire composition. Consequently, the mechanism of the metal transfer behavior was clarified.

2. Experimental Procedure

2.1. Materials and Welding Parameters

In the current work, welding experiments were produced on mild steel plates (SS400-JIS G 3101). The chemical compositions and mechanical properties of the base material are shown in Tables 1 and 2, respectively. The dimensions of a workpiece are 300 mm × 50 mm × 9 mm. The main purpose of this paper is to clarify the metal transfer process, so bead-on-plate welding is carried out for simplicity as in [17,19,20]. To clarify the metal transfer behavior, three types of wire electrodes with a diameter of 1.2 mm were investigated, including a solid commercial wire and two prototyped metal-cored wires. The solid wire (wire 1) is the JIS Z3312 YGW11 wire, which corresponds to the A5.18 ER70S-G wire of the AWS classification. Meanwhile, the two metal-cored wires correspond to the A5.20 E70T-1C wire of AWS classification, including a wire without any alkaline elements (wire 2) and a wire with a flux containing 0.084 mass% of sodium (wire 3). According to production limitations, only the amount of sodium was controlled as an experimental parameter. Wire 3 is same as the wire 4 in [19], which has the largest amount of sodium of the prototyped wires. The details of the chemical compositions of the three wires are listed in Table 3.

Table 1. Chemical compositions (mass%) of base material plates (JIS G 3101-2015).

Base Material	C	Mn	S	P
Mild steel SS400	–	–	$\leq$0.050	$\leq$0.050

Table 2. Mechanical properties of mild steel SS400 for 9 mm plates (JIS G 3101-2015).

Base Material	Yield Point (N/mm^3)	Tensile Strength (N/mm^2)	Elongation (%)	Bendability	
				Bending Angle	Inner Radius (mm)
Mild steel SS400	$\geq$245	400–510	$\geq$17	180°	13.5

Table 3. Chemical compositions (mass%) of wires.

Wire	Fe	C	Si	Mn	Cu	Al	Ti + Zr	Na
Wire 1	97.20	0.04	0.73	1.58	0.23	-	0.22	0
Wire 2	96.63	0.04	0.90	2.00	-	0.26	0.17	0
Wire 3	95.79	0.04	0.90	2.00	-	0.26	0.17	0.084

The welding process was conducted in direct current electrode positive (DCEP) mode using a DP-350 welding power source (OTC Daihen) equipped with an appropriate welding torch and wire feeder system. The metal transfer of three wires was investigated under two welding current levels at 280 A and 320 A. In this paper, we focused on a current range around the transition from globule (or project) transfer to streaming transfer for MCAW. The welding current was maintained among wires by varying the wire feeding speed for each wire. The welding voltage was adjusted to maintain a constant arc length of approximately 5.2 mm, for observing the metal transfer process clearly. During the welding, the welding torch was fixed, and the base metal was moved by an actuator. The details of the welding conditions are shown in Table 4. An experimental photo setup is depicted in Figure 1.

Table 4. Summary of welding conditions.

Parameters	Value/Unit
Welding current (output)	280 and 320 A
Welding voltage (output)	31.2–33.8 V
Shielding gas	Pure argon, 20 L/min
Contact tip to work distance	20 mm
Welding velocity	7 mm/s

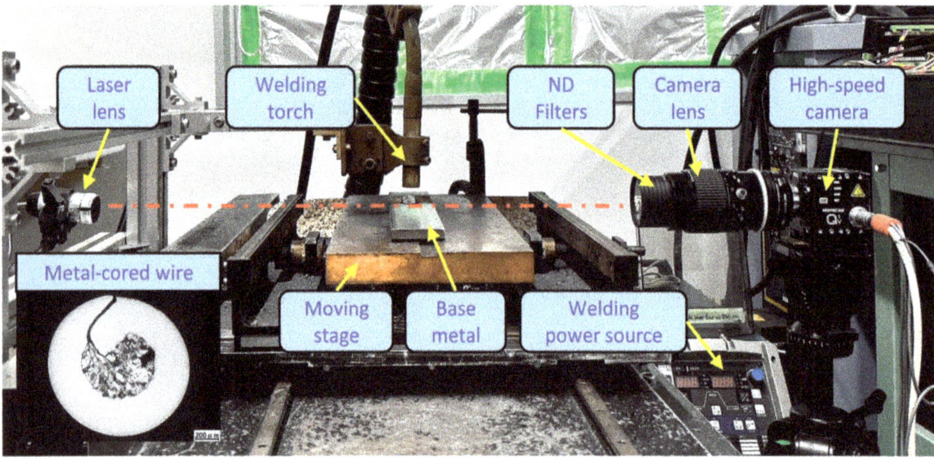

Figure 1. A photo of the experimental setup.

2.2. Visualization Procedure

In this study, a shadowgraph method was applied to observe the metal transfer behavior. The technique uses a radiation light source to illuminate an object from the opposite side of a camera, which allows the camera to obtain the shadow as the geometry of the object. The observation system consisted of a high-speed camera (Memrecam Q1v, Nac Image Technology, Tokyo, Japan), a camera lens (AF Micro-Nikkor, Nikon, Tokyo, Japan), and a laser system (Cavilux HF, Cavitar, Tampere, Finland). The laser system releases a laser from the laser lens to the camera lens, as shown in Figure 1. The camera was set to focus on the arc area, which obtains a clear image with a size of 640 × 480 pixels. The aperture value was f/4. Five neutral (ND) filters were utilized to reduce the intensive arc radiation. The observation conditions are summarized in Table 5.

Table 5. Summary of laser observation conditions.

Parameters	Value/Unit
Camera name	Nac, Memrecam Q1v
Frame rate	4000 fps
Shutter speed	20 μs
Aperture	f/4
Image	640 × 480 pixels
Laser wavelength	640 nm
The central wavelength of iron filter (Fe I filter)	540.0 nm
The central wavelength of sodium filter (Na I filter)	589.0 nm

In addition, a spectral observation technique was applied to investigate the arc properties in more detail. The camera was equipped with a bandpass filter, which allows light with a narrow range of wavelengths to go through. Two bandpass filters, appropriate to observe spectra of iron plasma for three wires and sodium plasma for wire 3, were applied, termed Fe I filter and Na I filter, respectively. The central wavelength of the two filters is presented in Table 2. The two filters had a full width at half maximum (FWHM) of 10.0 nm. In this experiment, the number of ND filters is three ND8, and the aperture was set at f/22.

3. Results and Discussions

3.1. Observation Results

The observation results of the metal transfer of the three investigated wires at a welding current of 280 A are shown in Figure 2 as time-sequential images. The molten metal transportation and arc light concentration were captured simultaneously, which supports an evaluation of the arc attachment behavior on the wire tip during welding. The figure shows one cycle of the molten metal movement from the wire electrode to the weld pool. In Figure 2a, the metal transfer of the solid wire (wire 1) shows a streaming transfer in which a long liquid column in a pencil shape forms at the tip of the wire. The molten droplet was separated at the end of the column to contact the surface of the weld pool. At this welding current, the metal transfer of wire 1 in pure argon gas was reported in previous studies to be a streaming transfer mode [20,21]. The duration of one transfer in this condition is short at around 3 ms. However, due to the molten metal coming into direct contact with the molten pool unevenly, short-circuiting can occur between the end of the liquid column and the top of the molten metal droplet, as depicted at a frame of 0 ms. Consequently, the short-circuiting phenomenon resulted in a spatter, as observed in the period from frame 0.5 to 2 ms.

Figure 2b,c shows the metal transfer of the metal-cored wire without sodium (wire 2) and the metal-cored wire with sodium (wire 3), respectively. The droplet was transferred in a projected spray transfer mode in both circumstances. It can be observed that the droplet diameter was similar to the wire diameter. The times for a cycle transfer for wires 2 and 3 are 5.75 and 6.5 ms, corresponding to droplet transfer frequencies of approximately 174 and 154 Hz, respectively. For these two wires, in the duration of the droplet growth, it can be observed that the arc was attached at the neck position between the droplet and wire tip to cover the entire droplet. It can be considered that when the arc attachment is moved overhead of the droplet, a large part of the current is conducted directly from the neck of the wire to the argon plasma, avoiding the lower part of the droplet, because the dense iron plasma under the droplet has a relatively low electrical conductivity compared to that of argon plasma [22], so the electromagnetic force will act on the neck position effectively to enhance the droplet detachment. Meanwhile, the arc pressure in welding with pure argon gas was considered to be low. Thus, the droplet in metal-cored wires was transferred smoothly in projected mode. The reason for the metal transfer of metal-cored wire not to be streaming transfer was explained by Trinh et al. [20], in which two commercial solid and metal-cored wires were compared in pure argon shielding gas. It should be noted that during the arcing time, the unmelted flux column of metal-cored wire prevented the liquid column formation necessary for streaming transfer mode, particularly at 280 A welding current.

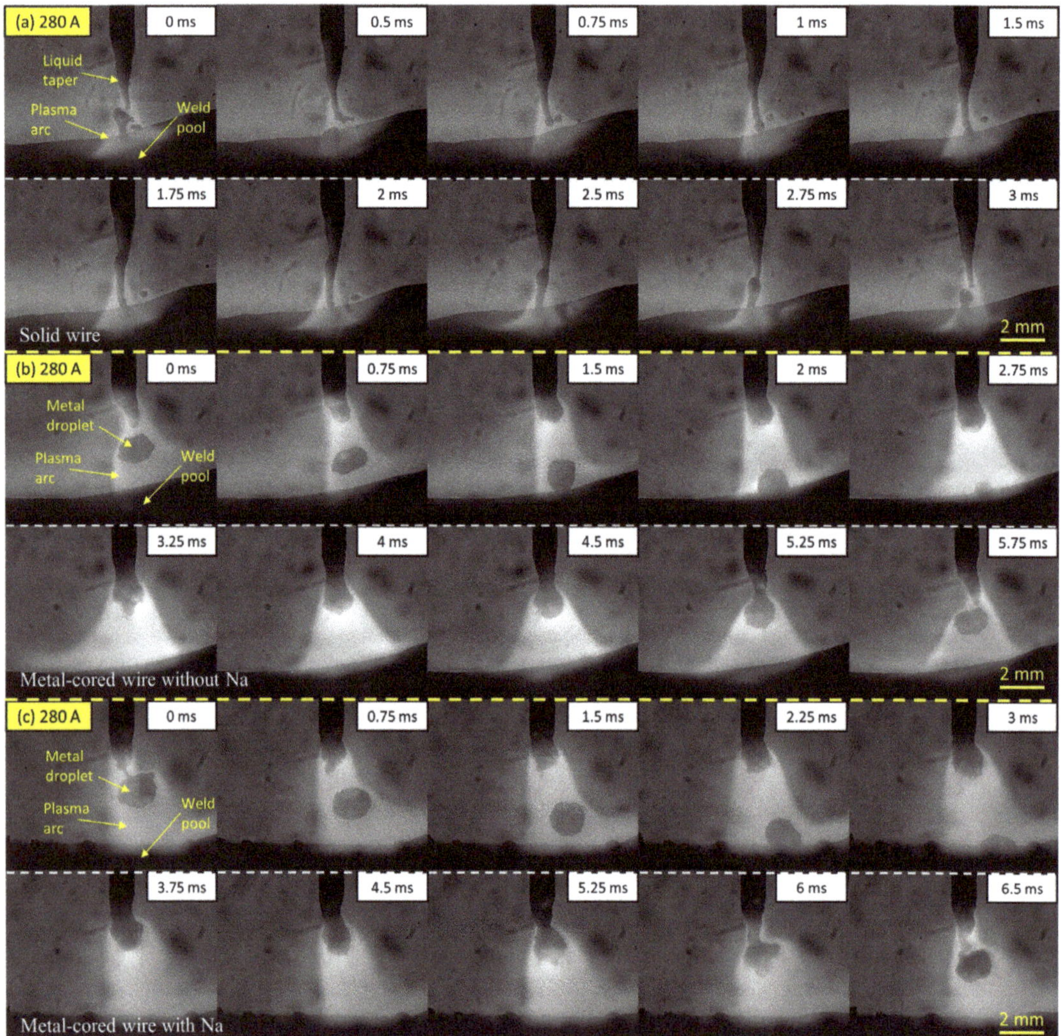

Figure 2. Metal transfer in wire 1 (**a**), wire 2 (**b**), and wire 3 (**c**) at 280 A welding current.

The metal transfer regime of the three wires at a high welding current of 320 A was compared in Figure 3. In Figure 3a, wire 1 shows a streaming transfer mode in the same manner as the transfer behavior in Figure 2a. At this welding current, the tip of the liquid column was extended to reach the surface of the base metal. The liquid column was thought to be related to the influence of electromagnetic force acting on the wire. To facilitate the droplet separation, the current following via molten metal on the wire tip must be sufficiently strong.

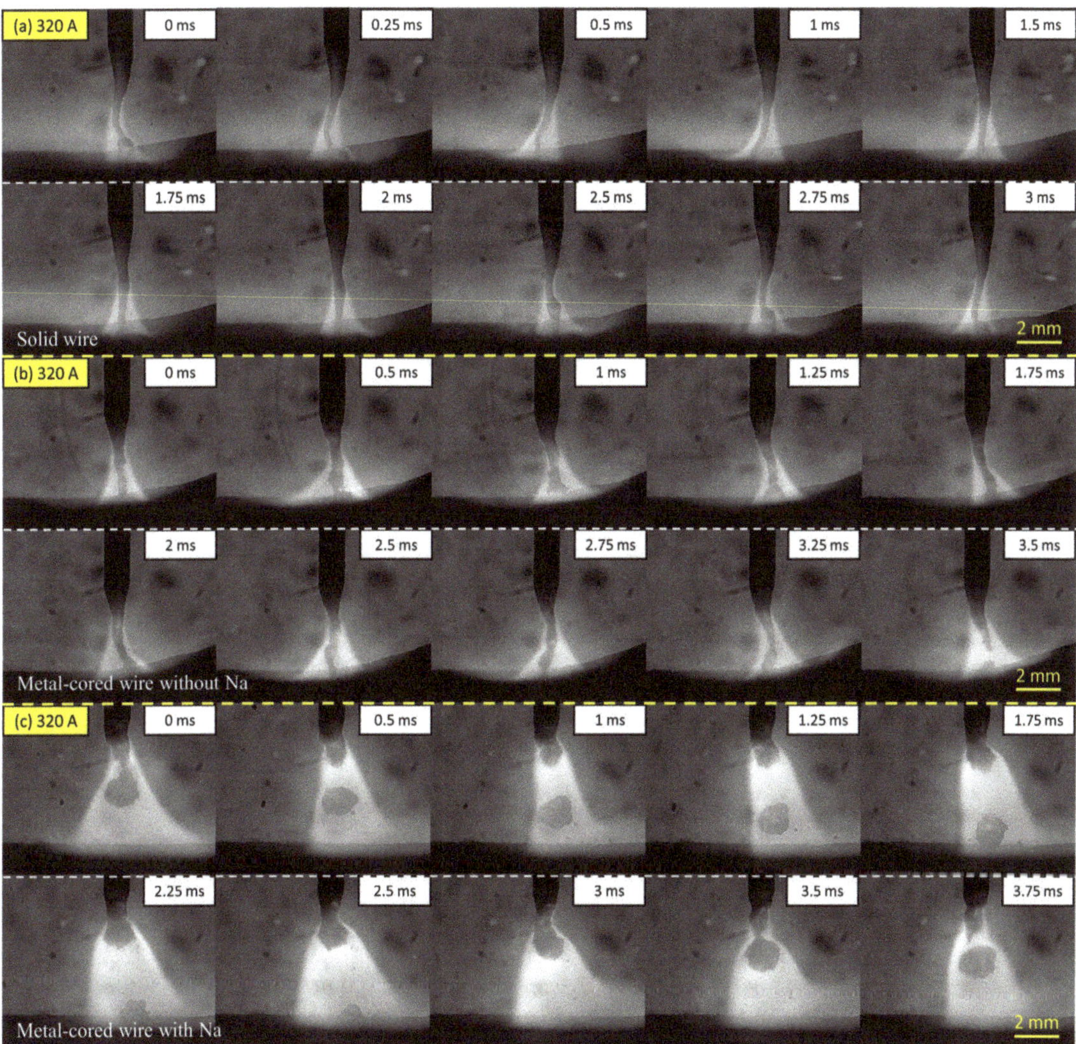

Figure 3. Metal transfer in wire 1 (**a**), wire 2 (**b**), and wire 3 (**c**) at 320 A welding current.

On the other hand, when a large proportion of current flows through the gas plasma rather than the molten droplet, the strong downward momentum of electromagnetic force squeezes molten metal on the wire tip to form a long liquid column [22]. When the welding current increases to 320 A, Joule heating becomes higher than 280 A, which increases the temperature to reduce the surface tension of the molten metal. As a result, the length of the column in 320 A was larger than that in 280 A. The tip of the column becomes slightly waved under the effect of driving forces, as shown in frames 0 to 1.5 ms. It was reported that when the current increases continually to around 400 A, the liquid column becomes unstable and rotates around the wire axis due to the strong electromagnetic force acting on the molten metal [23].

Figure 3b shows the images of metal transfer in wire 2 without sodium in the flux. The metal transfer mode is in streaming transfer, which is different from the projected transfer mode of this wire in 280 A, as shown in Figure 2b. A long liquid column was combined with a melted solid wire sheath and flux inside the wire for the streaming transfer in this wire. On the other hand, the metal transfer in wire 3 with 0.084% of sodium, shown in Figure 3c, was maintained in a projected transfer similar to 280 A. The cycle time for a droplet transfer is around 3.75 ms, corresponding to a transfer frequency of approximately 267 Hz. A molten droplet was generated and transferred under the arc attachment position during the projected transfer for this wire.

This study also investigated distributions of iron plasma and sodium plasma at 320 A to clarify the difference in metal transfer mechanism among the wires. Figure 4 shows the iron plasma distribution obtained by the high-speed camera with the Fe I filter. In Figure 4a, the iron plasma in wire 1 was located around the tip of the liquid column, which generated iron vapor plasma in a conical shape from a position higher than the detachment point to the surface of the weld pool. The result is consistent with a previous study by Trinh et al. [20], in which the iron and argon plasma in the streaming transfer was observed to separate to form a dual structure in the arc plasma. Thus, the argon plasma in this circumstance was expected to cover iron plasma at the arc center. In Figure 4b, the iron plasma distribution of wire 2 is similar to that of wire 1 in streaming transfer.

In contrast, the iron plasma distribution of wire 3 in Figure 4c shows a different behavior. During the projected transfer, the iron plasma attached under the droplet from the beginning of droplet formation, as observed from frame 0.5 to 3.25 ms. Due to the smaller size of the molten droplet, the entire droplet was enveloped by iron plasma.

Figure 5 shows time-sequential photos of metal transfer observation using the Na I filter for wire 3 at 320 A of welding current. It can be observed that the sodium plasma was distributed from the wire tip to the surface of the weld pool. In a previous study conducted by Trinh et al. [19], it was explained that very fine sodium particles were distributed in the wire flux before melting and injected into the arc intermittently to form sodium plasma after melting. At the droplet detachment moment, the sodium plasma under the wire tip was observed to be confined to a radially narrow region, as shown in frames 0 and 0.5 ms. During the period from 1 to 3.25 ms, sodium plasma was broadened from the wire tip and covered the droplet.

Figure 6 compares the iron and sodium plasma distribution of wire 3 at 320 A. Figure 6a shows the iron plasma when a droplet was completely detached from the wire tip, defined as time t_0, and Figure 6b shows the iron plasma at 2 ms after that moment. At the time t_0, the iron plasma has an arc root around the tip of the unmelted flux. The iron plasma gradually expanded to envelop the entire droplet surface, as observed at $t_0 + 2$ ms. In addition, Figure 6c,d show the sodium plasma distribution when another droplet detached from the wire tip, defined as time t_1, and that at $t_1 + 2$ ms, respectively. The sodium plasma was concentrated at the tip of the melted flux, as shown in Figure 6c. At the time t_1, the sodium vapor evaporated from the melted flux was observed to be distributed radially narrower than the iron vapor. The sodium vapor broadened at the wire tip when the droplet size increased at the moment $t_1 + 2$ ms, as shown in Figure 6d.

3.2. Metal Transfer Mechanism

Based on the experimental results, a schematic of the arc plasma distribution for wire 3 was suggested in Figure 7. Three plasma regions were considered, including sodium, iron, and argon plasma. In metal-cored wire 3, sodium is added only to the flux inside the wire, which is enveloped by the wire sheath. Even though sodium has a significantly lower boiling point than iron (1156 K compared to 3134 K), sodium could not evaporate at a higher position than the solid part of the wire sheath covering the flux around the wire tip because the solid wire sheath is thought to prevent the evaporation. In addition, the arc attachment was reported to move upward when the iron intensively evaporated because the arc temperature was lowered by a substantial radiation loss, causing a reduction in the electrical conductivity of iron plasma [22]. It should be noted that the highest position of arc attachment corresponds to the taper part of the wire sheath, where the wire surface begins to melt by the intensive heat flux from the high-temperature arc and the electron condensation [24,25]. The iron vapor is considered to start the evaporation from this taper part, unlike the sodium vapor.

Furthermore, with pure argon shielding gas, the argon plasma was observed to be located at the top of the molten metal, and the iron plasma was connected closely below the argon plasma for welding a commercial metal-cored wire [20]. The commercial metal-cored wire contains some alkaline elements, such as potassium and sodium, which may lead the arc phenomena similar to wire 3 in this study.

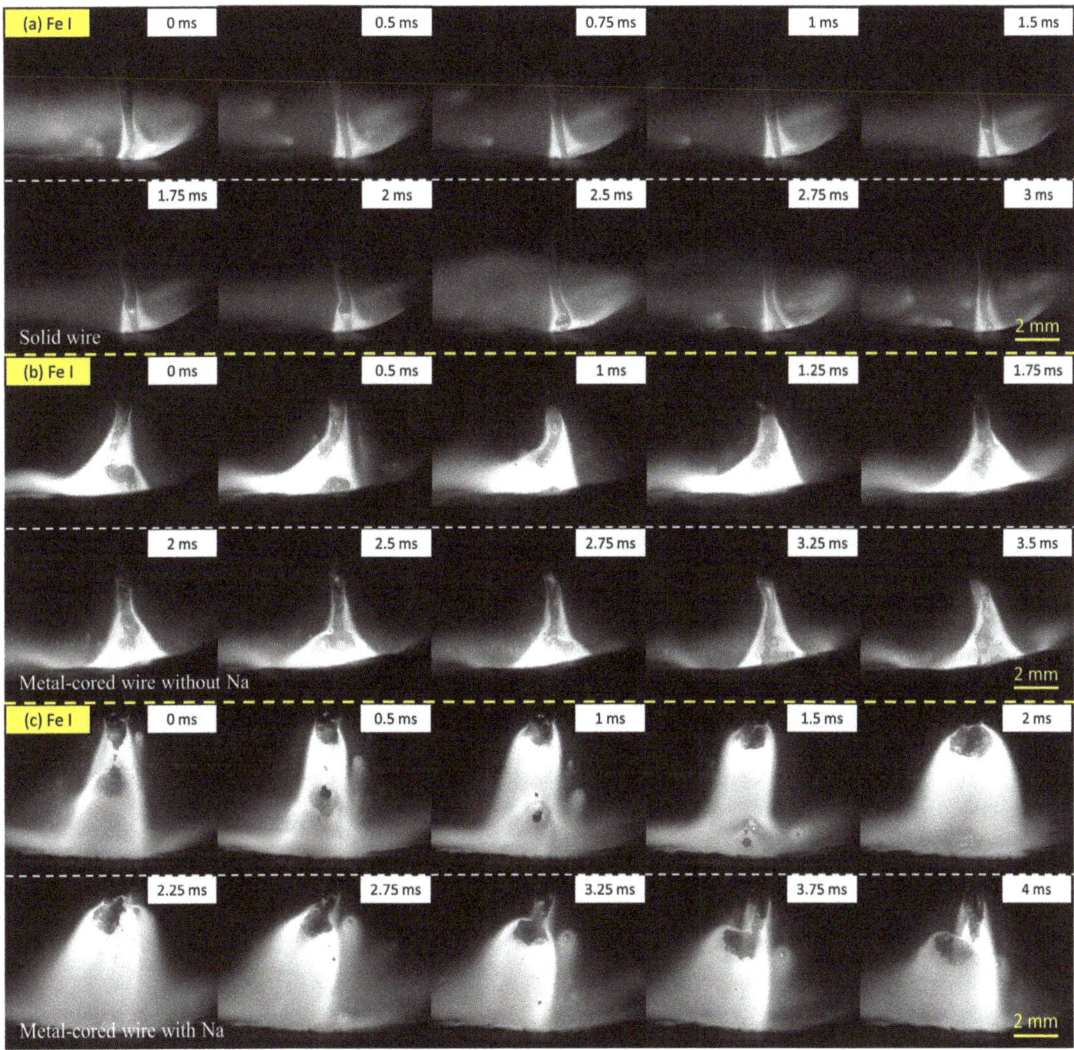

Figure 4. Metal transfer observation for wire 1 (**a**), wire 2 (**b**), and wire 3 (**c**) using the Fe I filter at 320 A of welding current.

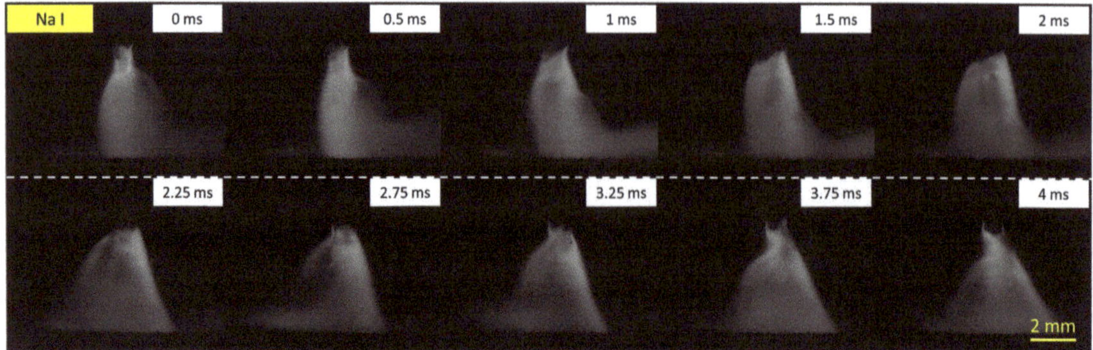

Figure 5. Metal transfer observation using the Na I filter for wire 3 at 320 A welding current.

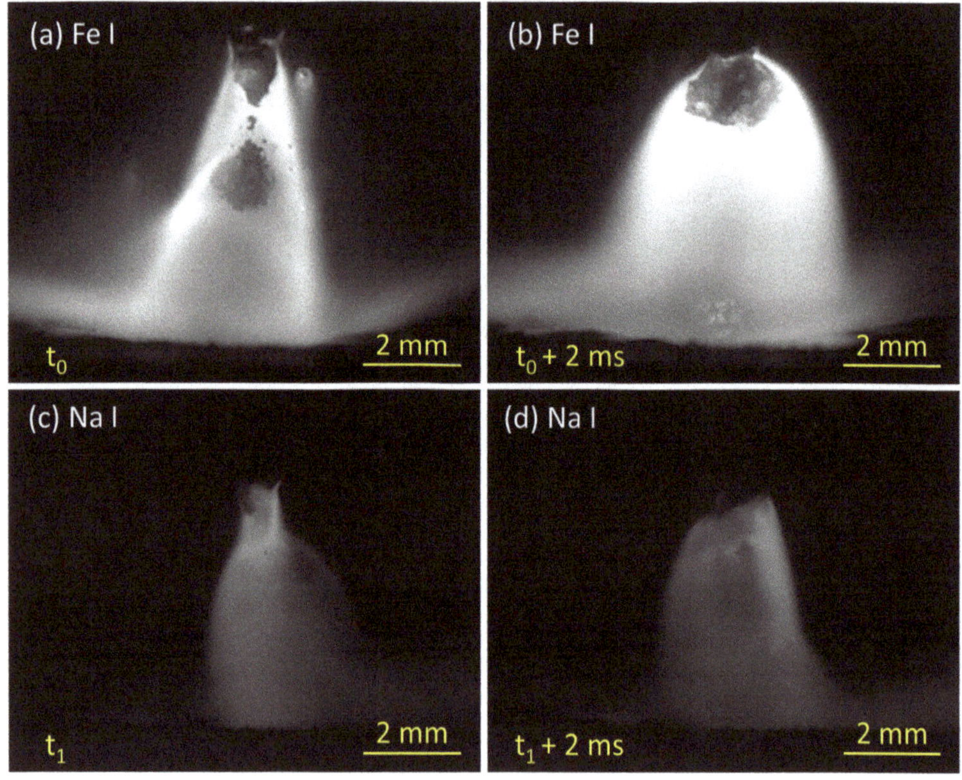

Figure 6. The typical images of iron vapor plasma (**a**,**b**) and sodium vapor plasma distribution (**c**,**d**) of wire 3 at 320 A of welding current.

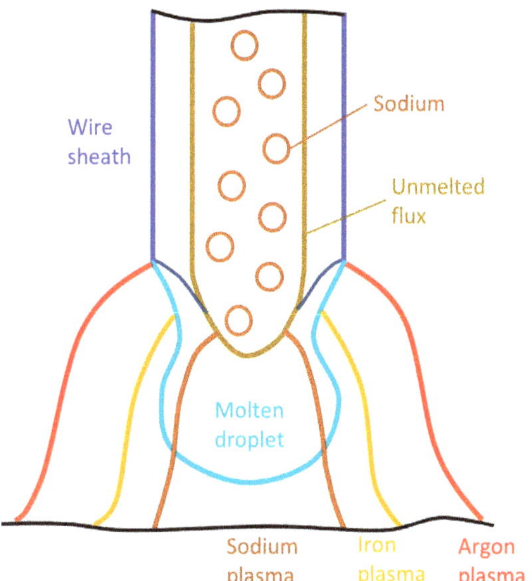

Figure 7. Position of plasma distribution for wire 3.

Trinh et al. [19] found that sodium inside the wire assisted a new current path from the wire tip to the base metal to bypass the molten droplet. In that experiment, sodium was thought to evaporate from a higher position than iron on the wire tip because the investigation was conducted at a low current of 220 A in argon +20% CO_2 shielding gas. Under that welding condition, the arc tended to constrict due to the high specific heat of CO_2. The iron plasma was especially concentrated underneath the bottom of the droplet, lower than the position where sodium evaporated. In this study, pure argon was used as a shielding gas. At a high current of 320 A, the droplet size in the projected transfer was small, which enabled the iron plasma to reach the overhead of the droplet to attach to a higher position than sodium plasma. The arc plasma can be separated into the outer argon plasma and inner metal vapor plasma, in which metal vapor consists of the iron vapor covering the sodium vapor. Therefore, the sodium, iron, and argon plasmas were distributed as discussed above.

Based on the evaluation in Figure 7, a mechanism comparing the metal transfer behavior of the three wires at 320 A welding current was suggested in Figure 8. In Figure 8a,b, the streaming transfer was observed in solid wire 1 and metal-cored wire 2 without sodium. In both circumstances, the argon plasma was located on the surface of the wire. It was reported that most of the current flowed in the argon plasma region owing to its high temperature. On the other hand, iron plasma has a lower temperature due to the strong radiation loss of iron vapor [21,22]. During the welding, the electromagnetic force acting on a molten droplet is significantly influenced by the current path that follows inside the droplet [26,27]. The direction of electromagnetic force in the wire is almost inward, because the axial component of the current is larger than the radial component. According to the separation into the argon and iron plasmas explained above, most of the electromagnetic force is considered to be applied above the iron plasma. When the current paths inside the droplet are insignificant, the electromagnetic force is not enough to separate the droplet from molten metal on the wire tip. As a result, the liquid metal at the wire tip was squeezed to form a long, tapered liquid column.

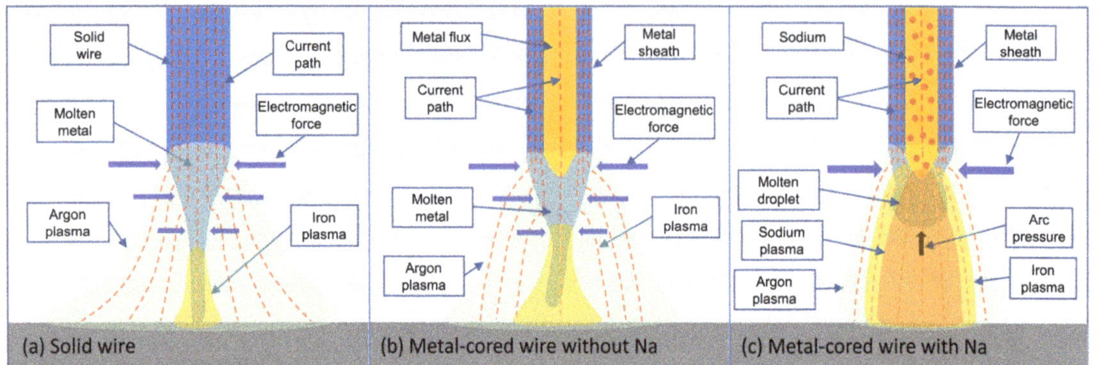

Figure 8. Mechanism of metal transfer (**a–c**) at a high welding current of 320 A.

In Figure 8a, the taper extends to come into contact with the surface of the weld pool. In GMAW, the molten metal at the wire tip was heated by combining the Joule heating effect, thermal conduction from the high-temperature arc, and electron condensation at the surface of the wire. The combination of these heat sources strongly affects the wire melting phenomena and configuration of the wire tip [3,28]. As explained previously, the length of the liquid column in 320 A was larger than that in 280 A. In Figure 8b, the streaming transfer behavior of wire 2 was slightly different. In metal-cored wire, the low electrical conductivity of metal flux leads to most of the current flows inside the solid metal sheath; as a result, the flux was heated mainly from the molten metal sheath, flux, and arc by the thermal conduction. The unmelted flux was ineffective in liquid column formation, which increased the transition current to the streaming transfer in metal-cored wire 2.

In Figure 8c, the metal transfer in the metal-cored wire shows a projected behavior. The droplet was transferred in a diameter less than the wire diameter. In addition, a comparison of the calculated electrical conductivities of the sodium, iron, and argon plasmas as a function of temperature is shown in Figure 9. The calculated value of iron and argon plasma was obtained, as discussed in [29,30]. At the same time, sodium vapor was applied in similar methods, with the momentum transfer for interactions between metal atoms and electrons [31]. The figure shows that sodium plasma has a higher electrical conductivity than iron, especially at a temperature range from 3000 to 7000 K. On the other hand, the electrical conductivity of argon plasma becomes significant at a high temperature above 10,000 K. Sodium and iron have ionization energies of 5.1 eV and 7.9 eV [32], respectively. The presence of sodium plasma in the arc will increase the current following through the metal vapor, thus increasing the current flowing out from the bottom of the droplet. As a result, the electromagnetic force applied on the molten metal and the arc pressure under the droplet increase. Consequently, the metal transfer mode of wire 3 was the projected transfer even in a high welding current.

3.3. Weld Bead Formation

In this study, weld beads were collected after completing experiments to investigate the effect of metal transfer on the weld bead formation. Figure 10 shows the weld bead appearance of three wires at 320 A welding current. In Figure 10a, the weld bead of solid wire 1 shows the variation at the toe of the weld bead along the welding direction. It can be explained by the instability of the liquid column on the wire tip forming in the streaming transfer [33,34]. Before welding, the surface of the base material was ground to remove the oxide layer, which reduced the presence of oxygen in the arc area. In the GMAW process with solid wire in pure argon shielding gas, the arc and metal transfers were unstable due to the movement of cathode spots. This phenomenon can be limited by adding small amounts of oxygen to the shielding gas [35].

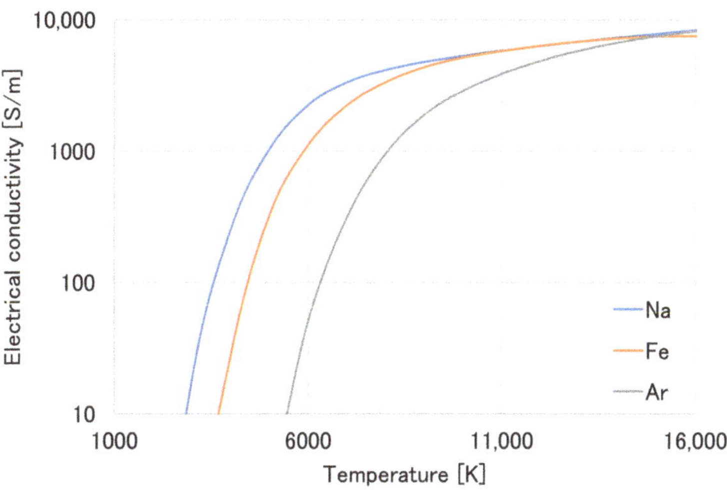

Figure 9. Electrical conductivity of sodium (Na), iron (Fe), and argon (Ar) plasma.

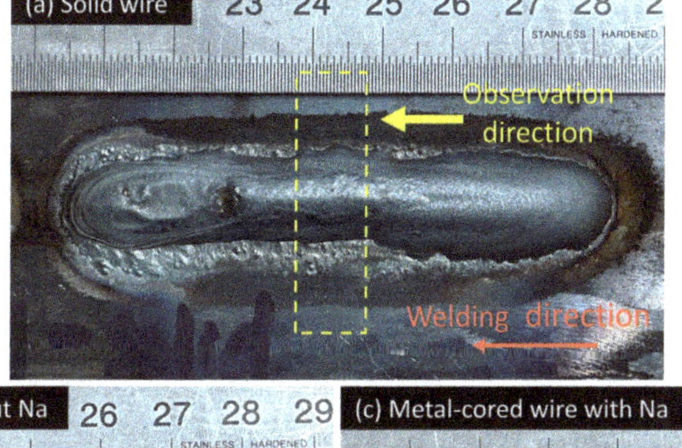

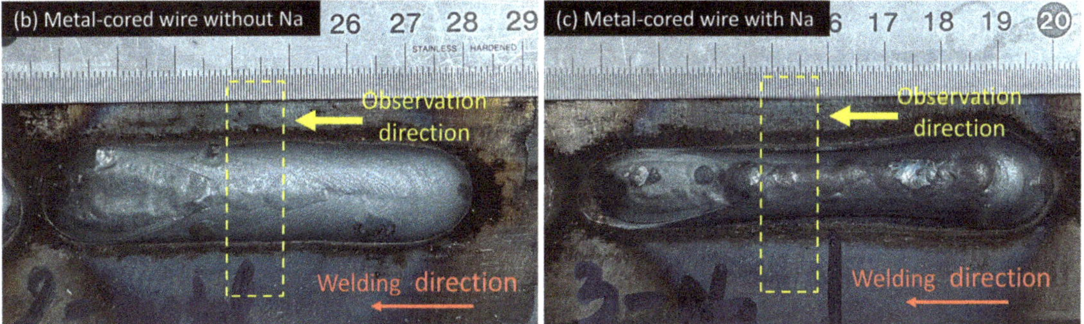

Figure 10. The photo of the weld bead appearance.

On the other hand, the weld bead of two metal-cored wires shows a smooth weld toe. In metal-cored wire, the flux contains oxides of iron and metal alloys, which has a role similar to oxygen in shielding gas to prevent cathode spot movement. In addition, a dashed rectangle in the photos implying a specimen position with observation directions for cross-section investigation is shown in Figure 10.

Figure 11 shows the cross-section of three weld beads presented in Figure 10. The weld bead geometry of solid wire showed a finger shape, as observed in Figure 11a. In pure argon gas, the finger shape is a typical geometry of streaming transfer mode [36,37]. The cross-section shows many porosities located at the root penetration of the weld pool. It is considered that the long liquid column at the wire tip leads to molten metal transferred under the weld pool surface in streaming mode at 320 A, as observed in Figures 3a and 4a. The submerged arc might cause the gasses to be trapped in the root of the weld bead after solidification. The cross-section of metal-cored wire 2 in Figure 11b shows a finger-like shape. The weld penetration in wire 2 was observed to be less than that in wire 1. It implies that the streaming transfer of wire 2 was less intensive than wire 1.

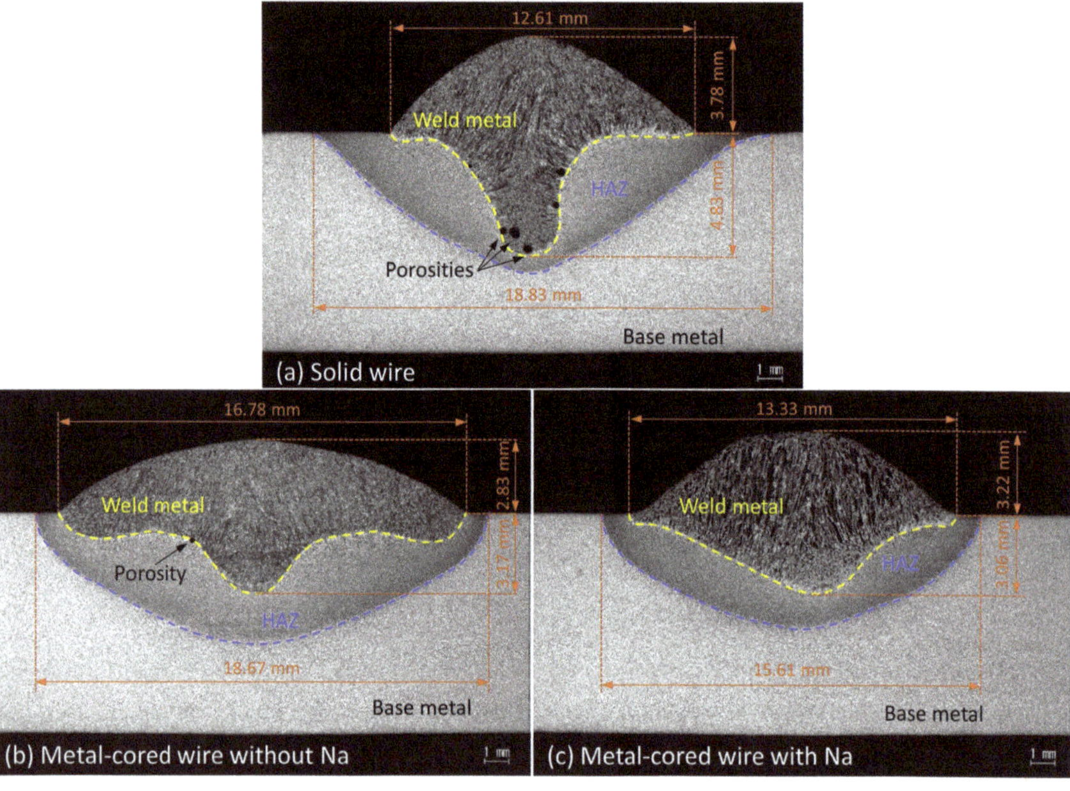

Figure 11. Weld bead cross-section.

On the other hand, the cross-section of wire 3 in Figure 11c shows a shallow geometry. The deep penetration in this circumstance was similar to the bead of wire 2; however, the fusion line is smoother, and the heat-affected zone is narrower than that of wires 1 and 2 (15.61 mm compared to 18.83 and 18.67 mm, respectively). The narrower heat-affected zone implies that the heat input from the arc to base metal is smaller for wire 3, which is considered to be caused by a lowering of the arc voltage (the arc voltages for wires 1, 2, and 3 were 32.2, 33.8, and 31.2 V, respectively). As presented in Figure 9, the addition of an alkali element increases the electrical conductivity of the arc, thus the arc voltage might decrease. This decrease in heat input also leads to the lower wettability of the weld bead for wire 3 in Figure 10. The result indicates that welding with wire 3 has better weld stability than with the other.

The finding in this study implies that the sodium or alkaline elements can improve the arc stability by preventing the metal transfer not streaming at the high welding current.

In addition, the impact of wire composition on the metal transfer is dominant other than the effect of the wire structure in GMAW with metal-cored wire.

4. Conclusions

This study investigated the effect of wire structure and wire composition of the metal-cored wire using high-speed camera observation. The comparison considered the metal transfer behavior for a solid wire, a metal-cored wire without alkaline elements, and another metal-cored wire with 0.084% mass of sodium in pure argon shielding gas under two welding currents of 280 and 320 A. The results lead to the following conclusions:

1. At 280 A, the solid wire showed a streaming transfer, and the other metal-cored wire showed a projected transfer mode. When the current increased to 320 A, the metal transfer of metal-cored wire without sodium changed to the streaming, while that of the metal-cored wire with sodium remained projected.
2. For metal-cored wire without sodium, the reason for the delayed transition to streaming transfer at 280 A corresponds to the wire structure. The unmelted flux with low electrical conductivity inside the wire limited the liquid column formation, which increased the transition current to streaming transfer.
3. In the metal-cored wire containing sodium, the iron plasma was observed to cover the sodium plasma. The presence of sodium plasma increases the current path through the metal vapor plasma region. As a result, the electromagnetic force acting on the neck of the droplet was more effective, and the arc pressure under the droplet increased. Consequently, the metal transfer is projected at 320 A welding current.
4. The results indicated that the effect on the welding stability of an alkaline element in the wire is larger than that of the wire structure for the metal-cored wire. Furthermore, the wire, including sodium, showed the best weld bead formation.

Our results provide strong evidence that the addition of sodium as an alkaline element to the wire is beneficial for metal-cored welding in argon at high currents.

Author Contributions: Conceptualization, methodology, and supervision, H.V.B.; experiment and writing—original draft preparation, N.Q.T.; writing—review and editing, S.T.; discussion, T.S., T.K., K.Y., A.L., A.B.M. and M.T. All authors have read and agreed to the published version of the manuscript.

Funding: This research received no external funding.

Institutional Review Board Statement: Not applicable.

Informed Consent Statement: Not applicable.

Data Availability Statement: Data can be made available based on the requirements to verify this work.

Acknowledgments: The authors are grateful for the support toward this study provided by a joint research agreement between the Joining and Welding Research Institute, Osaka University, Japan; Hanoi University of Science and Technology, Vietnam; Kobe Steel Ltd., Japan; and Thai Kobelco Welding Co., Ltd., Thailand.

Conflicts of Interest: The authors declare no conflict of interest.

References

1. Rhee, S.; Kannatey-Asibu, E. Observation of Metal Transfer during Gas Metal Arc Welding. *Weld. J.* **1992**, *71*, 381–387.
2. Liu, S.; Siewert, T.A. Metal Transfer in Gas Metal Arc Welding: Droplet Rate. *Weld. J.* **1989**, *68*, 52–58.
3. Hu, J.; Tsai, H.L. Heat and Mass Transfer in Gas Metal Arc Welding. Part I: The Arc. *Int. J. Heat Mass Transf.* **2007**, *50*, 833–846. [CrossRef]
4. Hu, J.; Tsai, H.L. Heat and Mass Transfer in Gas Metal Arc Welding. Part II: The Metal. *Int. J. Heat Mass Transf.* **2007**, *50*, 808–820. [CrossRef]
5. Lancaster, J.F. The Physics of Fusion Welding Part 2: Mass Transfer and Heat Flow. *IEE Proc. B Electr. Power Appl.* **1987**, *134*, 297–316. [CrossRef]
6. Huang, Y.; Zhang, Y.M. Laser-Enhanced GMAW. *Weld. J.* **2010**, *89*, 181S–188S.

7. Fan, Y.; Yang, C.; Lin, S.; Fan, C.; Liu, W. Ultrasonic Wave Assisted GMAW. *Weld. J.* **2012**, *91*, 91S–99S.
8. Talalaev, R.; Veinthal, R.; Laansoo, A.; Sarkans, M. Cold Metal Transfer (CMT) Welding of Thin Sheet Metal Products. *Est. J. Eng.* **2012**, *18*, 243–250. [CrossRef]
9. Scotti, A.; Ponomarev, V.; Lucas, W. A Scientific Application Oriented Classification for Metal Transfer Modes in GMA Welding. *J. Mater. Process. Technol.* **2012**, *212*, 1406–1413. [CrossRef]
10. Wu, C.S.; Zou, D.G.; Gao, J.Q. Determining the Critical Transition Current for Metal Transfer in Gas Metal Arc Welding (GMAW). *Front. Mater. Sci. China* **2008**, *2*, 397–401. [CrossRef]
11. Iordachescu, D.; Quintino, L. Steps toward a New Classification of Metal Transfer in Gas Metal Arc Welding. *J. Mater. Process. Technol.* **2008**, *202*, 391–397. [CrossRef]
12. Kah, P.; Latifi, H.; Suoranta, R.; Martikainen, J.; Pirinen, M. Usability of Arc Types in Industrial Welding. *Int. J. Mech. Mater. Eng.* **2014**, *9*, 15. [CrossRef]
13. Wang, W.; Liu, S.; Jones, J.E. Flux Cored Arc Welding: Arc Signals, Processing and Metal Transfer Characterization. *Weld. J.* **1995**, *74*, 369–377.
14. Valensi, F.; Pellerin, N.; Pellerin, S.; Castillon, Q.; Dzierzega, K.; Briand, F.; Planckaert, J.P. Influence of Wire Initial Composition on Anode Microstructure and on Metal Transfer Mode in GMAW: Noteworthy Role of Alkali Elements. *Plasma Chem. Plasma Process.* **2018**, *38*, 177–205. [CrossRef]
15. Bang, K.S.; Jung, H.C.; Han, I.W. Comparison of the Effects of Fluorides in Rutile-Type Flux Cored Wire. *Met. Mater. Int.* **2010**, *16*, 489–494. [CrossRef]
16. Yamamoto, E.; Yamazaki, K.; Suzuki, K.; Koshiishi, F. Effect of Flux Ratio in Flux-Cored Wire on Wire Melting Behaviour and Fume Emission Rate. *Weld. World* **2010**, *54*, R154–R159. [CrossRef]
17. Trinh, N.Q.; Tashiro, S.; Suga, T.; Kakizaki, T.; Yamazaki, K.; Morimoto, T.; Shimizu, H.; Lersvanichkool, A.; Van Bui, H.; Tanaka, M. Effect of Flux Ratio on Droplet Transfer Behavior in Metal-Cored Arc Welding. *Metals* **2022**, *12*, 1069. [CrossRef]
18. Starling, C.M.D.; Modenesi, P.J. Metal Transfer Evaluation of Tubular Wires. *Weld. Int.* **2007**, *21*, 412–420. [CrossRef]
19. Trinh, N.Q.; Tashiro, S.; Tanaka, K.; Suga, T.; Kakizaki, T.; Yamazaki, K.; Morimoto, T.; Shimizu, H.; Lersvanichkool, A.; Murphy, A.B.; et al. Effects of Alkaline Elements on the Metal Transfer Behavior in Metal Cored Arc Welding. *J. Manuf. Process.* **2021**, *68*, 1448–1457. [CrossRef]
20. Trinh, N.Q.; Tashiro, S.; Suga, T.; Kakizaki, T.; Yamazaki, K.; Lersvanichkool, A.; Van Bui, H.; Tanaka, M. Metal Transfer Behavior of Metal-Cored Arc Welding in Pure Argon Shielding Gas. *Metals* **2022**, *12*, 1577. [CrossRef]
21. Ogino, Y.; Hirata, Y.; Asai, S. Discussion of the Effect of Shielding Gas and Conductivity of Vapor Core on Metal Transfer Phenomena in Gas Metal Arc Welding by Numerical Simulation. *Plasma Chem. Plasma Process.* **2020**, *40*, 1109–1126. [CrossRef]
22. Hertel, M.; Trautmann, M.; Jäckel, S.; Füssel, U. The Role of Metal Vapour in Gas Metal Arc Welding and Methods of Combined Experimental and Numerical Process Analysis. *Plasma Chem. Plasma Process.* **2017**, *37*, 531–547. [CrossRef]
23. Lancaster, J.F. The Physics of Welding. *Phys. Technol.* **1984**, *15*, 73–79. [CrossRef]
24. Egerland, S. A Contribution to Arc Length Discussion. *Soldag. Insp.* **2015**, *20*, 367–380. [CrossRef]
25. Ogino, Y.; Hirata, Y.; Murphy, A.B. Numerical Simulation of GMAW Process Using Ar and an Ar–CO_2 Gas Mixture. *Weld. World* **2016**, *60*, 345–353. [CrossRef]
26. Nemchinsky, V.A. The Effect of the Type of Plasma Gas on Current Constriction at the Molten Tip of an Arc Electrode. *J. Phys. D Appl. Phys.* **1996**, *29*, 1202–1208. [CrossRef]
27. Haidar, J. An Analysis of the Formation of Metal Droplets in Arc Welding. *J. Phys. D Appl. Phys.* **1998**, *31*, 1233–1244. [CrossRef]
28. Modenesi, P.J.; Starling, C.M.D.; Reis, R.I. Wire Melting Phenomena in Gas Metal Arc Welding. *Sci. Technol. Weld. Join.* **2005**, *10*, 610–616. [CrossRef]
29. Murphy, A.B. The Effects of Metal Vapour in Arc Welding. *J. Phys. D Appl. Phys.* **2010**, *43*, 434001. [CrossRef]
30. Tanaka, M.; Yamamoto, K.; Tashiro, S.; Nakata, K.; Yamamoto, E.; Yamazaki, K.; Suzuki, K.; Murphy, A.B.; Lowke, J.J. Time-Dependent Calculations of Molten Pool Formation and Thermal Plasma with Metal Vapour in Gas Tungsten Arc Welding. *J. Phys. D Appl. Phys.* **2010**, *43*, 434009. [CrossRef]
31. Nakamura, Y.; Lucas, J. Electron Drift Velocity and Momentum Cross-Section in Mercury, Sodium and Thallium Vapours. II. Theoretical. *J. Phys. D Appl. Phys.* **1978**, *11*, 337–345. [CrossRef]
32. National Institute of Standards and Technology. NIST Atomic Spectra Database Ionization Energies Form. Available online: https://physics.nist.gov/PhysRefData/ASD/ionEnergy.html (accessed on 3 April 2023).
33. Waszink, J.H.; Van Den Heuvel, G.J.P.M. Heat Generation and Heat Flow in the Filler in Gma Welding. *Weld. J.* **1982**, *61*, 269S–282S.
34. Modenesi, P.J.; Nixon, J.H. Arc Instability Phenomena in GMA Welding. *Weld. J.* **1994**, *73*, 219s–224s.
35. Ushio, M.; Matsuda, F. Effect of Oxygen on Stabilization of Arc in 9% Ni-Steel GMA Welding. *Trans. JWRI* **1978**, *7*, 93–100.
36. Brien, A.O. *Welding Handbook*; American Welding Society: Doral, FL, USA, 1987; Volume 2, ISBN 0871717298.
37. Murray, P.E.; Scotti, A. Depth of Penetration in Gas Metal Arc Welding. *Sci. Technol. Weld. Join.* **1999**, *4*, 112–117. [CrossRef]

Article

Mechanical Properties and Microstructure of Austenite—Ferrite Duplex Stainless Steel Hybrid (Laser + GMAW) and SAW Welded Joint

Ryszard Krawczyk [1], Jacek Słania [1,2], Grzegorz Golański [3,*] and Tomasz Pfeifer [2]

[1] Faculty of Mechanical Engineering and Computer Science, Czestochowa University of Technology, Armii Krajowej 21, 42-201 Czestochowa, Poland
[2] Łukasiewicz Research Network Upper Silesian Institute of Technology, K. Miarki 12-14, 44-100 Gliwice, Poland
[3] Department of Material Engineering, Czestochowa University of Technology, Armii Krajowej 19, 42-201 Czestochowa, Poland
* Correspondence: grzegorz.golanski@pcz.pl; Tel.: +48-34-3250-721

Abstract: The purpose of the research was to develop a technology for producing thick-walled duplex steel welded joints. The material used in the research was X2CrNiMoN22 duplex steel in the form of a 15 mm thick plate. The welded joint was produced by the modern, high-performance Hybrid Laser Arc Welding (HLAW) method. The HLAW method involves welding a joint using a laser, the Gas Metal Arc Welding (GMAW) method and the Submerged Arc Welding (SAW) method. The HLAW method was used to make the root pass of the double butt welded joint, while the filler passes were made by the SAW method. The obtained welded joint was subjected to non-destructive and destructive testing. The non-destructive and macroscopic tests allowed the joint to be classified to the quality level B. Microscopic examinations revealed the presence of ferritic–austenitic microstructure in the base material and the weld, with different ferrite content in specific joint areas. The analysed joint had high strength properties (tensile strength (TS) ~ 790 ± 7 MPa) and high ductility of weld metal (~160 ± 4 J) heat-affected zone (~216 ± 26 J), and plasticity (bending angle of 180° with no macrocracks). At the same time, hardness on the cross-section of the welded joint did not exceed 280 HV10.

Keywords: duplex steel; HLAW + SAW welded joint; microstructure; mechanical properties

Citation: Krawczyk, R.; Słania, J.; Golański, G.; Pfeifer, T. Mechanical Properties and Microstructure of Austenite—Ferrite Duplex Stainless Steel Hybrid (Laser + GMAW) and SAW Welded Joint. *Materials* **2023**, *16*, 2909. https://doi.org/10.3390/ma16072909

Academic Editor: Raul D. S. G. Campilho

Received: 7 March 2023
Revised: 31 March 2023
Accepted: 4 April 2023
Published: 6 April 2023

1. Introduction

Duplex steels are corrosion-resistant steels with approx. 50% ferrite and 50% austenite. They have favourable mechanical properties (strength and toughness) in the service temperature range from −50 to 280 °C, good general corrosion and pitting resistance, stress corrosion cracking resistance and adequate weldability and formability. The performance of the duplex steel depends not only on the content of alloying elements, but mainly the content of ferrite and austenite in the microstructure. Compared to conventional austenitic steels, duplex steels have also a more attractive price as a construction material due to lower content of expensive nickel. Due to their valuable performance, duplex steels are used in various branches of modern industry, especially in the chemical, food, petrochemical and construction industries [1,2].

The welding of duplex steel structures and components entails the need to take into account not only the impact of chemical composition of the filler metal and base material, but also the thermal welding conditions on the composition of structure, the volume fraction of the phases—ferrite and austenite in the heat-affected zone (HAZ) and the weld. For welded joints, the important parameter that determines the welded joint structure is the cooling rate associated in a significant way with the welding conditions. Therefore, when welding duplex steel, particular attention should be paid to the heat input also in direct correlation with the shielding gas composition, the preparation of the joint for welding and the method for depositing beads [3–6].

The increase in the cooling rate of a duplex steel welded joint contributes to inhibition of the ferrite-to-austenite conversion, resulting in a higher volume fraction of ferrite. This metallurgical regularity is found in the weld and the HAZ of the welded joint where a coarse-grained microstructure with high ferrite content can be formed as a result of incorrectly selected parameters and too fast cooling after welding. This structure has low plasticity and low corrosion resistance. It is assumed that duplex steels demonstrate their beneficial properties when ferrite content is not lower than 25–30% [5,7].

When welding thick-walled duplex or superduplex steel components, a significant problem is the precipitation of, among others, harmful intermetallic phases, e.g., chi—or sigma—phases, within the weld volume or HAZ due to multi-pass welding [8–10]. The appearance of these phases contributes not only to the occurrence of brittleness, but also has a negative influence on the welded joint corrosion resistance. One of the methods to limit the likelihood of these harmful phases being precipitated is to minimise the amount of energy introduced during welding (arc energy) and use the appropriate interpass temperature [8,10]. The selection of the appropriate welding heat input is an important element for making a correct welded joint in duplex stainless steel. Welding with low arc energy provides rapid cooling, which, however, can result in the formation of a structure with too much ferrite in the weld and the HAZ. The increase in the proportion of ferrite in these areas can lead to deterioration of the plastic properties and decrease in the corrosion resistance of the welded joint. On the other hand, welding made with too high heat input can contribute to the precipitation and growth of unfavourable hard and brittle phases due to the extension of the HAZ residence time at above 280 °C [7,8,10,11].

In production conditions, there is a need for high-performance welding processes, especially for thick plates. This involves the introduction of a significant amount of heat into the joint. This heat may adversely affect the phase structure of the welded joint. The purpose of the non-destructive and destructive tests carried out in the work and the analysis of the results obtained was to confirm the usefulness of the high-performance combined process (hybrid welding and SAW) for welding thick-walled X2CrNiMoN22-5-3 plates.

2. Materials and Research Methodology

The analysis was performed on a thick-walled duplex steel welded joint. The prepared joint was a butt joint of X2CrNiMoN22-5-3 steel plates with a thickness of t = 15 mm. The chemical composition of the test base material determined with the Bruker XP spectrometer is presented in Table 1.

Table 1. The average chemical composition of X2CrNiMoN22-5-3 steel, wt. %.

Steel Grade	Average Chemical Composition, wt. %								
	C	Si	Cr	Ni	Mo	Mn	P	S	N
X2CrNiMoN22-5-3	0.011	0.53	22.8	7.21	3.14	1.29	0.022	0.002	0.11

The test butt joint was prepared with a double-side bevel (2Y), a 5 mm high threshold and a groove angle of 70°. The test joint was made under automated conditions using the hybrid welding process, i.e., HLAW and classic SAW welding (Figure 1).

The following equipment was used for the hybrid welding:

- KUKA KR30HA welding robot fitted with Trumpf TruDisk 12,002 laser with Trumpf D70 head for laser welding;
- EWM welding power supply with a push-pull feeder for MAG welding.

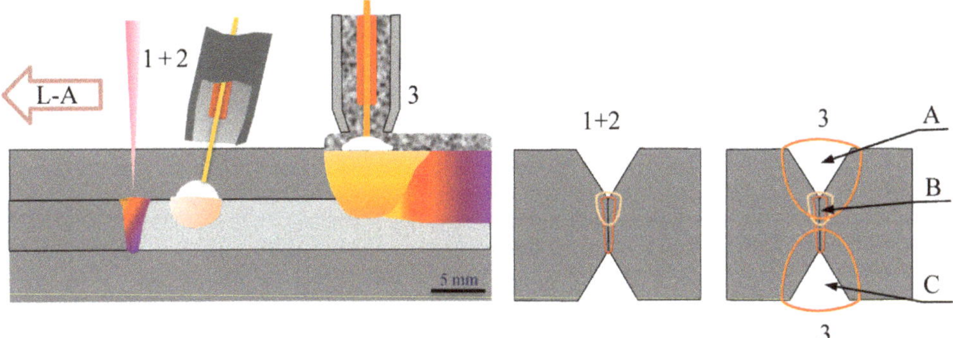

Figure 1. Diagram of the application of the hybrid + SAW process: (1 + 2) hybrid process (laser + GMAW), (3) SAW; A, B, C—microstructure investigation areas.

For submerged arc welding, the LAH 1001 DC power supply fitted with the A2 Multitrack welding tractor with A2/A6 PEK digital process controller was used. The filler material was OK Autrod 2209 welding electrode, whose chemical composition and mechanical properties based on the manufacturer's certificate are presented in Table 2.

Table 2. Characteristics of the filler metal type of OK Autrod 2209.

Material Grade	Chemical Composition of Weld Metal, % wt							Properties of Weld Metal			
	C	Mn	Si	Cr	Mo	Ni	N	Tensile Strength N/mm^2	Yield Strength N/mm^2	El. %	KV_{-20} J
OK Autrod 2209	<0.08	1.50	0.50	22.5	3.2	8.5	0.15	765	600	28	85

where: El.—elongation; KV—impact energy determined at −20 °C.

For the MAG method of the hybrid welding process, the ESAB OK Autrod 2209 welding electrode and M12 shielding gas (ArCo$_2$ with the active gas content of 2.5%) were used, while for the submerged arc welding, the ESAB OK Autrod 2209 welding electrode and ESAB OK FLUX 10.93 were used. The schematic diagram of the hybrid + SAW welding process used is presented in Figure 1.

The first bead was deposited in the central zone of the joint by the hybrid (laser + GMAW) method in the L-A system, while the second and third ones were made as single-pass filler welds by the SAW method. The heat input during welding by the hybrid (laser + GMAW) and SAW methods was 0.83 and 1.54 kJ/mm, respectively. Before welding the drying heating was applied at 300 °C for two hours and during welding the recommended interpass temperature of 110 °C was controlled.

Non-destructive testing, i.e., visual tests (VT), penetrant tests (PT) and radiographic tests (RT) were performed to evaluate the quality of the welded joint. The VT, PT and RT were made on both sides of the weld over the entire length of the joint in accordance with the requirements of the relevant standards, e.g., [12,13]. For evaluation of the experimental joint, the quality level B was adopted. The evaluation of the mechanical properties of the test joint included: macroscopic examinations, static tensile test, side bend test, impact test and hardness measurement.

The tensile and bend tests were carried out with the MTS 800 testing machine. The minimum value of 660 N/mm^2 was adopted for evaluation of the tensile strength (TS) and the bending angle of 180° was used for the bend test evaluations. The diameter of the bending mandrel was 50 mm. The impact energy test on standard Charpy V-notch samples taken from two areas, i.e., in the weld and the heat-affected zone (HAZ), was carried out with the WOLPERT W-15 impact testing machine. The impact energy was evaluated taken transversely from each of the zones of the welded joint, i.e., weld (VWT 0/2) and HAZ (VHT 0/2), based on their average values according to the adopted criterion of the minimum impact energy of KV_{min} 60 J. The impact test was carried out at room temperature. The Vickers hardness measurement was made with the indenter load of 10 kG (98.1 N)—HV10 using the QATM Qness 60 A+ hardness testing machine. The measurements were taken on the transverse microsection along three measurement lines determined next to the top and bottom face of the weld and in the central zone, respectively. The measurements were evaluated according to the adopted criterion of the maximum hardness in the area of the welded joint, $HV10_{max}$ 450. The results of mechanical tests, i.e., static tensile test, presented in this paper were the average of three measurements, while the bend test included four measurements.

The macroscopic examination was performed on the transverse microsection including all zones of the joint: the weld, both HAZs and adjacent areas of the base material in the welded joint. The surface of the microsection was etched with the Barah reagent for approx. 20 s. To evaluate the macrostructure of the welded joint, $10\times$ magnification was used in accordance with the adopted quality level B criterion.

The microscopic examinations were performed on transverse metallographic microsection etched with the Barah reagent. The duration of etching of the metallographic microsection prepared by grinding and polishing was approx. 10–15 s. The Barah reagent is commonly used to reveal the structure of duplex steels/cast steels [14]. The observation and recording of the images of the welded joint microstructure was carried out with the Keyence VHX 7000 digital microscope (DM). The volume percentage of austenite and ferrite in the test welded joint was determined by computer image analysis using the DM software.

3. Research Results and Analysis

3.1. Non-Destructive Tests

The visual tests confirmed that the test joint was produced properly, both on the top and bottom side of the weld. The evaluation of the joint meets the specified requirements in accordance with the adopted quality level B criterion. In turn, the penetrant tests revealed neither non-linear nor linear indications in both tested areas on the top and bottom side of the weld. The evaluation of the joint also confirmed the quality level B. Based on the radiographic test, the test joint was rated at the quality level B.

The positive results obtained during the non-destructive tests were the basis for further destructive testing to evaluate the properties of the test joint.

3.2. Macroscopic Examinations

The macroscopic image of the test X2CrNiMoN22-5-3 steel welded joint (Figure 2) shows that it has proper structure, i.e., correct penetration and regular fusion into both edges of the materials joined, proper arrangement of individual passes and beads, and small and mild weld face reinforcement. The particular weld fusion area of the first pass made by the hybrid method and remelted on both sides by the SAW method showed no deviations from the proper structure. The HAZ width was equal on both sides of the fusion line and amounted up to 2.5 mm at the bottom and top of the weld. In addition, no significant recorded welding imperfections were found on the cross-section of the joint in any of the assessed zones, i.e., in the base materials (BMs), HAZs and weld. The macroscopic examination of the X2CrNiMoN22-5-3 steel welded joint confirmed that the joint was made properly and with good quality.

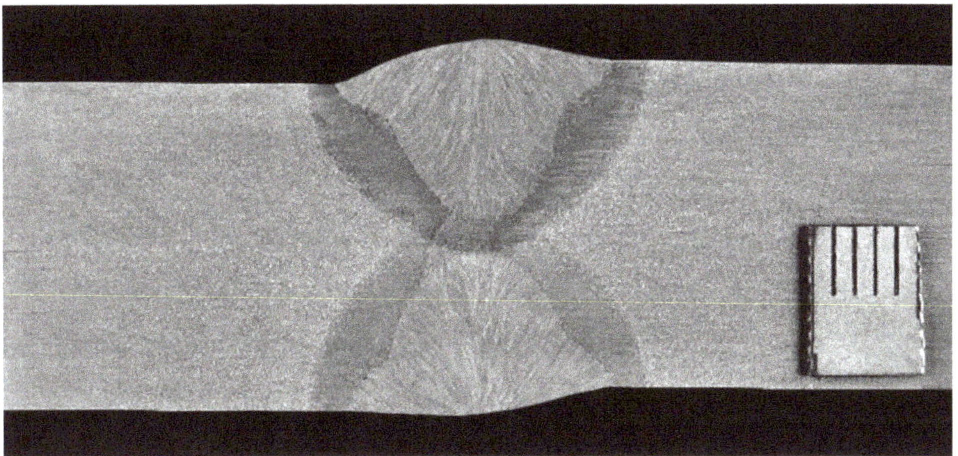

Figure 2. Macroscopic image of the cross-section of X2CrNiMoN22-5-3 steel welded joint.

3.3. Microscopic Examinations

Microstructure of the based material, i.e., X2CrNiMoN22-5-3 steel is presented in Figure 1. The test steel in the as-received condition had a two-phase microstructure consisting of banded dark-etched ferrite grains and bright-etched austenite grains (Figure 3). The austenite and ferrite grains were elongated in the rolling direction of plastic deformation. The microstructure of the test material was typical of this steel grade [4,11,14]. The volume percentage of ferrite in the base material was approx. 48%, and of austenite, approx. 52%.

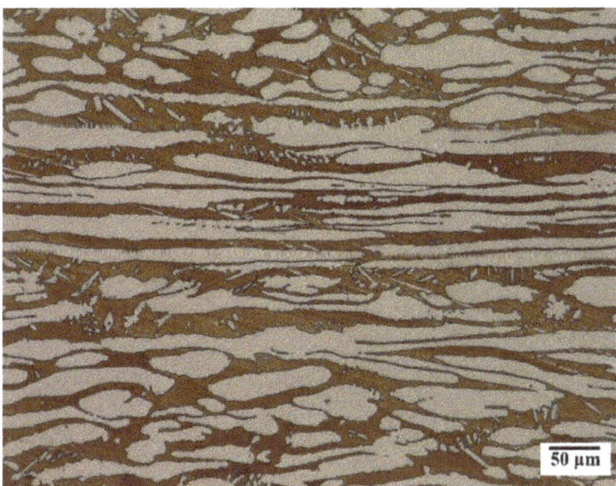

Figure 3. Microstructure of the base material.

At the fusion line (Figure 4), both on the face side and the fusion side of the welded joint, a section of the so-called incomplete fusion was visible where a smooth transition of austenite grains from the fusion area in the base material to the weld was observed. The weld in the analysed welded joint had a ferritic–austenitic structure typical of acicular/lamellar structures and Widmanstätten structures. No coarse-grained zone characteristic of this group of steels was observed close to the HAZ fusion line (Figure 5).

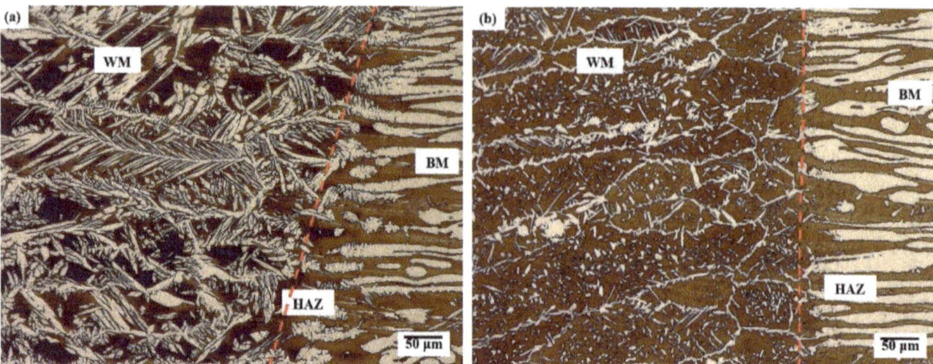

Figure 4. Microstructure in the fusion line on: (**a**) face side, (**b**) fusion side: WM—weld metal; HAZ—heat-affected zone; BM—base material, red dashed line—weld line.

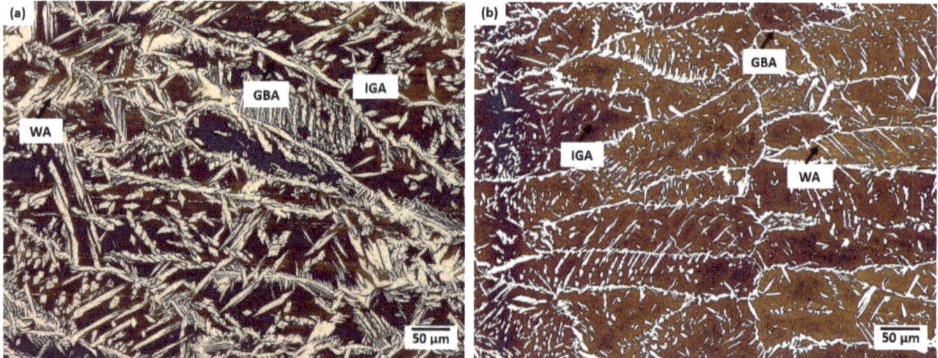

Figure 5. Microstructure of the weld in the area of: (**a**) face; (**b**) fusion, where: GBA—allotriomorphic austenite precipitation on grain boundary, WA—Widmansttaten-like austenite; IGA—intragranular austenite phase.

Generally, in the weld and the HAZ of the duplex steel, four austenite forms can be distinguished: allotriomorphic austenite precipitation on the grain boundary, side-plate Widmanstätten-like austenite, intragranular austenite phases and partially transformed austenite [5,7]. In the test joint, the following forms of austenite were observed both on the face side and the fusion side in the microstructure of this area: allotriomorphic austenite precipitation on the grain boundary (GBA), side-plate Widmanstätten-like austenite (WA) and intragranular austenite (IGA). The GBA and the WA were the predominant austenite phase forms in the analysed joint both on the face and fusion side (Figure 4). Similar observations of the microstructure of weld metal and HAZ were reported by other researchers [3,5,11,15]. The GBA nucleates and grows at the ferrite–ferrite boundary as energy-privileged sites. The precipitation of the GBA takes place at 1350 ÷ 800 °C [16]. Austenite grains grew along ferrite–ferrite boundaries forming thin layer of allotriomorphic austenite phase. The WA precipitates at 800 ÷ 650 °C, where diffusion is fast, nucleating directly on the GAB and/or partially transformed austenite. The WA formation requires a relatively small driving force and little undercooling [17]. The presence of the WA in the weld microstructure can be attributed to the presence of nitrogen in the chemical composition of the base material/filler metal [5,6]. In turn, the IGA precipitation takes place at 1000 ÷ 1100 °C in the chromium/molybdenum depleted zones and nickel/nitrogen-rich microareas at metastable ferrite [17].

The form of austenite nucleates heterogeneously on non-metallic inclusions, precipitates or dislocations and requires greater degree of undercooling [17,18]. On the face side of the weld, the ratio of the volume percentage of ferrite and austenite was around 55/45, while in the hybrid welding-affected area in the fusion zone it was 65/35 and in the remelting area—68/32. In the weld zone and HAZ (close to the fusion line), duplex steels solidify to a completely ferritic form, and the precipitation of austenite takes place in a solid form as a result of the following transformation: $L \rightarrow L + \delta \rightarrow \delta + \gamma$ [16]. The formation of austenite phase is conditioned by the diffusion of stabilising austenite elements to this phase, while ferrite is enriched with elements stabilising this phase. The volume percentage of the austenite formed depends on the chemical composition of the steel, the cooling rate after heat treatment and/or welding process and the diffusion rate of diffusing elements. The increase in the cooling rate in the weld zone and HAZ can lead to a higher content of metastable ferrite due to inadequate time for transformation of ferrite phase to austenite phase [11,16,19]. Numerous austenite particles in the form of the WA (Figure 5) also indicates a high cooling rate in these areas of the weld. However, according to [5], the presence of the GBA and WA in the structure of the joint requires smaller driving force and less undercooling. Due to performance, it is not acceptable for industrial uses that the austenite volume percentage is less than 25% or the ferrite phase content is greater than 75%. The volume percentage of austenite in the weld meets the austenite content recommended for duplex steel joints of at least 25–30% [7].

3.4. Mechanical Testing

Figure 6 shows the Vickers hardness distribution on the cross-section of the welded joint. The measurements revealed that the BM hardness was between 259 and 270 H10, for HAZ it was 261–275 HV10, and in the weld—261–280 HV10. The obtained hardness on the cross-section of the welded joint was lower than the $HV10_{max}$ 450 criterion. In general, the relatively small differences in hardness on the cross-section of the test joint (~260–280 HV10) were mainly related to the variation in ferrite-to-austenite ratio in the given area of the welded joint.

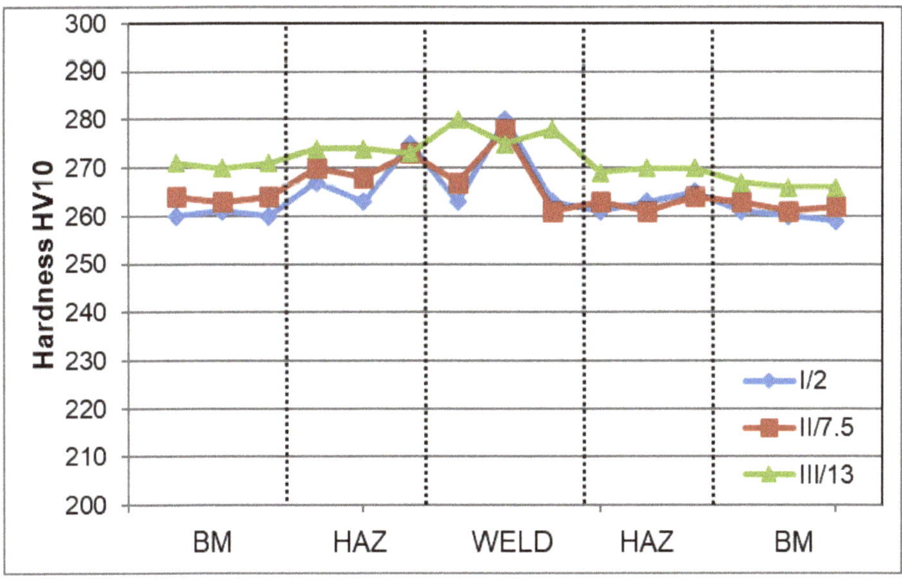

Figure 6. The results of the Vickers hardness measurement (HV10) on cross-section of the joint determined along lines from the top surface, where: I/2—measurement at a distance of up to 2; II/7.5—measurement line at a distance of 7.5 mm; III/13—measurement line at a distance of 13 mm.

The highest hardness observed in the weld was due to the higher proportion of ferrite in this area compared to that in the BM or HAZ. Additionally, small ferrite grain in the microstructure of the welded joint has a significant effect on its hardness according to [19,20]. Hardness in the area of the welded joint which is comparable to that in the BM also indicates the absence of secondary phase precipitates affecting the increase in the precipitation hardening, resulting, among other things, in a significant increase in hardness. Additionally, the higher ferrite content in the weld compared to other areas of the analysed joint resulted in a relatively high tensile strength (TS) of approx. 790 ± 7 MPa. This value was higher by approx. 20% than the minimum TS required for this grade of material which is 660 MPa. The similar high TS values for duplex steel welded joints were observed in, but not limited to, Refs. [5,11,15]. Metastable ferrite in duplex steels is responsible for strength properties; hence, its higher content will result in an increase in these properties. In turn, according to [19,20], the fine ferrite grain in the area of the welded joint has a favourable effect on its strength. The effect of ferrite content in the test joint was also observed in the obtained value of impact energy (KV). The impact energy for samples with a notch cut in the weld axis (VWT 0/2) was 163 ± 4 J, and for samples with a notch in the HAZ (VHT 0/2) KV = 216 ± 26 J. The test material was characterised by high ductility, expressed by impact energy of 371 ± 8 J. Therefore, the impact energy results obtained for the weld and HAZ were approx. 44 and 58% of the roughness of the base material, respectively. The obtained results for impact energy were higher not only than the minimum requirements for this steel grade, but also than the data presented in, but not limited to, [4,5,17,21] for duplex steel welded joints. The toughness of a duplex steel welded joint depends on its microstructure, which is associated with the optimisation of the cooling rate, and the quantity and size of ferrite grain—fine grain increases the impact energy. The higher impact energy in the HAZ to was probably associated with the higher volume percentage of plastic austenite phase in this zone of the microstructure compared to the weld. The ductile austenite phase significantly prevents crack growth in contrast to the relatively fragile ferrite. Fine austenite boundaries in the HAZ also significantly prevent brittle crack growth [5,22]. According to [19], the high impact energy in the area of the welded joint of duplex steel can also be associated with the increase in the number of low angle grain boundaries (LAGBs). High proportion of LAGBs also has a positive influence on the corrosion resistance of duplex steel [23]. In turn, the tests in [24] show that not only fine grain, but also the grain boundary orientation has a characteristic influence on the impact energy of the welded joint of the duplex steel. The increase in percentage of austenite $\Sigma3$ coincidence site lattice increases the grain boundary impact energy of the welded joint [24]. On the other hand, the decrease in impact energy in the duplex steel welded joint area can be related to precipitates at the grain boundaries and/or inadequate transformation of ferrite phase into austenite phase in rapid air cooling from elevated temperatures after welding process [21]. The high value of TS and impact energy can also be significantly affected by the optimally selected welding filler metal [4,5]. The obtained values of impact energy in the welded joint zone were several times higher than criterion of KV_{min} = 60 J assumed for the examined material. The high plasticity and ductility of the joint produced were also evidenced by the results of the side bend test. The bend test of the welded joint showed the absence of macrocracks at the bending angle of 180° (Figure 7). This indicates the beneficial effect of the ferrite to austenite ratio in the volume of the welded joint on its plastic properties.

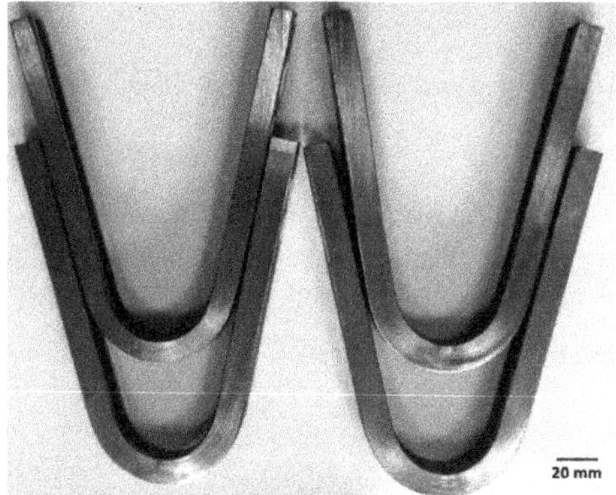

Figure 7. View of the samples after bend test.

4. Summary

The thick-walled X2CrNiMoN22-5-3 duplex steel joint welded by the combined hybrid (laser + GMAW) and SAW method was subjected to physical metallurgy analysis. The joint was made as a double-sided joint with 2Y bevel in the central fusion zone using the hybrid (laser + GMAW) method, while the rest of the weld groove was filled by the SAW method. The investigations allowed for the following conclusions:

1. The analysed welded joint has proper macro- and microscopic structure in all its areas, both in the fusion zone made by hybrid laser + GMAW method and in the filler welds made by the SAW method, which allowed the highest quality level B to be obtained for this joint.
2. The analysed joint had high strength properties (TS ~ 790 ± 7 MPa) with good toughness in the range of ~163 ± 4 ÷ 216 ± 26 J and also a very good plasticity (bending angle of 180° with no cracks).
3. The use of high-performance welding processes conducted under automated conditions using skilfully selected parameters allows to obtain a high-quality joint while ensuring high efficiency and beneficial reduction in production costs.
4. The combination of hybrid welding processes (laser + GMAW) and SAW used in the paper has a large potential in terms of controlling parameters that affect the amount of heat input. The application of the developed welding technology in industrial conditions allows thick-walled duplex steel welded joints with the required quality level to be obtained.

Author Contributions: Conceptualization, R.K., J.S. and G.G.; methodology, R.K., J.S., G.G. and T.P.; validation, G.G.; formal analysis, R.K., J.S. and G.G.; investigation, R.K., G.G. and T.P.; writing—original draft preparation, R.K., J.S. and G.G.; writing—review and editing, R.K., J.S. and G.G.; supervision, G.G. All authors have read and agreed to the published version of the manuscript.

Funding: This research received no external funding.

Institutional Review Board Statement: Not applicable.

Informed Consent Statement: Not applicable.

Data Availability Statement: Not applicable.

Conflicts of Interest: The authors declare no conflict of interest.

References

1. Vinoth Jebaraj, A.; Ajaykumar, L.; Deepak, C.R.; Aditya, K.V.V. Weldability, machinability and surfacing of commercial duplex stainless steel AISI2205 for marine applications—A recent review. *J. Adv. Res.* **2017**, *8*, 183–199. [CrossRef]
2. Charles, J. Composition and properties of duplex stainless steel. *Weld. World* **1995**, *36*, 43–54.
3. Mohammed, G.R.; Ishak, M.; Aigda, S.N.; Abdulhadi, H.A. Effect of heat input on microstructure, corrosion and mechanical characteristics of welded austenitic and duplex stainless steels: A review. *Metals* **2017**, *7*, 39. [CrossRef]
4. Li, L.; Du Zh Sheng, X.; Zhao, M.; Song, L.; Han, B.; Li, X. Comparative analysis of GTAW + SMAW and GTAW welded joints of duplex stainless steel 2205 pipe. *Int. J. Press. Vessel. Pip.* **2022**, *199*, 104748. [CrossRef]
5. Devendranath Ramkumar, K.; Mishra, D.; Ganesh Gaj, B.; Vignesh, M.K.; Thiruvengatam, G.; Sudharsan, S.P.; Arivazhagan, N.; Sivashanmugam, N.; Rabel, A.M. Effect of optimal weld parameters in the microstructure and mechanical properties of autogeneous gas tungsten arc weldments of super-duplex stainless steel UNS S32750. *Mater. Des.* **2015**, *66*, 356–365. [CrossRef]
6. Varbai, B.; Majlinger, K. Physical and theoretical modelling of the nitrogen content of duplex stainless steel weld metal: Shielding gas composition and heat input effects. *Metals* **2019**, *9*, 762. [CrossRef]
7. Hsieh, R.I.; Liou, H.-Y.; Pan, Y.-T. Effect of cooling time and alloying elements on the microstructure of Gleeble-simulated heat-affected zone of 22%Cr duplex stainless steels. *J. Mater. Eng. Perform.* **2001**, *10*, 526–536. [CrossRef]
8. Valiente Bermejo, M.A.; Eyzop, D.; Hurtig, K.; Karlsson, L. Welding of large thickness super duplex stainless steel: Microstructure and Properties. *Metals* **2021**, *11*, 1184. [CrossRef]
9. Słania, J.; Krawczyk, R.; Wójcik, S. Quality requirements put on the Inconel 625 austenite layer used on the sweet pile walls of the boiler's evaporator to utilize waste thermally. *Arch. Metall. Mater.* **2015**, *60*, 677–685. [CrossRef]
10. Valiente Bermejo, M.A.; Eyzop, D.; Hurtig, K.; Karlsson, L. New approach to the study of multi-pass welds–microstructure and properties of welded 20 mm-thick superduplex stainless steel. *Appl. Sci.* **2019**, *9*, 1050. [CrossRef]
11. Pecly PH, R.; Almeida, B.B.; Perez, G.; Pimenta, A.R.; Tavares, S.S.M. Microstructure, corrosion resistance and hardness of simulated heat-affected zone of duplex UNS S32205 and superduplex UNS 32750 stainless steel. *J. Mater. Eng. Perform.* **2023**. [CrossRef]
12. *EN ISO 15614-1:2017-08*; Specification and Qualification of Welding Procedures for Metallic Materials—Welding Procedure Test. Part 1: Arc and Gas Welding of Steels and Arc Welding of Nickel and Nickel Alloys. ISO: Geneva, Switzerland, 2017.
13. *EN ISO 5817:2014-05*; Welding—Fusion-Welded Joints in Steel, Nickel, Titanium and Their Alloys (Beam Welding Excluded)—Quality Levels for Imperfections. ISO: Geneva, Switzerland, 2014.
14. Varbai, B.; Majlinger, K. Optimal etching sequence for austenite to ferrite ratio evaluation of two lean duplex stainless steel weldments. *Measurement* **2019**, *147*, 106832. [CrossRef]
15. Mourad, A.-H.I.; Khourshid, A.; Sharef, T. Gas tungsten arc and laser beam welding processes effects on duplex stainless steel 2205 properties. *Mater. Sci. Eng.* **2012**, *549*, 105–113. [CrossRef]
16. Chen, L.; Tan, H.; Wang Zh Jiang, Y. Influence of cooling rate on microstructure evolution and pitting corrosion resistance in the simulated heat-affected zone of 2304 duplex stainless steels. *Corros. Sci.* **2012**, *28*, 168–174. [CrossRef]
17. Zhang, Z.; Jing, H.; Xu, L.; Han, Y.; Zhao, L. Investigation on microstructure evolution and properties of duplex stainless steel joint multi-pass welded by using different methods. *Mater. Des.* **2016**, *109*, 670–685. [CrossRef]
18. Geng, S.; Sun, J.; Guo, L.; Wang, H. Evolution of microstructure and corrosion behavior in 2205 duplex stainless steel in GTA-welding joint. *J. Manuf. Process.* **2015**, *19*, 32–37. [CrossRef]
19. Kose, C.; Topal, C. Texture, microstructure and mechanical properties of laser beam welded AISI 2507 super duplex stainless steel. *Mater. Chem. Phys.* **2022**, *289*, 126490. [CrossRef]
20. Zhang, Z.; Jing, H.; Xu, H.; Han, Y.; Zhao, L.; Lv, X.; Zhang, J. Influence of heat input in electron beam process on microstructure and properties of duplex stainless steel welded interface. *Appl. Surf. Sci.* **2018**, *435*, 352–366. [CrossRef]
21. Tavares, S.S.M.; Pardal, J.M.; Lima, L.D.; Bastos, I.N.; Nascimento, A.M.; de Souza, J.A. Characterization of microstructure, chemical composition, corrosion resistance and toughness of a multipass weld joint of superduplex stainless steel UNS S32750. *Mater. Charact.* **2007**, *58*, 610–616. [CrossRef]
22. Haghdadi, N.; Cizek, P.; Hodgson, P.D.; Beladi, H. Microstructure dependence of impact toughness in duplex stainless steel. *Mater. Sci. Eng.* **2019**, *745*, 369–378. [CrossRef]
23. Zhang, Z.; Jing, L.; Xu, L.; Han, Z.; Gao, Z.; Zhao, L.; Zhang, J. Microstructural characterization and electron backscatter diffraction analysis across the welded interface of duplex stainless steel. *Appl. Surf. Sci.* **2017**, *413*, 327–343. [CrossRef]
24. Cui, S.; Yu, Y.; Tian, F.; Pang, S. Morphology, microstructure and mechanical properties of S32101 duplex stainless steel joints in K-TIG welding. *Materials* **2022**, *15*, 5432. [CrossRef] [PubMed]

materials

MDPI

Article

Molecular Dynamics Study on the Welding Behavior in Dissimilar TC4-TA17 Titanium Alloys

Peng Ou [1,2], Zengqiang Cao [1,*], Ju Rong [3] and Xiaohua Yu [3]

1   School of Mechanical Engineering, Northwestern Polytechnical University,127 Youyi Ave, West, Xi'an 710072, China
2   Kunming Precision Machinery Research Institute, P.O. Box No.26, Kunming 650118, China
3   Faculty of Materials Science and Engineering, Kunming University of Science and Technology, Kunming 650093, China
*   Correspondence: czq66326@nwpu.edu.cn

Abstract: Titanium alloys have become the material of choice for marine parts manufacturing due to their high specific strength and excellent resistance to seawater corrosion. However, it is still challenging for a single titanium alloy to meet the comprehensive specifications of a structural component. In this study, we have applied a molecular dynamics approach to simulate the aging phase transformation, K-TIG welding process, and mechanical properties of the TC4-TA17 (Ti6Al4V-Ti4Al2V) alloy. The results show that during the aging phase transformation process, changes in the structure of the titanium alloys are mainly manifested in the precipitation of a new phase from the sub-stable β-phase, and after the state stabilization, the α-phase content reaches 45%. Moreover, during the melting and diffusion process of TC4-TA17, aluminum atoms near the interface diffuse, followed by titanium atoms, while relatively few vanadium atoms are involved in the diffusion. Finally, the results of tensile simulations of the TC4-TA17 alloy after welding showed that stress values can reach up to 9.07 GPa and that the mechanical properties of the alloy in the weld zone are better than those of the single alloys under the same conditions. This study will provide theoretical support for the optimization of process parameters for TC4-TA17 alloy welding.

Keywords: molecular dynamics; TC4-TA17 alloy; radial distribution function; phase transition; diffusion coefficient

Citation: Ou, P.; Cao, Z.; Rong, J.; Yu, X. Molecular Dynamics Study on the Welding Behavior in Dissimilar TC4-TA17 Titanium Alloys. *Materials* **2022**, *15*, 5606. https://doi.org/10.3390/ma15165606

Academic Editors: Raul D. S. G. Campilho and Antonio Riveiro

Received: 17 May 2022
Accepted: 7 July 2022
Published: 16 August 2022

Publisher's Note: MDPI stays neutral with regard to jurisdictional claims in published maps and institutional affiliations.

1. Introduction

The exploration and exploitation of marine resources impose special requirements on seawater corrosion resistance along with the strength of engineering materials, and the selection of appropriate materials is crucial [1,2]. Titanium alloys have recently been demonstrated to be favorable materials for marine parts manufacturing due to their advantages of high specific strength and excellent seawater corrosion resistance [3,4]. However, it remains challenging for a single type of titanium alloy to satisfy the composite indicator of structural members in terms of performance and cost [5]. Specifically, considering a pressure-resistant hull of marine parts with bulged flanges, the flanges do not require as high a strength as the pressure-resistant hull because the pressures on either side of them are equivalent. Therefore, parts serving in the marine environment tend to require composite connections of dissimilar titanium alloys so as to achieve a property balance between the structures [6–8].

To fully exploit the performance characteristics of different types of titanium alloys and to reduce costs, it is necessary to adopt appropriate dissimilar titanium alloys for the pressure-resistant hull and flanges. The Ti6Al4V (TC4) titanium alloy displays great comprehensive performance, and it is suitable for various press-forming processes by virtue of its high strength and fine process plasticity, which is extensively applied in the pressure-resistant hulls of marine parts [9–11]. Ti4Al2V (TA17) belongs to a class of medium-strength titanium alloys with promising weldability, corrosion resistance, and impact toughness, and is particularly beneficial for the manufacture of loaded components, such as flanges and skeletons employed in the marine environment [12–14]. Actually, these two typical titanium alloys have shown great promise in marine parts manufacturing, but there is still a shortcoming regarding long-term service structural integrity when simply joining the two together. This effect is caused by an undesirable incompatibility at the connection interface owing to different features of TC4 and TA17. Based on the above analysis, some researchers consider that fusion welding (including K-TIG welding, electron beam welding, laser beam welding, and so on) with high precision, low distortion, and reliable connections offers encouraging potential [15,16]. Currently, it is difficult to investigate the narrow linker region of fusion welding from a macro-scale perspective, whereas the application of molecular dynamics to study the welding behavior of titanium alloys from a micro-scale perspective has been favored. Zhu et al. [17] examined the diffusion process of V atoms in titanium alloys, as a result of which the influence laws of V atom concentration on the diffusion distance and diffusion coefficient were obtained. Shimono et al. [18] performed molecular dynamics simulations of the recrystallization process of Ti–Al alloys in the amorphous state to derive the critical cooling rate of Ti–Al alloys; in addition, the way in which the Al content affects the amorphous formation ability of Ti–Al alloys was established. Semiatin et al. [19] experimentally probed the self-diffusion coefficients of each atom in TC4 alloy and expressed the Arrhenius function for the variation of diffusion coefficient with diffusion temperature. The studies mentioned above suggest that molecular dynamics is an effective simulation tool in material calculations. Nevertheless, most studies have focused on structural transformation processes in alloys and diffusion linkage processes between single-crystal dissimilar metals, whereas molecular dynamics simulations of diffusion linkage processes between dissimilar multi-alloy systems have been less well studied. Consequently, the aging transformation, fusion welding process, and mechanical properties of the TC4-TA17 (Ti6Al4V-Ti4Al2V) alloy need to be further investigated.

Here, a molecular dynamics study of K-TIG welding behavior between two ternary titanium alloys, TC4 and TA17, has been carried out. The initial configurations of TC4, TA17, and TC4-TA17, matching the corresponding atomic contents, were established by a random substitution method based on the LAMMPS (large-scale atomic/molecular massively parallel simulator) software package [20]. Precipitation of the α phase from the sub-stable β phase was simulated, and the stable $\alpha + \beta$ biphase structure was modeled by aging pretreatment of the TC4 configuration. On this basis, the effects of process parameters on the diffusion linker region width and Ti atom diffusion coefficients of TC4-TA17 dissimilar titanium alloys were evaluated. Furthermore, molecular dynamics simulations of tensile deformation behavior in the diffusion linker region were performed to acquire stress–strain curves with different process parameters. Molecular dynamics simulation results have revealed that TC4-TA17 exhibits satisfactory welding behavior in dissimilar titanium alloys. It is expected that this study will provide theoretical support for the optimization of process parameters for fusion welding and will further promote the practical application of dissimilar titanium alloys in marine parts manufacturing.

2. Computational Methods

2.1. Selection of Alloy Potential Function

A molecular dynamics approach was used to simulate the aging phase change behavior of the two Ti–Al–V ternary alloys and the welding melting process between them. In molecular dynamics, the state of motion of any particle can be obtained by integrating over Newton's equations of motion [21], and the differential equation of Newton's second law is as follows:

$$m_i \frac{d^2 r_i}{dt^2} = -\nabla_i V(r_1, r_2 \ldots r_N) \tag{1}$$

where $V(r_1, r_2 \ldots r_N)$ is the potential function of the atom, and m_i and r_i are the mass and position of the i-th atom, respectively. By solving for the state quantities (velocity and position) of the particles, the trajectory of the system can be described. By averaging the results for the system using statistical methods, the required macroscopic physical quantities, such as temperature, pressure, stress, etc., can be obtained to investigate the equilibrium thermodynamic properties of the system and the microstructure of the particles.

In molecular dynamics simulations, the interaction potential between atoms is defined by a mathematical function of atomic potential energy versus coordinates [21,22]. The embedded atom method (EAM) is a semi-empirical theory, proposed by Daw and Baskes in 1983, based on density generalized theory using effective medium and quasi-atomic approximations [23–26]. EAM theory treats each atom in a system as an exotic impurity of the substrate under study, and the system energy is expressed as the sum of the embedding energy and the interaction potential energy, so that polyatomic interactions can be included in the embedding energy. EAM is applicable to metallic atomic systems and provides a better description of the way in which bonds are formed between atoms compared to other potential functions. The potential function is generalized from the two-body potential to take into account the free electron gas effect in the local region of each atom, using the gel model to simplify the description of the complex environment surrounding the atom. Its functional form is as follows [24]:

$$U_{total} = \sum_i F(p_i) + \frac{1}{2} \sum_i \sum_{j \neq i} \varnothing(r_{ij}) \tag{2}$$

$$p_i = \sum_{j \neq i} f(r_{ij}) \tag{3}$$

where F is the atomic embedding energy as a function of p_i, $\varnothing$ is the short-range pair potential, and p_i is a function of atomic electron density. The first term on the right-hand side of Equation (2) is the embedding energy, which represents the energy at which the atom is embedded at the electron density of the substrate, and the second term is the conventional pair potential, which describes the repulsion between atoms in a solid system. The EAM model parameters are determined by fitting the binding energy, single vacancy formation energy, lattice constant, elastic constant, and structural energy difference of the elements. From Equations (2) and (3), the force between atoms can be deduced as:

$$F_{ij} = -\left\{ \left[F'(p_i) + F'(p_j) \right] f'(r_{ij}) + \varphi'(r_{ij}) \right\} \tag{4}$$

Zhou and co-workers developed a library of alloy potentials for 16 commonly used metal elements, including Ti, Al, and Cu, and were able to derive alloy potential function files for any combination through a fitting procedure, but the alloy library did not contain the element V [25]. In 2021, Zhao et al. added V to the existing alloy library [25]. Based on the above work, a fitting procedure is applied here to generate the required Ti–Al–V alloy potential function files. Under the above conditions, the Nosé–Hover temperature control method was used to simulate the molecular dynamics of the welding melting process of TC4 and TA17 alloys. The calculated melting point of TC4 alloy is about 1780 K, as compared to the actual melting point of about 1950 K. The simulated value is close to the actual value, which shows the suitability of the selected potential function.

2.2. The Establishment of an Atomic Model

The initial three-dimensional configurations of the two alloys were set as bcc structures, and the system in which they were located used NVT system synthesis. Considering the scale effect, periodic boundary conditions were set in the x and y directions, and contractive boundary conditions were set in the z direction. The box size was set to $20 \times 20 \times 80$ (bcc structured cubic cell, Ti lattice constant 0.328 nm), with the TC4 alloy occupying the lower part of the box with a size of $20 \times 20 \times 35$ and the TA17 alloy occupying the upper part of the box with the same dimensions. The total number of atoms was 56,800. In the TC4 alloy model, corresponding Ti atoms were replaced with 6 at% Al atoms and 4 at% V atoms at random, whereas in the TA17 alloy model, the corresponding Ti atoms were replaced with 4 at% Al atoms and 2 at% V atoms at random. The resulting initial atomic configuration is shown in Figure 1. For ease of observation, the configurations above and below the contact surface are distinguished by different colors. The Ti atoms are colored gray and red, the Al atoms are colored pink and blue, and the V atoms are colored brown and light green.

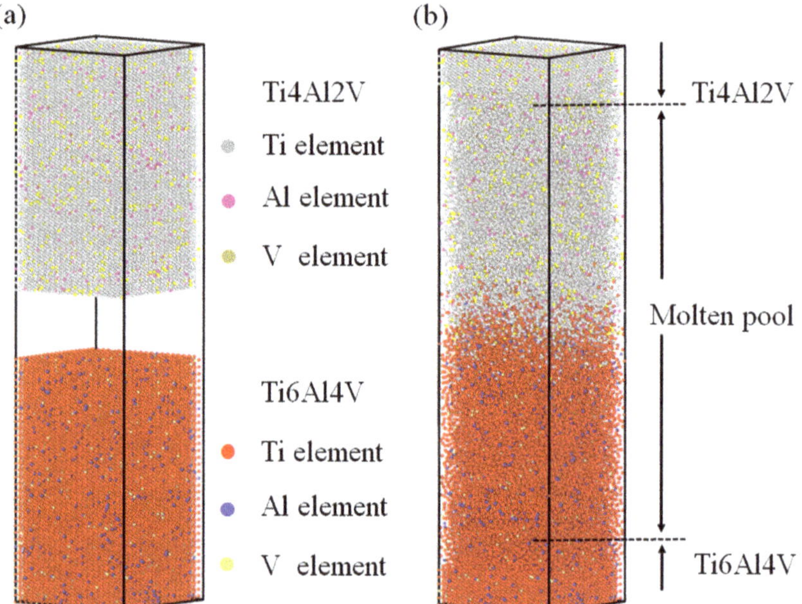

Figure 1. Welding model of TC4-TA17 dissimilar titanium alloy: (**a**) molecular dynamics calculation model before welding, in which the upper part is TC4 titanium alloy and the lower part is TA17 alloy; (**b**) molecular dynamics structure after welding, in which the middle is the molten pool area. The alloys form an obvious metallurgical bond.

2.3. Phase Change, Welding and Stretching Process Simulation

Having established the atomic model, the phase change simulation process was first carried out with LAMMPS to obtain the initial state of the material. This was followed by a simulation of the melting and welding process, and finally, the tensile mechanical properties of the alloy were tested. All simulation parameter settings are basically the same as in previous references [17–19] to ensure that the simulation can be repeated.

The phase change temperature of the general $\alpha + \beta$ titanium alloy was 900 K. The sub-stable phase formed by high-temperature quenching of the general $\alpha + \beta$ titanium alloy was followed by precipitation of a new phase as a result of a phase change process in the lower phase temperature region, significantly strengthening the alloy. The phase synthesis was run for 6000 steps, and atomic information was recorded at intervals of 20 steps. In a simulation of the aging treatment of the TC4 and TA17 alloys, the three atomic boundary layers above and below the model were fixed and stiffened, and a z-directional load was applied to each atom in this boundary layer, similar to rigid die pressurization, so that a uniform pressure was applied. The simulation step was set to 0.01 ps (1 ps = 10–12 s).

Before simulating the welding melting process, the initial temperature of the model was set to 1200 K. After running 80,000 steps, the temperature was raised to 2000 K to melt the alloy. The model was then held at 2000 K for 100 ps to obtain the diffusion of each atom throughout the melting and diffusion processes.

Finally, molecular dynamics simulations were used to obtain the changes in the mechanical properties of the alloy before and after melting, as well as its stress–strain curves at different temperatures and after different holding times. In order to better repeat our simulation, the detailed input file can be downloaded from the supporting materials.

3. Results and Discussion

3.1. Simulation of the Aging Phase Transition

The microscopic details associated with atoms can be captured more intuitively by molecular dynamics simulation methods. Figure 2a shows a schematic diagram of the dot arrangement on the (100) crystal plane after completion of the aging phase transformation of the TC4 (as an example; the TA17 alloy has similar properties) primary protocell model. From the figure, it can easily be seen that the TC4 protocell model shows a twin structure after the precipitation phase, as well as some regions of stacking layer dislocations, where the red region is the hcp structure in the form of lamellae and the green region is the fcc structure in the form of stripes and small lamellae. Pinsook et al. [27] also observed striped fcc structures in their simulations of the bcc–hcp martensite phase transformation process of Zr and attributed this phenomenon to the fact that phase transformation stresses changed the stacking order of the hcp structure, resulting in the formation of fcc structures in some regions, and that the formation of stacking layer dislocations and twin boundaries could reduce some of the stresses. In addition, the structure of multiple atomic layers may also be formed as transition regions at the intersection of multiple grains of different orientations.

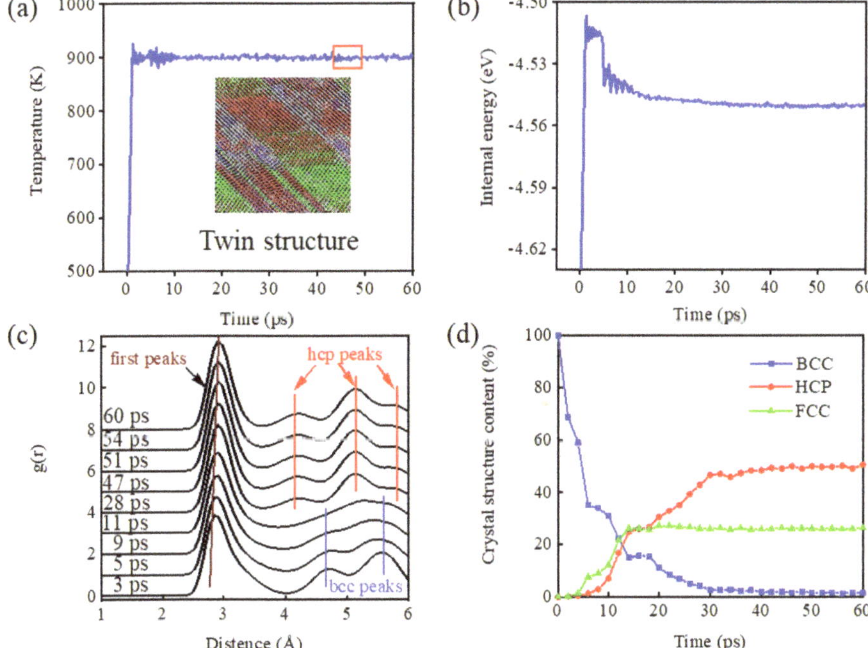

Figure 2. Microstructure of TC4 alloy after heat treatment and return to room temperature: (**a**) the relaxation process of the transformation from high-temperature face-centered cubic phase to low-temperature close-packed hexagonal phase finally forms a partial twin structure; (**b**) the change in binding energy during phase transformation; (**c**) the radial distribution function during phase transformation; and (**d**) analysis of the relative contents of each phase.

Figure 2b shows the trend of the internal energy of the Ti–Al alloy system during the simulation. It can easily be seen that the internal energy shows a trend of significant decrease with increasing simulation time, followed by a slow decrease until it reaches a plateau. This is due to the fact that the β-phase of the Ti–Al alloy is sub-stable under the set simulated temperature conditions, and shifts to the thermodynamically more stable α-phase when aging at this temperature to reduce the system energy. In molecular dynamics simulations, the radial distribution function (RDF) is commonly used to characterize the degree of ordering of the dotted structure [28–30]. Figure 2c shows RDF plots of the TC4 ternary alloy at different moments during the aging phase transformation. As can be seen in the figure, the first peak of each plot incrementally shifts to the right with time, and its width increases. This change is due to the fact that the atoms of the bcc structure have eight nearest-neighbor atoms at 0.284 nm and six next-nearest-neighbor atoms at 0.322 nm. It is also evident from Figure 2c that the wave crest becomes wider due to the atomic vibrational shift. However, as the model undergoes a phase transition and the atoms are relatively displaced, the intensity of the characteristic peak corresponding to the initial bcc structure gradually decreases and the peak profile gradually becomes flat. The bcc structure gradually becomes an hcp structure with 12 nearest-neighbor atoms, the position of the first peak of the RDF curve is shifted to 0.295 nm, and the height and width of the peak increase due to the increase in coordination number. TC4 shows a shift from the characteristic peak of the bcc structure to the characteristic peak of the hcp structure at 11 ps.

Figure 2d shows plots of the relative contents of the bcc and hcp structures and the fcc structure versus simulation time. It can be seen from the figure that, during the first 14 ps, the model bcc structure content decreased sharply and the hcp and fcc structure contents gradually increased. The rates of change of the bcc and hcp structure contents slowed down during the period 15–30 ps, while the fcc structure content stabilized after 15 ps. After 30 ps, the contents of the crystal structures all stabilized, with the hcp structure content at about 50%, the fcc structure content at about 27%, and the bcc structure content close to 0.

For a better visualization of changes in the crystal structure during phase transformation of the model, the co-nearest-neighbor angle distribution analysis method proposed by Ackland et al. [30] was used for processing, and then the phase transformation of the TC4 alloy was visualized with the aid of OVITO software [31,32]. Figure 3 depicts the evolution of the atomic dot structure of the (011) crystal plane with simulation time. The blue atomic region in the figure shows the bcc dot structure, the red atomic region shows the hcp dot structure, and the green atomic region shows the fcc dot structure. As can be seen from the figure, by 5 ps, most of the atoms in the model had deviated from the structural dot matrix due to the displacement generated by vibrations; by 11 ps, hcp atomic clusters started to form in the crystal structure and the new phase started to nucleate. This result is consistent with the plateau period in Figure 2c. As shown in Figure 3c, the new phase nucleated at the very beginning increased significantly, indicating that the process of new phase growth mainly occurred in the late stage of the TC4 aging phase transformation process; by 47 ps, the crystal structure had stabilized.

3.2. Simulation of K-TIG Welding of TC4-TA17 Alloy

Figure 4 shows distribution plots of root-mean-square deviation (RMSD) with holding time for Ti, Al, and V at holding temperatures of 1500–2183 K. In Figure 4a–c, the RMSDs for the Ti, Al, and V atoms show linear variations with holding time. The magnitude of the RMSD gradually increases with increasing holding temperature. This indicates that the higher the temperature, the higher the frequency of Ti, Al, and V atoms jumping in the connection interface [33–36]. Atomic diffusion coefficients were calculated by applying the mean-square difference formula. Mean-square displacement yielded diffusion coefficients of the corresponding elements, with a slope of the simulation time distribution plot of 1/6 [32]. Linear fitting of the RMSD data was carried out using Origin software. When the holding temperature was below 1983 K, the slopes for the Ti, Al, and V atoms were smaller. This implies that their atomic diffusion coefficients were small. When the holding temperature was higher than 1983 K, sharp increases in the slopes of the diffusion coefficients were observed, implying larger atomic diffusion coefficients. From Figure 4d, it can be seen that at the same holding temperature, Ti and V atomic diffusion coefficients are comparable in magnitude, whereas that of Al is larger.

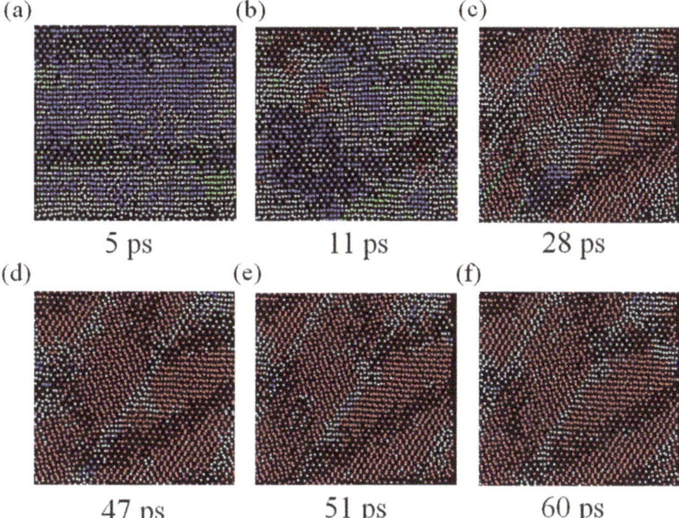

Figure 3. Structural evolution of TC4 alloy from high-temperature phase to room-temperature phase: (**a–f**) the corresponding structures at 5 ps, 11 ps, 28 ps, 47 ps, 51 ps, and 60 ps, respectively.

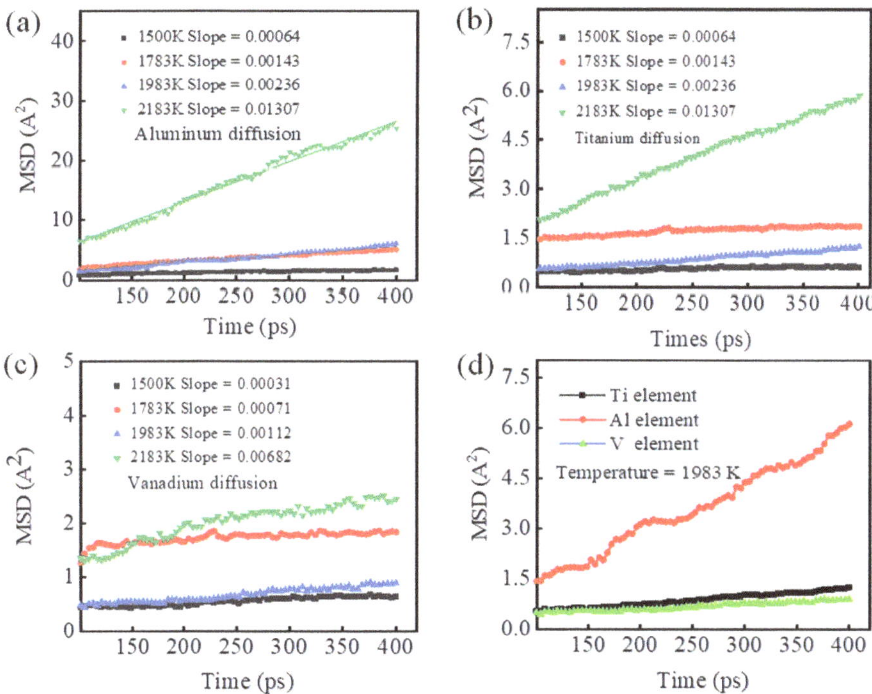

Figure 4. Diffusion mechanism of Al, Ti, and V atoms during welding: (**a–c**) root-mean-square displacements of Al, Ti, and V atoms at different temperatures; (**d**) root-mean-square displacements of Al, Ti, and V atoms at 1983 K.

Figure 5 shows the atomic distributions of Ti, Al, and V with the holding time and pressure fixed at 400 ps and 3.5 MPa, respectively, at a melting temperature of 1983 K after complete diffusion. For ease of observation, the Ti, Al, and V atoms of the TA17 model above the contact surface are colored gray, yellow, and purple, respectively. The Ti, Al, and V atoms of the TC4 model below the contact surface are colored red, blue, and light yellow, respectively. It can be seen that at constant holding time and pressure, when the system temperature is 1983 K, it is mainly the Al atoms near the interface that diffuse, followed by the Ti atoms, whilst the V atoms have the lowest diffusion capacity; this is consistent with the results in Figure 4d [37,38].

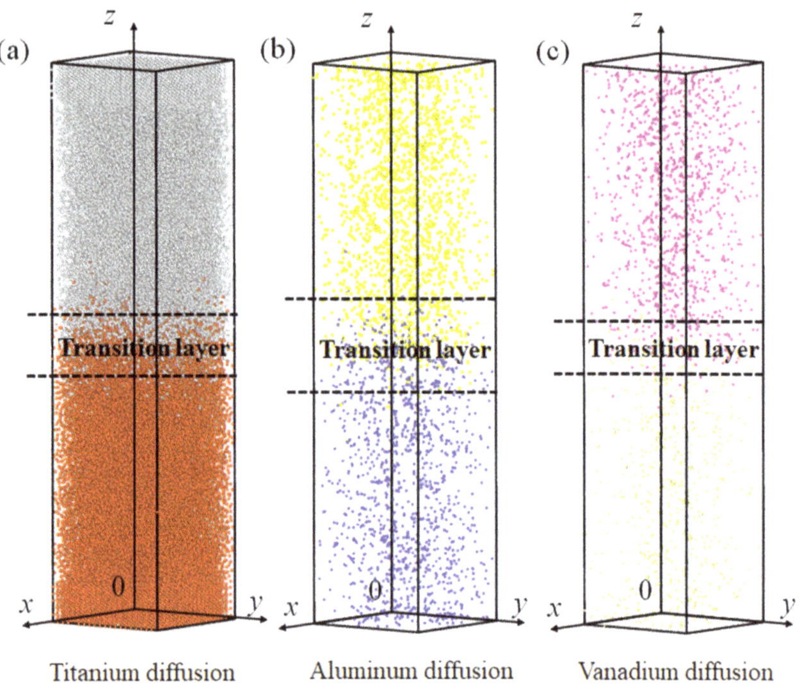

Figure 5. Analysis of the TC4-TA17 dissimilar titanium alloy welding diffusion layer: (**a–c**) diffusion layer structures of Al, Ti, and V atoms, respectively.

The main factors affecting the results of TC4-TA17 alloy diffusion bonding are holding temperature, pressure, holding time, and cooling rate. We proceeded by examining the effect of holding temperature on the diffusion bonding process of the TC4-TA17 alloy. Figure 6 shows the distributions of Ti–Al–V atomic concentrations at different holding temperatures. In the figure, Ti_V1 and Al1 are the atoms of the model below the contact surface, and Ti_V2 and Al2 are the atoms of the model above the contact surface. Since the initial content of V atoms was not high and the content in the diffusion region was even lower, to facilitate calculation processing and observation, we summed the atomic concentrations of Ti_V and did not calculate them separately. The areas between the dashed lines in Figure 6 represent the diffusion transition layers. It is generally stipulated that the area in which each atom concentration exceeds 5% in the contact surface and the lower part is the transition layer, so that the value of the diffusion connection width can be determined. It can be seen from the figure that as the heat preservation temperature was increased, the width of the diffusion connection increased significantly. A higher holding temperature can promote plastic yielding of the diffusion bonding interface, thereby expanding the contact area and reducing the probability of voids.

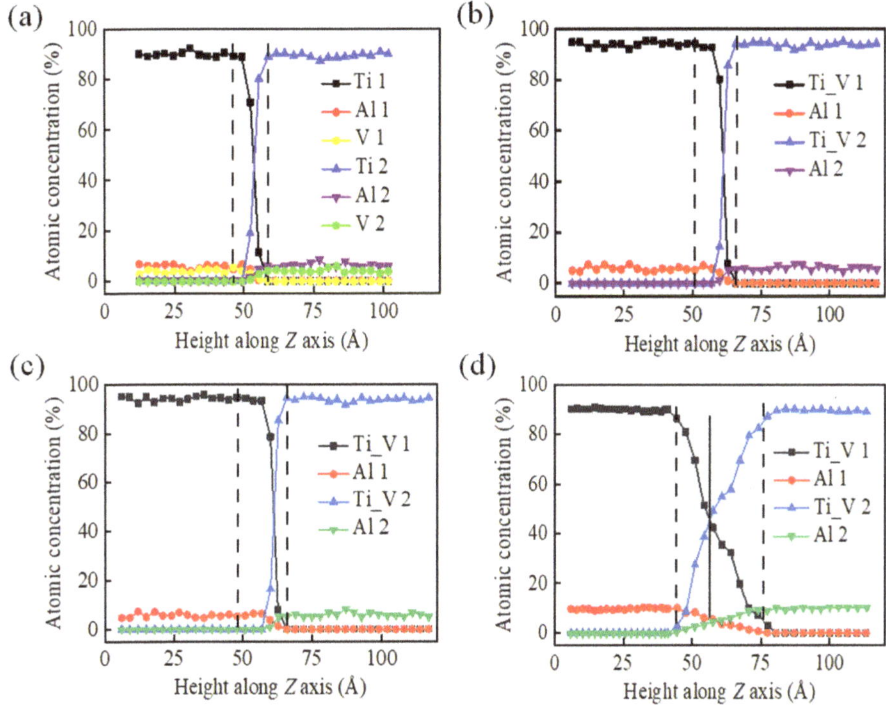

Figure 6. Fine structures of diffusion layers at different holding temperatures, as calculated from atomic positions using MATLAB software: (**a–d**) 1000 K, 1200 K, 1400 K, and 1600 K, respectively.

3.3. Simulation of Tensile Mechanical Properties

Figure 7 shows tensile stress–strain curves for TC4-TA17 with different parameters. As can be seen in Figure 7a, stress in TC4-TA17 increases with strain in stage I ($\varepsilon = 0$–0.07) and in stage II ($\varepsilon = 0.05$–0.12), similar stages to those of elastic deformation in the macroscopic materials. In stage III ($\varepsilon = 0.12$–0.18), a peak stress value of 9.07 GPa is reached at $\varepsilon = 0.16$ [39]. Thereafter, the stress decreases due to the continuous fracture of metal bonds; at the same time, the atoms are displaced by plastic deformation and then form new bonds with other metal atoms, causing the stress to increase. In stage IV ($\varepsilon > 0.18$), with increasing strain, the rate of metal bond breakage exceeds the rate of formation of new metal bonds and the stress begins to decrease; the TC4-TA17 shows the onset of necking and eventually fractures. In Figure 7b, with the increase in tensile force, the model of the upper and lower sides of the atomic movement of different degrees reaches tensile fracture, while the welding position is relatively solid. This indicates that the tensile mechanical properties of the TA17 and TC4 welding areas are good, exceeding those of TA17 and TC4 individually. It is worth noting that the tensile curve and fracture locations calculated by us are basically consistent with the experimental results, which proves that the simulation results are satisfactory [40–44].

As shown in Figure 7c, we simulated tensile experiments of TC4-TA17 welding samples under different holding times. It was found that as the holding time was increased, the initial elastic phase of the TC4-TA17 welded samples was prolonged and the tensile strength was increased. As can be seen in Figure 7d, with the increase in holding temperature, the yield strength of the TC4-TA17 welded samples also increased. At holding temperatures of 1000 K and 1200 K, the elastic phase was short, the yield point was at $\varepsilon = 0.07$, and the yield strengths were low. At holding temperatures of 1400 K and 1600 K, there was a significant extension of the resilience phase, the yield point appeared at $\varepsilon = 0.15$, and the peak yield strength was significantly higher than those at 1000 K and 1200 K. This indicates that the tensile mechanical properties of the TC4-TA17 welded samples are significantly improved with the increase in holding temperature.

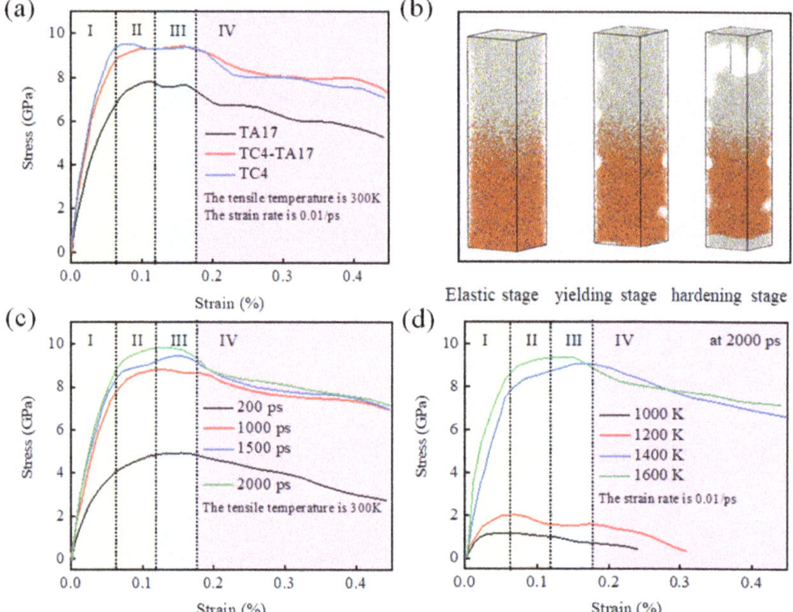

Figure 7. Fracture characteristics and fracture mechanism analysis of TC4-TA17 dissimilar titanium alloy after welding: (**a**) fracture characteristics of TC4-TA17 dissimilar titanium alloy at the tensile temperature is 300 K and the strain rate is 0.01/ps; and (**b**) fracture mechanism; (**c**) strain rate (the tensile temperature of 300 K); and (**d**) effect of holding temperature on tensile properties (the strain rate is 0.01/ps).

4. Conclusions

(1) During the phase transformation process, crystal defects such as lamination and twin boundary are prone to occurring at the junction of grain boundaries, so as to alleviate stress caused by the distortion. Increasing the content of Al, as an α-phase-stabilizing element, facilitates nucleation and precipitation of this phase and facilitates the phase change.

(2) A higher holding temperature promotes the plastic yield of the diffusion connection interface, thus enlarging the contact area and reducing the probability of void formation.

(3) The tensile process in TC4-TA17 simulation is similar to those of macroscopic materials, involving elastic deformation and a plastic deformation stage, and showing a necking phenomenon. When the system enters plastic deformation, the stress–strain curve shows a spike, and high yield strength is acquired.

Author Contributions: P.O. writing—original draft preparation; X.Y. and J.R. software; Z.C. review, editing, methodology and supervision. All authors have read and agreed to the published version of the manuscript.

Funding: This research received no external funding.

Institutional Review Board Statement: Not applicable.

Informed Consent Statement: Not applicable.

Data Availability Statement: The data that support the findings of this study are available upon reasonable request from the authors.

Conflicts of Interest: The authors declare no conflict of interest.

References

1. Malahy, K.A.; Hodgkiess, T. Comparative studies of the seawater corrosion behaviour of a range of materials. *Desalination* **2003**, *158*, 35–42. [CrossRef]
2. Fu, Y.Q.; Zhou, F.; Wang, Q.Z.; Zhang, M.D.; Zhou, Z.F. Electrochemical and tribocorrosion performances of CrMoSiCN coating on Ti-6Al-4V titanium alloy in artificial seawater. *Corros. Sci.* **2020**, *165*, 108385. [CrossRef]
3. Zhang, J.; Zhang, M.; Cui, W.C.; Tang, W.X.; Wang, F.; Pan, B.B. Elastic-plastic buckling of deep sea spherical pressure hulls. *Mar. Struct.* **2018**, *57*, 38–51. [CrossRef]
4. Wagner, H.N.R.; Hühne, C.; Niemann, S. Robust knockdown factors for the design of spherical shells under external pressure: Development and validation. *Int. J. Mech. Sci.* **2018**, *141*, 58–77. [CrossRef]
5. Li, L.Z.; Wang, S.G.; Huang, W.; Jin, Y. Microstructure and mechanical properties of electron beam welded TC4/TA7 dissimilar titanium alloy joint. *J. Manuf. Process.* **2020**, *50*, 295–304. [CrossRef]
6. Liu, H.; Dai, X.; Shi, M.H. EBSD Analysis on Tube-Plate Assembly Welding Line of TA 16 and TA17 Alloys. *Rare Metal Mat. Eng.* **2012**, *41*, 1756–1760.
7. Luo, Q.; Wang, L.; Chen, X.; Liu, S.W. Studies on the corrosion behavior of TA16 and TA17 titanium alloys in high temperature and high pressure water. *Light Met.* **2012**, *2*, 56–59.
8. Wang, C.M.; Guo, Q.L.; Shao, M.H.; Zhang, H.; Wang, F.F.; Song, B.Y.; Ji, T.J.; Li, H.X. Microstructure and corrosion behavior of linear friction welded TA15 and TC17 dissimilar joint. *Mater. Charact.* **2022**, *187*, 111871. [CrossRef]
9. Liu, Z.Y.; He, B.; Lyu, T.Y.; Zou, Y. A review on additive manufacturing of titanium alloys for aerospace applications: Directed energy deposition and beyond Ti-6Al-4V. *JOM* **2021**, *73*, 1804–1818. [CrossRef]
10. Dong, L.L.; Lu, J.W.; Fu, Y.Q.; Huo, W.T.; Liu, Y.; Li, D.D.; Zhang, Y.S. Carbonaceous nanomaterial reinforced Ti-6Al-4V matrix composites: Properties, interfacial structures and strengthening mechanisms. *Carbon* **2020**, *164*, 272–286. [CrossRef]
11. Li, Z.; Wang, J.; Dong, Y.Z.; Xu, D.K.; Zhang, X.H.; Wu, J.H.; Gu, T.Y.; Wang, F.H. Synergistic effect of chloride ion and Shewanella algae accelerates the corrosion of Ti-6Al-4V alloy. *J. Mater. Sci. Technol.* **2021**, *71*, 177–185. [CrossRef]
12. Yan, B.; Kong, N.; Zhang, J.; Li, H.B. High temperature formability of Ti–4Al–2V titanium alloy under hot press forming process. *Mater. Res. Express* **2019**, *6*, 126501. [CrossRef]
13. Fu, C.; Wang, Y.Q.; He, S.L.; Zhang, C.L.; Jing, X. Microstructural characterization and mechanical properties of TIG weld joint made by forged Ti–4Al–2V alloy. *Mater. Sci. Eng. A* **2021**, *821*, 141604. [CrossRef]
14. Huang, D.M.; Yang, X.F.; Wei, Q.F.; Chen, Y.; Guo, H.; Liang, S.L.; Wang, Y. Causes of surface stripe cracks of Ti–4Al–2V alloy cold-rolled sheet. *Rare Met.* **2014**, *33*, 522–526. [CrossRef]
15. Du, S.; Wang, S.; Ding, K. A novel method of friction-diffusion welding between TiAl alloy and GH3039 high temperature alloy. *J. Manuf. Process.* **2020**, *56*, 688–696. [CrossRef]
16. Karfoul, M.K.; Tatlock, G.J. Interfacial processes during diffusion welding of titanium alloy/aluminium couples under ambient atmosphere. *Weld. World* **2019**, *63*, 841–849. [CrossRef]
17. Zhu, L.L.; Zhang, Q.F.; Chen, Z.Q.; Wei, C.D.; Cai, G.M.; Jiang, L.; Jin, Z.P.; Zhao, J.C. Measurement of interdiffusion and impurity diffusion coefficients in the bcc phase of the Ti–X (X= Cr, Hf, Mo, Nb, V, Zr) binary systems using diffusion multiples. *J. Mater. Sci.* **2017**, *52*, 3255–3268. [CrossRef]
18. Shimono, M.; Onodera, H. Molecular dynamics study on formation and crystallization of Ti–Al amorphous alloys. *Mater. Sci. Eng. A* **2001**, *304*, 515–519. [CrossRef]
19. Semiatin, S.L.; Knisley, S.L.; Fagin, P.N.; Barker, D.R.; Zhang, F. Microstructure evolution during alpha-beta heat treatment of Ti-6Al-4V. *Metall. Mater. Trans. A* **2003**, *34*, 2377–2386. [CrossRef]
20. Plimpton, S. Fast parallel algorithms for short-range molecular dynamics. *J. Comput. Phys.* **1995**, *117*, 1–19. [CrossRef]
21. Harish, M.S.; Patra, P.K. Temperature and its control in molecular dynamics simulations. *Mol. Simulat.* **2021**, *47*, 701–729. [CrossRef]
22. Wang, H.; Zhang, L.F.; Han, J.Q.; Dee, W.N.E. PMD-kit: A deep learning package for many-body potential energy representation and molecular dynamics. *Comput. Phys. Commun.* **2018**, *228*, 178–184. [CrossRef]
23. Daw, M.S.; Baskes, M.I. Semiempirical, quantum mechanical calculation of hydrogen embrittlement in metals. *Phys. Rev. Lett.* **1983**, *50*, 1285. [CrossRef]
24. Baskes, M.I. Modified embedded-atom potentials for cubic materials and impurities. *Phys. Rev. B* **1992**, *46*, 2727–2742. [CrossRef] [PubMed]
25. Zhou, X.W.; Johnson, R.A.; Wadley, H.N.G. Misfit-energy-increasing dislocations in vapor-deposited CoFe/NiFe multilayers. *Phys. Rev. B* **2004**, *69*, 144113. [CrossRef]
26. Zhao, S.J.; Xiong, Y.X.; Ma, S.H.; Zhang, J.; Xu, B.; Kai, J.J. Defect accumulation and evolution in refractory multi-principal element alloys. *Acta Mater.* **2021**, *219*, 117233. [CrossRef]
27. Pinsook, U. Molecular dynamics study of vibrational entropy in bcc and hcp zirconium. *Phys. Rev. B* **2002**, *66*, 024109. [CrossRef]

28. Ackland, G.J.; Jones, A.P. Applications of local crystal structure measures in experiment and simulation. *Phys. Rev. B* **2006**, *73*, 4104. [CrossRef]

29. Zhao, F.; Zhang, J.; He, C.W.; Zhang, Y.; Gao, X.L.; Xie, L. Molecular dynamics simulation on creep behavior of nanocrystalline TiAl alloy. *Nanomaterials* **2020**, *10*, 1693. [CrossRef]

30. Song, C.B.; Lin, T.S.; He, P.; Jiao, Z.; Tao, J.; Ji, Y.J. Molecular dynamics simulation of linear friction welding between dissimilar Ti-based alloys. *Comp. Mater. Sci.* **2014**, *83*, 35–38. [CrossRef]

31. Stukowski, A. Structure identification methods for atomistic simulations of crystalline materials. *Model. Simul. Mater. Sci. Eng.* **2012**, *20*, 045021. [CrossRef]

32. Falconi, S.; Ackland, G.J. Ab initio simulations in liquid caesium at high pressure and temperature. *Phys. Rev. B* **2006**, *73*, 184204. [CrossRef]

33. Cui, Z.Q.; Zhou, X.L.; Meng, Q.B. Atomic-scale mechanism investigation of mass transfer in laser fabrication process of Ti-Al alloy via molecular dynamics simulation. *Metals* **2020**, *10*, 1660. [CrossRef]

34. Bai, D.S.; Sun, J.F.; Chen, W.Y.; Du, D.X. Molecular dynamics simulation of the diffusion behaviour between Co and Ti and its effect on the wear of WC/Co tools when titanium alloy is machined. *Ceram. Int.* **2016**, *42*, 17754–17763. [CrossRef]

35. Li, P.; Wang, L.S.; Yan, S.L.; Meng, M.; Xue, K.M. Temperature effect on the diffusion welding process and mechanism of B2–O interface in the Ti2AlNb-based alloy: A molecular dynamics simulation. *Vacuum* **2020**, *173*, 109118. [CrossRef]

36. Liu, C.H.; Zhu, X.J.; Li, X.M.; Shi, Q.S.J. Investigation on sintering processes and mechanical properties of Ti–Ta alloys by molecular dynamics simulation. *Powder Technol.* **2021**, *11*, 117069. [CrossRef]

37. Xu, Z.F.; Rong, J.; Yu, X.H.; Song, Y.M.; Zhan, Z.L.; Liu, J.X.; Meng, K. Preparation of Al-based coatings of titanium ingot and its high temperature oxidation resistance. *Rare Met. Mat. Eng.* **2017**, *46*, 1961–1965.

38. Rong, J.; Wang, Y.; Li, R.Y.; Yu, X.H.; Meng, K. Interface structure between titanium substrate and high temperature oxidation resistant aluminum-base coatings. *Rare Met. Mat. Eng.* **2018**, *47*, 682–686.

39. Zhang, B.; Zhang, X.Y.; Li, C.; Zhou, K.C. Molecular dynamics simulation on phase transformation of Ti-Al alloy with low Al content. *Rare Met. Mat. Eng.* **2012**, *41*, 1010–1015. [CrossRef]

40. Wang, S.Q.; Li, W.Y.; Jing, K.; Zhang, X.Y.; Chen, D.L. Microstructural evolution and mechanical properties of electron beam welded dissimilar titanium alloy joints. *Mat. Sci. Eng. A* **2017**, *697*, 224–232. [CrossRef]

41. Wang, S.Q.; Liu, J.H.; Chen, D.L. Effect of strain rate and temperature on strain hardening behavior of a dissimilar joint between Ti–6Al–4V and Ti17 alloys. *Mater. Des.* **2014**, *56*, 174–184. [CrossRef]

42. Wang, S.Q.; Liu, J.H.; Lu, Z.X.; Chen, D.L. Cyclic deformation of dissimilar welded joints between Ti–6Al–4V and Ti17alloys: Effect of strain ratio. *Mat. Sci. Eng. A* **2014**, *598*, 122–134. [CrossRef]

43. Xu, W.F.; Ma, J.; Luo, Y.X.; Fang, Y.X. Microstructure and high-temperature mechanical properties of laser beam welded TC4/TA15 dissimilar titanium alloy joints. *Trans. Nonferrous Met. Soc. China* **2020**, *30*, 160–170. [CrossRef]

44. Zhao, P.K.; Fu, L. Strain hardening behavior of linear friction welded joints between TC11 andTC17 dissimilar titanium alloys. *Mat. Sci. Eng. A* **2015**, *621*, 149–156. [CrossRef]

Article

FEM Numerical and Experimental Work on Extrusion Welding of 7021 Aluminum Alloy

Dariusz Leśniak [1,*], Wojciech Libura [1], Beata Leszczyńska-Madej [1], Marek Bogusz [1], Jacek Madura [1], Bartłomiej Płonka [2], Sonia Boczkal [2] and Henryk Jurczak [3]

[1] Faculty of Non-Ferrous Metals, AGH University of Science and Technology, 30-059 Krakow, Poland; libura@agh.edu.pl (W.L.); bleszcz@agh.edu.pl (B.L.-M.); bogusz@agh.edu.pl (M.B.); madura@agh.edu.pl (J.M.)

[2] Łukasiewicz Research Network—Institute of Non-Ferrous Metals, Light Metals Division, 44-100 Gliwice, Poland; bartlomiej.plonka@imn.lukasiewicz.gov.pl (B.P.); sonia.boczkal@imn.lukasiewicz.gov.pl (S.B.)

[3] Albatros Aluminum Corporation, 61-102 Poznan, Poland; h.jurczak@albatros-aluminum.com

* Correspondence: dlesniak@agh.edu.pl; Tel.: +48-12-617-31-96

Abstract: Extrusion welding of AlZnMg alloys encounters great technological difficulties in practice associated with high shaping forces and the low quality of longitudinal welds. Three different chemical compositions of 7021 aluminum alloy, differing in terms of Zn and Mg contents, were used in the first stage of the research. The laboratory device modelling the behavior of metal in welding chambers of the porthole die was applied to examine the ability of 7021 alloys to produce high-quality joints. The weldability tests were carried out for different welding temperatures—400, 450 and 500 °C—and for a fixed welding pressure of 300 MPa. The microstructural effects in pressure-welds were evaluated with the use of OM and SEM/EDS. The temperature–pressure parameters in the welding chambers were analyzed by using the FEM method for original porthole dies while extruding tubes with dimensions of Ø50 × 2 mm. Finally, the industrial extrusion trials were performed with examination of the structure and strength of the seam welds. It was found that it is possible to produce high-quality high-strength welds in tubes extruded from AlZnMg alloys in industrial conditions (the strength of welds in the range of 96–101% of the strength of the basic non-welded material) through properly matched alloy chemical composition of the alloy, construction of the porthole dies and temperature–speed conditions of deformation.

Keywords: AlZnMg alloys; extrusion welding; porthole die geometry; welding conditions; seam weld quality

Citation: Leśniak, D.; Libura, W.; Leszczyńska-Madej, B.; Bogusz, M.; Madura, J.; Płonka, B.; Boczkal, S.; Jurczak, H. FEM Numerical and Experimental Work on Extrusion Welding of 7021 Aluminum Alloy. *Materials* 2023, 16, 5817. https://doi.org/10.3390/ma16175817

Academic Editors: Chih-Chun Hsieh and Raul D. S. G. Campilho

Received: 5 July 2023
Revised: 12 August 2023
Accepted: 16 August 2023
Published: 24 August 2023

1. Introduction

The hot extrusion of profiles with the use of porthole dies is a common production technique. In this process, which is also called welding extrusion, the heated billet is divided into inlet channels of the die, and then the separate streams of metal enter the welding chamber, and they finally flow through the die cavity in the form of a hollow section. The obtained profile contains the longitudinal welds, the number of which depends on the die construction. The specific microstructure of the welds may be the reason for the poor strength of the profile.

The process is traditionally applied to extrude aluminum alloys, e.g., the 6xxx series. It is known from industrial practice that extrusion welding of high-strength alloys such as AlCuMg, AlMg3–5 (Mg3–5 in aluminum alloy represents the Mg chemical composition in mass percentage), AlZnMg and AlZnMg (Cu) causes serious difficulties in terms poor weldability and high deformation resistance. The process parameters, such as temperature and pressure in the welding chamber, extrusion speed and particularly porthole die design play an important role in effective welding.

Khan et al. [1] observed the critical role of welding chamber geometry on weld quality in 6082 and 7008 alloys. Formation of voids containing oxygen under the bridge is responsible for the poor strength of the welds. Similar observations were revealed by Oosterkamp [2]. Duplancic [3] investigated the quality of welds in relation to the porthole die geometry and process parameters. For the AlZnMg1 alloy to guarantee the good strength of the welds, the normal pressure in the welding chamber should be 6-fold higher than the flow stress of the alloy, whereas this factor should be 10-fold greater for AlZnMgCu1.5.

Donati and Tomesani [4] studied weld quality in connection with the porthole die geometry and process parameters for 6060 and 6082 alloys. The inlet channels should be wide for good weld quality and high exit speed. They proposed a new kind of die, a "butterfly" type, which has been implemented in industrial practice and increased the exit speed by 20%.

The results of FEM simulations of porthole die extrusion are presented in many works [5–9]. The main research stream therein is focused on the distribution of hydrostatic pressure and temperature in a welding chamber of the die. The data presented indicate that the pressure value in the welding chamber is responsible for the good quality of the welds. The pressure in turn is determined by the geometry of the inlet channels, the dimensions of the welding chamber, the length of the bearing land, the temperature and extrusion speed as well as the deformation value in the process. Akeret [10] proposed a qualitative criterion for welding the materials in the extrusion process. For the first time, he described the weldability as the relationship between the normal pressure σ_n in a welding area and the yield stress k of the metal. He provided limited values for different AlMgSi alloys leading to successful welding. Plata and Piwnik [11] presented an energetic method for longitudinal extrusion welding in which the contact surface, the contact time and the normal pressure to the yield stress ratio play the most important roles. A similar criterion presented by Donati and Tomesani [12] introduced a factor correcting the real time of contact.

Theoretical criteria for extrusion welding can be used in process optimization in connection with a numerical simulation of the process which allows predicting fundamental process parameters and conditions for welding. Ceretti [13] investigated the criterion of Plata and Piwnik based on the simulation of the extrusion of the 6061 alloy.

Many authors performed studies on the weldability of alloys by using hot working processes, e.g., upsetting, rolling or hot extrusion [14]. However, there are no tests that fully control the welding conditions that occur in extrusion including the fact that the welding process is run without contact with air atmosphere.

The effects of different geometries of the weld chamber and the processing conditions on the quality of the weld seam are investigated by Baker [15]. Through computer simulations, conditions related to weld seam formation were modelled and compared with the experimental results. The experimental results demonstrate that metal flow controlled by the die geometry causes defects, leading to the inferior mechanical performance of the extrudate. A comprehensive microstructural characterization of welds is presented. The authors of the study [16] introduced the analysis of different quality criteria of longitudinal seam welds that appear in aluminum profiles during extrusion using porthole dies. A new improved dimensionless welding quality criterion is proposed by the authors. The quantitative correspondence between welding types and the values of the developed criterion is also defined. In the study [17], porthole extrusion using different depths of welding chambers were designed and manufactured. Profiles extruded with different depths of the welding chambers were obtained by performing extrusion experiments. The welding quality of the extruded profiles was characterized by microstructure observation, a tensile test and fracture analysis. In the study [18], the welding quality of a 6063 aluminum alloy hollow square tube extruded by a porthole die was studied. The K criterion was introduced to evaluate the welding quality. The DRX fraction, grain size, and its influence on the welding quality were analyzed. They found that the uniform microstructure near the welding line would also affect the welding quality. Porthole die extrusion of Mg-Al-Zn alloy was conducted using the as-cast, as-homogenized and as-extruded billets [19]. The effects of the initial microstructure on the texture and mechanical properties of the extruded profiles were investigated. The results showed that complete dynamic recrystallization (DRX) occurred in the welding zone of all profiles, and the profile extruded from the as-cast billet had the smallest DRXed grains. Fine nanostructures of the bonding interfaces of weld seams formed by porthole die extrusion in the absence/presence of a gas pocket behind the bridge of the extrusion die were studied in [20] to understand interfacial bonding mechanisms. It was found that the formation of adverse gas pockets can be avoided by increasing the depth of the welding chamber.

An experimental setup is proposed [21] in order to analyze the welding condition of aluminum specimens under different loading conditions and temperatures. Based on backward cup extrusion, two specimens were compressed together within a heated container. Further, microstructural analysis of the welded specimens was conducted to observe if solid-state bonding occurred. The results of the trials are used to analyze the quality criteria for the seam welds during extrusion processes.

In the research [22], the commercial FEM software package, DEFORM v13.1, was used to simulate experiments in order to understand the relationship between the bridge design and the thermal mechanical history. The bridge can be divided into two parts: the lower part, close to the welding chamber, and the upper part, which initially split the billet into metal streams. The results showed that the geometry of the lower bridge influenced extrusion loads and profile exit temperatures. In contrast, changes to the geometry of the upper bridge had little effect on the porthole die extrusion process.

Different types of welding were found during porthole die extrusion by varying the extrusion die structure and process parameters [23]. It was found that the welding quality is determined by metal flow behavior, the solid-state bonding process and microstructural evolution. The formation of the unbonded macrodefects in the extruded profile is affected by improper metal flow behavior, while the formation of the microvoids on the bonding interface is caused by insufficient solid-state bonding. The influence of billet heating temperature and extrusion speed on the microstructure and mechanical properties of welding seams was studied in [24]. It was found that, in porthole die extrusion of aluminum alloy profiles, fine or coarse grains and microvoids can be formed in welding seams. The hardness, strength and ductility of the extruded profiles can be improved by increasing billet heating temperature and extrusion speed.

The metal flow law in the unsteady stage during the porthole die extrusion process was studied by Wang et al. [25]. They proposed the interfacial bonding mechanism of the longitudinal welds in the unsteady zones of the profiles and a method for evaluating the length of the unsteady zones of the profiles. The study by Annadurai et al. on extrusion die design using FEM simulations allowed for a reduction in experimental trials by 2/3 and a doubling of the lifespan of the die [26]. The die design and FEM simulations show the potential to predict extrusion defects and final profile microstructure, texture and mechanical properties [27,28]. Yu et al. used FEM model predictions to understand the effect of the bridge shape on the textures that formed along the weld seam during porthole die extrusion [29].

In the study [30], the dynamic recrystallization (DRX) behavior of an Al-Zn-Mg alloy during the porthole die extrusion process was studied. Higher deformation temperature and lower strain rate are favorable for the occurrence of DRX. The volume fraction of DRX at the zones close to bridge and porthole wall is much higher than that in the other zones. The objective of the study [31] was to develop a new porthole extrusion die for improving the welding pressure in the welding chamber by using numerical analysis. Through numerical analysis, the welding pressures in the welding chamber between the new porthole die and the conventional porthole die were compared with each other. The shape of the porthole die bridges was studied by FEM [32]. They found that dies that have curved bridges offer optimum process conditions. A Taguchi method and ANOVA were used to optimize the geometry of the welding chamber and other process parameters [33].

An optimal combination of the investigated parameters was determined for all the process outputs according to specific process needs.

The effect of process parameters on longitudinal welding quality was studied by Chen et al. [34]. The morphology of the poor longitudinal weld seam is observed by optical microscopy and scanning electron microscopy. Simulation results show that a lower extrusion speed, higher billet temperature and larger billet diameter are beneficial to improve longitudinal welding quality. To improve the flow balance in the die, a design approach was introduced to find an appropriate die structure that includes the porthole and pocket geometry correction, the bearing length adjustment, and the bridge structure modification [35]. Using the proposed die, the predicted velocity relative difference (VRD) and the maximum velocity difference (ΔV) of the extrudate were significantly lower than those of an initial die.

This work presents an innovative way for predicting seam weld quality in a tube of $\varnothing 50 \times 2$ mm from 7021 alloy. The method employs a patented laboratory device to investigate the welding process, allowing full modelling of conditions in a welding chamber of the porthole die [36]. This method allows reducing the time-consuming and costly industrial extrusion trials. Generally, the AlZnMg and AlZnMgCu alloys are known as difficult to weld during extrusion through porthole dies, so the proper choice of the alloy composition and the proper design of the porthole die construction are very challenging for these aluminum alloys. The FEM simulations as well as the industrial verification complete the research procedure. In the first stage, the Akeret indicator $\frac{\sigma_n}{k}$ was determined in relation to the chemical composition of the alloy. The Akeret indicator σ_n/k is a welding parameter for the susceptibility of the material to extrusion welding. The lower the value of this indicator, the higher the susceptibility to extrusion welding. For example, for easily deformable and easily weldable AlMgSi alloys, the value of this parameter is 2–3, while for difficult-to-deform alloys with lower weldability, this parameter is in the range of 6–10 (σ_n is the normal stress affecting the weld in the welding chamber of porthole die; k is the flow stress of the deformed material). The second stage comprised FEM simulations and provided information where the required value of the indicator above was obtained in the welding chambers of different geometries. The third stage comprised microstructural analysis of the welds and mechanical testing of samples including welds; and for comparison, these were taken from the base material without welds.

2. Materials and Methods

2.1. Characterization of AlZnMg Alloys

The billets, with the chemical composition presented in Table 1 and a diameter of 178 mm, were DC cast in semi-industrial conditions. Three alloys were investigated at a EN AW-7021 grade. In the case of all examined alloys, the low-melting microstructure components were dissolved during homogenization by soaking to a degree sufficient in practice—no incipient melting peaks on the DSC curves are noted. As a result, the significant increase in solidus temperature was achieved, and the obtained values are within the range from 559.2 °C for alloy 3 to 613.2 °C for alloy 1 (Table 2).

Table 1. The chemical composition of investigated EN AW-7021 alloys, mass percentage.

Alloy Denotation	Si	Fe	Cu	Mg	Cr	Zn	Ti	Zr
7021 alloy 1	0.09	0.22	0.00	1.20	0.00	5.27	0.01	0.15
7021 alloy 2	0.08	0.21	0.00	2.12	0.00	5.47	0.01	0.15
7021 alloy 3	0.09	0.22	0.00	2.06	0.00	8.02	0.02	0.15

2.2. Method and Device for Testing Ability to Extrusion Welding

A new method and device for weldability testing was carried out using a developed authorial patented device (Figure 1a,b, [1]). This device enables repeating conditions of joining of a metal which are expected to exist in a welding chamber of the porthole die during extrusion of hollow sections (processes of shearing and compression without the air admission). The test of weldability consists of cutting of two samples from the tested alloy (separated by steel counter-samples) and then axial pressing (welding) at the assumed temperature and under the assumed unitary pressure (hydrostatic pressure). The cutting of the samples heated in the heating chamber (1) and placed in the shearing and welding cartridge tool (2) is conducted by the upper shearing punch (6). The axial pressure is conducted by the hydraulically driven compression stem (3), which affects the welding surface of the samples perpendicularly.

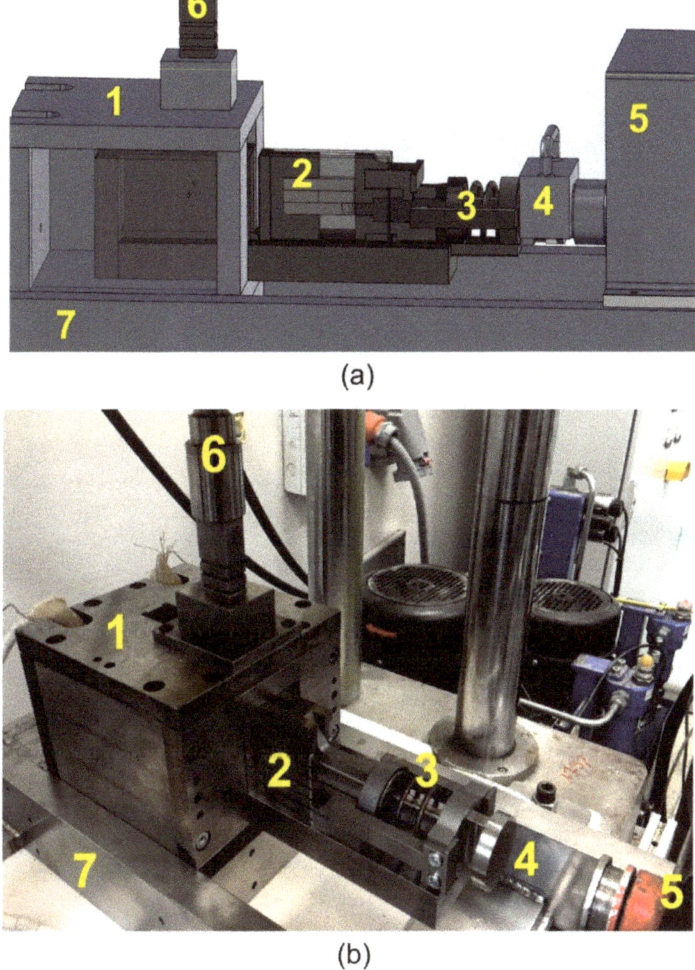

Figure 1. Device for weldability tests of metals and alloys: 3D model visualization (**a**) and research laboratory device (**b**); 1—heating chamber, 2—shearing and welding cartridge tool, 3—compression hydraulically driven stem with presser, 4—adapter of hydraulic cylinder and tool assembly holder, 5—hydraulic cylinder, 6—upper shearing punch, and 7—main construction ram (Patent No PL230273B1, AGH Krakow, 2018 [36]).

Figure 2 presents the subsequent stages of work of the device for weldability tests: a—the tool cassette is inserted into the heating chamber and the presser presses the sample packet; b—the samples are sheared with the upper punch while the presser is still held; c—after cutting, the shear punch stops and the compression punch hydraulically applies the specific pressure force on the welded samples. In this way, the situation is similar to that of the real porthole die, where separate streams of metal move on the channel surface and are welded in the welding chamber. Thus, the steel plate "plays role" of the 1 inlet channel.

Table 2. DSC test results of as-cast and homogenized EN AW-7021 alloys.

Alloy	Solidus Temperature, °C	Incipient Melting Heat, J/g
7021 alloy 1	611.8	-
7021 alloy 2	478.1	0.07
7021 alloy 3	478.2	0.68
7021 alloy 1 (homogenized)	613.2	-
7021 alloy 2 (homogenized)	572.1	0.29
7021 alloy 3 (homogenized)	559.2	-

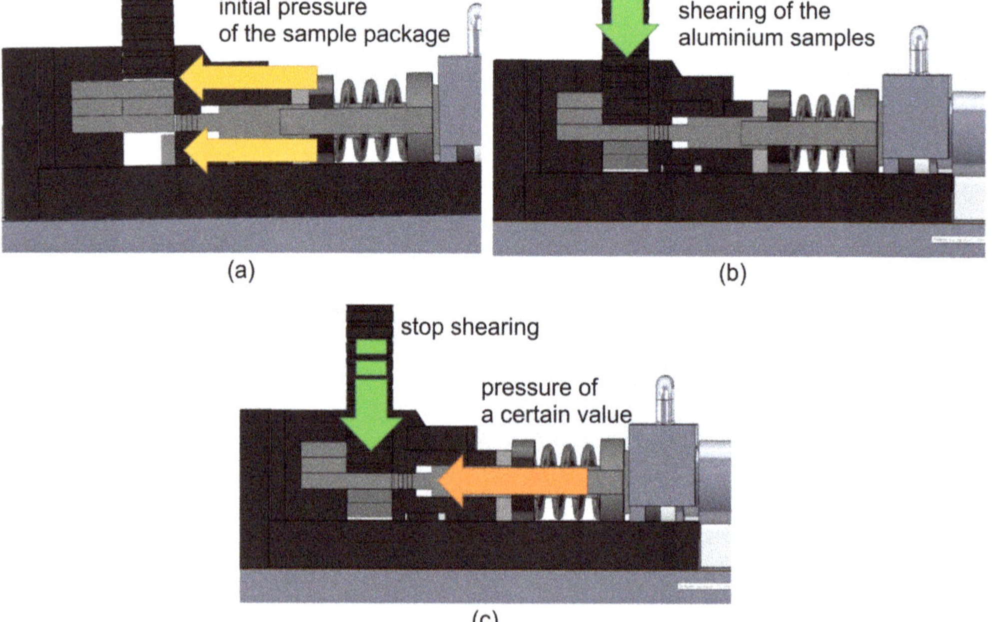

Figure 2. Individual stages of the weldability test: (**a**)—the tool cassette is inserted into the heating chamber and the presser presses the sample package, (**b**)—the samples are sheared with the upper punch while holding the presser continuously, and (**c**)—the shearing punch stops and the pressing punch applies a certain pressure to the welded samples.

Figure 3 shows the arrangement of the samples for the alloy tested (the orange and green ones) and steel counter-samples (the blue light and blue dark ones) in the tool cassette, in the starting position (left view) and in the end position (right view). The initial sample dimensions were of 60 × 10 × 10 mm. Dimensions of samples and counter-samples in the starting position are shown in Figure 4.

2.3. Methodology of the Microstructural and Mechanical Examination

The method of taking samples from the extruded tube for further examination of the microstructure and mechanical properties is presented in Figure 5. Welding extrusion is the process of extruding hollow profiles using a bridge–porthole die. In this process, the metal from the input material (ingot) is cut on the die bridge, then flows through the inlet channels, successively flows into the welding chambers located under the bridge and finally flows out through the working opening of the die. The finished product has longitudinal welds in the cross-section (seam welds) located along the entire profile. The samples for microscopic examination were mounted in resin, mechanically ground with sandpaper with appropriate gradation, and then mechanically polished in two stages using a diamond paste suspension and a colloidal silicon oxide suspension for finishing polishing. To reveal the microstructure of the samples for observation with a light microscope, the samples were anodized in Barker reagent with the composition of 100 mL H_2O + 2 mL HBF_4. The microstructure of the samples was examined by means of light microscopy (OLYMPUS GX51 microscope, Tokyo, Japan) and scanning electron microscopy (Hitachi SU 70 microscope, Tokyo, Japan). Additionally, the chemical composition in the microareas was analyzed using the energy-dispersive X-ray spectroscopy (EDS) method (Thermo Fisher Scientific, Waltham, MA, USA). An analysis of the chemical composition within the grains was performed to determine the content of individual alloying elements. In each case, a minimum of 20 spot analyses were performed. The test was carried out at an acceleration voltage of 15 kV. In addition, the average grain diameter was determined in the weld area and outside of this area (average chord method), and the average grain elongation coefficient in the weld area was determined. Usually, the samples with a weld in the center are prepared for static tensile tests, allowing for an assessment of the quality of the joint by comparing the mechanical properties of samples with the weld and reference samples (without weld). Studies of the basic mechanical properties—yield strength (YS), ultimate tensile strength (UTS) and percentage elongation (A%)—were carried out using the INSPECT 100 strength machine (with a maximum tensile strength of 100 kN).

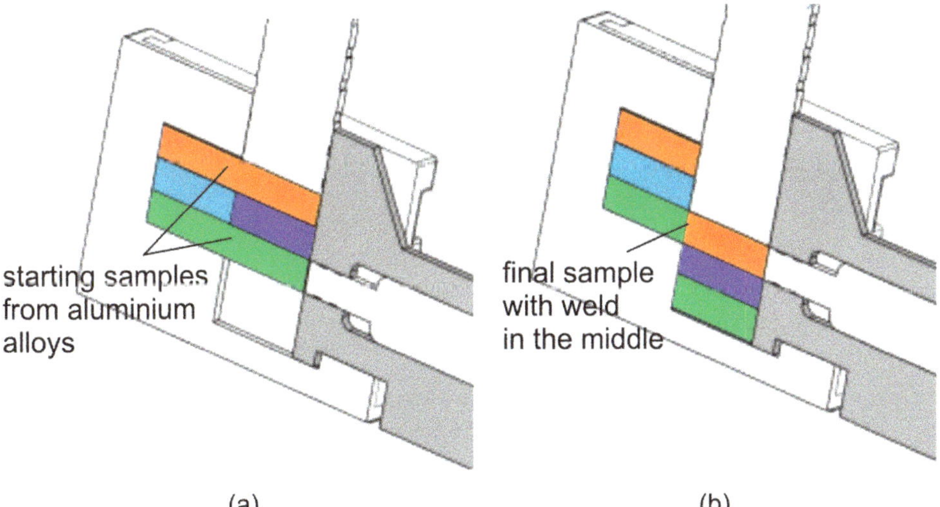

Figure 3. Arrangement of samples and counter-samples in the toolbox in the starting position (**a**) and in the final position after the end of the tests (**b**); orange and green colours mean samples from aluminium alloy; blue and purple colours mean counter-samples from steel.

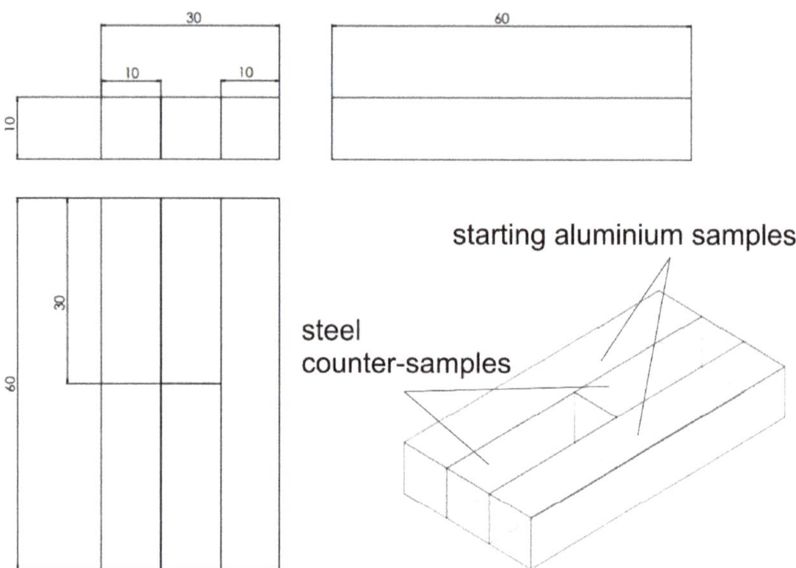

Figure 4. Dimensions of samples and counter-samples in the starting position.

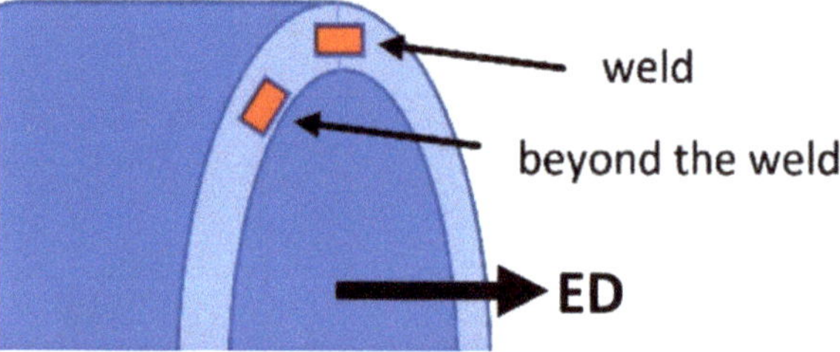

Figure 5. The method of taking samples from the extruded tube for further examination of the structure and mechanical properties.

Analysis of the crystallographic orientation of grains was performed under a high-resolution INSPECT F50 FEI scanning electron microscope with attachments for chemical analysis by EDS and a camera for EBSD. The EBSD analysis was performed using EDAX APEX^{TM} EBSD software. The cross-section samples were cut, ground and polished mechanically. In the last stage of preparation, the samples were polished using the Leica RES101 Ion Milling System. Maps were scanned at a resolution of 800 × 800 mm and a step of 1 mm.

2.4. FEM Numerical Modeling of Extrusion Welding

The extrusion process of tubes of Ø50 × 2 mm from aluminum 7021 alloy 2 was FEM modeled by using porthole dies of various geometries: conventional porthole die for 6xxx alloys based on the local 3-armed bridges (die 1) and die 2 with maximal broad inlet channels, shaped pockets, bearings of varied length and proper geometry of the central baffle and mandrels. Moreover, the thickness of the bridges for die 2 was increased by 4 mm, whereas their length was increased by 40 mm in relation to die 1. The dimensions of the porthole dies discussed are shown in Figure 6.

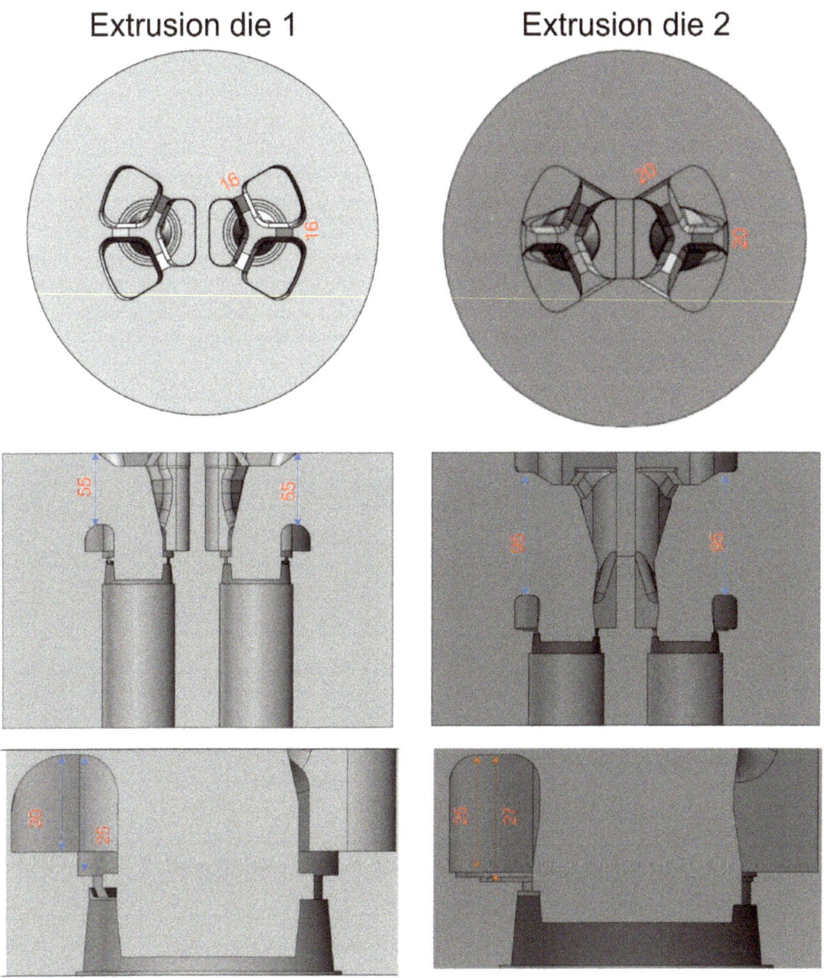

Figure 6. Dimensions of different porthole dies for extrusion of tubes of Ø50 × 2 mm from 7021 alloy 2—top view and the cross-sectional view.

The overall dimensions for both dies are the same. The diameter of the dies is Ø310 mm and the thickness of the tool set assembly is 210 mm. In the case of die 1, the bridge part is 65 mm, the length of the bridges is 55 mm, and the thickness is 16 mm; the height of the welding chambers is 20 mm. The die plate with the backer that completes the set is 145 mm thick.

In the case of die 2, the bridge part is 103 mm, longer than that of die 1 by 38 mm. Longer bridges with a length of 95 mm and a thickness of 20 mm were used. The height of the welding chambers of the complex set is also greater, 25 mm. The die plate is 107 mm thick; and in this case, no backer is used. A chamfered contact surface was used at an angle of 20° instead of the parallel connection of the porthole and die plate as in die 1, providing stiffness and support without the use of a backer.

Conducting numerical simulations required a number of preparatory works to be performed, e.g., development of a digital three-dimensional CAD model of both the die and the entire toolkit enabling installation in an industrial press (Figure 7). Three-dimensional tool models were developed using SolidWorks Premium v2019 SP0.0 CAD/CAM software.

Preparatory work and numerical simulations were carried out using QForm Extrusion v10.1.7 3D software. A special calculation module enables importing a CAD model of tools and generating a mesh of finite elements on their surface and in the volume of the model. The numerical simulation module consists of a pre-processor and a post-processor that provides visualization of the results [1]. The numerical model of the finite element simulations is defined based on the flow formula proposed by Zienkiewicz and Pittman, where the deformed material is treated as an incompressible and rigid viscoplastic continuum, while elastic deformations are neglected [2]. The calculations are based on the Euler–Lagrange model, which uses finite elements to simultaneously relate the material flow with the deformation and temperature distribution of the tool. This means that the elastic deformation of the die affects the way the metal flows, while the deformation of the tool itself is determined by the pressure of the metal on its surface. The software is based on two discrete models. The first of them, the Lagrange model, is designed to simulate the transient state of the initial stage of the process when the metal fills the die, while the second, the combined Euler–Lagrange model, is used to simulate the steady-state phase [3].

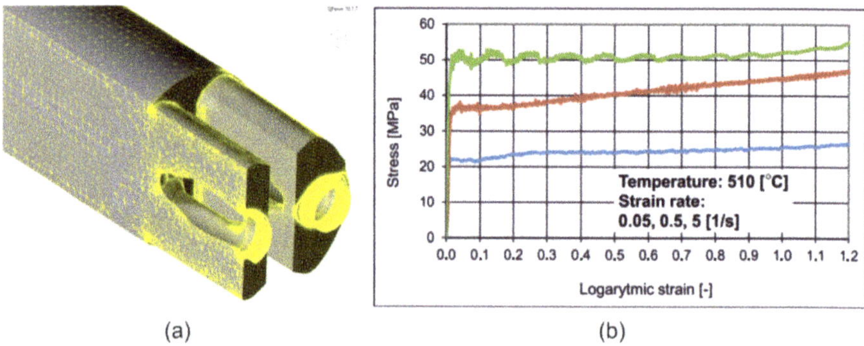

Figure 7. The geometry of the extruded material (**a**) and FEM material model indicating the dependence of plastic stress on logarithmic strain for different strain rates for alloy 2 (**b**); blue colour for 0.05 1/s, red colour for 0.5 1/s and green colour for 5 1/s [37].

The mesh of finite elements, both surface and volume die set and workpiece, is made of elements with the geometry of triangles of various sizes connected by nodes. The mesh of the computational domain-aluminum filling the interior of the die as well as the die set was created automatically after defining the appropriate geometrical elements of the tool. The size of the local elements was selected using individually selected factors for each of the die areas. The default size factor is 1, the coefficient of the adaptation of the container is 1.0, the coefficient of the adaptation of the welding chamber is 0.95, the coefficient of the adaptation of the pocket and profile geometry is 0.45, and the coefficient of the adaptation of bearing is 0.80.

In the case of die 1, the number of nodes on the surface and in the volume of the workpiece is 220,146 and 261,203 for the die set—a total of 481,355 nodes. The number of mesh elements on the surface and in the volume of the workpiece is 1,149,535 and 1,392,619 for the die set—a total of 2,542,154 finite elements. In the case of die 2, the number of nodes on the surface and in the volume of workpiece is 225,440 (due to larger inlet channels and the volume of metal filling the die) and 232,893 for the die set—a total of 458,333 nodes. The number of mesh elements on the surface and in the volume of the workpiece is 1,173,504 and 1,243,138 for the die set—a total of 2,416,642 finite elements.

FEM mesh verification was carried out in a special QShape module for mesh preparation. The prepared surface and volume mesh are checked by the algorithm used in QShape for further reconstruction and potential errors—starting numerical calculations in the main calculation 3D solver is not possible for an unverified mesh.

Obtaining high-accuracy simulation results requires a detailed definition of the rheological properties of the deformed material. The Hensel–Spittel constitutive Equation (1) was used to describe the deformation characteristics of the alloy:

$$\sigma_p = A \cdot e^{m_1 T} \cdot T^{m_9} \cdot \varepsilon^{m_2} \cdot e^{\frac{m_4}{\varepsilon}} \cdot (1 + \varepsilon)^{m_5 T} \cdot e^{m_7 \varepsilon} \cdot \dot{\varepsilon}^{m_3} \cdot \dot{\varepsilon}^{m_8 T} \tag{1}$$

where

σ—plastic stress,
ε—plastic strain,
$\dot{\varepsilon}$—strain rate, and
T—temperature of deformation.

The concept of describing the yield stress with one curve makes it possible to determine empirically the rheological properties of the material in laboratory tests, e.g., in the high-temperature compression test. The effect of temperature during plastic deformation on the decrease in yield stress value should be taken into account. For this reason, the determination of the characteristics of the selected material using a wide range of temperatures and strain rates is crucial for the description of rheological properties used in material extrusion tests [4]. The coefficients from Equation (1) are shown in Table 3.

Table 3. The coefficients from the Hensel–Spittel equation.

A	m_1	m_2	m_3	m_4	m_5	m_7	m_8	m_9
1090	-0.0675	-0.055	0.21	-0.015	0.019	-0.026	-0.0019	0.056

In the simulations, the friction model defined by Levanov (2) was adopted on the contact surface of the deformed metal and the tool.

$$F_t = m \frac{\overline{\sigma}}{\sqrt{3}} \left(1 - exp\left(-1.25\frac{\sigma_n}{\overline{\sigma}}\right)\right) \tag{2}$$

where m is the coefficient of friction, σ_n is the perpendicular contact pressure and σ is the actual stress. The equation can be considered as a combination of the constant friction model and the Coulomb friction model. The second term in parentheses of Equation (2) takes into account the effect of normal pressure on surface contact. For high pressure values, the expression approximates the conditions defined by the constant friction model. On the other hand, for low pressure values, it defines a linear approximation depending on the normal stress at the surface contact. All the defined extrusion process parameters are presented in Table 4.

Table 4. The defined parameters of the FEM modelled extrusion process of tubes of Ø50 × 2 mm from 7021 alloy 2.

Alloy	7021 alloy 2
Billet dimensions	Ø178 × 700 mm
Billet temperature	480 °C
Container/Die temperature	465 °C
Extrusion ratio	42
Stem velocity	1.58 mm/s
Metal exit speed	3–4.5 m/min
Friction coefficient	m = 1

2.5. Extrusion Trials of Round Tubes

The extrusion trials of the tube of Ø50 × 2 mm from 7021 alloy 2 by using the double-hole porthole die 1 and porthole die 2 (Figure 8) were carried out on the 7-inch hydraulic press of 25 MN capacity. The billet dimensions and the conditions of the trials were identical with these in the numerical simulation. Die 1 was based on the variant usually used during extrusion of the 6000 series alloys, whereas die No 2 was the modified version elaborated in FEM simulation of the investigated process. Figure 8 shows the layout of the inlet channels in both the tested dies. Despite the similarity in the inlet channel layout, the channels in die 2 are wider while the bridges are higher and thicker to withstand the increased extrusion force and to avoid elastic deflection of the die. The height of the welding chamber is higher in die 2. The die insert was equipped with pockets, which regulate metal flow in the region close to the die opening. In such a case, the metal flow is more uniform and this improves the geometrical stability and dimensional accuracy of the extruded profile. Alloy 2 was used in the experiments. The extruded tubes were air cooled on the run-out table and next submitted to ageing. During the extrusion trials, the ram speed, exit speed, extrusion pressure and profile temperature were recorded. The cracking on the extrudate surface means that the maximal extrusion speed was exceeded. The maximum exit speed during extrusion of the 7000 series alloys is usually very low (below 1.5 m/min), so each achievement in this field considerably improves process efficiency.

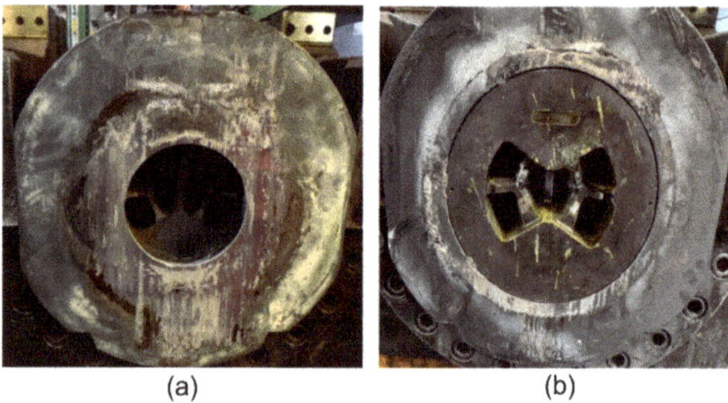

Figure 8. Porthole extrusion dies used in experimental trials of the extrusion of tubes of Ø50 × 2 mm from aluminum 7021 alloy 2 by using the 7-inch 2500 T hydraulic press: (**a**)—die 1; (**b**)—die 2.

3. Results

3.1. Weldability Tests

Figure 9 shows the stress–strain curves recorded during static tensile testing of samples welded from alloys 1, 2 and 3 at 450 °C and 500 °C. The influence of the welding temperature on the mechanical properties of the welds produced in the weldability tests at a given unit pressure $p = 300$ MPa for 7021 alloys with different contents of Mg and Zn (7021 alloy 1, 2 and 3) was determined (Figure 10). The highest strength and plastic properties were recorded at a welding temperature of 450 °C for all three analyzed alloys. The relative strength of the welds was 86% (alloy 1), 93% (alloy 2) and 85% (alloy 3). Slightly lower values of the relative strength of the welds were obtained for a welding temperature of 500 °C—but only for 7021 alloys 2 (61%) and 3 (66%). In the case of alloy 1, the relative strength of the welds dropped dramatically from 86% to 12%.

The highest UTS in welding trials was achieved for samples from alloy 3 with the highest content of main alloy additions (Figure 10b). However, the increase in the content of alloying elements also translates into a reduction in the plastic flow resistance of the metal during heat extrusion as well as a higher extrusion force and higher mechanical loads acting on the dies. Therefore, in order to ensure an metal flow rate from the die (process efficiency), extrusion force and dimensional tolerances resulting from a small elastic deflection of the die, alloy 2 was adopted for further analysis.

Due to the temperature and speed conditions of the industrial process of the extrusion of hollow sections from 7021 alloy, the temperature in the welding area of

550 °C was assumed for further analysis. For the alloy defined in this way and the welding temperature, the Akeret weldability index was determined, which in this case is $6_n/k$ = 300 MPa/51 MPa = 5.88.

Microstructural observations of the samples after the welding process using light microscopy show that, regardless of the process conditions used outside the weld, the grains have a regular, near-axial shape, while in the weld area the grains are elongated (Figure 11, Figures S1 and S2 in Supplementary File). In most cases, no discontinuities were found in the weld area, except for samples of alloys 1 and 2 welded at the lowest temperature of 400 °C, where discontinuities were found along the entire length of the weld (see arrows) (Figure 11a,b). A characteristic feature of the microstructure of alloys 2 and 3 was the occurrence of numerous precipitates within the grains, regardless of the welding process parameters used (Figure 11b,c, Figures S1 and S2 in Supplementary File).

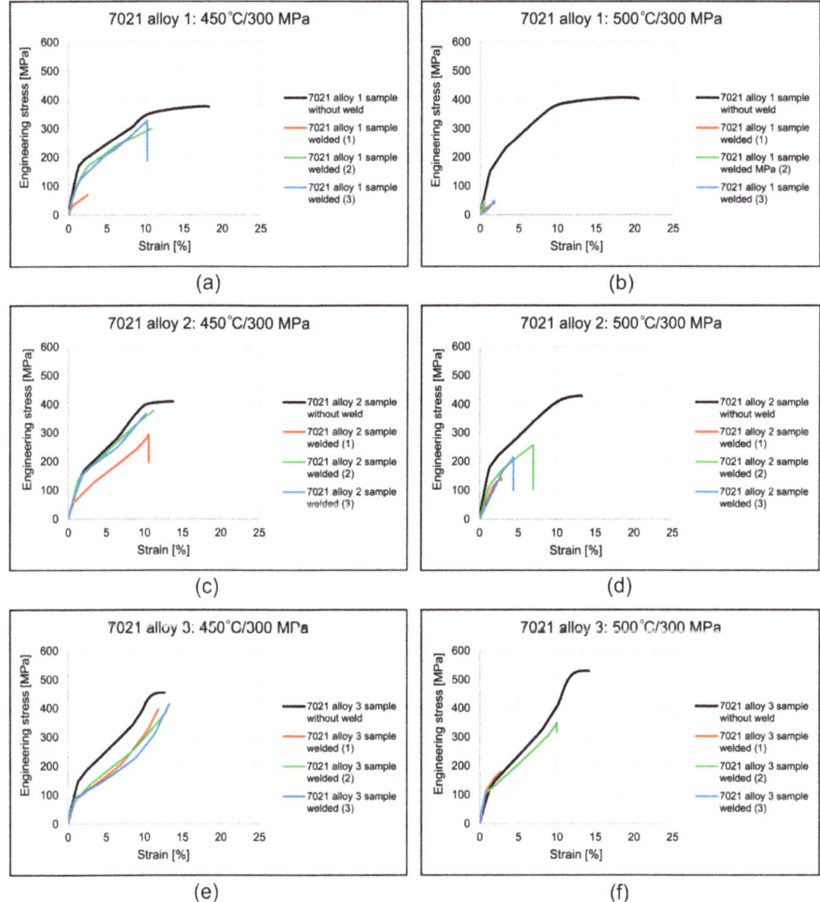

Figure 9. Stress–strain curves recorded during static tensile testing of samples welded from alloys 1, 2 and 3 at 450 °C and 500 °C; 7021 alloy 1: welding temperature 450 °C/welding pressure 300 MPa (**a**) 7021 alloy 1: welding temperature 500 °C/welding pressure 300 MPa (**b**) 7021 alloy 2: welding temperature 450 °C/welding pressure 300 MPa (**c**) 7021 alloy 2: welding temperature 500 °C/welding pressure 300 MPa (**d**) 7021 alloy 3: welding temperature 450 °C/welding pressure 300 MPa (**e**) 7021 alloy 3: welding temperature 500 °C/welding pressure 300 MPa (**f**).

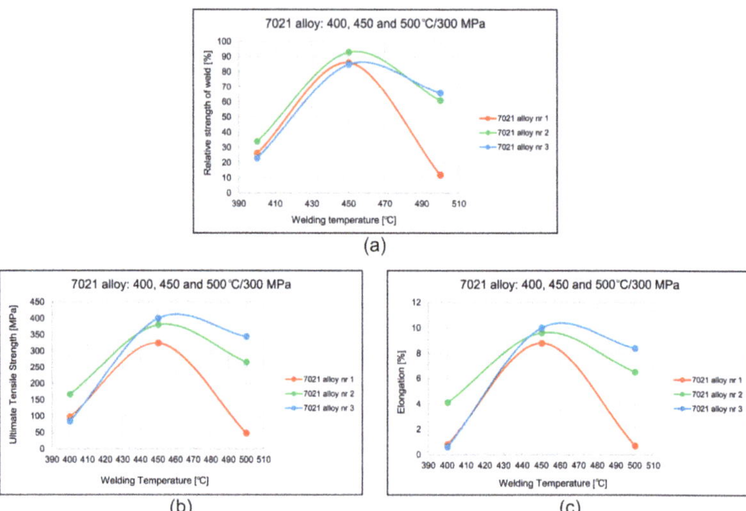

Figure 10. Influence of welding temperature on the relative strength of welds (**a**), the ultimate tensile strength (UTS) of samples (**b**) and elongation (**c**) for 7021 alloys with different chemical compositions.

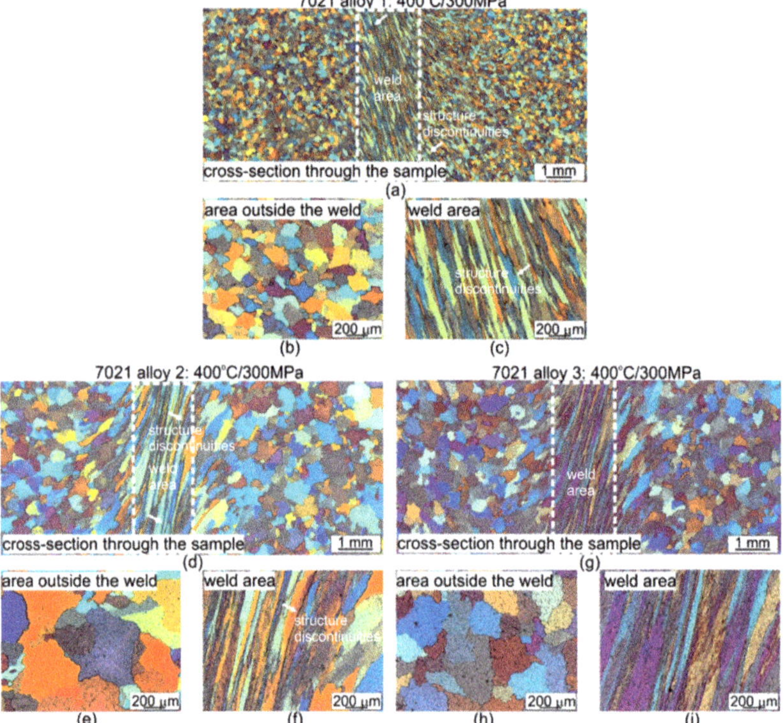

Figure 11. Microstructure of welded 7021 alloy in relation to chemical composition; welding was performed under identical process parameters: T = 400 °C, p = 300 MPa; (**a–c**) 7021 alloy 1: 1.20% Mg, 5.27% Zn; (**d–f**) 7021 alloy 2: 2.12% Mg, 5.47% Zn; (**g–i**) 7021 alloy 3: 2.12% Mg, 8.02% Zn; light microscopy.

Measurements of the width of the weld zone proved that for alloy 1, the width of the welding area measures approximately 2.5 mm, regardless of the welding temperature. For alloy 2, the width of the welding area measures approximately 1.7 mm for welding temperatures of 400 °C and 450 °C. On the other hand, it is much wider at approximately 2.8 mm after welding at 500 °C. In the case of alloy 3, the welding width was found to increase with increasing temperature; the width of the welding area was 1.77 mm, 2.53 mm and 3 mm for welding temperatures of 400 °C, 450 °C and 500 °C, respectively.

The grain size measurements proved that, in the alloy 1 samples, the average grain diameter outside the weld was 147–196 µm, with the samples welded at 400 °C having the largest grain size. The average diameter of the grains outside the weld area of the samples welded at 450 °C and 500 °C was comparable and was approximately 150 µm (Figure 12a). The determined average grain size in the welding area is 228–260 µm and, similarly to the outside of the weld, is the largest for samples welded at 400 °C; for the other welding variants, it is comparable (Figure 12a). The coefficient of grain elongation of alloy 1 samples in the weld area is 7.7–9.1 and is the smallest for samples welded at the lowest temperature of 400 °C. In this alloy, fine dispersive precipitations are presented in the microstructure after welding at 450 °C and 500 °C, which inhibited grain growth.

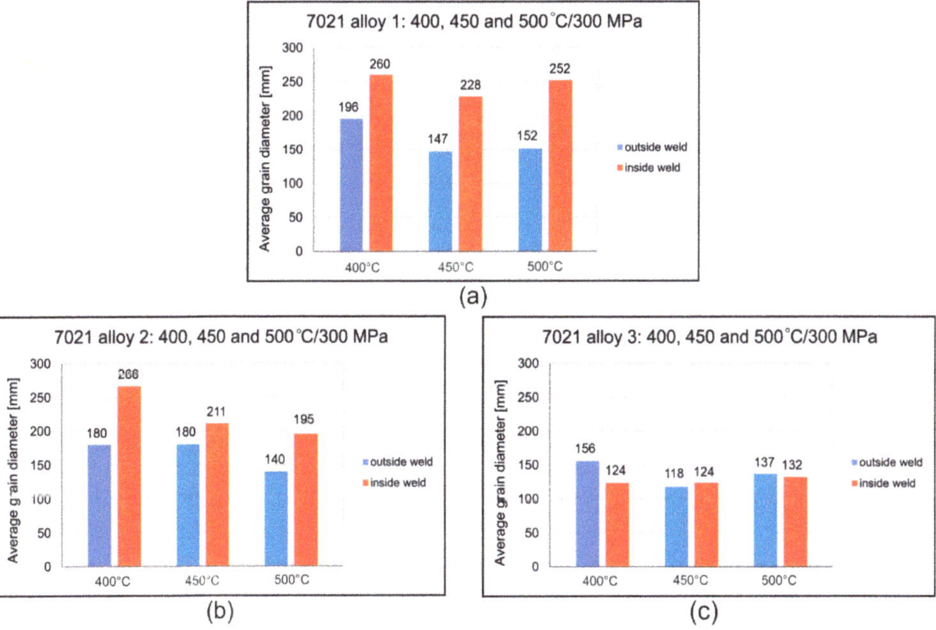

Figure 12. Average grain size outside and inside the weld; (**a**) 7021 alloy 1: 1.20% Mg, 5.27% Zn; (**b**) 7021 alloy 2: 2.12% Mg, 5.47% Zn; (**c**) 7021 alloy 3: 2.12% Mg, 8.02% Zn.

In the alloy 2 samples, the average grain diameter outside the weld is 140–180 µm, with the smallest grain size for samples welded at 500 °C. The determined average grain size in the welding area is 195–266 µm and, similarly to the outside of the weld, is the smallest for samples welded at 500 °C (Figure 12b). The elongation coefficient is 8.5–14.2 and is the highest for samples welded at the lowest temperature of 400 °C. After welding at 450 °C and 500 °C, the grains in the weld area were wider and shorter; characteristically, in the microstructure, the number of dispersive precipitates on the grain cross-section decreased, and the number of obstacles that inhibit the movement of grain boundaries decreased.

Samples from alloy 3 are characterized by the finest grain size, with the average grain diameter outside the weld area being 118–156 µm, while in the welding area, it is comparable for all variants of the welding process and is 124–132 μm (Figure 12c). The elongation coefficient determined is at the level of 6.5–9.2. The microstructure of alloy 3 has the highest number of dispersive precipitates, which inhibit grain growth when exposed to high temperature and pressure during the welding process. The numerous precipitates did not adversely affect the welding process and no discontinuities were found in the welding area of this alloy, regardless of the welding process conditions used (Figure 11, Figures S1 and S2 in Supplementary File).

Figure 13 and Figures S3 and S4 in Supplementary File) show the results of the chemical composition analysis in micro-areas for alloy 2—welding at 450 °C (Figure 13: outside weld and Figure S3 in Supplementary File: in weld) and 500 °C (Figure S4 in Supplementary File: outside weld and Figure S5 in Supplementary File: in weld). Figure 14 shows the Mg content for the alloys tested when welded at 450 °C (Figure 14a) and 500 °C (Figure 14b). Figure S6 in Supplementary File, in turn, shows the Zn content for all alloys welded at 450 °C and 500 °C (Figure S6 in Supplementary File). The mean Mg and Zn content for samples welded at 450 °C was comparable in the welding and non-welding areas and averaged 1.75 wt.%. Mg and 6.5 wt.%. Zn (Figure 14a,b). Examination of the chemical composition of the grain cross-section locally indicates the presence of Ti and Zr, with these elements being present in the dispersoids observed inside the grains. Similarly, in samples welded at a higher temperature of 500 °C, the average content of Mg and Zn in the welding area and outside the weld was comparable, being higher than in samples welded at 450 °C and averaging 1.79 wt.% Mg and 6.56 wt.%, Zn (Figures S5 and S6 in Supplementary File).

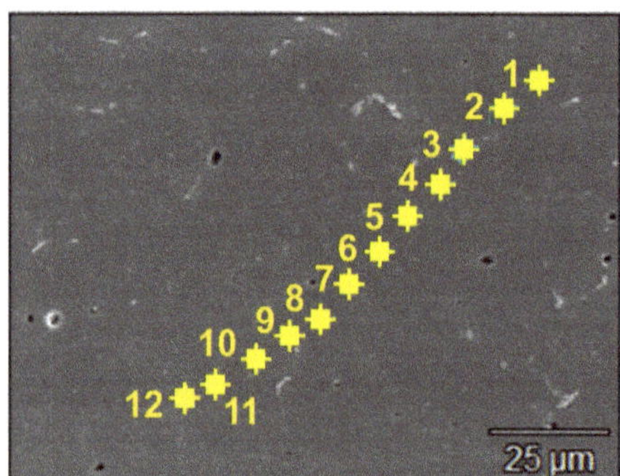

Elements concentration, wt.%						
Area	Mg	Al	Ti	Fe	Zn	Zr
1	1.85	91.45	0.06		6.64	
2	1.73	91.44		0.09	6.56	0.18
3	1.77	91.53	0.05		6.65	0.00
4	1.73	91.67			6.58	0.02
5	1.79	91.76			6.44	
6	1.73	91.70	0.04	0.07	6.45	
7	1.84	91.54	0.08		6.44	0.09
8	1.77	91.36	0.06		6.71	0.10
9	1.77	91.60			6.58	0.05
10	1.73	91.71	0.12		6.38	0.07
11	1.71	91.79	0.07	0.08	6.35	
12	1.69	91.55	0.01	0.02	6.62	0.11

Figure 13. Microstructure of alloy 2 in the area outside the weld and results of the chemical composition test on the grain cross-section; welding process conditions: T = 450 °C and p = 300 MPa; SEM/EDS.

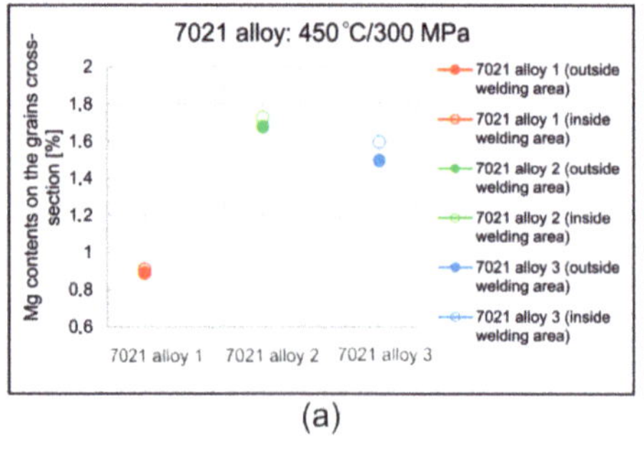

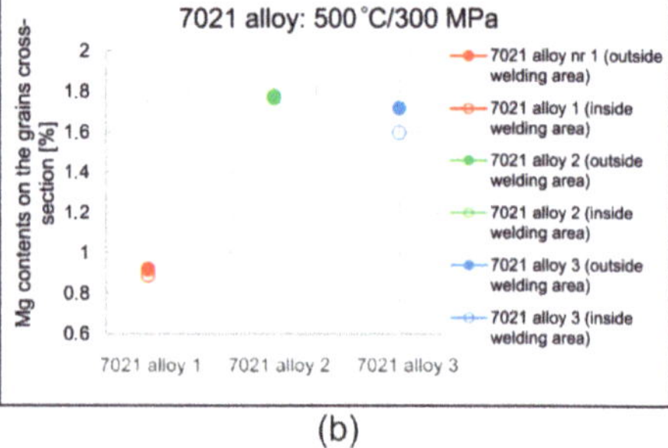

Figure 14. Average Mg content on the grain cross-section in the welding area and outside the welding area for the 7021 alloy 2 tested: (**a**) T = 450 °C, p = 300 MPa; (**b**) T = 500 °C, p = 300 MPa.

After welding at 450 °C, the largest variation in Mg content across the grain cross-section was found for alloy 3 (Figure 14). The average Mg content in the non-welded area is 1.48 wt.%, while in the welding area it is 1.59 wt.% and Zn 8.4 wt.% and 9.1 wt.%, respectively. Welding at a higher temperature of 500 °C resulted in an increase in the inhomogeneity of the distribution of the main alloying elements on the cross-sectional area, especially for alloy 3 (Figures S4–S6 in Supplementary File). The average Mg content in the area outside the weld for this alloy is 1.6 wt.%, while in the welding area it is 1.72 wt.%, Zn, respectively—9.6 wt.%. in the non-welded area and 9.15 wt.% in the welding area.

Figure 15 and Figure S7 in Supplementary File show representative images of the fractures of specimens welded at 450 °C and 500 °C, after uniaxial tensile tests. Fractures of samples of all the alloys welded at 450 °C show features of plastic fracture, the proportion depending on the type of alloy. In the case of alloy 1, the characteristic rounded hollows are only locally visible. In the case of alloys 2 and 3, rounded hollows and elevations, characteristic of plastic fracture, are visible on the entire surface. The presence of dimples in the material shows that plastic deformation occurs. Decohesion occurs in the successive parallel slip planes of favorable orientation. As a result of this process, new free surfaces are formed in materials. The shape and size of the dimples are determined by the size and distribution of microstructure discontinuities, such as the micropores, disperse particles and microcracks, plastic properties of the material, and the acting stresses.

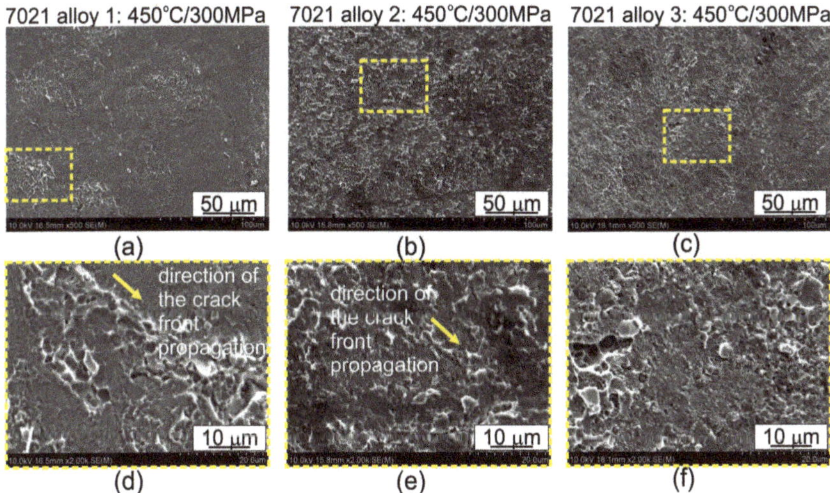

Figure 15. Fracture surfaces after uniaxial the tensile test; welding was performed under identical process parameters: $T = 450$ °C, $p = 300$ MPa; (**a,d**) 7021 alloy 1: 1.20% Mg, 5.27% Zn; (**b,e**) 7021 alloy 2: 2.12% Mg, 5.47% Zn; (**c,f**) 7021 alloy 3: 2.12% Mg, 8.02% Zn. Figures (**d–f**) are the enlarged areas indicated by the dashed rectangles in Figures (**a–c**).

When the specimens were welded at 500 °C, the fracture of the specimens definitely differed from alloy to alloy (Figure S7 in Supplementary File). In the case of alloy 1, surface decohesion under shear occurred during the uniaxial tensile test, and no signs of plastic deformation are observed. In the case of alloys 2 and 3, holes and elevations characteristic of plastic fracture are visible on the entire surface, with their proportion being greater in alloy 2 (covering the entire surface). The arrows in Figure 15 and Figure S7 in Supplementary File show the direction of the crack front propagation. The fractures of the specimens after the uniaxial tensile test shown in Figure 15 and Figure S7 in Supplementary File are consistent with the results of the microstructure observations. Samples showing a predominance of plastic fracture contribution were characterized by good weld quality.

3.2. FEM Numerical Calculations

Figure 16 shows the mean stress distributions in the inlet channels of the porthole die, the welding chambers, and the die opening for die variant 1 and for die variant 2 of the die (left) and the enlarged mean stress distributions in the welding chambers and the die opening of the porthole die (right) during extrusion of the Ø50 × 2 mm tube from EN AW-7021 alloy. Differences between dies can be seen, especially for the inner welding chambers. In general, in the upper regions of the welding chambers (just below the bridge), maximum compressive stress values favorable for material welding are observed in the range 247–267 Mpa depending on the die design, with slightly higher values recorded for die 1. It can be concluded that the maximum values of the average stress in the welding chambers of the porthole dies correspond to the compressive stress value occurring in the laboratory weldability tests (σ_w = 300 Mpa). Moving towards the die opening, these values decrease to reach mean stress values close to zero in the die opening itself. In addition, die 2 has lower compressive stresses compared to die 1 that goes from the extrusion axis towards the width of the welding chamber. This level and distribution of the average stress indicate slightly more favorable conditions for the joining of the metal strands in the welding chambers in the case of die 1. In the metal case of the temperature distribution at the height of the welding chamber (Figure 17), comparable temperature values in the range 488–503 °C with a maximum in the die opening are observed for both dies. These values correspond to the welding temperature in the laboratory weldability test (500 °C), for which a moderate relative weld strength of 62% was obtained for 7021 alloy 2. In the case of plastic stress distributions in the welding chambers of porthole die, significantly higher values of this parameter were obtained for die 2: more than 30 for die 2 compared to 20 for die 1 (Figure 18). Additionally, for die 2, there are larger areas with maximum plastic stress values in the volume of the welding chamber, which may indicate more favorable bonding conditions for the metal planes resulting from more intensive metal mixing in the welding chambers.

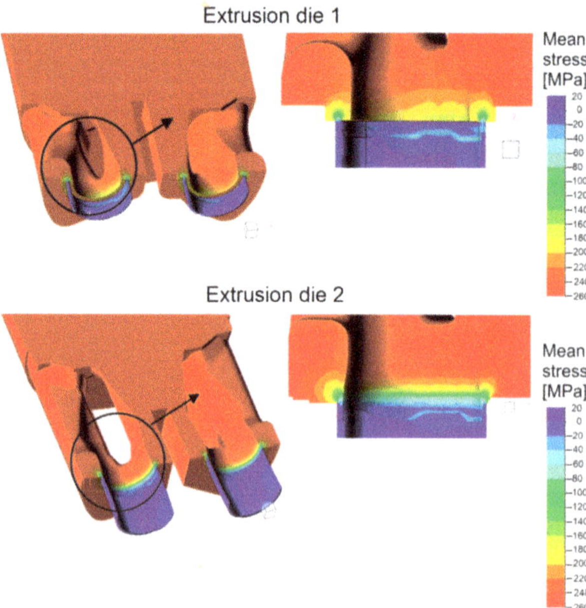

Figure 16. Distribution of mean stress during extrusion of tubes of Ø50 × 2 mm from 7021 aluminum alloy through porthole dies of different geometries—extrusion die 1 and extrusion die 2 (results of FEM calculations).

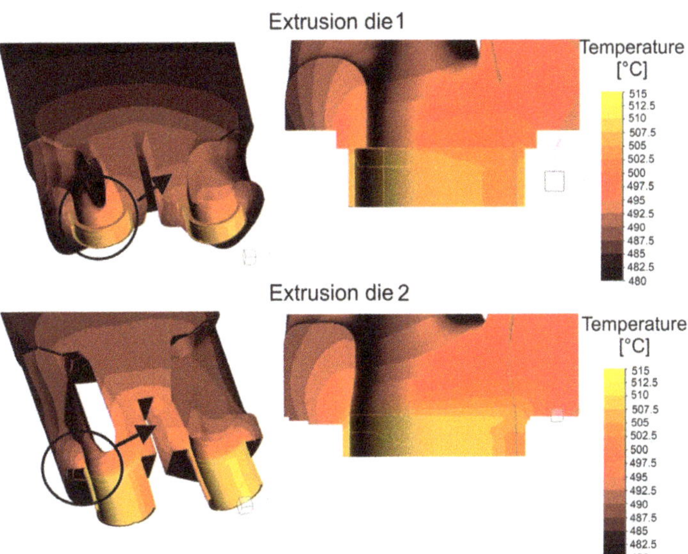

Figure 17. Distribution of temperature during extrusion of tubes of Ø50 × 2 mm from 7021 aluminum alloy through porthole dies of different geometries—extrusion die 1 and extrusion die 2 (results of FEM calculations).

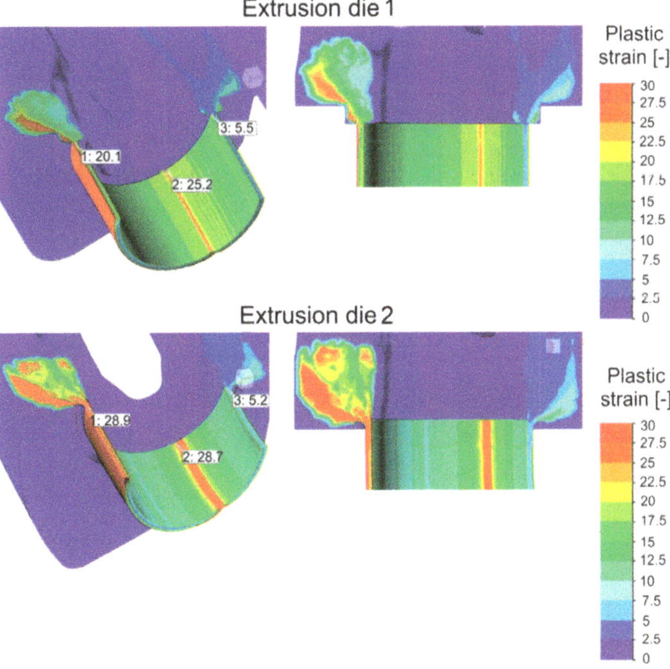

Figure 18. Distribution of plastic strain during extrusion of tubes of Ø50 × 2 mm from 7021 aluminum alloy through porthole dies of different geometries—extrusion die 1 and extrusion die 2 (results of FEM calculations).

Figure 19 presents the distributions described above in numerical notation. Figure 19, bottom right, shows the distribution of the weldability parameter σ_m/σ_i at the height of the welding chambers of die 1 and die 2, i.e., the ratio of the average stress σ_m to the stress intensity σ_i, a parameter that informs about the material welding conditions in the extrusion process through the porthole dies. The higher value of this parameter, the higher probability of producing a high-quality joint in the extruded hollow profile. For comparison, this graph also shows with a dashed line the level of the minimum weldability index from the laboratory weldability tests ($\sigma_w/k = 5.88$), which already guaranteed the production of a relatively good quality weld for the analyzed 7021 alloy 2. As can be seen, at the prevailing height of the welding chambers from the sub-bridge space toward the die opening, there are sufficiently favorable welding conditions (determined numerically by FEM) for both dies (σ_m/σ_i of 6–7), which ensure relatively good material susceptibility to extrusion welding—exceeding the minimum weldability index determined by laboratory weldability tests. However, the closer to die opening, the greater the deterioration of the metal welding conditions.

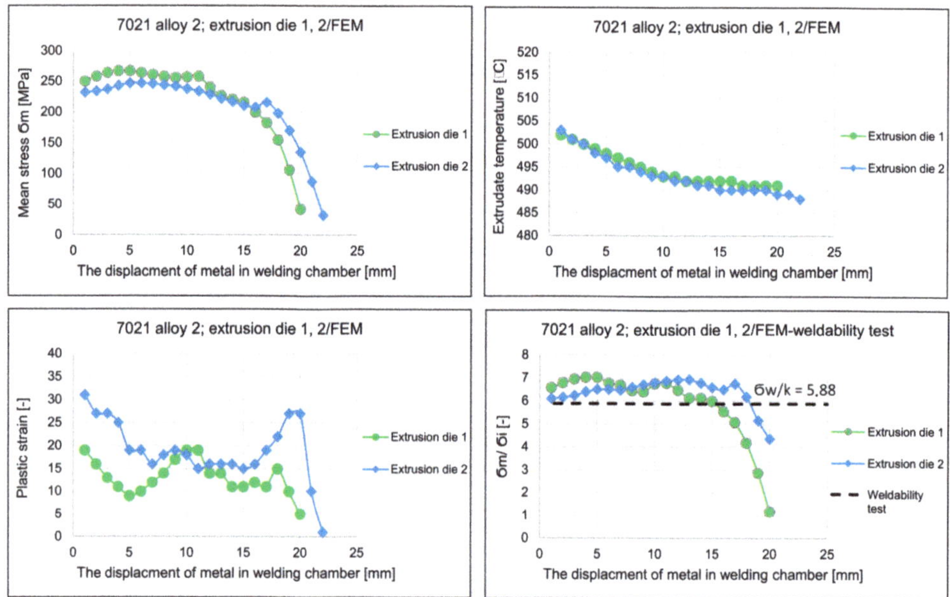

Figure 19. FEM numerically predicted the relationship between the displacement of metal in the welding chamber and the mean stress σ_m, extrudate temperature, plastic strain and weldability index σ_m/σ_i during extrusion of Ø50 × 2 mm tubes from 7021 aluminum alloy 2.

Based on this distribution, it can be concluded that in the first stage of metal welding, up to approximately half of the height of the welding chamber, more favorable welding conditions occur for die 1, while in the second stage of welding, more favorable welding conditions occur for die 2. In summary, it can be concluded that the influence of die design on the conditions for joining metal strands in the welding chambers is complex. Die 1 provides higher compressive stress values in the welding chambers in absolute terms, while die 2 guarantees better material mixing due to higher plastic stress values in the welding chambers. The weldability parameter indicates more favorable conditions for the metal strands to fuse together for die 1 in the first welding stage and the more favorable conditions for the metal strands to fuse together for die 2 in the second welding stage. Experimental verification under industrial conditions will ultimately indicate which factors that affect the weldability of the metal in the extrusion process are dominant.

3.3. Extrusion Trials

Figure 20 shows photos of Ø50 × 2 mm tubes extruded from 7021 alloy 2 using 2-hole die 1 and 2-hole die 2. The top and middle rows show photos of tubes on the press run out immediately after the metal exits the die opening (Figure 20 top and middle). The bottom row, on the other hand, shows cross-sectional photos of tubes on the cooling table (Figure 20 bottom). From these results, the good surface quality and dimensional accuracy of the extruded products can be inferred. However, detailed studies of the geometric stability of tubes extruded by die 1 (which is the subject of another publication) indicated relatively large dimensional deviations of wall thickness, going beyond the values allowed by the relevant standard [37].

Extrusion die 1

Extrusion die 2

Figure 20. Extruded tubes of Ø50 × 2 mm from 7021 alloy 2 on the press run-out for dies of different geometries: die 1; $T_0 = 480\,°C$, $V_1 = 3.5\,m/min$, die 2; $T_0 = 480\,°C$, $V_1 = 4.5\,m/min$.

The process parameters recorded during extrusion of the tubes in question indicate a metal discharge velocity from the die hole in the range 3.5–4.5 m/min, depending on the geometry of the porthole die used (Figure S8 in Supplementary File). Higher values for the metal exit speed were recorded for die 2, which also provided high dimensional accuracy of the extruded tubes. The temperature measured immediately after the metal exited the die opening was in the range of 570–580 °C (Figure S8 in Supplementary File). Figure 21 shows example macrostructures of the extruded tube using die 1 with 3 longitudinal welds visible in the cross-section.

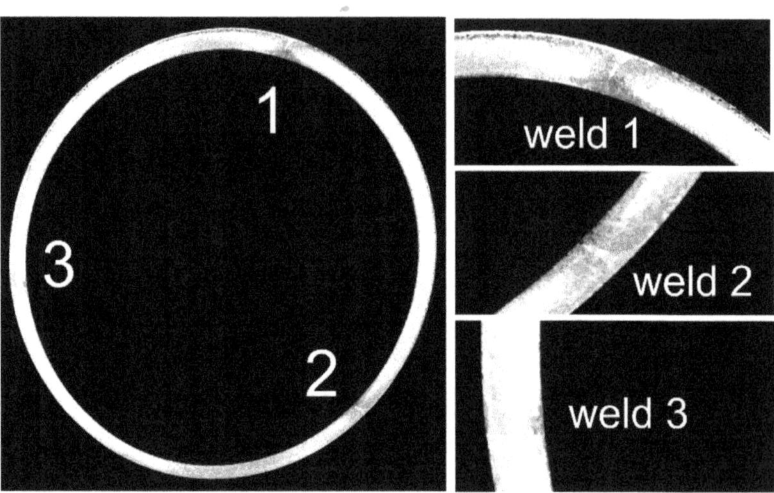

Figure 21. Example macrostructures of a tube of Ø50 × 2 mm extruded through die 1 from 7021 alloy 2 revealing the locations of welds in the cross-section.

Figure 22 shows the microstructures and analyses of the crystallographic orientation images taken from the cross-sectional view of the tube from the weld area and outside the weld area for die 1 and die 2. From observations in the weld area, it can be seen that the shape and grain size do not differ much from those outside the weld area. This is true for both die 1 and die 2 extruded tubes, and the differences are visible in the inverse-pole figure image. In samples taken from tubes extruded on die 1 and die 2, a maximum was found in the <001> direction, indicating an axial texture typical of the extruded products. Minor differences in texture blurring are also evident in the images in the sample taken from the extruded tube on die 1 between the weld and outside the weld. The image of the orientation distribution map in the sample extruded on die 2 differs from the others in local grain growth. At the weld area of this sample, recrystallized grains are visible, probably formed by friction processes during welding. The microstructure of the sample extruded on die 2 also differs in the shape of the grain outside the weld. The grains in this case are elongated.

Grain size distributions were also made from orientation images (Figure 23). Although the grain size is similar in the tested materials and oscillates at approximately 15 μm, the histograms of the grain size distributions are of a different nature. Finer grains were found in the extruded tubes in die 1 at the weld site. In contrast, there are more grains in the 15 to 30 μm range outside the weld. The tube extruded in die 2 showed a particular tendency toward grain growth in the welding. It differed from the others in having a slightly higher average grain size and a greater spread of the grain size range outside the welding area.

Figure 24 shows the mechanical properties of Ø50 × 2 mm tube extruded from 7021 alloy 2 through porthole die 1 (left) and porthole die 2 (right). Mechanical properties were determined in a static tensile test for 3 specimens containing a weld and 1 specimen without a weld. The results indicate high mechanical properties for both dies analyzed—slightly higher tensile properties YS and UTS were obtained for the tube extruded by die 1, while slightly higher plastic properties were obtained for the tube extruded by die 2. Tensile strengths UTS for the unwelded material were obtained at 472 Mpa (die 1) and 476 Mpa (die 2). The average tensile strength Rm for welded specimens is noteworthy at 474 Mpa for the die 1 extrusion variant and at 454 Mpa for the die 2 extrusion variant. This indicates a high susceptibility to welding of 7021 alloy 2 and the production of high-strength longitudinal welds in the extruded tubes—the relative strength of the welds is approximately 100.5% (die 1) and 95.4% (die 2). In the case of the die 2 extrusion variant, highly plasticized material was also obtained in the extruded tubes, both for the samples with a weld (percentage elongation in the range of 17.5–20%) and for the non-welded material (percentage elongation of approximately 17.5%). In the case of the extruded tube using die 1, the percentage of elongation was below 15%.

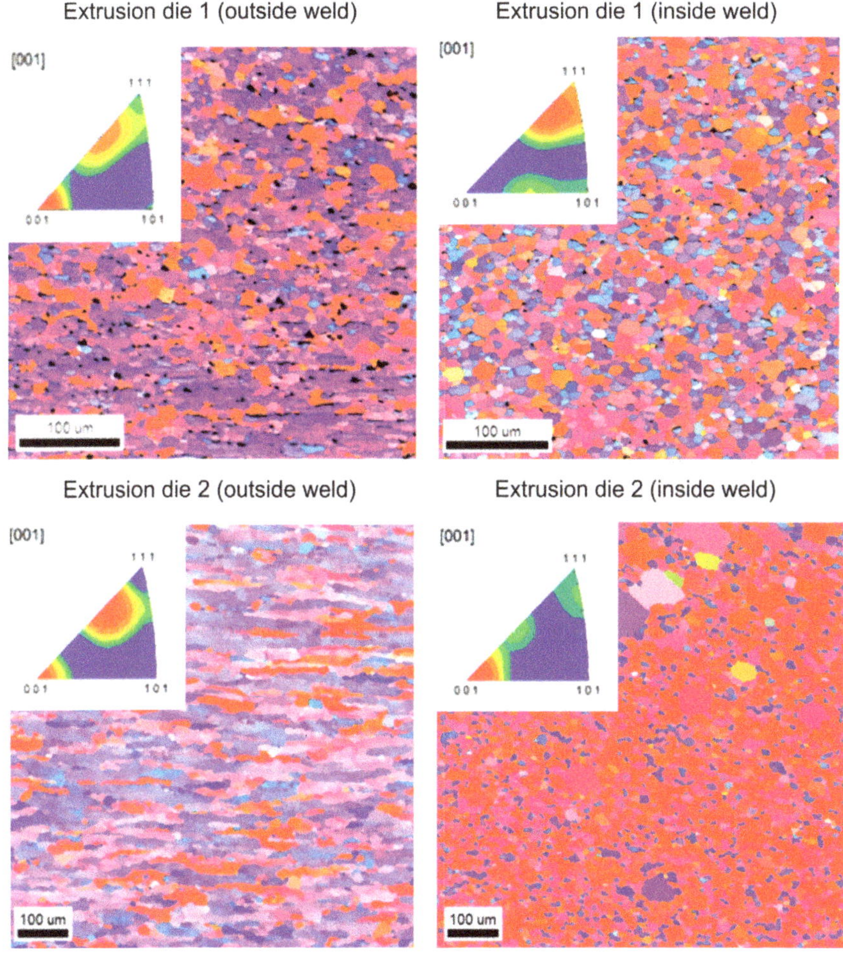

Figure 22. Microstructures and crystallographic orientation images taken from the cross-sectional view of the tube of Ø50 × 2 mm extruded from 7021 alloy 2 from the weld area and outside the weld area for die 1 and die 2.

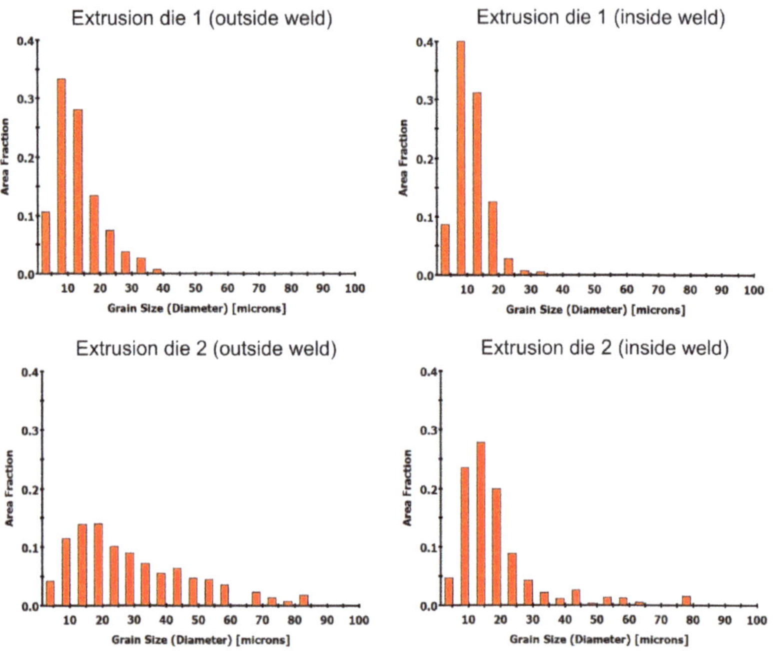

Figure 23. Grain size distributions for material outside the weld and inside the weld for tubes extruded by using die 1 and die 2 (tube of Ø50 × 2 mm extruded from 7021 alloy 2).

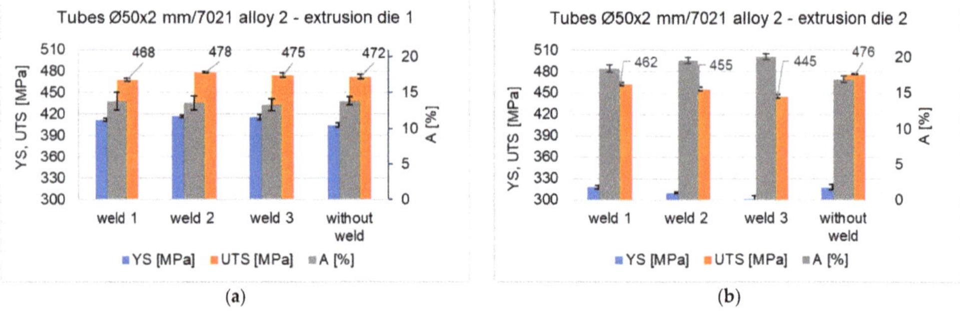

Figure 24. Mechanical properties of Ø50 × 2 mm tube extruded from 7021 alloy 2 through porthole die 1 (**a**) and porthole die 2 (**b**) determined in the static tensile test.

4. Discussion

In this work, an innovative way for predicting weld seam weld quality in round tubes from the 7021 alloy is presented. The method is based on employment of a patented laboratory device for investigation of welding phenomena, allowing full modelling of conditions which are expected to exist in a welding chamber of the porthole die. The 7021 copper-free alloy was chosen because it is relatively easier to extrude with the use of the porthole die—the extrusion force is lower than that for copper-containing alloys, e.g., the 7075 alloy, and the permissible exit speed is higher. The performed investigations include comprehensive tests, beginning from the homogenization of the billets, through laboratory testing of weldability and the industrial test performed on the basis of the preceding FEM simulation of the extrusion process. The three alloys of different contents of Mg and Zn were tested in the welding tests. The increase in the solidus temperature of the investigated alloys as a result of homogenization enables applying a higher heating temperature of the billets that, in turn, decreases the extrusion force and increases the exit speed from the die and, in this way, the productivity of the process increases. The welding tests with the use of the original authors' device enable determination of the temperature–pressure conditions necessary to obtain good-quality welds.

The billets from the tested alloys of 7021 grade were first submitted to a homogenization procedure. In all the examined alloys, the low melting components were dissolved during homogenization to a degree sufficient for extrusion practice. As a result, a significant increase in the solidus temperature was achieved, with values within the range from 559.2 °C for alloy 3 to 613.2 °C for alloy 1. The higher solidus temperature enables increasing the billet heating temperature, and so the extrusion force can be decreased. In addition, the exit speed from the die and the resulting effectiveness of the extrusion process can be effectively increased.

In the welding test, three temperatures of 400 °C, 450 °C and 500 °C were applied, whereas the compression pressure was assumed to be stable at the level of 300 MPa. The influence of the welding parameters for alloys with different contents of Mg and Zn (1, 2 and 3) on the mechanical properties of the welds was determined. The highest strength and plastic properties were recorded for a welding temperature of 450 °C for all three analyzed alloys. The relative strength of the welds was 86% (alloy 1), 93% (alloy 2) and 85% (alloy 3). Slightly lower values of the relative strength of the welds were obtained for a welding temperature of 500 °C but only for the alloys 2 (61%) and 3 (66%). In the case of alloy 1, the relative strength of the welds dropped dramatically from 86% to 12%.

Measurements of the width of the weld zone (Figure 11 and Figure S1 and in Supplementary File) proved that for alloy 1, the width of the welding area measures approximately 2.5 mm, regardless of the welding temperature. For alloy 2, the width of the welding area measures approximately 1.7 mm for welding temperatures of 400 °C and 450 °C. On the other hand, it is much wider at approximately 2.8 mm after welding at 500 °C. In the case of alloy 3, the welding width was found to increase with increasing temperature; the width of the welding area was 1.77 mm, 2.53 mm and 3 mm for welding temperatures of 400 °C, 450 °C and 500 °C, respectively.

The grain size measurements proved that, in alloy 1 samples, the average grain diameter outside the weld was 147–196 µm, with the samples welded at 400 °C having the largest grain size. The average diameter of the grains outside the weld area of the samples welded at 450 °C and 500 °C was comparable and was approximately 150 µm (Figure 12a). The determined average grain size in the welding area is 228–260 µm and, similarly to the outside of the weld, is the largest for samples welded at 400 °C; for the other welding variants, it is comparable (Figure 12a). In the case of this alloy, fine dispersive precipitations are present in the microstructure after welding at 450 °C and 500 °C, which inhibited grain growth.

In the alloy 2 samples, the average grain diameter outside the weld is 140–180 µm, with the smallest grain size for samples welded at 500 °C. The determined average grain size in the welding area is 195–266 µm and, similarly to the outside of the weld, is the smallest for samples welded at 500 °C (Figure 12b). After welding at 450 °C and 500 °C, the grains in the weld area were wider and shorter; characteristically, in the microstructure, the number of dispersive precipitates on the grain cross-section decreased, and the number of obstacles that inhibit the movement of grain boundaries decreased.

Samples from alloy 3 are characterized by the finest grain size, with the average grain diameter outside the weld area being 118–156 µm, while in the welding area, it is comparable for all variants of the welding process and is 124–132 µm (Figure 12c). The microstructure of alloy 3 has the highest number of dispersive precipitates, which inhibit grain growth when exposed to high temperature and pressure during the welding process.

The fracture of the welded samples was also submitted to microscopic observations (Figure 15 and Figure S7 in Supplementary File). The figures show representative images of the fractures of specimens welded at 450 °C and 500 °C, after uniaxial tensile tests. Fractures of samples of all the alloys welded at 450 °C show features of plastic fracture, the proportion depending on the type of alloy. In the case of alloy 1, the characteristic rounded hollows are only locally visible. In the case of alloys 2 and 3, rounded hollows and elevations, characteristic of plastic fracture, are visible on the entire surface. The presence of dimples in the material shows that plastic deformation occurs. When the specimens were welded at 500 °C, the fracture of the specimens definitely differed from alloy to alloy. In the case of alloy 1, surface decohesion under shear occurred during the uniaxial tensile test, and no signs of plastic deformation are observed. In the case of alloys 2 and 3, holes and elevations characteristic of plastic fracture are visible on the entire surface, with their proportion being greater in alloy 2 (covering the entire surface). The fractures of the specimens after the uniaxial tensile test are consistent with the results of the microstructure observations. Samples showing a predominance of plastic fracture contribution were characterized by good weld quality.

The FEM simulations were performed for two porthole dies, which differ in the geometry of the inlet channels and welding chambers (Figures 16–18). The calculations indicated differences in the distributions of predicted welding parameters, such as mean stress, temperature and plastic strain. In general, in the upper regions of the welding chambers (just below bridge), the maximum compressive stress values are observed in the range 247–267 MPa depending on the die design. The slightly higher values were obtained for die 1. It can be seen that the maximum values of the mean stress in the welding chambers of the porthole dies correspond to the compressive stress value occurring in the laboratory weldability tests (σ_w = 300 MPa). In the case of temperature distribution at the height of the welding chamber (Figure 19), comparable values within the range of 488–503 °C with a maximum in the die opening are observed for both the dies tested. These values are slightly higher than that where the maximal relative strength was obtained in the laboratory test and correspond to a welding temperature of 500 °C, for which a moderate relative weld strength of 62% was obtained for 7021 alloy 2. In the case of plastic strain distributions in the welding chambers of the porthole die, significantly higher values were obtained for die 2: above 30 for die 2 compared to 20 for die 1. Additionally, in the case of die 2, there is a larger area of maximum plastic strain values in the volume of the welding chamber, which may indicate more favorable bonding conditions resulting from the more intensive mixing of the metal in the welding chambers.

The distribution of the welding parameters σ_m/σ_i in the welding chambers is also important. In the laboratory weldability tests, an σ_w/k value equal to 5.88 guaranteed the production of a relatively good quality weld for alloy 2. As can be seen, at the prevailing height of the welding chambers from the sub-bridge space toward the die opening, there are sufficiently favorable welding conditions (determined numerically by FEM) for both dies (σ_m/σ_i of 6–7), which ensures relatively good material susceptibility to extrusion welding. The influence of die design on the conditions for joining metal streams in the welding chambers is inconclusive. Die 1 provides higher compressive stress values in the welding chambers, whereas die 2 guarantees better material mixing due to higher plastic strain values in the welding chambers. The weldability parameter indicates more favorable conditions for the metal streams to join together for die 1 in the first welding stage and the more favorable conditions for joining together for die 2 in the second welding stage (Figure 19).

Industrial extrusion tests were performed using alloy 2. Round tubes of Ø50 × 2 mm were extruded through porthole dies 1 and 2 as described in the FEM simulations. The simulation provided useful information for proper selection of extrusion parameters and die geometry. The exit temperature and exit speed were recorded during the experiment. The higher permissible exit speed was obtained for die 2 at the level of 4.5 m/min compared to 3 m/min for die 1. The temperature measured immediately after the metal exited the die opening was in the range of 570–580 °C. The measurements of the dimensional accuracy of the extruded tubes indicated better behavior of die 2.

The FEM simulation and the industrial trials indicated the important role of the porthole die geometry when extruding the 7000 series alloys. Die 1 was based on the variant usually used during extrusion of the 6000 series alloys, whereas die No 2 was the modified version elaborated in FEM simulation of the process. Finally, it can be noticed that the die design for the 7000 series alloys and particularly for the 7021 alloy differs considerably from that of die 1. Despite the similarity in the inlet channel layout, the channels should be wider while the bridges must be higher and thicker to withstand the increased extrusion force and to avoid elastic deflection of the die. The height of the welding chamber should be higher. The die insert should be equipped with pockets, which regulate metal flow in the region close to the die opening. In such a case, the metal flow is more uniform and this is what improves the geometrical stability and dimensional accuracy of the extruded profile.

EBSD studies of the microstructure of the industrially extruded tubes proved the variation in grain size and shape at the weld site and outside the weld for both die 1 and die 2 extruded tubes, with die 1 extruded tubes having a more homogeneous microstructure in the cross-section (Figure 22). The average grain size of the tubes tested is approximately 15 μm. For tubes extruded by die 1, grains of less than 20 μm predominate, with grains in the 20–40 μm range more frequently observed in the area outside the weld (Figure 23). In the case of tube extruded by die 2, when using a higher extrusion speed (4.5 m/min), the grains have a varied shape in the area outside the weld. In the microstructure, there are both evenly spaced and elongated grains with clearly bulged boundaries, indicating the occurrence of structure renewal processes. In the weld area, the grains are evenly spaced, with locally large recrystallized grains, which were probably formed during the deformation process due to significant heating of the material (Figure 22). In the microstructure of the tube extruded by die 2, grains up to 80 μm are present, with grains of less than 30 μm predominating. In addition, the histogram showing the grain distribution in the area outside the weld is more flattened than that showing the weld area (Figure 23).

The mechanical properties of the extruded tubes were determined in the tensile test. The results indicate high mechanical properties for both the dies analyzed–higher strength properties $R_{0.2}$ and R_m were obtained for the tube extruded with die 1, while slightly higher plastic properties were obtained for the tube extruded with die 2. The R_m for the unwelded material was obtained at 472 MPa (die 1) and 476 MPa (die 2). The average R_m for welded specimens is as high as 474 MPa for die 1 and 454 MPa for die 2. This indicates a very high susceptibility to welding of 7021 alloy 2 that guaranties production of high-strength longitudinal welds in the extruded tubes—the relative strength of the welds is approximately 100.5% (die 1) and 95.4% (die 2). The relative strength of the tubes is higher in comparison to that obtained in the laboratory test but it can be explained by more intensive plastic deformation in the welding chambers during the extrusion process. The slightly lower Rm values and higher A values for tubes extruded by die 2 can be explained by the local grain growth in the weld area, as well as the larger grain size compared to tubes extruded by die 1 (Figure 24).

Comparison of the strength properties of samples with a longitudinal weld produced in the extrusion process of the AlZnMg alloy 2 and the numerically predicted FEM level of compressive stress in the die welding chamber, determined by the die construction, shows a clear relationship. An approximately 8–10% higher level of compressive stress in the welding chamber for die 1 translates into an approximately 5% higher UTS for the product extruded on this die. Moreover, FEM predicted higher values of the plastic strain in the welding chamber of die 2 resulting in a higher percentage elongation for samples with the weld. This allows us to conclude that the FEM results and the results of extrusion experiments are in high agreement, at least on a qualitative level. A high quantitative consistency can be found between the weldability index determined for alloy 2 in weldability tests (σ_w/k – 5.88) and numerically determined by FEM for both dies (σ_w/k of 6–7). Both dies ensured obtaining this indicator, which was confirmed by the results of the static tensile test (high-quality seam welds of high strength). Higher values of the weldability index for matrix 1 were obtained in the first stage of metal welding in the welding chamber, which proves that this initial stage of welding determines the quality/mechanical strength of the welds.

5. Conclusions

1. It is possible to produce the high-quality weld seams in the hollow profiles extruded from the AlZnMg alloys through the porthole dies. The UTS in the T5 temper for extruded tubes of Ø50 × 2 mm from the 7021 alloy 2 with weld seams in relation to the die used was as high as 454–474 MPa, whereas the relative strength of the welds was within the range of 95 ÷ 100.5 MPa. Worth noting is the high plasticity of samples with the welds—the percentage elongation for die 2 was within the range of 17.5–20%. The optimum welding temperature that produces the best mechanical strength is obtained at 450 °C. This is possible due to proper selection of the chemical composition of the 7021 alloy (Zn content at the level of 5.47% and Mg of 2.12%) as well as appropriate selection of stress–temperature conditions in the extrusion process, e.g., a unit pressure at the level of 300 MPa resulting from the die design and a temperature of 550 °C determined by process conditions.

2. Predicted through FEM, higher values of compression stress in the welding chamber of the porthole die 1 resulted in higher UTS for samples with the weld from 7021 alloy 2. In turn, FEM predicted higher values of the plastic strain in the welding chamber of die 2, resulting in a higher percentage elongation for samples with the weld.

3. The proposed method for predicting the quality of welds in the extruded hollow profiles from the aluminum alloys is based on the laboratory tests of weldability with the use of patented device, which simulates bonding conditions in the welding chamber of the porthole dies. At first, the weldability factor $6_n/k$ of Akeret is determined, and next, it is verified numerically for the extrusion process. The performed analysis confirmed the possibility of obtaining a factor value of $6_n/k = 5.88$ for both the dies examined. This result was verified positively in the industrial trials of the extrusion of tubes of Ø50 × 2 mm from alloy 2. The obtained industrial trial values of the relative strength of welds are higher compared to those predicted in the laboratory tests and can be explained by the positive influence of the intensive plastic mixing of the material within the inlet channels and in welding chambers.

4. Examination of the microstructure of specimens welded in a device for weldability tests has shown that, regardless of the parameters of the welding process and the chemical composition of the alloy, the grains in the area of welding have an elongated shape (heavy sheared regions where the samples undergo welding), while grains have a regular, near-axial shape outside the area of welding. The width of the weld area depends on the alloy type and welding process conditions and is approximately 2.5 mm for alloy 1, 1.7–2.8 mm for alloy 2 and 1.77–3 mm for alloy 3. In most of the samples tested, no discontinuities in the welding area were found. Investigations of the chemical composition in the welding area and outside the welding area also showed no differences in the content of the main alloying elements Mg and Zn.

5. Studies of the microstructure of tubes made of alloy 2 extruded in industrial conditions showed greater microstructure differentiation in the weld area and outside the weld area in the case of extrusion through porthole die 2. For tubes extruded by die 2, in the weld area, the grains are similar in shape to parallel, with locally large, recrystallized grains. The variation in microstructure on the cross-section of the tube extruded by die 2 influenced the mechanical properties. The UTS for the unwelded material was as high as 472 MPa for die 1 and 476 MPa for die 2, whereas, for welded specimens, is as high as 474 MPa for die 1 and 454 MPa for die 2. The slightly lower UTS values and higher elongation values for tubes extruded by die 2 can be explained by the local grain growth in the weld area, as well as by the slightly larger grain size compared to tubes extruded by die 1.

Future research will focus on the possibilities of further maximizing the metal exit speed during extrusion of AlZnMg alloys, as these alloys are generally difficult to deform. Solutions will be sought that will ensure a consensus between the quality of longitudinal welds in extruded products (mechanical strength), its dimensional tolerances and the efficiency of the extrusion process.

Supplementary Materials: The following supporting information can be downloaded at: https://www.mdpi.com/article/10.3390/ma16175817/s1, Figure S1: Microstructure of welded 7021 alloy in dependence of chemical composition; welding was performed under identical process parameters: T = 450 °C, p = 300 MPa; (a–c) 7021 alloy 1: 1.20%Mg, 5.27%Zn, (d–f) 7021 alloy 2: 2.12%Mg, 5.47%Zn, (g–i) 7021 alloy 3: 2.12%Mg, 8.02%Zn; light microscopy; Figure S2: Microstructure of welded 7021 alloy in dependence of chemical composition; welding was performed under identical process parameters: T = 500 °C, p = 300 MPa; (a–c) 7021 alloy 1: 1.20%Mg, 5.27%Zn, (d–f) 7021 alloy 2: 2.12%Mg, 5.47%Zn, (g–i) 7021 alloy 3: 2.12%Mg, 8.02%Zn; light microscopy; Figure S3: Microstructure of Alloy 2 in the area inside the weld and results of the chemical composition test on the grain cross-section; welding process conditions: T = 450 °C, p = 300 MPa; SEM/EDS; Figure S4: Microstructure of Alloy 2 in the area inside the weld and results of the chemical composition test on the grain cross-section; welding process conditions: T = 500 °C, p = 300 MPa; SEM/EDS; Figure S5: Microstructure of Alloy 2 in the area outside the weld and results of the chemical composition test on the grain cross-section; welding process conditions: T = 500 °C, p = 300 MPa; SEM/EDS; Figure S6: Average Zn content on the grain cross-section in the welding area and outside the welding area for the alloys tested: (a) T = 450 °C, p = 300 MPa, (b) T = 500 °C, p = 300 MPa; Figure S7: Fracture surfaces after uniaxial tensile test; welding was performed under identical process parameters: T = 500 °C, p = 300 MPa; (a,d) 7021 alloy 1: 1.20%Mg, 5.27%Zn, (b,e) 7021 alloy 2: 2.12%Mg, 5.47%Zn, (c,f) 7021 alloy 3: 2.12%Mg, (c,d) 8.02%Zn; Figure S8: Registered exampled technological parameters of extrusion process of tubes of $\varnothing 50 \times 2$ mm from 7021 alloy no 2 for extrusion die 1 and die 2: metal exit speed (on the left) and extrudates temperature (on the right).

Author Contributions: Conceptualization, D.L., W.L. and B.P.; methodology, B.L.-M., J.M., M.B. and S.B.; software, J.M.; validation, H.J., formal analysis, D.L.; investigation, B.L.-M., J.M., M.B. and S.B.; writing—original draft preparation, B.L.-M., J.M., M.B. and S.B.; writing—review and editing, D.L. and W.L.; visualization, J.M. and M.B.; supervision, D.L.; project administration, D.L. All authors have read and agreed to the published version of the manuscript.

Funding: This research was funded by THE NATIONAL CENTRE FOR RESAERCH AND DEVELOPMENT, grant number TECHMATSTRATEG2/406439/10/NCBR/2019 "Extrusion welding of high-strength shapes from aluminum alloys 7xxx series".

Institutional Review Board Statement: Not applicable.

Informed Consent Statement: Not applicable.

Data Availability Statement: Not applicable.

Conflicts of Interest: The authors declare no conflict of interest.

References

1. Khan, Y.A.; Valberg, H.; Irgens, I. Joining of metal streams in extrusion welding. *Int. J. Mater. Form.* **2009**, *2*, 109–112. [CrossRef]
2. Oosterkamp, A.; Djapic, L.; Nordeide, A. Kissing Bond phenomena in solid-state welds of aluminium alloys. *Weld. J.* **2004**, *83*, 225–231.
3. Duplancic, I.; Prgin, J. Determination of parameters required for joining process in hollow dies. In Proceedings of the 6th International Aluminium Extrusion Technology Seminar, Chicago, IL, USA, 14–17 May 1996; pp. 225–230.
4. Donati, L.; Tomesani, L.; Minak, G. Characterization of seam weld quality in AA6082 extruded profiles. *J. Mater. Process. Technol.* **2007**, *191*, 127–131. [CrossRef]
5. Kloppenborg, T.; Schwane, M.; Ben Khalifa, N.; Tekkaya, A.E.; Brosius, A. Experimental and numerical analysis of material flow in porthole die extrusion. *Key Eng. Mater.* **2012**, *491*, 97–104. [CrossRef]
6. Filice, L.; Gagliardi, F.; Ceretti, E.; Fratini, L.; Giardini, C.; La Spisa, D. Analysis of joint quality along welding plane. *Key Eng. Mater.* **2010**, *424*, 79–86.
7. Tang, D.; Zhang, Q.; Li, D.; Peng, Y. Numerical and experimental study on seam welding behavior in extrusion of micro-channel Tube. *Key Eng. Mater.* **2012**, *491*, 189–195. [CrossRef]

8. Selvaggio, A.; Segatori, A.; Guzel, A.; Donati, L.; Tomesani, L.; Tekkaya, A.E. Extrusion Benchmark 2011: Evaluation of different design strategies on process conditions, die deflection and seam weld quality in hollow profiles. *Key Eng. Mater.* **2012**, *491*, 1–10. [CrossRef]
9. Reggiani, B.; Segatori, A.; Donati, L.; Tomesani, L. Finite element modelling of the charge welds evolution in a porthole die. *Key Eng. Mater.* **2012**, *491*, 19–26. [CrossRef]
10. Akeret, R. Extrusion welds—Quality aspects are now centre stage. In Proceedings of the 5th Aluminium Extrusion Technology Seminar, Chicago, IL, USA, 19–22 May 1992; pp. 319–336.
11. Plata, M.; Piwnik, J. Theoretical and experimental analysis of seam weld formation in hot extrusion of aluminum alloys. In Proceedings of the 7th International Aluminium Extrusion Technology Seminar, Chicago, IL, USA, 16–19 May 2000; pp. 205–211.
12. Donati, L.; Tomesani, L. The prediction of seam weld quality in aluminum extrusion. *J. Mater. Process. Technol.* **2004**, *153–154*, 366–373. [CrossRef]
13. Ceretti, E.; Mazzoni, L.; Giardini, C. Simulation of metal flow and welding prediction in porthole die extrusion: The influence of the geometrical parameters. *Int. J. Mater. Forum* **2009**, *2*, 101–104. [CrossRef]
14. Gagliardi, F.; Ambrogio, G.; Filice, L. Optimization of Aluminium Extrusion by Porthole Die Using a Down Scaled Equipment. *Key Eng. Mater.* **2012**, *491*, 173–180. [CrossRef]
15. Bakker, A. Weld Seams in Aluminium Alloy Extrusions: Microstructure and Properties. Ph.D. Thesis, Delft University, Delft, The Netherlands, 2016.
16. Kniazkin, I.; Vlasov, A. Quality prediction of longitudinal seam welds in aluminium profile extrusion based on simulation. *Procedia Manuf.* **2020**, *50*, 433–438. [CrossRef]
17. Yu, J.; Zhao, G.; Chen, L. Analysis of longitudinal weld seam defects and investigation of solid-state bonding criteria in porthole die extrusion process of aluminum alloy profiles. *J. Mater. Process. Technol.* **2016**, *237*, 31–47. [CrossRef]
18. Li, S.; Li, L.; Liu, Z.; Wang, G. Microstructure and Its Influence on the Welding Quality of 6063 Aluminum Alloy Porthole Die Extrusion. *Materials* **2021**, *14*, 6584. [CrossRef] [PubMed]
19. Chen, L.; Zhang, J.; Zhao, G.; Wang, Z.; Zhang, C. Microstructure and mechanical properties of Mg-Al-Zn alloy extruded by porthole die with different initial billets. *Mater. Sci. Eng. A* **2018**, *718*, 390–397. [CrossRef]
20. Yu, J.; Zhao, G. Interfacial structure and bonding mechanism of weld seams during porthole die extrusion of aluminum alloy profiles. *Mater. Charact.* **2018**, *138*, 56–66. [CrossRef]
21. Crosio, M.; Hora, D.; Hora, P. *Experimental and Numerical Analysis of Bonding Criteria for Seam Welds during Aluminium Extrusion Processes of Hollow Profiles*; Research Collection; ETH Zurich: Zürich, Switzerland, 2019.
22. Wang, Y.; Wells, M.A. The Effect of the Bridge's Angle during Porthole Die Extrusion of Aluminum AA6082. *Metals* **2023**, *13*, 605. [CrossRef]
23. Yu, Y.; Zhao, G. Study on the welding quality in the porthole die extrusion process of aluminum alloy profiles, Science Direct. *Procedia Eng.* **2017**, *207*, 401–406. [CrossRef]
24. Yu, J.; Zhao, G.; Cui, W.; Zang, C.; Chen, L. Microstructural evolution and mechanical properties of welding seams in aluminum alloy profiles extruded by a porthole die under different billet heating temperatures and extrusion speeds. *J. Mater. Process. Technol.* **2017**, *247*, 214–222.
25. Wang, Y.; Zhao, G.; Zhang, W.; Sun, L.; Wang, X.; Lu, Z. Interfacial bonding mechanism and length evaluation method of the longitudinal welds in the unsteady deformation process of porthole die extrusion of aluminum alloy profiles. *J. Mater. Res. Technol.* **2022**, *20*, 1624–1644. [CrossRef]
26. Annadurai, S.; Mulla, I.; Mohammed, M. Die correction and die design enhancement using profile simulation in extrusion industry. In Proceedings of the Twelfth International Aluminum Extrusion Technology Seminar and Exposition, Orlando, FL, USA, 3–5 May 2022.
27. Kolpak, F.; Schulze, A.; Dahnke, C.; Tekkaya, A.E. Predicting weld-quality in direct hot extrusion of aluminium chips. *J. Mater. Process. Technol.* **2019**, *274*, 116294. [CrossRef]
28. Yu, J.; Zhao, G.; Cui, W.; Chen, L.; Chen, X. Evaluating the welding quality of longitudinal welds in a hollow profile manufactured by porthole die extrusion: Experiments and simulation. *J. Manuf. Process.* **2019**, *38*, 502–515. [CrossRef]
29. Wang, Y.; Zang, A.; Mahmoodkhani, Y.; Wells, M.; Poole, W.; Parson, N. The Effect of Bridge Geometry on Microstructure and Texture Evolution During Porthole Die Extrusion of an Al–Mg–Si–Mn–Cr Alloy. *Metall. Mater. Trans. A* **2021**, *52*, 3503–3516. [CrossRef]
30. Chen, G.; Chen, L.; Zhao, G.; Zhang, C.; Cui, W. Microstructure analysis of an Al-Zn-Mg alloy during porthole die extrusion based on modeling of constitutive equation and dynamic recrystallization. *J. Alloys Compd.* **2017**, *2017*, 80–91. [CrossRef]
31. Lee, S.Y.; Lee, I.K.; Jeong, M.S.; Ko, D.C.; Kim, B.M.; Lee, S.K. Development of Porthole Extrusion Die for Improving Welding Pressure in Welding Chamber by Using Numerical Analysis. *Trans. Mater. Process.* **2017**, *26*, 115–120. [CrossRef]
32. Ayer, O.; Özmen, B.G.; Karakaya, I. An Optimization Study for Bridge Design of a Porthole Extrusion Die. *Adv. Sci. Technol. Res. J.* **2019**, *13*, 270–275. [CrossRef] [PubMed]
33. Gagliardi, F.; Ciancio, C.; Ambrogio, G. Optimization of porthole die extrusion by Grey-Taguchi relation al analysis. *Int. J. Adv. Manuf. Technol.* **2018**, *94*, 719–728. [CrossRef]
34. Chen, W.; Xu, C.; Pan, P.; Ruanc, X.; Ji, H. Effect of process parameters on longitudinal weld seam quality of aluminum alloy profile for an automobile fuel tank protector. *Sci. Procedia Manuf.* **2020**, *50*, 159–167. [CrossRef]

35. Truong, T.; Hsu, Q.; Tong, V.; Sheu, J. A Design Approach of Porthole Die for Flow Balance in Extrusion of Complex Solid Aluminum Heatsink Profile with Large Variable Wall Thickness. *Metals* **2020**, *10*, 553. [CrossRef]
36. Lesniak, D.; Rekas, A.; Libura, W.; Zasadzinski, J.; Latos, T.; Jurczak, H. Method for Testing Weldability of Metals or Alloys and the Device for Producing Welded Samples for Testing Weldability of Metals or Alloys. Patent No PL230273B1, 31 October 2018.
37. Lesniak, D.; Zasadzinski, J.; Libura, W.; Zaba, K.; Puchlerska, S.; Madura, J.; Balcerzak, M.; Plonka, B.; Jurczak, H. FEM Numerical and Experimental Study on Dimensional Accuracy of Tubes Extruded from 6082 and 7021 Aluminium Alloys. *Materials* **2023**, *16*, 556. [CrossRef]

 *materials*

Article

A Finite-Difference Based Parallel Solver Algorithm for Online-Monitoring of Resistance Spot Welding

Tomas Teren *, Lars Penter * , Christoph Peukert and Steffen Ihlenfeldt

Chair of Machine Tools Development and Adaptive Control, Institute of Mechatronic Engineering, Technische Universität Dresden, 01069 Dresden, Germany
* Correspondence: tomas.teren@protonmail.com (T.T.); lars.penter@tu-dresden.de (L.P.)

Abstract: Although resistance spot welding (RSW) was invented at the beginning of the last century, the online-monitoring and control of RSW is still a technological challenge and of economic and ecological importance. Process, material and geometry parameters of RSW are stored in the database of the process control system. Prospectively, these accumulated data could serve as the base for data-driven and physics-based models to monitor the spot weld process in real-time. The objective of this paper is to present a finite-difference based parallel solver algorithm to simulate RSW time-efficiently. The Peaceman–Rachford scheme was combined with the Thomas algorithm to compute the electrical–thermal interdependencies of the resistance spot welding process within seconds. Finally, the electric–thermal model is verified by a convergence analysis and parameter study.

Keywords: resistance spot welding; finite difference method; real-time simulation; digital twin

 check for updates

Citation: Teren, T.; Penter, L.; Peukert, C.; Ihlenfeldt, S. A Finite-Difference Based Parallel Solver Algorithm for Online-Monitoring of Resistance Spot Welding. *Materials* **2022**, *15*, 6348. https://doi.org/10.3390/ma15186348

Academic Editor: Raul D.S.G. Campilho

Received: 25 June 2022
Accepted: 6 September 2022
Published: 13 September 2022

Publisher's Note: MDPI stays neutral with regard to jurisdictional claims in published maps and institutional affiliations.

1. Introduction

Between 7000 and 12,000 spot welds hold a car body together depending on its size [1] and 91.7 million cars were produced in the automotive industry worldwide [2]. One major technological and economical advantage of resistance spot welding over other joining technologies is the ease of integration in automated production lines. Furthermore, it is a lucrative process in technological and organizational terms. The process setup is simple, and the process cycle is, in the order of milliseconds, very short. After stacking one metal sheet on the top of another, pneumatic driven cylinders move electrode caps to clamp the metal sheet stack. Subsequently, the metal sheets are pressed together and a load of up to several hundred MPa is applied (squeeze time). Then, the electrodes are connected to a voltage, and thus an electric current crosses the metal sheets, in which Joule heat initiates weld nugget growth and sheet metal fusion (welding time). At the end, the electrodes rest briefly upon the metal sheets (hold time) before they are moved to the next weld spot (off time) and the process cycle restarts. The most important quality criterion for spot welds is its tensile strength. It can be determined by destructive test methods such as chisel or tensile test. However, after the test the weld is destroyed, and further deployment is impossible. Alternatively, non-destructive test methods can evaluate the weld quality and warrant further use of the assembly after testing. A widespread non-destructive method is ultrasonic testing which aims at detecting the effective contact area size of the weld joint. If this area exceeds a minimal threshold, the joint weld is accepted and dismissed otherwise. The equipment is expensive, requires qualified staff, and ultrasonic testing results scatter broadly. It tends to underestimate the welding spot diameter by approximately up to 2.5 mm [3]. Other non-destructive methods are numerical methods that simulate nugget growth and geometric parameters, such as nugget diameter and penetration depth, which—due to their correlation with weld strength—allow an indirect assessment of the weld joint quality. However, current RSW models are too slow for integration into real-time monitoring and control systems, as discussed in the state-of-the-art section. An online monitoring system for RSW bears the potential of adjusting welding process parameters from one manufactured spot weld to the next in the assembly line, and, as a consequence, reducing the number of NOK welds and save time, costs, and energy. In the early stages, refs [4,5] applied the finite-difference method (FDM) to electro-thermal models of RSW. With the assumption of a constant electric current, ref [4] proposed a model to predict the temperature distribution as a function of time and space, allowing for variations in the mechanical properties of the sheet metal. Ref [5] presents the temperature-dependent electrical potential distribution in the base metal and the interfaces for various electrical currents. In both studies, the nugget diameter and penetration depth derived from the computed temperature field showed good agreement with experimental data. Ref [6] used a control volume formulation and central differences to model the dimensionless temperature field, the nugget growth for different welding currents, electrode tip shapes and thickness ratios of work pieces. The enthalpy-temperature relation was capitalized to account for the phase change. The simulation results in terms of weld nugget growth, nugget thickness and shape were consistent with experimental results. In [7] the finite volume method (FVM) was adopted to build a complex RSW simulation model, which considered—among other aspects—the electric current density, the magnetic field intensity, the temperature, and the velocity field for work pieces with flat faced or truncated electrodes. The effects of the electrode face radius and cone angle on transport mechanism, for example, mass transfer, and various other non-linear phenomena were clearly demonstrated; simulation results agreed well with experimental data. In [8], the mass, momentum, heat and species transport, as well as the magnetic field intensity, were discretized by a control-volume formulation to compute the dynamic electrical resistance during RSW. The simulation result suggest that the dynamic resistance of AISI 1008 steel can be divided into four distinct stages, in which the contact resistance and the bulk resistance contributions vary over time. Several years later, ref [9] developed a control volume based finite difference code for the electrical and thermal field and combined it with a commercial code that provided the mechanical model. Based on this hybrid-approach, the computed nugget size deviated from the experimental data by merely three percent. Many simulations of RSW are based on the finite-element method (FEM), which derives model equations from integration over the finite-element domain. For example, the general-purpose simulation program ABAQUS© (Version 5.7) was used to conduct a parametric study on different electrode shapes, welding currents, and electrode forces for Al-alloys in [1]. Ref [10] analyzes the influence of electrode-water cooling on welding of aluminum alloys AA5182. It utilizes LS-DYNA© (R11.0) to build a thermal-electrical-mechanical model. The simulation results indicated that water cooling affects the temperature distribution in the sheets only slightly, and thus, does not influence nugget growth at all. However, it has a significant effect on the electrode cooling during hold-time. Another study on aluminum alloys for RSW processes was conducted in [11], where a calibrated contact resistance model for AA5182 was presented. The underlying electric–thermal–mechanical FE model could reproduce weld nugget diameters deviating from real experiments by four percent. FEM-based, SORPAS© 3D is a special purpose simulation program with a multi-physics model for RSW [12]. It was applied to investigate short-pulse welding on aluminum alloy AA6016-T4 to reduce the required energy for producing sound welds in thin sheets [13]. The nugget formation was found to happen in two distinct phases: the nucleation, in which 60–80% of the final diameter evolves due to high contact resistance, and the growth stage, when further nugget growth is induced by heat conduction. Savings of approximately 50% regarding energy and time were achieved. SORPAS©.2D provides simulation results with high accuracy, but it requires approximately an hour to run an RSW-simulation with a resolution of 1000 finite elements on a conventional desktop computer [14].

In view of the development labor in past decades, it becomes clear that in the framework of appropriately set model assumptions and on the fundament of suitable material and process data, the nugget growth of RSW can be simulated with sufficient accuracy. Hitherto, numerical analyses of RSW in academia and commercial special purpose programs paid attention to simulation accuracy rather than computational speed; none of the previously cited research papers indicated information on the simulation run-time. In terms of a model-based real-time monitoring and quality assessment system, two essential requirements can be formulated: it should be capable of differentiating between OK and NOK spot welds (sufficient model accuracy) and the time window for the quality assessment is to be shorter than the time between two consecutive spot welds (computation speed). In numerical simulation, these requirements contradict each other, that is, increased simulation accuracy comes along with increased computational time. In the sense of this optimization problem, it is reasonable to constrain the simulation model to physical phenomena, which are predominantly relevant to RSW, to reduce the computation cost. RSW is based on Joule heating, i.e., the transformation of electrical to thermal energy, to join metal stacks by fusion. Thus, the electric-thermal model is considered as the core of any multi-physics model for RSW. The finite difference method, for heat transfer in solids elaborately described in [15], derives model equations from replacing partial derivative terms by finite differences. In this paper, this numerical method is applied to develop an electric-thermal model for RSW by means of the Peaceman–Rachford scheme. It leads to a set of linear equations with tridiagonal band matrices which are solved by the Thomas Algorithm rapidly. Based upon this solving algorithm, the electric-thermal model is verified and investigated on its suitability for real-time simulations.

2. Resistance Spot Welding Model

This paragraph describes all aspects necessary for implementing the electric-thermal model presented in this paper. It includes the model geometry, the boundary conditions (Section 2.1), the electric model (Section 2.2), the thermal model (Section 2.3), the material model (Section 2.4), and the solution methodology (Section 2.5).

2.1. Model Geometry and Boundary Conditions

According to DIN EN ISO 5821-C0-16-23 the electrode cap geometry was defined. Axial symmetry along the faying surface is supposed and, thus, it suffices to model the upper electrode cap and sheet of the weld joint. Furthermore, the electric-thermal field is assumed to be constant in circumferential direction, which allows the cylindrical coordinate system of the plane (r, z) to represent the cylindrical coordinate system of the three-dimensional space (r, ϕ, z). The modelled plane consists of 1007 nodes and can be considered as the entity of three connected rectangle regions I, II, and III. They are meshed equidistantly and consist of 6×20 (I), 19×37 (II), and 23×8 (III) nodes along the r- and z-axis, respectively. The model geometry as well as its dimensions are depicted in Table 1. To determine the space increment, the criterion for explicit methods Equation (1) is applied. Rearranging it leads to

$$\Delta r = \Delta z \geq \sqrt{2 \cdot a_{max} \cdot \Delta t_{max}}. \tag{1}$$

In the study at hand, the maximum thermal diffusivity is associated with the copper electrode $a_{Cu} = \lambda \cdot c_p^{-1} \cdot \rho^{-1} = 1.01748 \times 10^{-4} \mathrm{m^2 s^{-1}}$ and the largest time increment is $\Delta t_{max} = 6 \times 10^{-4}$ s. Thus, the space increments are selected as $\Delta r = \Delta z = 0.4$ mm.

In order to solve for the temperature and electric potential fields, the boundary conditions of the electric–thermal model must be specified. The electric current streams unidirectional, i.e., a direct current is adopted. Except from the top of the upper electrode, Equation (2), and the faying surface, Equation (3), all system border nodes are electric isolators Equations (4) and (5).

$$\Phi|_{CD} = \Phi_{Electrode} \tag{2}$$

$$\Phi|_{OG} = 0 \text{ V} \tag{3}$$

$$\left.\frac{\partial \Phi}{\partial r}\right|_{OA} = \left.\frac{\partial \Phi}{\partial r}\right|_{BC} = \left.\frac{\partial \Phi}{\partial r}\right|_{DE} = \left.\frac{\partial \Phi}{\partial r}\right|_{FG} = 0 \tag{4}$$

$$\left.\frac{\partial \Phi}{\partial z}\right|_{AB} = \left.\frac{\partial \Phi}{\partial z}\right|_{EF} = 0 \tag{5}$$

Table 1. Geometry features and dimensions.

Geometry Feature	Dimension (mm)
sheet radius r_s	8.8
sheet thickness h_s	2
inner electrode radius r_i	6
outer electrode radius r_o	8
electrode height h_e	23
cooling recess h_f	10.5
space increments $\Delta r = \Delta z$	0.4
height of contact layer	10^{-2}

The initial temperature of the sheet and electrode, modelled by Equation (6), corresponds to the assumed ambient temperature of 20 °C. Convection between the cooling water and the electrode is modelled as a Dirichlet boundary condition by loading a constant temperature of 20 °C onto separating border nodes, Equation (6). Due to Joule heating, each node acts as a volumetric heat source, and heat is transferred to cooler adjacent nodes. The interfacial heat generation between the sheets is accounted for by Equation (7). Convective heat transfer to the surrounding air is negligible [10,16]. Therefore, the nodes contacting the surrounding air, Equation (8), and the remaining ones, Equation (9), simulate adiabatic system borders.

$$\left. T \right|_{t=0} = \left. T \right|_{AB} = \left. T \right|_{BC} = 20\,°C \tag{6}$$

$$\left. Q \right|_{OG} = \frac{1}{\sigma_{contact}} \left(\frac{\partial \Phi}{\partial z} \right)^2 \tag{7}$$

$$\left. \frac{\partial T}{\partial z} \right|_{DE} = \left. \frac{\partial T}{\partial z} \right|_{EF} = 0 \tag{8}$$

$$\left. \frac{\partial T}{\partial r} \right|_{OA} = \left. \frac{\partial T}{\partial r} \right|_{CD} = \left. \frac{\partial T}{\partial r} \right|_{FG} = \left. \frac{\partial T}{\partial r} \right|_{OG} = 0 \tag{9}$$

Material properties of Cu are assigned to the electrode–sheet interface. The estimation of heat transfer across the interfaces is uncertain. Hence it is simplified and treated as heat conduction in a solid body.

2.2. Electric Model

Equation (10) is a second order partial differential equation of elliptic type and the model equation for the electrical field. It is solved for the electrical potential, which is used to calculate the Joule heating in Equation (14).

$$\frac{\partial}{\partial r} \left(\frac{1}{\sigma} \frac{\partial \Phi}{\partial r} \right) + \frac{1}{\sigma r} \left(\frac{\partial \Phi}{\partial r} \right) + \frac{\partial}{\partial z} \left(\frac{1}{\sigma} \frac{\partial \Phi}{\partial z} \right) = 0. \tag{10}$$

Finite difference methods discretize partial differential equations by replacing derivatives with finite differences, which are obtained by a Taylor series approximation. Equation (10) factors in resistance as function of the space coordinates (r, z) and the temperature T. Applying the finite difference method to Equation (10) yields the finite-difference form of the partial derivatives:

$$\frac{\partial}{\partial r} \left(\frac{1}{\sigma} \frac{\partial \Phi}{\partial r} \right) \approx \frac{1}{\Delta r^2} \left(\frac{\Phi_{i-1,j} - \Phi_{i,j}}{\sigma_{i-1,j} + \sigma_{i,j}} - \frac{\Phi_{i,j} - \Phi_{i+1,j}}{\sigma_{i,j} + \sigma_{i+1,j}} \right), \tag{11}$$

$$\frac{1}{\sigma r}\left(\frac{\partial \Phi}{\partial r}\right) \approx \frac{1}{2(i-1)\Delta r^2}\left(\frac{\Phi_{i-1,j}-\Phi_{i,j}}{\sigma_{i-1,j}+\sigma_{i,j}}+\frac{\Phi_{i,j}-\Phi_{i+1,j}}{\sigma_{i,j}+\sigma_{i+1,j}}\right), \tag{12}$$

$$\frac{\partial}{\partial z}\left(\frac{1}{\sigma}\frac{\partial \Phi}{\partial z}\right) \approx \frac{1}{\Delta z^2}\left(\frac{\Phi_{i,j-1}-\Phi_{i,j}}{\sigma_{i,j-1}+\sigma_{i,j}}-\frac{\Phi_{i,j}-\Phi_{i,j+1}}{\sigma_{i,j}+\sigma_{i,j+1}}\right). \tag{13}$$

Substituting the partial derivatives in Equation (10) with Equations (11)–(13) leads to the finite difference model equation, which is used to compute the potential distribution in the electrode-sheet configuration.

2.3. Thermal Model

2.3.1. Heat Diffusion Equation

The heat diffusion equation is a second order partial differential equation of parabolic type. For a solid or motionless fluid volume unit, it states that the rate of change of thermal energy stored equals the net rate of in- and outgoing conductive energy transfer and the rate of thermal energy generation. The partial differential equation of the heat diffusion is defined by:

$$\rho \cdot c_p \cdot \frac{\partial T}{\partial t} = \frac{\partial}{\partial r}\left(\lambda\frac{\partial T}{\partial r}\right) + \frac{\lambda}{r}\frac{\partial T}{\partial r} + \frac{\partial}{\partial z}\left(\lambda\frac{\partial T}{\partial z}\right) + Q. \tag{14}$$

After discretizing (14) by the finite difference method, the terms are as follows:

$$\frac{\partial}{\partial r}\left(\lambda\frac{\partial T}{\partial r}\right) \approx \left(\frac{\lambda_{i-1,j}+\lambda_{i,j}}{2\Delta r^2}T_{i-1,j}-\frac{\lambda_{i+1,j}+2\lambda_{i,j}+\lambda_{i,j}}{2\Delta r^2}T_{i,j}+\frac{\lambda_{i+1,j}+\lambda_{i,j}}{2\Delta r^2}T_{i+1,j}\right) \tag{15}$$

$$\frac{\lambda}{r}\frac{\partial T}{\partial r} \approx \lambda_{i,j}\left(\frac{T_{i+1,j}-T_{i-1,j}}{2\Delta r^2(i-1)}\right) \tag{16}$$

$$\frac{\partial}{\partial z}\left(\lambda\frac{\partial T}{\partial z}\right) \approx \left(\frac{\lambda_{i,j-1}+\lambda_{i,j}}{2\Delta z^2}T_{i,j-1}-\frac{\lambda_{i,j+1}+2\lambda_{i,j}+\lambda_{i,j-1}}{2\Delta z^2}T_{i,j}+\frac{\lambda_{i,j+1}+\lambda_{i,j}}{2\Delta z^2}T_{i,j+1}\right) \tag{17}$$

$$\frac{\partial T}{\partial t} \approx \frac{T^{n+1}-T^n}{\Delta t}. \tag{18}$$

Substituting the partial derivative terms in Equation (14) with Equations (15)–(18) leads to the discretized heat diffusion equation, which is applied to compute the temperature field in the electrode-sheet configuration.

2.3.2. Joule Heating

Joule heating connects the electrical potential to the thermal model. It is implemented as the source term in the heat diffusion Equation (14). The formula for Joule heating is defined as Equation (19) and can be discretized by central differences Equations (20) and (21) for the gradient of the electrical potential. After inserting Equations (20) and (21) into Equation (19), the discretized source term emerges and can be embedded in Equation (14).

$$Q = \frac{1}{\sigma}\left(\left(\frac{\partial \Phi}{\partial r}\right)^2 + \left(\frac{\partial \Phi}{\partial z}\right)^2\right) \tag{19}$$

$$\frac{\partial \Phi}{\partial r} \approx \frac{\Phi_{i+1,j}-\Phi_{i-1,j}}{2\Delta r} \tag{20}$$

$$\frac{\partial \Phi}{\partial z} \approx \frac{\Phi_{i,j+1}-\Phi_{i,j-1}}{2\Delta z} \tag{21}$$

2.3.3. Contact Resistance Model

The basics of electrical contacts were studied and published by Holm [17] and Greenwood [18]. The contact resistance can be decomposed into constriction and film resistance.

However, the applied equation in this model ignores the distinction and considers both components as one entity for the sake of simplicity. The formula that relates resistance to resistivity is:

$$R = \frac{\sigma \cdot l}{A},$$ (22)

where the length l and the area A are the size of a three-dimensional electrical resistor. The model of linear variation of voltage within the contact zone is adopted from [5]. Hence, the electric current density Equation (23), and the interfacial heat generation at the faying surface Equation (24) can be computed by

$$J_{contact} = \frac{1}{\sigma_{contact}} \left(\frac{\partial \Phi}{\partial z} \right)_{contact},$$ (23)

$$Q_{contact} = \frac{1}{\sigma_{contact}} \left(\frac{\partial \Phi}{\partial z} \right)_{contact}^2.$$ (24)

Equations (23) and (24) indicate that the electric current flows perpendicular through the faying surface. The voltage drop at the faying surface is discretized by the forward difference according to:

$$\frac{\partial \phi}{\partial z} \approx \frac{\Phi_{contact,j+1} - \Phi_{contact,j}}{\Delta z_{contact}}.$$ (25)

The contact layer height between the sheets is set to $\Delta z_{contact} = 0.01$ mm and is regarded as the average roughness of the faying surface, which in practice deforms more the heavier the electrode load is. According to the literature it ranges between 0.01 and 0.05 mm [5,19].

2.3.4. Phase Change Model

The phase change from solid to liquid is of particular importance. Neglecting the effect would lead to an unrealistically high temperature field beyond the melting point. While the sheet melts, the temperature remains constant and energy—the specific latent heat H—is absorbed to break down the lattice structure of the solid elementary cells. The specific latent heat for Al amounts to 397 kJ/kg. The specific heat capacity of the phase change $c_{phase} = 1.14$ kJ/(kg · °C) results from the arithmetic mean of the specific heat capacity at the solidus and liquidus temperature. For RSW simulations, the latent heat can be transformed into an equivalent temperature difference

$$\Delta T = \frac{H}{c_{phase}}.$$ (26)

It can be considered as an artificial temperature reservoir, which can be used to differentiate between solid, solid–liquid and liquid phase. In the solid state ($T_{sheet} < T_{solidus}$), $\Delta T = 348.25$ °C remains constant; the sheet temperature increases. As the sheet enters the solid–liquid phase ($T_{liquidus} > T_{sheet} > T_{solidus}$), the difference temperature $T_{sheet} - T_{solidus}$ is subtracted from ΔT. As long as the sheet remains in the solid–liquid phase ($0 < \Delta T < 348.25$ °C), the temperature increase is suppressed. When $\Delta T = 0$ °C is reached, the liquid phase ($T_{sheet} > T_{liquidus}$) begins and the model continues to increase the temperature.

2.4. Material Model

The material data set encompasses specific heat capacity c_p, thermal conductivity λ, electric resistance σ and density ρ, all as functions of temperature. The authors in [20] provide data on specific heat capacity for Al and Cu in all three phases; on electric resistance only for phase change and liquid stage. These electrical resistance values were inserted into the Wiedemann–Franz Law Equation (27) to obtain the thermal conductivity.

$$\lambda = \frac{T}{\sigma} \cdot L$$ (27)

For solid state, ref [21] provides values for the thermal conductivity of Al and Cu. These served the calculation of electrical resistance by the Wiedmann–Franz Law. The density for phase change and liquid state stem from [20] as well. The solid density values for Cu and Al were determined by

$$\rho(T, \alpha) = \frac{\rho_{25}}{(1 + \alpha(T - 25\,^{\circ}C))^3}. \tag{28}$$

The temperature dependent thermal expansion factor α, the density value of the sheet $\rho_{25}^{Al} = 2700\ kg/m^3$ and the electrode $\rho_{25}^{Cu} = 8960\ kg/m^3$ are given in [22]. Depending on temperature intervals, the material data were averaged and are summarized in Table 2.

Table 2. Material data set.

	Aluminum Sheet			Copper Electrode Cap
	Solid **(T < 660 °C)**	**Phase Change** **(T = 660 °C)**	**Liquid** **(T > 660 °C)**	**Solid**
$\rho\ (kg/m^3)$	2663	2385	2323	8874
$c_p\ (J/kg \cdot {}^{\circ}C)$	1041	1194	1085	412
$\lambda(W/(m \cdot {}^{\circ}C))$	231.5	209	102	372
$\sigma(n\Omega m)$	56	110	270	38.7

The solidus temperature for Al is defined at 660 °C. In the simulation, the temperature of the electrode cap never came near solidus temperature. Therefore, material parameters of copper were restricted to solid state exclusively. For each phase a contact resistance was defined. The contact resistances for the solid and solid–liquid phase were aligned with the bulk material resistance of the aluminum sheet as the faying surface and the bulk material are assumed to possess the same consistency beyond solidus temperature.

2.5. Solution Methodology

D.W. Peaceman and H. H. Rachford introduced an alternating-direction implicit scheme for finite difference methods—the so-called Peaceman–Rachford scheme [23]. It originates a set of linear equations with tridiagonal band matrices which can be solved by the Thomas Algorithm efficiently. To the author's best knowledge, this solution methodology was used to simulate the melting during the RSW for the first time. By using the discretized heat diffusion Equation (14), the application of the Peaceman–Rachford scheme is demonstrated in this section. The approach is analogously viable to the discretized form of Equation (10).

2.5.1. Peaceman–Rachford Scheme

The Peaceman–Rachford scheme is an unconditionally stable method permitting an arbitrary large time step size. Locally, it is second order accurate in space and time $O\left(\Delta r^2, \Delta z^2, \Delta t^2\right)$. The heat diffusion equations casted into the Peaceman–Rachford scheme leads to following equations:

$$\left(1 - \frac{\mu_r}{2}\left(\delta_r^2 + \frac{\delta_r}{(i-1)}\right)\right)T_{i,j,n+0.5} = \left(1 + \frac{\mu_z}{2}\delta_z^2\right)T_{i,j,n} + \frac{\Delta t}{2}Q_{i,j,n+0.5} \tag{29}$$

$$\left(1 - \frac{\mu_z}{2}\delta_z^2\right)T_{i,j,n+1} = \left(1 + \frac{\mu_r}{2}\left(\delta_r^2 + \frac{\delta_r}{(i-1)}\right)\right)T_{i,j,n+0.5} + \frac{\Delta t}{2}Q_{i,j,n+0.5}. \tag{30}$$

The alternating-direction character of the PR-Scheme is clarified by Equations (29) and (30). At first, the known temperature distribution $T_{i,j,n}$ is used to compute the intermediate temperature distribution $T_{i,j,n+0.5}$ in radial direction by Equation (29). Afterwards, this intermediate solution serves as the input for the subsequent calculation carried out by Equation (30), which outputs the temperature distribution $T_{i,j,n+1}$ in axial direction. The solution of Equation (30) constitutes the solved temperature field of the resistance spot weld nugget. The difference operators $\delta_r T_{i,j}$, $\delta_r^2 T_{i,j}$, and $\delta_z^2 T_{i,j}$ are defined as follows:

$$\delta_r T_{i,j} = \lambda_{i,j} \frac{T_{i+1,j} - T_{i-1,j}}{2} \tag{31}$$

$$\delta_r^2 T_{i,j} = \frac{\lambda_{i-1,j} + \lambda_{i,j}}{2} T_{i-1,j} - \frac{\lambda_{i+1,j} + 2\lambda_{i,j} + \lambda_{i-1,j}}{2} T_{i,j} + \frac{\lambda_{i+1,j} + \lambda_{i,j}}{2} T_{i+1,j} \tag{32}$$

$$\delta_z^2 T_{i,j} = \frac{\lambda_{i,j-1} + \lambda_{i,j}}{2} T_{i,j-1} - \frac{\lambda_{i,j+1} + 2\lambda_{i,j} + \lambda_{i,j+1}}{2} T_{i,j} + \frac{\lambda_{i,j+1} + \lambda_{i,j}}{2} T_{i,j+1}. \tag{33}$$

The Equations (29) and (30) must be arranged according to $a_i T_{i-1,j,n+0.5} + b_i T_{i,j,n+0.5} + c_i T_{i+1,j,n+0.5} = d_{i,n}$ and $a_j T_{i,j-1,n+1} + b_j T_{i,j,n+1} + c_j T_{i,j+1,n+1} = d_{j,n+0.5}$, respectively, in order to be formatted appropriately for the application of the Thomas-Algorithm.

2.5.2. Thomas Algorithm and Code Implementation

The Peaceman–Rachford scheme leads to a system of algebraic equations, one for each of Equations (29) and (30). The Thomas algorithm as a direct solver for tridiagonal system of algebraic equations, treats Equations (29) and (30) indifferently, i.e., it is applied to both matrix equations equally. Thus Equations (29) and (30) can be unified in a general format:

$$\begin{bmatrix} b_1 & c_1 & & & & \\ a_2 & b_2 & c_2 & & & \\ & a_3 & b_3 & c_3 & & \\ & & \ddots & \ddots & \ddots & \\ & & & a_{N-1} & b_{N-1} & c_{N-1} \\ & & & & a_N & b_N \end{bmatrix} \begin{bmatrix} T_1 \\ T_2 \\ T_3 \\ \vdots \\ T_{N-1} \\ T_N \end{bmatrix} = \begin{bmatrix} d_1 \\ d_2 \\ d_3 \\ \vdots \\ d_{N-1} \\ d_N \end{bmatrix}. \tag{34}$$

The Thomas Algorithm consists of two phases. First, the matrix equation is brought into an upper diagonal shape by zeroing a_k, and substituting b_k and d_k by

$$b_k = b_k - \frac{a_k}{b_{k-1}} c_{k-1} \tag{35}$$

$$d_k = d_k - \frac{a_k}{b_{k-1}} d_{k-1}. \tag{36}$$

for $k = 2, \dots, N$. Second, the temperature field is solved by backward substitution, based on

$$T_k = \frac{d_k - c_k T_{k+1}}{b_k}, \tag{37}$$

beginning in the last row with $T_N = d_N / b_N$ towards the first row with T_1. To provide an overview of the coupling between the electric and the thermal model and the underlying program structure, a pseudo-code is depicted in the Appendix A. After initialization of the model geometry and fixed material properties, the welding time t_w can be defined as the product of the time step Δt and the number of time steps n arbitrarily. The number of time steps n also determines how often the temperature dependent material properties are updated as well as how many times the electric and thermal fields are calculated. The electrical field is computed iteratively until the residuum and the difference of succeeding solutions fall below given predetermined break conditions, respectively. By combining the Peaceman–Rachford Scheme with the Thomas algorithm, the temperature field is calculated by two main sequences, which are referred to as sweeps or scans in pertinent literature. During the first sweep, intermediate temperature values are calculated for each row grid point wise from left to right by Equation (29). Analogously to the first sweep, the second sweep calculates the final temperature values for each column grid point wise from bottom to top by Equation (30). For further details on the theory and implementation of the

Peaceman–Rachford scheme and the Thomas Algorithm, the books [24,25] can be consulted. This solver methodology operates line by line, which makes parallel computing feasible and promotes real-time simulation. The simulation model was run on an Intel Core i5-6500 CPU (3.2–3.6 GHz) and in MATLAB© (R2018b).

3. Results

Paragraphs 3 and 4 aim at verifying the implemented model from a numerical and physical point of view by a convergence analysis and a parameter study, respectively. It demonstrates that the electric-thermal model meets general expectations on the behavior of numerical methods and on the physics of RSW. Despite the Joule heating in the electrode has also been part of the computation and influenced the simulation run-time, it is not analyzed in detail in the forthcoming paragraphs due to its negligible low temperature increase.

3.1. Convergence Analysis and Computation Speed

In order to examine the electro–thermal model's run time t_{sim} and convergence behavior, three spot welds with welding times $t_w = 40$ ms, $t_w = 50$ ms und $t_w = 60$ ms were simulated. The voltage between the electrode cap and the faying surface drops by 0.5 V and the contact resistance between the sheets amounts to 400 $\mu\Omega$m for all variants. All other simulation parameters are known from preceding paragraphs and are identical for all simulation variants as well. For each welding time $t_w = n \cdot \Delta t$, the number of simulations runs n and the time steps Δt were combined twelve times, see Tables 3–5. The number of variants per simulated welding time was chosen to be twelve so as to ensure that the mean sheet temperature $\overline{T}$ remains constant when the time step Δt is further decreased (or simulations run n is further increased). Therefore, the twelfth variant of each Tables 3–5 is the closest approximation of the assumed exact solution for the corresponding welding time and set spatial grid. The discretization errors ϵ are calculated by referring to the mean temperature $\overline{T}$ of the twelfth variants in all three tables, i.e., $\varepsilon = \left(\overline{T} - \overline{T}_{variant12}\right)/\overline{T}_{variant12}$. The influence of the simulations runs n and the influence of the time step Δt on the convergence behavior were to be analyzed separately. Thus, across the Tables 3–5 the simulation runs n were held constant with varying time steps Δt for the variants 1, 2, 4, 5, 7, 8, 10, and 11; while the simulations runs n varied with constant time steps Δt for the variants 3, 6, 9, and 12. The arithmetic mean of the sheet temperature $\overline{T}$ over all nodes and its standard deviation $SD(\overline{T})$ as well as the minimum and maximum values of the sheet temperature T_{min} and T_{max} were determined. Finally, the run-time t_{sim} of each simulation variant was measured manually with a stopwatch and is therefore subject to slight measurement errors.

Table 3. Characterization of sheet temperature field depending on (n,Δt)-variants (welding time $t_w = 40$ ms).

Variant	n	Δt (μs)	$\overline{T}$ (°C)	$SD(\overline{T})$ (°C)	T_{max} (°C)	T_{min} (°C)	ϵ (%)	t_{sim} (s)
1	100	400	332	229	1000	79	7.4	3
2	200	200	326	218	941	82	5.5	3.5
3	400	100	320	210	900	82	3.6	4
4	1000	40	315	203	862	82	1.9	5
5	2000	20	312	200	845	82	1.0	6
6	4000	10	311	198	832	82	0.7	8
7	10,000	4	310	196	824	82	0.3	9
8	20,000	2	310	196	821	82	0.3	14
9	40,000	1	309	195	819	82	0.0	20
10	100,000	0.4	309	195	818	82	0.0	36
11	200,000	0.2	309	195	818	82	0.0	66
12	400,000	0.1	309	195	817	82	0.0	125

Table 4. Characterization of sheet temperature field depending on (n,Δt)-variants (welding time $t_w = 50$ ms).

Variant	n	Δt (μs)	$\overline{T}$ (°C)	SD($\overline{T}$) (°C)	T_{max} (°C)	T_{min} (°C)	ε (%)	t_{sim} (s)
1	100	500	419	302	1456	118	9.7	6
2	200	250	409	286	1368	115	7.1	8
3	500	100	398	268	1270	113	4.2	10
4	1000	50	392	260	1226	112	2.6	10
5	2000	25	388	255	1194	112	1.6	10
6	5000	10	385	250	1169	111	0.8	12
7	10,000	5	384	249	1160	111	0.5	14
8	20,000	2.5	383	248	1155	111	0.3	18
9	50,000	1	383	247	1152	111	0.3	27
10	100,000	0.5	382	247	1151	111	0.0	43
11	200,000	0.25	382	247	1150	111	0.0	73
12	500,000	0.1	382	247	1150	110	0.0	162

Table 5. Characterization of sheet temperature field depending on (n,Δt)-variants (welding time $t_w = 60$ ms).

Variant	n	Δt (μs)	$\overline{T}$ (°C)	SD($\overline{T}$) (°C)	T_{max} (°C)	T_{min} (°C)	ε (%)	t_{sim} (s)
1	100	600	526	408	2001	150	10.0	11
2	200	300	515	392	1918	146	7.7	13
3	600	100	501	371	1805	141	4.8	15
4	1000	60	496	364	1768	139	3.8	15
5	2000	30	490	356	1732	137	2.5	16
6	6000	10	483	348	1695	136	1.0	19
7	10,000	6	481	346	1685	136	0.6	22
8	20,000	3	480	345	1677	135	0.4	25
9	60,000	1	479	343	1671	135	0.2	37
10	100,000	0.6	479	343	1670	135	0.2	50
11	200,000	0.3	478	343	1669	135	0.0	79
12	600,000	0.1	478	343	1668	135	0.0	199

3.2. Parameter Study

For qualitatively verifying the simulation model, n = 10,000, as this choice leads to results with acceptable low discretization error ($\epsilon \leq 1\%$). As process parameters, the welding time, the applied voltage, and the electric current affect the amount of thermal energy produced in the metal sheets. Aside from above mentioned process parameter, the electric contact resistance at the faying surface also influences the Joule heating. Thus, the parameter study is performed by varying the electrode voltage, the welding time and the contact resistance one by one while all other parameters are held constant. The results of the parameter study are shown in Figure 1 with corresponding data in Table 6. The reference spot weld is depicted separately in the first row of Table 6 and referenced by Figure 1a.

According to Section 2.3.4, a grid point is solid (T < 660 °C), mushy (T = 660 °C) or liquid (T > 660 °C). Thus, the upper limit of the temperature scale in Figure 1 is set to 1000 °C to permit the distinction between these three phases. The simulation of the reference weld (Figure 1a) was conducted with the parameters U = 0.5 V, σ = 400 μΩm, and $t_w = 40$ ms. Its temperature field is described with $\overline{T} = 310$ °C, SD($\overline{T}$) = 196 °C, $T_{min} = 82$ °C, $T_{max} = 824$ °C and marked with a thin molten and mushy zone along the faying surface; the rest of the sheet is solid. Compared to Figure 1a, the spot weld in Figure 1b exhibits lower temperature values $\overline{T}$, SD($\overline{T}$), T_{min}, and T_{max} due to a reduced electrode voltage of 0.45 V. It is solid except for a mushy area at the contact layer. The Figure 1c indicates a molten and mushy phase along the faying surface, both shaped like a flat ellipse. On account of an increased voltage of 0.55 V, the temperature field of Figure 1c is overall higher than in the reference weld. An examination of the welds in Figure 1a–c shows that the temperature increases with voltage. In Figure 1d near the left-bottom corner, a slight molten pool can be observed. Furthermore, a thin mushy zone evolves along the faying surface. Compared to the reference weld, the contact resistance $\sigma = 300\ \mu\Omega m$ and the values of $\overline{T}$, $SD(\overline{T})$, T_{min}, and T_{max} are lower. An increase of the contact resistance up to $\sigma = 500\ \mu\Omega m$ (Figure 1e) induces higher temperatures than in the reference spot weld. As a result, an increase/decrease of the contact resistance causes higher/lower spot weld temperatures. Finally, the welding time of the reference spot weld was varied by ± 10 ms to verify the model. While Figure 1f depicts merely the onset of a fusion area, Figure 1g shows clearly an elliptically shaped weld spot. By contrast to Figure 1a, the weld temperatures in Figure 1f,g are decreased and increased, respectively. As expected, the weld spot temperature increases the longer the welding time lasts.

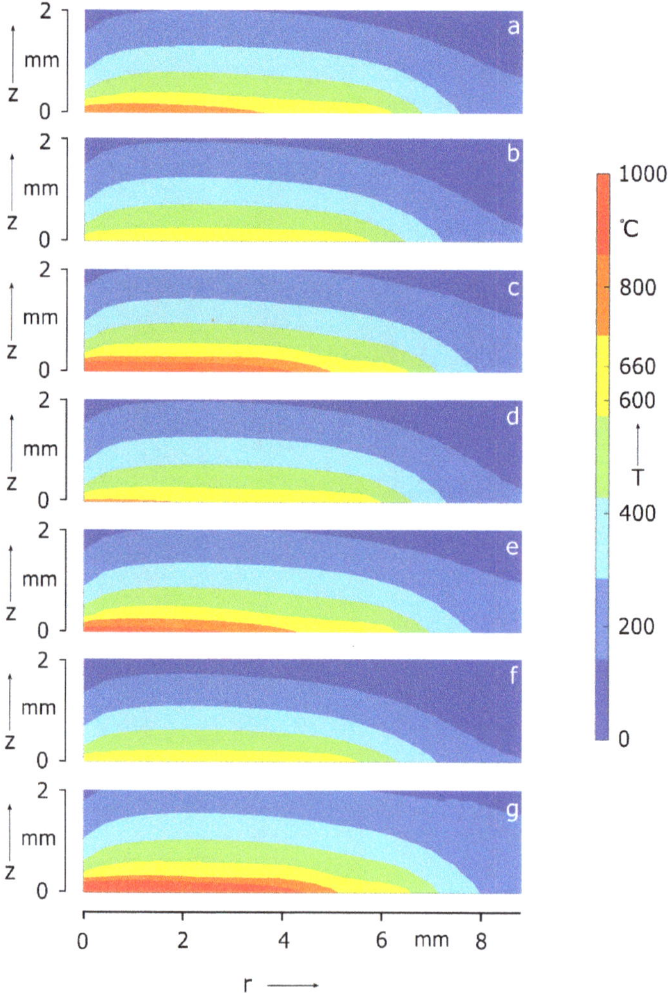

Figure 1. reference spot weld (**a**) and spot welds for varying voltage (**b**,**c**), contact resistance (**d**,**e**) and welding time (**f**,**g**).

Table 6. simulation parameters and results of parameter study.

	U/(V)	σ/($\mu\Omega$m)	t_w/(ms)	$\overline{T}$/($^\circ$C)	$SD(\overline{T})$/($^\circ$C)	T_{min}/($^\circ$C)	T_{max}/($^\circ$C)
a	0.5	400	40	310	196	82	824
b	0.45	400	40	281	181	68	660
c	0.55	400	40	358	245	93	1094
d	0.5	300	40	291	185	71	724
e	0.5	500	40	328	209	88	901
f	0.5	400	30	254	186	44	660
g	0.5	400	50	384	249	111	1160

4. Discussion

4.1. On Computation Speed and Convergence Analysis

In Table 3, the simulation variant 4 with $\Delta t = 40$ ms (n = 1000) exhibits a discretization error of 1.9 %, while variant 7 with $\Delta t = 4$ μs (n = 10,000) leads to a discretization error of 0.3 %. It exemplifies the expected relation between the time step and the discretization error, i.e., the discretization error decreases with decreasing time step. Moreover, the run-time increases with shrinking time step as discussed hereinafter. The run-time for the variant 4 with $\Delta t = 40$ μs and variant 7 with $\Delta t = 4$ μs in Table 4 are 5 s and 9 s, respectively. In addition, the run-time relates to the discretization error conversely, i.e., the run-time increases with decreasing discretization error. By comparing variants 4 and 7 of Table 3 again, it is discernible that the run-time increases from 5 s to 9 s while the discretization error decreases from 1.9 % to 0.3 %. These relations apply to all (n,t_w)-variants in Tables 3–5 equally. From a numerical point of view, it has been shown that the relations among the time step, discretization error and run-time meet common expectations on the general behavior of finite difference methods. Another observation is that the solution converges from top to bottom for decreasing time step size. For example, in Table 5 the variants 3, 6, 9, and 12 possess mean sheet temperatures $\overline{T}$ and time steps Δt of ($\overline{T}$ = 398 $^\circ$C, $\Delta t = 100$ μs), ($\overline{T}$ = 385 $^\circ$C, $\Delta t = 10$ μs), ($\overline{T}$= 383 $^\circ$C, $\Delta t = 1$ μs) and ($\overline{T}$ = 382, $\Delta t = 0.1$ μs). The standard deviation $SD(\overline{T})$, T_{min}, and T_{max} behave analogously. However, an unexpected effect can be observed by comparing the variants 4 in Tables 3–5. The run-time increases with increasing time step Δt for a fixed number of simulation runs. For example, the run-time amounts to 5 s, 10 s and 15 s for corresponding time steps of $\Delta t = 40$ μs, $\Delta t = 50$ μs and $\Delta t = 60$ μs, although the simulation runs n = 1000 are identical in all three variants. It can be explained by the fact that an increased time step leads to higher temperature values of the intermediate simulation result. High temperature values require more computer bits of the central processor unit than lower temperature values and this coincides with higher computational costs.

The demand for fast computation models for the precise prediction of process simulation has been growing since the beginning of Industry 4.0. The idea of integrating real-time-capable digital twins into production processes to increase productivity, by reducing waste or increase quality, and use them as monitoring and control units is receiving increasing attention in industry and research. In [26] a digital twin for RSW is presented, which visualizes the temperature field. It consists of an interpolation model based on experiments and FE computations. The digital twin delivers almost identical results to the simulation model (deviation < 1%) and takes only 10 s instead of one hour (FEM model). In [27] the inherent strain and deformation method are applied to predict the total deformation of 23 resistance spot welds in a vehicle part within around 90 min. The resulting deformation, the so-called inherent deformation, is achieved by calculating the difference between the total and elastic deformation. The total deformation is identified by experiments; the elastic deformation is calculated by an FEM tool. In [28], an equivalent parametric methodology for modelling multi-pass longitudinal welds on planar structures, such as plates and rectangular hollow sections, is introduced. The so-called welding equivalence model consists of a single and multi-layered shell, connection and beam elements and is generated by an automatic sub-program that acts on already existing FE shell model. It uses transient thermal and steady-state structural analysis to identify residual stresses and local distortions typical of multi-pass welds. Compared to classic numerical 3D models, it reduces the computing time by a factor of ten. Although the results of the studies [26–28] are remarkable in terms of the shortened computing time, the extended preparation time is critical; namely carrying out experiments and/or numerical simulations. It is not evident how the computational time savings compare to the additional preparation time, or whether this translates into an overall time saving. The computation speed of numerical models remains a bottleneck in the above-mentioned studies. It underlines the importance of time-efficient simulation models.

For the design of numerical programs exist at least three aspects that have a significant impact on the computing speed of transient thermal conduction problems. This includes the choice between explicit vs. implicit difference methods, iterative vs. direct solvers and structured vs. unstructured grids. Explicit methods are conditionally stable, so that a given stability criterion must be met. It limits the size of the time step to be selected and leads to long run times for simulations. Implicit methods are unconditionally stable and are not subject to this restriction. Because a larger time step may be selected for implicit methods, the simulation run time can be shorter. The disadvantage of these methods is the increased effort during implementation. Iterative solvers, such as the Gauss–Seidel process, require numerous repetitions to converge. Here, the iteration runs are canceled if the residual and/or the difference of two consecutive solution values fall below a pre-specified tolerance. Direct solvers only need one time step to reach the solution for a certain point in time. The Thomas algorithm, an example of a direct solver, represents a recursion formula that uses the boundary conditions to indicate exactly the result of the difference equation at a point in space. The prerequisite for the application of the Thomas algorithm is the presence of a linear equation system with a tridiagonal coefficient matrix. Grid-based discretization methods (FDM, FVM, FEM), distinguish between structured and unstructured grids or meshes. Unstructured meshes, which are usually used in FEM [29], have irregularly distributed nodes and their cells do not need to have a standard shape. Therefore, they are the preferred method for generating meshes in areas with complex geometries. However, the use of unstructured meshes complicates the numerical algorithm due to the inherent data management problem, which requires a special program to number and organize the nodes, edges, surfaces, and cells of the grid. In addition, linearized difference schema operators on unstructured meshes are not usually band matrices, making it difficult to use implicit schemes. The numerical algorithms based on unstructured grids are the most costly in terms of computing time and memory. Structured grids, which are the basis of the FDM, implicitly contain the order of the grid elements in their solution; the application of a program for the management of the grid elements is omitted. As a result of the ordered structure of structured grids, model equations with band matrices are created, which allow the application of time-efficient solvers. A major disadvantage of primitive FDM is that complex geometry, characterized by curves, sharp/obtuse angles, can only be modelled if losses in computational accuracy are acceptable. However, the use of elliptical grids can level out this disadvantage [30]. Regarding these three major aspects of the numerical software design, the solver algorithm was designed in favor of a short simulation run time, i.e., an implicit finite-difference method with a direct solver on a structured mesh was developed.

4.2. On Parameter Study

In aluminum alloy spot welds two types of nugget development have been observed in studies based on practical experiments. In one of them, melting starts as a circle around the center at the contact area. Gradually, from all sides the melting continues inwards until a complete nugget is formed [31]. In the other type initial melting is located at the center of the faying surface, before it spreads in vertical and horizontal direction outwards [13,32]. The nugget growth in the present study starts in the center and extends outwards and, thus, can be assigned to the latter type. Hereby proof for the qualitative correct simulation of the nugget development is given. Table 6 points out that the sheet temperature and the weld spot of the aluminum sheet grow with the applied electrode cap voltage, the contact resistance and welding time. These observations align with general expectations of RSW [33].

5. Conclusions

As mentioned at the beginning, the requirements for a model-based real-time monitoring and control system for RSW are sufficient model accuracy and computation speed. In terms of the computation speed, the simulation run-time falls below 20 s for a discretization error $\leq 1\%$, whereas the process cycle (squeezing-, welding-, holding-, off-phase) lasts up to a couple of seconds depending on process conditions. If the real-time simulation should run to RSW-Process simultaneously, the simulation run-time must undercut the cycle time of the spot weld process. This requires additional effort to reduce the run-time below the cycle time. In terms of software optimization, technics of high-performance computing, for example, vectorization and parallel computing as well as efficient programming possess the potential to accelerate the computation speed additionally. Moreover, hardware with more or higher processing power can support real-time simulation of RSW. After all, in combination with an adequately chosen solving algorithm the finite difference method seems to be a feasible approach for computing resistance spot welding close to real-time. Investigations to come will include the validation of the model by means of experiment.

Author Contributions: Conceptualization, T.T. and L.P.; methodology, T.T. and L.P.; software, T.T.; formal analysis, T.T.; investigation, T.T.; resources, S.I.; data curation, C.P.; writing—original draft preparation, T.T.; writing—review and editing, L.P and C.P.; visualization, C.P.; supervision, S.I.; project administration, S.I.; funding acquisition, S.I. All authors have read and agreed to the published version of the manuscript.

Funding: This research was funded by the Deutsche Forschungsgemeinschaft (DFG, German Research Foundation) within the project "Basic investigations for the in-situ simulation of resistance spot welding processes" under the grant numbers FU 307/15-1 resp. IH 124/17-1, Project-no.: 389519796.

Institutional Review Board Statement: Not applicable.

Informed Consent Statement: Not applicable.

Conflicts of Interest: The authors declare that they have no known competing financial interests or personal relationships that could have appeared to influence the work reported in this paper.

Nomenclature

Symbol	Quantity	Unit	Symbol	Quantity	Unit
T	temperature	$^\circ$C	r, φ, z	cylindrical coordinates	m
Q	volumetric heat source	W/m^3	t	time coordinate	s
H	latent heat	J/kg	$\Delta z/\Delta r$	spatial step in z/r-direction	m
L	Lorenz constant	[-]	Δt	time step	s
c_p	specific heat	J/(kg·K)	i/j	radial/axial index	[-]
Φ	electric potential	V	n	time and iteration index	[-]
α	thermal expansion factor	K^{-1}			
λ	thermal conductivity	W/(m·K)	∇	Nabla-operator	
μ	Fourier number	[-]	$\delta(\cdot)$	1st derivative difference operator	
ρ	density	kg/m^3	$\delta(\cdot)^2$	2nd derivative difference operator	
σ	specific electric resistance	μΩm			

Appendix A

```
%initialization:
- geometry of sheet and electrode
- fixed material properties of electrode
- time step Δt
- number time steps n              % simulations runs
- abs error = 0.001                % first break condition
- tolerance = 0.001                % second break condition
% start electric-thermal model
for i = 1 : n
- update temperature dependent material properties
- stop 1 = 0; % initial value of first break condition

% start electric model
while (stop1 == 0 and stop2 == 0)
        - 1st sweep of Peacheman-Rachford scheme (from left to right / r-direction):
                - upper diagonalizing (1st part of Thomas algorithm)
                - backward substitution (2nd part of Thomas algorithm)
        - 2nd sweep of Peacheman-Rachford scheme (from bottom to up / z-direction)
                - upper diagonalizing (1st part of Thomas algorithm)
                - backward substitution (2nd part of Thomas algorithm)
        - calculate residuum and relative error
        - if (residuum < abs error) then stop1 = 1 else stop1 = 0 end
        - if (difference successive solutions < tolerance) then stop2 = 1 else stop1 = 0 end
        - update electric boundaries
    end
% end electric model
% start thermal model
        - 1st sweep of Peacheman-Rachford scheme (from left to right / r-direction)
                - upper diagonalizing (1st part of Thomas algorithm)
                - backward substitution (2nd part of Thomas algorithm)
        - 2nd sweep of Peacheman-Rachford scheme (from bottom to up / z-direction)
                - upper diagonalizing (1st part of Thomas algorithm)
                - backward substitution (2nd part of Thomas algorithm)
        - determine phase state of sheet (optional)
        - update thermal boundaries
% end thermal model
end
% end electric-thermal model
```

References

1. Khan, J.A.; Xu, L.; Chao, Y.-J. Prediction of nugget development during resistance spot welding using coupled thermal-electrical-mechanical model. *Sci. Technol. Weld. Join.* **2013**, *4*, 201–207. [CrossRef]
2. Wagner, I. 2020. Available online: https://www.statista.com/statistics/262747/worldwide-automobile-production-since-2000/ (accessed on 24 November 2020).
3. Mathiszik, C.; Zschetzsche, J.; Füssel, U. Nondestructive characterization of the attachment surface in resistance pressure welding by remanence flux density imaging analysis. Final report. Zerstörungsfreie Charakterisierung der Anbindungs-fläche beim Widerstandspressschweißen durch bildgebende Analyse der Remanenzflussdichte. Schlussbericht IGF-Nr. 19.208 BR/DVS-Nr. 04.058: Technische Universität Dresden, Professur für Fügetechnik und Montage. *Tech. Rep.* **2019**. Available online: https://tu-dresden.de/ing/maschinenwesen/if/fue/ressourcen/dateien/ag_thermisches_fuegen/abschlussberichte/ Schlussbericht_IGF_19208?lang=de (accessed on 24 November 2021).
4. Han, J.; Orozco, J.E.; Indacochea; Chen, C.H. Resistance Spot Welding: A Heat Transfer Study Z. *Weld. J.* **1989**, *86*, 363s–371s.
5. Cho, H.S.; Cho, Y.J. A Study of the Thermal behavior in Resistance Spot welds. *Weld. J.* **1989**, *6*, 236s–244s.
6. Wei, P.; Ho, C.Y. Axisymmetric Nugget Growth During Resistance Spot Welding. *J. Heat Transf.* **1990**, *112*, 309. [CrossRef]

7. Wei, P.S.; Wang, S.C.; Lin, M.S. Transport Phenomena During Resistance Spot Welding. *J. Heat Transf.* **1996**, *112*, 309–316. [CrossRef]
8. Wang, S.; Wei, P.S. Modeling Dynamic Electrical Resistance During Resistance Spot Welding. *J. Heat Transf.* **2001**, *123*, 576–585. [CrossRef]
9. Khan, J.A.; Xu, L.; Chao, Y.-J.; Broach, K. Numerical Simulation of Resistance Spot Welding Process. *Numer. Heat Transf. Part A Appl.* **2010**, *37*, 425–446. [CrossRef]
10. MPiott, A.; Werber, L.; Schleuse, N.; Doynow, R.; Ossenbrink, V.G. Michailow Numerical and experimental analysis of heat transfer in resistance spot welding process of aluminum alloy AA5182. *Int. J. Adv. Manuf. Technol.* **2020**, *111*, 1671–1682. [CrossRef]
11. Piott, M.; Werber, A.; Schleuse, L.; Doynow, N.; Ossenbrink, R.; Michailov, V.G. Electrical Contact Resistance Model For Aluminum Resistance Spot Welding. *Math. Model. Weld Phenom.* **2019**, *12*, 1–18. [CrossRef]
12. Nielsen, C.V.; Zhang, W.; Perret, W.; Martins, P.A.; Bay, N. Three dimensional simulation of resistance spot welding. *Proc. Instiution Mech. Eng. Part D J. Automob. Eng.* **2015**, *229*, 885–897. [CrossRef]
13. ESchulz, M.; Wagner, H.; Schubert, W.; Zhang, B.; Balasubramanian, L.N. Brewer. Short-Pulse Resistance Spot Welding of Aluminum Alloy 6016-T4—Part 1. *Weld. J.* **2019**, *100*, 41–51. [CrossRef]
14. SWANTEC Software and Engineering ApS. 2020. Available online: https://www.swantec.com/technology/numerical-simulation/ (accessed on 24 November 2021).
15. Croft, D.R.; Lilley, D.G. *Heat Transfer Calculations Using Finite Difference Equations*; Applied Science Publishers Ltd.: London, UK, 1977.
16. Chang, B.H.; Li, M.V.; Zhou, Y. Comparative study of small scale and 'large scale' resistance spot welding. *Sci. Technol. Weld. Join.* **2004**, *6*, 273–280. [CrossRef]
17. Holm, R. *Electric contacts. Theory and Application*; completely rewritten edition; Springer: Berlin/Heidelberg, Germany, 1967; Volume 4.
18. Greendwood, J.A. Constriction resistance and the real area of contact. *Br. J. Appl. Phys.* **1966**, *17*, 1621. [CrossRef]
19. Wang, J.; Wang, H.P.; Lu, F.; Carlson, B.E.; Sigler, D.R. Analysis of Al-steel resistance spot welding process by developing a fully coupled multi-physics simulation mode. *Int. J. Heat Mass Transf.* **2015**, *89*, 1061–1072. [CrossRef]
20. Valencia, J.J.; Quested, P.N. *ASM Handbook, Volume 15: Casting*; Chapter: Thermophysical Properties; ASM International: Almere, The Netherlands, 2008. [CrossRef]
21. Incropera, F.P.; DeWitt, D.P. *Fundamentels of Heat and Mass Transfer*, 6th ed.; John Wiley & Sons: Hoboken, NJ, USA, 2007.
22. Smithells, C.J. *Smithells Metals Reference Book*, 8th ed.; Elsevier: Amsterdam, The Netherlands, 2004.
23. Peaceman, D.W.; Rachford, H.H. The numerical solution of parabolic and elliptic differential equations. *J. Soc. Indust. Appl. Math.* **1955**, *3*, 28–41. [CrossRef]
24. Thomas, J.W. *Numerical Partial Difference Equations: Finite Difference Methods*, 2nd ed.; Springer: New York, NY, USA, 1998.
25. Strikwerda, J.C. *Finite Difference Schemes and Partial Differential Equations*, 2nd ed.; Siam: Philadelphia, PA, USA, 2004.
26. Ren, S.; Ma, Y.; Ma, N.; Chen, Q.; Wu, H. Digital Twin for the Transient Temperature Prediction During Coaxial One-Sided Resistance Spot Welding of Al5052/CFRP. *J. Manuf. Sci. Eng.* **2022**, *144*, 031015. [CrossRef]
27. Chino, T.; Kunugi, A.; Kawashima, T.; Watanabe, G.; Can, C.; Ma, N. Fast Prediction for Resistance Spot Welding Deformation Using Inherent Strain Method and Nugget Model. *Materials* **2021**, *14*, 7180. [CrossRef]
28. Trupiano, S.; Belardi, V.G.; Fanelli, P.; Gaetani, L.; Vivio, F. A novel modeling approach for multi-passes butt-welded plates. *J. Therm. Stresses* **2021**, *44*, 829–849. [CrossRef]
29. Sherepenko, O.; Kazemi, O.; Rosemann, P.; Wilke, M.; Halle, T.; Jüttner, S. Transient Softening at the Fusion Boundary of Resistance Spot Welds: A Phase Field Simulation and Experimental Investigations for Al-Si-coated 22MnB5. *Metals* **2019**, *10*, 10. [CrossRef]
30. Liseikin, V.D. *Grid Generation Methods, Scientific Computation*; Springer International Publishing AG: Berlin/Heidelberg, Germany, 2017. [CrossRef]
31. Rashid, M.; Medley, J.B.; Zhou, Y. Nugget formation and growth during resistance spot welding of aluminum alloy. *Can. J. Metall. Mater. Sci.* **2013**, *50*, 61–71. [CrossRef]
32. Zhen, L.; Fuyu, Y.; Yang, L.; Yang, B.; Qi, Y.; Hui, T. Numerical and Experimental Study on Nugget Formation Process in Resistance Spot Welding of Aluminum Alloy. *Trans. Tianjin Univ.* **2015**, *21*, 35–139. [CrossRef]
33. Hamedi, M.; Atashparva, M. A review of electrical contact resistance modeling in resistance spot welding. *Weld World* **2017**, *61*, 269–290. [CrossRef]

materials

Article

Multi-Objective Welding-Parameter Optimization Using Overlaid Contour Plots and the Butterfly Optimization Algorithm

Rehan Waheed *, Hasan Aftab Saeed [ID] and Bilal Anjum Ahmed [ID]

Department of Mechanical Engineering (CEME), National University of Sciences and Technology (NUST), Sector H-12, Islamabad 4600, Pakistan; hasan.saeed@ceme.nust.edu.pk (H.A.S.); bilal.anjum@ceme.nust.edu.pk (B.A.A.)
* Correspondence: rehan.waheed@ceme.nust.edu.pk

Abstract: Distortion and residual stress are two unwelcome byproducts of welding. The former diminishes the dimensional accuracy while the latter unfavorably affects the fatigue resistance of the components being joined. The present study is a multi-objective optimization aimed at minimizing both the welding-induced residual stress as well as distortion. Current, voltage, and welding speed were the welding parameters selected. It was observed that the parameters that minimize distortion were substantially different from those that minimized the residual stress. That is, enhancing dimensional accuracy by minimizing distortion results in an intensification of residual stresses. A compromise between the two objectives was therefore necessary. The contour plots produced from the response surfaces of the two objectives were overlaid to find a region with feasible parameters for both. This feasible region was used as the domain wherein to apply the novel butterfly optimization algorithm (BOA). This is the first instance of the application of the BOA to a multi-objective welding problem. Weld simulation and a confirmatory experiment based on the optimum weld parameters thus obtained corroborate the efficacy of the framework.

Keywords: BOA; multi-objective optimization; residual stresses; response surface method; welding distortion

check for updates

Citation: Waheed, R.; Saeed, H.A.; Ahmed, B.A. Multi-Objective Welding-Parameter Optimization Using Overlaid Contour Plots and the Butterfly Optimization Algorithm. *Materials* **2022**, *15*, 4507. https://doi.org/10.3390/ma15134507

Academic Editor: Raul D.S.G. Campilho

Received: 17 May 2022
Accepted: 21 June 2022
Published: 27 June 2022

Publisher's Note: MDPI stays neutral with regard to jurisdictional claims in published maps and institutional affiliations.

1. Introduction

In the welding process, the parts being produced experience a thermal gradient. The temperature of the material under the welding torch is at or above the melting point of the base metal. In the same base metal, some regions are at ambient or room temperature, while others are in between the solidus and the ambient temperature. This temperature difference produces shrinkage forces in the base metal larger than the yield point of the material. These shrinkage forces during the heating and cooling cycle of the welding process produce distortion and locked-in stresses or residual stresses. The distortion and residual stresses both need to be minimized. The distortion produced during welding perturbs the dimensional requirements of the part being manufactured and causes fitment problems during assembly. The residual stresses need to be avoided for parts under fluctuating or cyclic loading. Parts of automobile frame structures, rotating machinery, airframes, and ship structures are all subjected to fatigue loading during their useful life. In all of these and similar cases, welding distortion and residual stress need to be addressed concurrently. The factors that play a vital role in producing welding distortion and residual stress need to be examined and a multi-objective optimization study is required to find the optimum values of the parameters that produce minimum distortion and residual stress in the welded structures. The experimental procedures for performing multiple test runs are expensive; therefore, welding simulations through FEA are utilized with the design of experiments (DOE) approach to find the optimum ranges of values. The response surface method (RSM) is a useful tool for this purpose. In the present work, contour plots of welding-induced distortion and residual stresses were produced from response surfaces. These contour plots were overlaid to find a feasible region of optimum weld parameters. From this feasible region is derived the solution domain for BOA optimization. The optimum welding process parameters were obtained after multiple iterations of the BOA.

2. Literature Review

In the past, the response surface method has been used as a tool to optimize the weld parameters. Prasada et al. [1] employed the response surface method to optimize the ultimate strength of Inconel sheets. They used a central composite rotatable design matrix and checked the influence of peak current, back current, pulse, and pulse width on the ultimate tensile strength of the base metal. Srivastava and Garg [2] used the response surface method with a Box–Behnken design. The responses they studied were the weld bead width, bead height, and the depth of penetration. Vasantharaja and Vasudevan [3] optimized the activated tungsten inert gas (TIG) welding process parameters using the response surface method. In their research, they applied the desirability approach for the optimization of the weld process parameters. Vidyarthi et al. [4] optimized the weld process parameters (i.e., welding current, speed, and flux coating density). They used the response surface method with a central composite design and studied the response of bead width, depth to width ratio of bead, weld fusion zone area, and depth of penetration. Lai et al. [5] applied the response surface method for the optimization of resistance spot welding. The response parameters they selected for the study were the electrode diameter and the effect on the electrode during the cooling process. Their study provided a useful technique for the design of resistance spot welding electrodes. Korra et al. [6] optimized the activated TIG welding process parameters (i.e., welding current, speed, and arc gap) by employing the response surface method with a central composite design. They studied the response of the process parameters on various weld bead geometry aspects and used the desirability approach for multi-objective optimization. Joseph et al. [7] used the Taguchi method for the design of experiments. In their work, they used the response surface method as a tool to develop a mathematical relation between the weld process parameters through regression analysis. This mathematical relation was used by them for further optimization of the welding process through the genetic algorithm (GA). Waheed et al. [8] used the response surface method and artificial intelligence to optimize welding induced distortion. Gunaraj and Murugan [9] used the response surface method for the optimization of weld process parameters for submerged arc welding. They used a central composite design that is rotatable and used the welding parameters (i.e., speed, arc voltage, wire feed rate, and nozzle-to-plate distance) to optimize the quality of the weld. The effect of welding residual stress during cyclic loading has been studied by many researchers. The control and optimization of residual stress are essential for welded structures under fatigue loading. Mochizuki [10] examined the problem of the minimization of the residual stress produced during welding. He concluded that the welding residual stress should be controlled during the welding process rather than relying on the post-weld heat treatment procedures.

Lee and Kyong [11] studied the fatigue crack growth rate under welding residual stress. They calculated the stress intensity factor in the presence of residual stress and used the linear elastic fracture mechanics (LEFM) technique to predict crack growth under fatigue loading. Hensel et al. [12] studied the fracture resistance of welded structures under fatigue loading. They concluded that the welding residual stress significantly affected the overall fatigue strength, crack growth rate, and fatigue life of the welded part. Farajian [13] observed in his work that weld residual stresses of magnitude equal to yield strength were present in large-welded structures. He also observed that the residual stresses present at the weld centerline were of higher magnitude than the stresses present near the toe of the weld bead. However, since the crack initiation usually starts from the toe of the weld bead during cyclic loading, this area also needs to be considered. Cui et al. [14] studied the deck-to-rib stresses in automobile bodies. The effect of stresses produced due to vehicle movement in the presence of residual stresses was the main interest of their work. They concluded that the welding residual stresses had a marked negative effect on the overall fatigue resistance. Chang [15] studied the softening of high tensile residual stresses through heat treatment procedures. He showed that high tensile stresses can be changed to compressive stresses by ultrasonic impact treatment. Barsoum and Barsoum [16] simulated fatigue crack propagation in the presence of welding residual stresses. First, they found the welding residual stresses through the finite element method (FEM) simulation. The welding residual stresses were mapped in the next simulation as an input load and the LEFM technique was then employed to predict the propagation of crack in the presence of weld residual stresses. In another study, Barsoum [17] studied the weld residual stresses near the weld root and toe in plate-to-tube joints. He performed a 2D FEM analysis and verified the simulated results with experimental data. Caruso and Imbrogno [18] performed the finite element modeling of AISI 441 steel plates and developed a user subroutine to predict the grain size variation and hardness of the steel plates. Murat and Ozler [19] developed a finite element model of friction stir welding. They predicted that the ratio of tool rotational speed and tool feed is critical for avoiding defects in friction stir welding. Moslemi et al. [20] developed a systematic procedure for calibrating heat source parameters before simulating the welding process for GMAW welding. Zhang and Dong [21] studied the possibility of brittle fracture in welded structures. They observed how residual stresses could decrease the plastic deformation capability of metals, thus decreasing the fatigue life of the welded structures.

Narwadkar and Bhosle [22] studied the angular distortion produced in the welded structures. They used the design of experiments approach. They observed the effect of welding current, voltage, and gas flow rate on the angular distortion of welded parts. Zhang et al. [23] used FEM simulations to observe the overall distortion in a large vacuum vessel, cutting down the cost of constructing a prototype of the vessel. Lorza et al. [24] built a thermomechanical model to simulate the TIG welding process. They observed the effect of welding voltage, current, speed, and torch parameters on the distortion produced during welding. Chen et al. [25] studied the distortion produced during welding in panel structures with stiffeners. The fillet joint configuration was simulated using FEM. The weld parameters of the welding current, voltage, and speed were used as the governing factors to control distortion. Additionally included in the study was the effect of the welding sequence. Multi-objective studies were adopted by several researchers to optimize the different weld responses. Rong et al. [26] optimized the longitudinal residual stress and transverse tensile stresses. Romero-Hdz et al. [27] used the GA for their multi-objective optimization of welding induced residual stresses and distortion. They used FE simulations to calculate the weld distortion and residual stresses for their study. Shao et al. [28], in a similar study, used multi-objective particle swarm optimization (MOPSO) to optimize the welding residual stress and distortion. They used welding current, speed, and voltage as the main parameters that affected the objective function.

The butterfly optimization algorithm (BOA) is a nature-inspired algorithm proposed by Arora and Singh [29]. They demonstrated the efficacy of the BOA over other nature-inspired metaheuristics by solving three classical benchmark engineering problems. They compared BOA with other nature-based optimization techniques such as artificial bee colony, cuckoo search, differential evolution, firefly algorithm, genetic algorithm, particle swarm, and the modified butterfly optimization algorithm. They found the BOA to be more efficient than the other metaheuristic algorithms. Yildiz et al. [30] applied the BOA to obtain the optimum shape of automobile suspension components, achieving a weight reduction of 32.9%. In previous research, the response surface method was used to formulate the objective function, but its outcome (i.e., the contour plots) was not utilized in the optimization process. In the present research, the boundaries of the solution domain were constrained through overlaid contour plots, which shrinks the solution domain. This helps in the implementation of BOA as a multi-objective optimization technique to find the optimum residual stress and welding induced distortion. The BOA has thereby been further enhanced to be used as a multi-objective optimization technique to obtain optimum weld parameters that produce minimum welding-induced distortion and residual stresses in the welded structures. This is the first instance of the application of the BOA to a multi-objective welding problem.

3. Methods

The research methods used in this study and their connection with each other are shown in Figure 1. First, a welding simulation was performed using temperature dependent material properties on a 2 mm thick ASTM A36 steel plate butt joint. The finite element analysis of the welding process requires the exact amount of heat to be applied on the base plate to obtain accurate results. For this purpose, the heat source needs to be modeled carefully. In the next step, welding process parameters that affect welding induced distortion and residual stress were identified. These parameters need to be modified to obtain the optimum values of welding induced distortion and residual stress. To observe the results of varying process parameters, a number of welding simulations are required. The DOE approach is used to minimize the number of experiments. Response surfaces and contour plots are generated as an outcome of DOE application. In addition, regression analysis is performed and the responses (i.e., the welding induced distortion and residual stress) are formulated in the form of equations. These equations and the contour plots are used to run the final optimization using BOA. The two equations serve as objective functions while the solution domain in BOA is defined by the overlaid contour plots. To validate the results of this multi-objective optimization, a welding simulation was performed using the optimum welding process parameters. Finally, a test sample was prepared to compare the results of the welding simulation with the actual values. The methods outlined above are further explained in the following subsections.

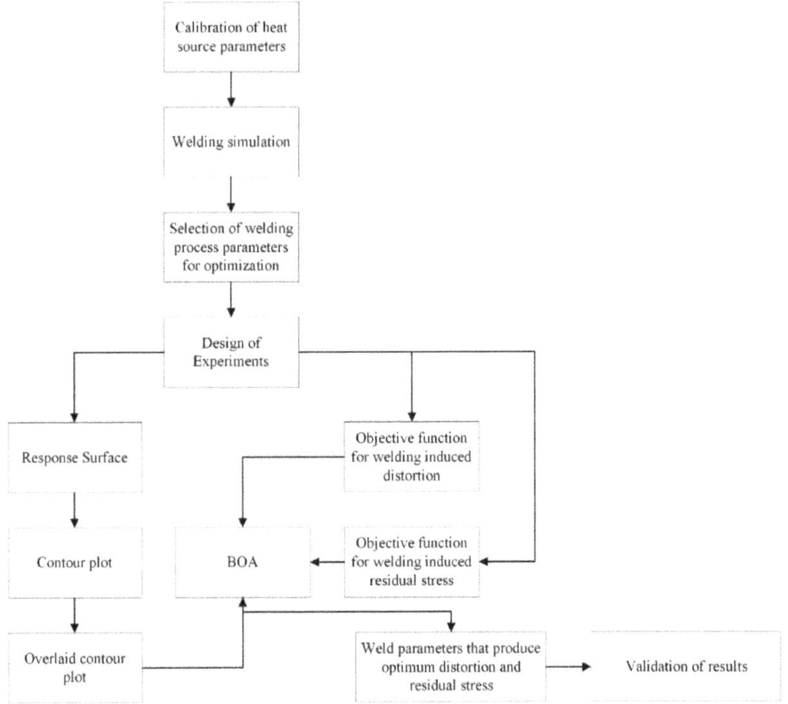

Figure 1. The framework of the multi-objective optimization study.

3.1. Welding Simulation

In this work, 2 mm thick ASTM A36 steel plates in a butt joint configuration were considered. To simulate the welding process, an FEM model of the base metal and the weld metal were prepared. The mesh of the FEM model is shown in Figure 2. The base material and filler material for welding process was the same (i.e., ASTM A36). The chemical composition of the base and filler material were taken from the work of Sajid and Kiran [31], as given in Table 1.

Figure 2. The FE model of the butt joint.

Table 1. The chemical composition of ASTM A36.

Carbon (C)	Manganese (Mn)	Phosphorous (P)	Sulfur (S)	Silicon (Si)	Copper (Cu)	Chromium (Cr)	Nickle (Ni)	Molybdenum (Mo)	Vanadium (V)	Titanium (Ti)	Niobium (Nb)	Iron (Fe)
0.15	0.69	0.018	0.004	0.18	0.24	0.15	0.088	0.0195	0.0048	0.0012	0.0024	98.4521

A couple-field thermomechanical analysis was used to simulate the welding process. The welding simulation was performed on ANSYS. A moving heat source was developed through a user sub-routine to provide heat input to the weld area. The heat was provided through a double ellipsoidal heat source as proposed by Goldak et al. [32]. The value set for the double ellipsoidal heat source parameters is given in Table 2. In thin steel plates, the change in the color of the steel due to the temperature rise is visible after the welding process. This phenomenon helps in adjusting the parameters of the double ellipsoidal heat source to match the temperature profile of the base material. The temperature profile during weld simulation is shown in Figure 3.

Table 2. The double ellipsoidal heat source parameters.

Double Ellipsoidal Heat Source Parameter	Length (mm)	
a_f	2	
a_r	6	
b	3	
c	2	

The density of elements in the heat-affected zone was increased by partitioning and mapped meshing. The mesh shown in Figure 2 was hex-dominant with couple-field elements to take up heat and stress simultaneously. The numerical setting included the application of load and boundary conditions. In the welding simulation, load is the amount of heat supplied through the welding torch. To apply thermal load at the element level, a new coordinate system was defined at the starting point of the welding process. This coordinate system was moved in each time step of analysis. The coordinates of each element in the double ellipsoid region were found through a user subroutine. Heat was applied to each element in this region. As the coordinate system moved forward in the next analysis step, heat was applied to new elements. The movement of the coordinate system was linked to the welding speed. The time step during the welding process was 0.5 s. After welding, the cooling phase started. In the cooling phase, the time step-size was increased exponentially to decrease the computational cost. To obtain the distortion and residual stress, a multilinear isotropic model was used. The stresses at the melting point temperature were ignored to avoid excessive plastic strains. These plastic strains pose difficulties in the convergence of solutions; however, their impact on the accuracy of results was negligible.

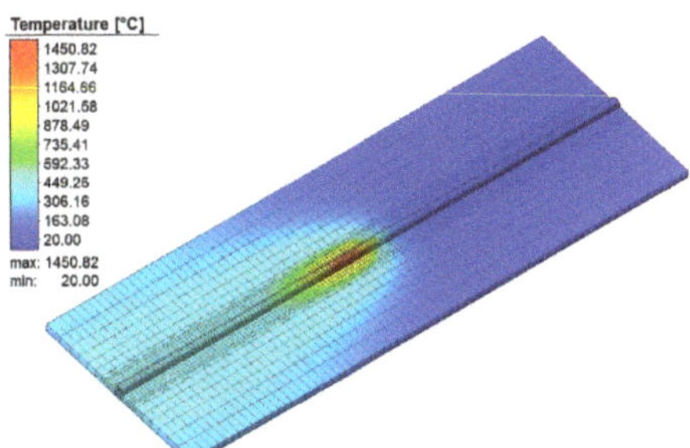

Figure 3. The temperature profile during the weld simulation.

Welding induced residual stress and distortion were caused by shrinkage forces generated as a result of nonuniform cooling and heating cycles. The heat generated in the welding process has a direct bearing on these shrinkage forces. The heat generated per unit length is given in Equation (1).

$$Welding\ heat\ per\ unit\ length = \frac{Current \times Voltage}{Welding\ speed} \tag{1}$$

It is evident from Equation (1) that the current, voltage, and welding speed are the three parameters that contribute to heat input in the welding process, and are therefore intimately linked with the phenomena of residual stress and distortion. Therefore, the input parameters in the present work were the welding current, voltage, and speed. All other factors were kept constant in the thermomechanical analysis.

3.2. Response Surface

The response surface method is a useful tool to observe the response of a system to multiple influencing factors. It is a graphical tool to represent the response of two factors at a time in the 3D plot. If three factors are under study, then three response surfaces are required to represent their effect. In optimization problems, the objective typically is to find the maximum or minimum of a function. If a graph of objective function is plotted, then the trend of that function can be observed. The expected response $E(y)$ is represented by Equation (2).

$$E(y) = f(x_1, x_2) = \eta \tag{2}$$

In Equation (2), η is the response surface for factors x_1 and x_2. In the design of an experiment approach, the most important goal is to find the outcome or response in a limited number of experiments. As the welding simulations require time and computational cost, they necessitate the use of the design of experiments. The experiments should be designed in such a manner that each contributing factor should have an equal representation. The experiments are arranged in a matrix or table. In designing experiments for the simulation of the welding process, the extreme values of a factor must not be combined in a single experiment; for example, the highest values of current, voltage, and the fastest welding speed must not be combined because that leads to unfeasible solutions. To avoid such unfeasible solutions, in the present work, the design of experiments table was prepared using the Box–Behnken design [33] with three factors, as listed in Table 3. It can be observed from Table 3 that each experiment has at least one mid-range value; for example, in experiment number 4, the welding current and voltage were 80 A and 13 V respectively, which corresponded to their higher value, while the welding speed was 5 mm/s, which was a mid-range value. The responses were the welding-induced residual stress and the distortion produced during welding. The two responses (i.e., residual stress and distortion) were recorded simultaneously during each run of the experiment (i.e., the thermomechanical simulation). The residual stress and distortion were measured on the weld axis.

Table 3. The DOE matrix.

Sr #	Current (A) X_1	Voltage (V) X_2	Welding Speed (mm/s) X_3
1	70	11	5
2	70	13	5
3	80	11	5
4	80	13	5
5	70	12	4
6	70	12	6
7	80	12	4
8	80	12	6
9	75	11	4
10	75	11	6
11	75	13	4
12	75	13	6
13	75	12	5
14	75	12	5
15	75	12	5

The last three experiments (i.e., #13, #14, and #15) had identical weld inputs consisting of central values for each parameter. Experiments based on these inputs are called the central run in the Box–Behnken design. The values of the parameters for the central run were obtained using the average of the respective parameter values.

The overall distortion observed during the simulation of experimental run #3 is shown in Figure 4. In the figure, the overall distortion of the base metal is shown. This type of distortion is classified as bowing or buckling of the plate. In the butt joint configuration, buckling with a slight amount of twisting or angular distortion was observed. The plate was deformed in the negative z-direction (i.e., along with the thickness).

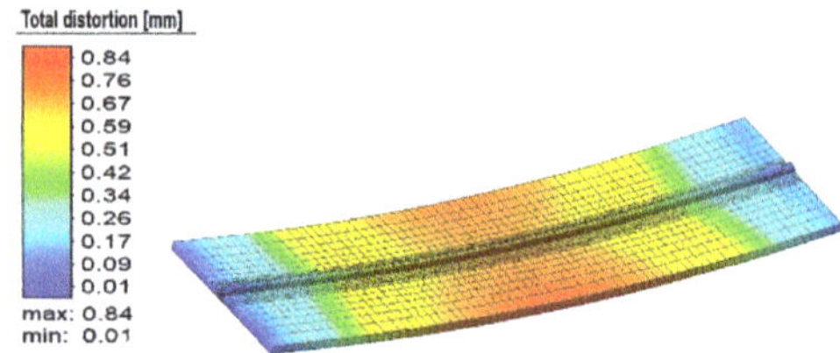

Figure 4. The total distortion in experiment number 3.

In Figure 5, the welding-induced residual stress of experiment run #10 is shown. The residual stress was tensile along with the weld bead. There was compressive residual stress away from the weld centerline.

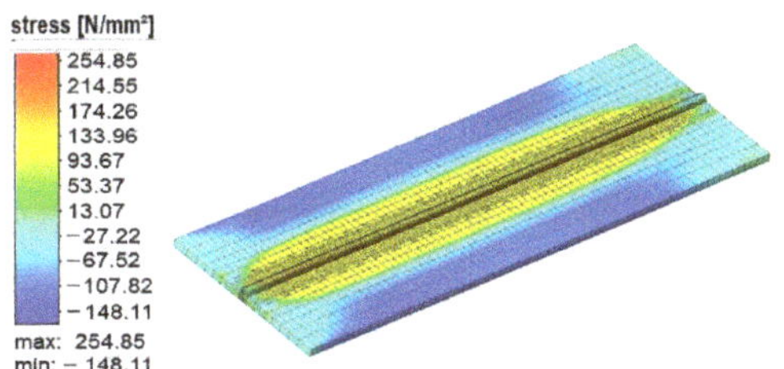

Figure 5. The welding induced residual stress.

Through the FEM simulations, the welding residual stress and distortion for all of the experimental runs were similarly calculated. In the next step, the data collected were used to run a regression analysis. Regression analysis was used to find the relation between the weld parameters and their effect on the respective responses (i.e., welding distortion and residual stress). The total welding distortion R_1 (in mm) and the maximum tensile residual stress R_2 (in MPa) are expressed in terms of welding current (X_1), voltage (X_2), and welding speed (X_3) in Equations (3) and (4), respectively.

$$R_1 = 32.75 - 1.197X_1 + 4.486X_2 - 6.035X_3 + 0.006160X_1{}^2 - 0.1835X_2{}^2 + 0.1690X_3{}^2 + 0.00050X_1X_2 + 0.058X_1X_3 - 0.0075X_2X_3 \qquad (3)$$

$$R_2 = 1580 - 33.28X_1 - 4.7X_2 - 16.5X_3 + 0.1987X_1{}^2 + 0.22X_2{}^2 + 0.97X_3{}^2 + 0.15X_1X_2 + 0.3X_1X_3 - 1.75X_2X_3 \qquad (4)$$

Equations (3) and (4) show the contribution of each factor and the combination of their products. A full quadratic analysis was run, which is evident from the quadratic terms in both the equations. The effect of an individual factor on the response variable can be observed in the surface plots. The surface plots of distortion against the three influencing factors are shown in Figure 6.

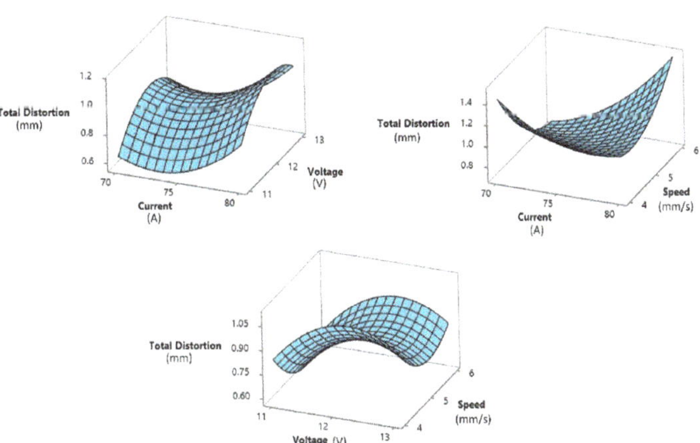

Figure 6. The surface plot of the distortion vs. welding current, voltage, and speed.

The surface plot of distortion vs. the welding current and voltage was a saddle-type plot. In this surface plot, the total distortion decreased at the midpoint or 75 A current, but in terms of voltage, the distortion increased at the midpoint voltage of 12 V. In the surface plot of distortion vs. voltage and speed, the same behavior of voltage was observed. The distortion first increased at low voltage; it then increased at the midpoint voltage and again decreased at high voltage. In the same surface plot, the behavior of welding speed could be observed. At low- and high-welding speeds, the distortion value increased, with a minimum distortion value at the midpoint of 5 mm/s.

The surface plots of residual stress values for the different weld parameters are shown in Figure 7. Welding speed had a different impact on the residual stress. The residual stress decreased with increasing welding speed. In terms of voltage, a lower value of residual stress was observed at minimum voltage. In terms of current, it was the midpoint value of 75 A that produced the minimum residual stress.

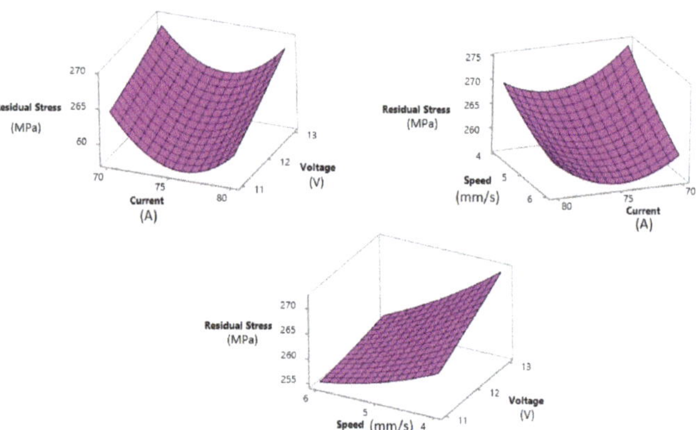

Figure 7. The surface plot of the distortion vs. welding current, voltage, and speed.

3.3. Overlaid Contour Plots

Contour plots are the 2D graphical representation of the response or interest to influencing factors on the *x*- and *y*-axis of the plot. In the case of multi-objective optimization studies, two contour plots can be overlaid to find optimum regions for both objective functions. It is a visual technique [34] to find the optimum regions for two or more objectives, but with only two influencing factors. If the number of factors increases to three, the overlaid plots become 3D and hence very difficult to study. An overlaid contour plot for the effect of welding current and voltage on the distortion and residual stress is shown in Figure 8.

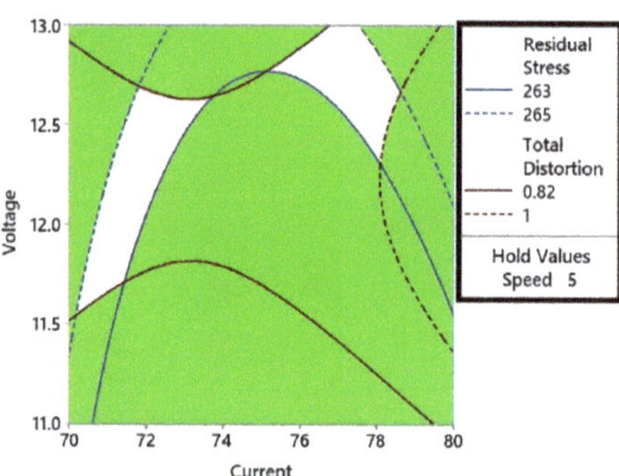

Figure 8. The overlaid contour plot of thee total distortion and residual stress with a speed of 5 mm/s.

In Figure 8, the residual stress was plotted with an upper limit of 265 MPa and a lower limit of 263 MPa. These values of residual stress represent the minimum residual stress range that intercepts the total distortion range on the welding current and voltage graph. The total distortion values intercepted on the graph were 0.82 mm and 1 mm. It can be observed in the contour plot that the lower value of distortion (0.82 mm) intercepted the lower value of residual stress (263 MPa) at three points. In the left feasible region shown in white, the minimum distortion value (0.82 mm) corresponded to a current value of 73 A.

This current value was kept constant and a contour plot was constructed for the welding voltage and speed as shown in Figure 9. These current and voltage values produced the optimum residual stress and distortion in the welded base metal. The advantage of using contour plots is that a range of optimum values for the current, voltage, and welding speed can be obtained. In the surface plots of residual stresses presented in Figure 7, it can be observed that the higher the welding speed, the lower will be the residual stresses produced. To check the optimum range of weld parameters with a combination of higher welding speed, another overlaid contour plot was constructed with the welding speed kept constant at 6 mm/s. This contour plot is shown in Figure 10.

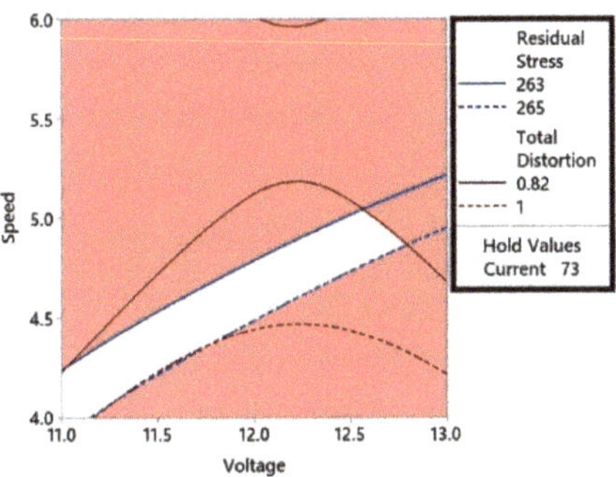

Figure 9. The overlaid contour plot of the total distortion and residual stress with the current at 73 A.

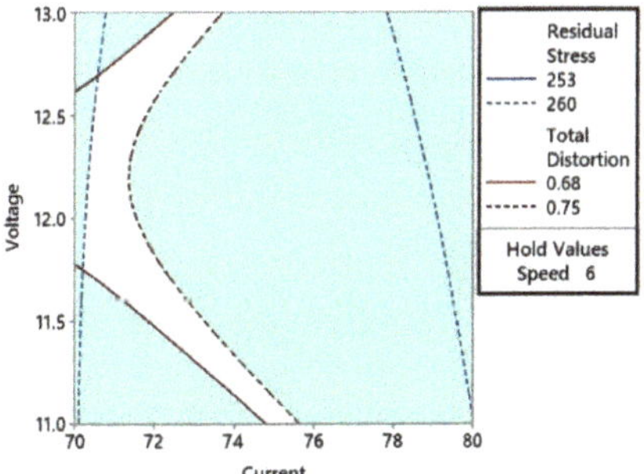

Figure 10. The overlaid contour plot of the total distortion and residual stress with a welding speed of 6 mm/s.

At higher welding speed values, new curve intercepts of welding-induced distortion and residual stress were found. The welding distortion was reduced to 0.68 mm in comparison to the previous minimum value of 0.82 mm. The new optimum value of residual stress was 260 MPa, which is lower than the previous value of 263 MPa. In Figure 10, there was a narrow band of the feasible region. This region corresponded to a range of total distortion (0.68–0.75 mm) and residual stress (253–260 MPa). Several combinations for the welding current and voltage can be selected by picking a point in the optimum region. Starting from the bottom of the contour plot in the middle of the optimum region, the current of 75 A corresponded to a voltage of 11 V. The first feasible set of weld parameters will therefore be 75 A and 11 V at a constant speed of 6 mm/s. A combination of 12 V with 71 A was found in the middle of the *y*-axis. A third combination was found at the top of the contour plot, as shown in Figure 10. This corresponded to the maximum voltage (i.e., 13 V with 73 A of current). It could also be observed from Figure 10 that the feasible region covered the entire range of voltage (from 11–13 V). In the case of current, the feasible range was between 71 A and 75 A.

3.4. The Butterfly Optimization Algorithm

The BOA is based on the food searching strategy of butterflies. Capable of differentiating between different fragrances and their intensities, butterflies search for their food source and mating partner by generating and detecting the smell in the air. When a butterfly senses food, it generates a fragrance that can be sensed by other butterflies in the vicinity. In BOA, the intensity of the fragrance is termed as the fitness value of the objective function. The coordinated movement of butterflies toward food is the global search. The butterflies will move randomly if they do not sense any fragrance. This step is defined as the local search. The three phases of the BOA are the initialization phase, the iteration phase, and the final phase. In the initialization phase, the objective function constraints and solution space are defined. In the present work, the solution space has been determined through overlaid contour plots. The initial population of butterflies is also set in the initial phase. The iteration phase is a combination of local and global search. The position of the butterflies changes during this phase. The movement of butterflies is controlled by the number of iterations set during this phase. To switch between the local and global search, a probability p. is defined. In the final phase, a stopping criterion is defined based on the number of iterations. The perceived magnitude of fragrance described by Arora and Singh [29] is presented in Equation (5).

$$f = cI^a \tag{5}$$

where f is the perceived fragrance; I is the stimulus intensity; c is the sensory modality; and a is the power exponent. I is dependent on the objective function. Its value is related to the amount of stimulus generated by a butterfly due to its position in the solution space. How this stimulus is perceived by other butterflies in the solution space is described by f. In the actual scenario, there are many physical hindrances such as the direction of air and obstacles between butterflies. These hindrances are interpreted through c and a. In the present work, the variables and their values used for the butterfly optimization were as follows:

Population size: $n = 30$
Sensory modality: $c = 0.01$
Power exponent: $a = 0.1$
Probability: $p = 0.8$
Number of iterations: $N = 100$

The variable I here is a combination of R_1 (welding distortion) calculated through Equation (3) and R_2 (residual stress) calculated through Equation (4). To rationalize the R_2 values, they are divided by 1000. Additionally, the minimization of welding distortion is given a weight of 60% while the minimization of residual stress is given a weight of 40%. The value of I is calculated for each butterfly within the solution space. The best solution is selected from all of the available solutions. In the next step, all butterflies move toward the best solution. This movement is initiated after a probability check. For a probability greater than the given value, a global search is initiated; otherwise, a local search is continued. The stopping criterion for the optimization algorithm is the number of iterations, which was set at 100.

4. Results

The optimum solution obtained from the BOA is given in Table 4.

Table 4. The optimum solution.

	Current (A)	Voltage (V)	Total Distortion (mm)	Residual Stress (MPa)
Optimum Solution	70.2	11.75	0.68	260

A welding simulation was performed using these optimum weld parameters to verify that the residual stress and welding-induced distortion fell inside the optimum range. Figure 11 shows the total distortion produced in the optimum solution. The residual stresses produced with the same set of welding parameters are shown in Figure 12.

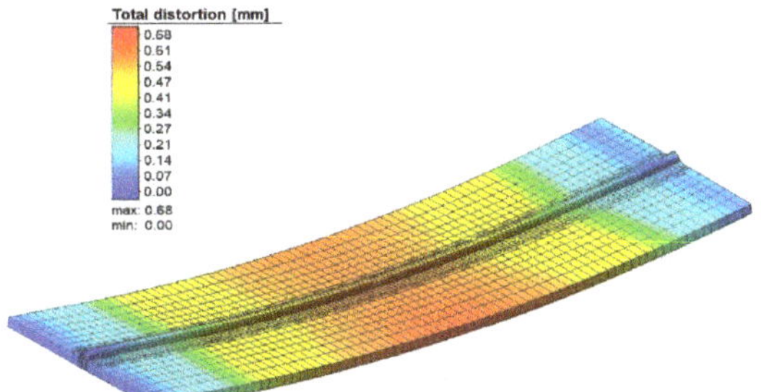

Figure 11. The total distortion at the optimum point.

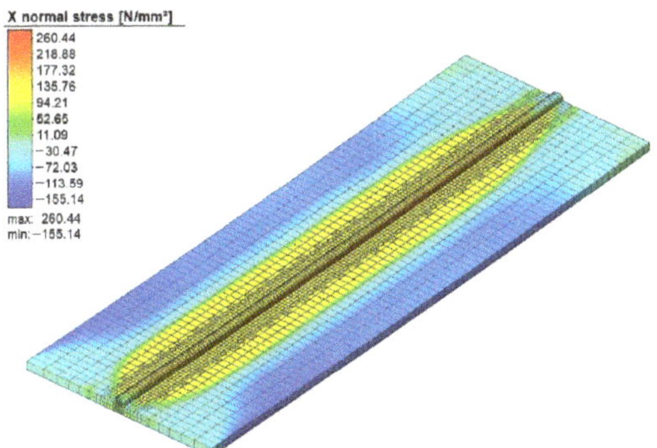

Figure 12. The residual stress at the optimum point.

To validate the results obtained by the welding simulation, a test sample was prepared by setting the weld parameters of the optimum point of Table 4. The weld sample is shown in Figure 13. The welding-induced distortion was measured along the weld line at 15 equally spaced points. A comparison of the actual and simulated distortion is shown in Figure 14.

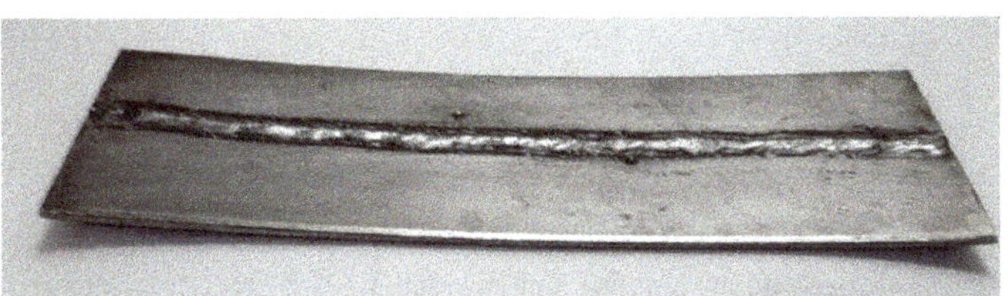

Figure 13. The weld sample for validation.

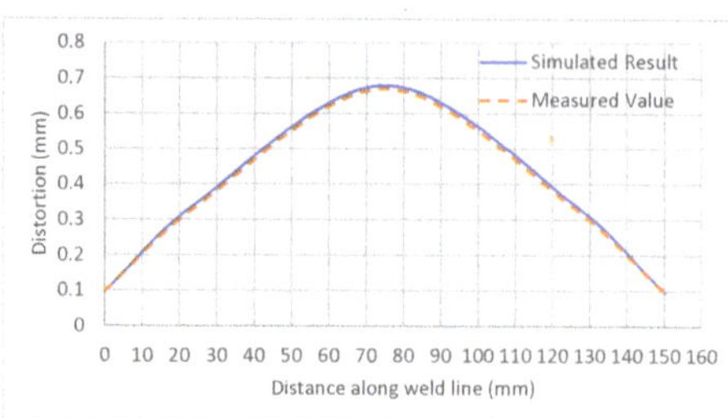

Figure 14. The validation of the distortion results at the optimum point.

In Table 5, the minimum, maximum, and optimum values of both quantities (i.e., distortion and residual stress) are shown.

Table 5. A summary of the results.

	Minimum		Optimum		Maximum	
	Residual Stress 258 MPa	Distortion 0.6 mm	Residual Stress 260 MPa	Distortion 0.68 mm	Residual Stress 275 MPa	Distortion 1.4 mm
Current (A)	76	73	70.2	70.2	70	80
Voltage (V)	11	11	11.75	11.75	13	12.25

It can be concluded from Table 5 that the minimum values of both the residual stress and distortion cannot be achieved at the same values of the welding current and voltage. To obtain the optimum values for both, the welding current was decreased while the voltage was increased to the values shown under 'Optimum'. The resulting welding-induced distortion and residual stress were plotted against the welding process parameters in Figure 15.

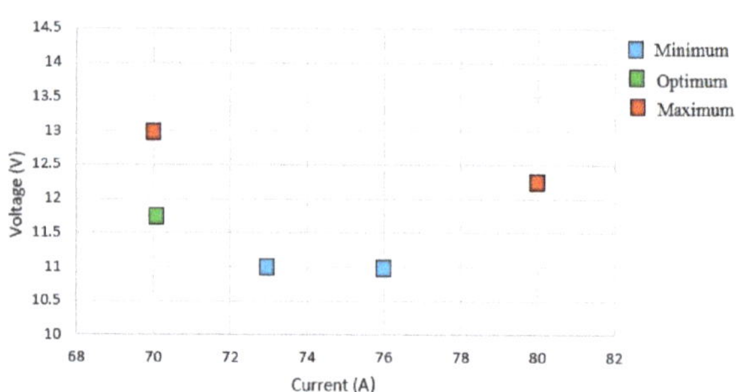

Figure 15. The minimum, optimum, and maximum values of the residual stress and distortion.

In Figure 15, the blue dots show the minimum values of the residual stress and welding-induced distortion. The values can be verified from Table 5. It should be noted that the minimum values of residual stress and distortion were realized at different values of the welding parameters. The left blue dot represents the welding distortion while the right blue dot represents the residual stress. In a similar manner, red dots show the other extreme (i.e., the maximum values). The green dot represents the optimum values that coincide with each other because they were realized at the same settings of voltage and current.

5. Conclusions

A multi-objective study was performed to simultaneously optimize the welding-induced distortion and residual stresses in thin metal plates for a butt joint-weld configuration. The optimum weld parameters to minimize welding-induced distortion and residual stresses were selected by using the response surface method. The response surfaces were utilized to generate a series of contour plots. The contour plots for the minimum distortion and residual stress were overlaid to determine an optimum region. This region contained the values of the optimum weld parameters within a range having upper and lower bounds.

The butterfly optimization algorithm (BOA) was applied to obtain the optimum value of the weld parameters from this solution space. These weld parameters provided the optimum values for the welding-induced distortion and residual stress. In this study, the BOA, as a multi-objective optimization technique, was successfully applied for the first time to a welding problem.

Author Contributions: R.W.: Research methodology, FEM modeling and welding simulation, programming of the butterfly optimization. H.A.S.: Selection of the welding parameters, design of the experiments, manuscript preparation and editing. B.A.A.: Conducted the experiments, surface, and contour plots for the weld parameters. All authors have read and agreed to the published version of the manuscript.

Funding: This research received no external funding.

Conflicts of Interest: The authors declare no conflict of interest.

References

1. Prasada, K.; Rao, C.; Rao, D. Optimizing pulsed current micro plasma arc welding parameters to maximize ultimate tensile strength of Inconel625 Nickel alloy using response surface method. *Int. J. Eng. Sci. Technol.* **2012**, *3*, 226–236. [CrossRef]
2. Srivastava, S.; Garg, R. Process parameter optimization of gas metal arc welding on IS:2062 mild steel using response surface methodology. *J. Manuf. Process.* **2017**, *25*, 296–305. [CrossRef]
3. Vasantharaja, P.; Vasudevan, M. Optimization of A-TIG welding process parameters for RAFM steel using response surface methodology. *Proc. Inst. Mech. Eng. Part L J. Mater. Des. Appl.* **2015**, *232*, 121–136. [CrossRef]
4. Vidyarthy, R.S.; Dwivedi, D.K.; Muthukumaran, V. Optimization of A-TIG process parameters using response surface methodology. *Mater. Manuf. Process.* **2017**, *33*, 709–717. [CrossRef]
5. Lai, X.M.; Luo, A.H.; Zhang, Y.S.; Chen, G.L. Optimal design of electrode cooling system for resistance spot welding with the response surface method. *Int. J. Adv. Manuf. Technol.* **2008**, *41*, 226–233. [CrossRef]
6. Korra, N.N.; Vasudevan, M.; Balasubramanian, K.R. Multi-objective optimization of activated tungsten inert gas welding of duplex stainless steel using response surface methodology. *Int. J. Adv. Manuf. Technol.* **2014**, *77*, 67–81. [CrossRef]
7. Joseph, J.; Muthukumaran, S. Optimization of activated TIG welding parameters for improving weld joint strength of AISI 4135 PM steel by genetic algorithm and simulated annealing. *Int. J. Adv. Manuf. Technol.* **2015**, *93*, 23–34. [CrossRef]
8. Waheed, R.; Saeed, H.A.; Butt, S.U.; Anjum, B. Framework for Mitigation of Welding Induced Distortion through Response Surface Method and Reinforcement Learning. *Coatings* **2021**, *11*, 1227. [CrossRef]
9. Gunaraj, V.; Murugan, N. Application of response surface methodology for predicting weld bead quality in submerged arc welding of pipes. *J. Mater. Process. Technol.* **1999**, *88*, 266–275. [CrossRef]
10. Mochizuki, M. Control of welding residual stress for ensuring integrity against fatigue and stress–corrosion cracking. *Nucl. Eng. Des.* **2007**, *237*, 107–123. [CrossRef]
11. Lee, C.H.; Kyong, H.C. Finite Element Computation of Fatigue Growth Rates for Mode I Cracks Subjected to Welding Residual Stresses. *Eng. Fract. Mech.* **2011**, *78*, 2505–2520. [CrossRef]
12. Hensel, J.; Nitschke-Pagel, T.; Ngoula, D.T.; Beier, H.-T.; Tchuindjang, D.; Zerbst, U. Welding residual stresses as needed for the prediction of fatigue crack propagation and fatigue strength. *Eng. Fract. Mech.* **2018**, *198*, 123–141. [CrossRef]
13. Farajian, M. Welding residual stress behavior under mechanical loading. *Weld. World* **2013**, *57*, 157–169. [CrossRef]
14. Cui, C.; Zhang, Q.; Luo, Y.; Hao, H.; Li, J. Fatigue reliability evaluation of deck-to-rib welded joints in OSD considering stochastic traffic load and welding residual stress. *Int. J. Fatigue* **2018**, *111*, 151–160. [CrossRef]
15. Cheng, X. Residual stress modification by post-weld treatment and its beneficial effect on fatigue strength of welded structures. *Int. J. Fatigue* **2003**, *25*, 1259–1269. [CrossRef]
16. Barsoum, Z. Residual Stress Effects on Fatigue Life of Welded Structures Using LEFM. *Powder Diffr.* **2009**, *16*, 449–467. [CrossRef]
17. Barsoum, Z. Residual stress analysis and fatigue of multi-pass welded tubular structures. *Eng. Fail. Anal.* **2008**, *15*, 863–874. [CrossRef]
18. Caruso, S.; Imbrogno, S. Finite element modelling and experimental validation of microstructural changes and hardness variation during gas metal arc welding of AISI 441 ferritic stainless steel. *Int. J. Adv. Manuf. Technol.* **2022**, *119*, 2629–2637. [CrossRef]
19. Murat, T.; Ozler, K. Numerical modeling of defect formation in friction stir welding. *Mater. Today Commun.* **2022**, *31*, 103539.

20. Moslemi, N.; Gohari, S.; Abdi, B.; Sudin, I.; Ghandvar, H.; Redzuan, N.; Hassan, S.; Ayob, A.; Rhee, S. A novel systematic numerical approach on determination of heat source parameters in welding process. *J. Mater. Res. Technol.* **2022**, *18*, 4427–4444. [CrossRef]
21. Zhang, J.; Dong, P. Residual Stresses in Welded Moment Frames and Implications for Structural Performance. *J. Struct. Eng.* **2000**, *126*, 306–315. [CrossRef]
22. Narwadkar, A.; Bhosle, S. Optimization of MIG Welding Parameters to Control the Angular Distortion in Fe410WA Steel. *Mater. Manuf. Process.* **2015**, *31*, 2158–2164. [CrossRef]
23. Zhang, W.; Jiang, W.; Zhao, X.; Tu, S.T. Analysis of the Effect of Tungsten Inert Gas Welding Sequences on Residual Stress and Distortion of Cfetr Vacuum Vessel Using Finite Element Simulations. *Metals* **2018**, *8*, 912. [CrossRef]
24. Lorza, R.L.; García, R.E.; Martinez, R.F.; Calvo, M.; Ángeles, M. Using Genetic Algorithms with Multi-Objective Optimization to Adjust Finite Element Models of Welded Joints. *Metals* **2018**, *8*, 230. [CrossRef]
25. Chen, Z.; Chen, Z.; Shenoi, R.A. Influence of Welding Sequence on Welding Deformation and Residual Stress of a Stiffened Plate Structure. *Ocean. Eng.* **2015**, *106*, 271–280. [CrossRef]
26. Rong, Y.; Huang, Y.; Xu, J.; Zheng, H.; Zhang, G. Numerical simulation and experiment analysis of angular distortion and residual stress in hybrid laser-magnetic welding. *J. Mater. Process. Technol.* **2017**, *245*, 270–277. [CrossRef]
27. Romero, J.; Gengis, T.-R.; Baidya, S. Deformation and Residual Stress Based Multi-Objective Genetic Algorithm for Welding Sequence Optimization. *Res. Comput. Sci.* **2017**, *132*, 155–179. [CrossRef]
28. Shao, Q.; Xu, T.; Yoshino, T.; Song, N. Multi-objective optimization of gas metal arc welding parameters and sequences for low-carbon steel (Q345D) T-joints. *J. Iron Steel Res. Int.* **2017**, *24*, 544–555. [CrossRef]
29. Arora, S.; Singh, S. Butterfly optimization algorithm: A novel approach for global optimization. *Soft Comput.* **2018**, *23*, 715–734. [CrossRef]
30. Yıldız, B.S.; Yıldız, A.R.; Albak, E.İ.; Abderazek, H.; Sait, S.M.; Bureerat, S. Butterfly optimization algorithm for optimum shape design of automobile suspension components. *Mater. Test.* **2020**, *62*, 365–370. [CrossRef]
31. Sajid, H.U.; Kiran, R. Influence of high stress triaxiality on mechanical strength of ASTM A36, ASTM A572 and ASTM A992 steels. *Constr. Build. Mater.* **2018**, *176*, 129–134. [CrossRef]
32. Goldak, J.; Chakravarti, A.; Bibby, M. A new finite element model for welding heat sources. *Metall. Trans. B* **1984**, *15*, 299–305. [CrossRef]
33. Box, G.E.; Behnken, D.W. Some new three level designs for the study of quantitative variables. *Technometrics* **1960**, *2*, 455–475. [CrossRef]
34. Montgomery, D.C. *Design and Analysis of Experiments*, 8th ed.; John Wiley and Sons: Hoboken, NJ, USA, 2013; pp. 478–544.

Article

Ultrasonic Welding of PEEK Plates with CF Fabric Reinforcement—The Optimization of the Process by Neural Network Simulation

Vladislav O. Alexenko [1], Sergey V. Panin [1,2,*], Dmitry Yu. Stepanov [1], Anton V. Byakov [1], Alexey A. Bogdanov [1,2], Dmitry G. Buslovich [3], Konstantin S. Panin [4] and Defang Tian [2]

1　Laboratory of Mechanics of Polymer Composite Materials, Institute of Strength Physics and Materials Science of Siberian Branch of Russian Academy of Sciences, 634055 Tomsk, Russia
2　Department of Materials Science, Engineering School of Advanced Manufacturing Technologies, National Research Tomsk Polytechnic University, 634050 Tomsk, Russia
3　Laboratory of Nanobioengineering, Institute of Strength Physics and Materials Science of Siberian Branch of Russian Academy of Sciences, 634055 Tomsk, Russia
4　Department of Chemical Physics, Institute for Laser and Plasma Technologies, National Research Nuclear University MEPhI, 115409 Moscow, Russia
*　Correspondence: svp@ispms.ru

Abstract: The optimal mode for ultrasonic welding (USW) of the "PEEK–ED (PEEK)–prepreg (PEI impregnated CF fabric)–ED (PEEK)–PEEK" lap joint was determined by artificial neural network (ANN) simulation, based on the sample of the experimental data expanded with the expert data set. The experimental verification of the simulation results showed that mode 10 (t = 900 ms, P = 1.7 atm, τ = 2000 ms) ensured the high strength properties and preservation of the structural integrity of the carbon fiber fabric (CFF). Additionally, it showed that the "PEEK–CFF prepreg–PEEK" USW lap joint could be fabricated by the "multi-spot" USW method with the optimal mode 10, which can resist the load per cycle of 50 MPa (the bottom HCF level). The USW mode, determined by ANN simulation for the neat PEEK adherends, did not provide joining both particulate and laminated composite adherends with the CFF prepreg reinforcement. The USW lap joints could be formed when the USW durations (*t*) were significantly increased up to 1200 and 1600 ms, respectively. In this case, the elastic energy is transferred more efficiently to the welding zone through the upper adherend.

Keywords: machine learning; neural network simulation; carbon fiber fabric; ultrasonic welding; lap joint; PEEK; prepreg; interface; adhesion; structural integrity

Citation: Alexenko, V.O.; Panin, S.V.; Stepanov, D.Y.; Byakov, A.V.; Bogdanov, A.A.; Buslovich, D.G.; Panin, K.S.; Tian, D. Ultrasonic Welding of PEEK Plates with CF Fabric Reinforcement—The Optimization of the Process by Neural Network Simulation. *Materials* **2023**, *16*, 2115. https://doi.org/10.3390/ma16052115

Academic Editor: Raul D. S. G. Campilho

Received: 4 February 2023
Revised: 25 February 2023
Accepted: 2 March 2023
Published: 6 March 2023

1. Introduction

Laminated polymer composites reinforced with continuous carbon fibers (CFs), so-called laminates, are widely applied in the aerospace and other high-tech industries. Zhang et al. [1] reviewed many global industrial applications of laminated composites in aerospace, wind energy, machine building, high pressure vessels, civil engineering, etc. Typically, thermoset resins were used to fabricate such materials, but thermoplastic ones have actively replaced them nowadays [2,3]. As evidence, a concise review of out-of-autoclave prepregs is presented in [4]. In addition, some new methods to produce CF-reinforced composites, including additive manufacturing, are being actively developed [5–8].

To this end, one of the important challenges is joining the CF-reinforced composites [9–12]. Based on the thermoplastic binders, the latter is solved by ultrasonic welding (USW) [13–16].

When developing the USW procedures for laminates, the influence of various technological parameters on the structure and functional properties of such joints is investigated. The research studies on such phenomena and the issues concerning optimization of the USW parameters include examining both heat generation and its transfer [17–19], the transmission of ultrasonic (US) vibrations [20,21], applied power and sonotrode displacement [22], adding a consolidator to a welding setup [23], effects of heating and shapes of energy directors (ED) [24–26], the possibility of USW joining of thermoplastics with thermosets [27]; principles of controlling the USW process [28]; the material crystallinity in a fusion zone [29]; both distribution and evolution of temperature fields [30]; the strain behavior under various test conditions [31], etc.

Even though few knowledge gaps remain in the USW of laminates, great interest in this topic still exists among researchers. In particular, two important areas should be noted: (i) the development of USW procedures for the particulate composites rather than the layered ones [32], and (ii) the production of the prepregs [33] or the US-assisted fabrication of multilayered CFRC [34].

Concerning USW joining the CFRC, some relevant reports of the scientific groups should be pointed out [35–38]. In addition, papers [39–41] are focused on developing repairing procedures. In terms of US-assisted prepreg fabrication, the automated placement of the CFs should be considered a very promising issue to discuss [42,43]. An alternative to this method is consolidating a composite pipe by in situ USW processing of a thermoplastic matrix composite tape [44]. One of the key issues to be solved regarding the optimization of the USW parameters is to consider both the viscosity and fluidity of the thermoplastics, as well as the impregnation of the CFFs [45].

Bonmatin et al. proposed an approach to the USW of laminates that included two layers of polyetherimide (PEI) between two polyetheretherketone (PEEK)/CF adherends, in addition to the PEI-based ED [46]. Considering the development of additive manufacturing of both polymers and composites, Khatri et al. investigated the efficiency of the USW of 3D-printed parts from both neat PEEK and its CF-reinforced composites [47]. It was suggested that such USW procedures, developed for the 3D-printed high-performance thermoplastics, might be implemented for manufacturing both non- and load-bearing structures.

It should also be noted that there is a distinction between continuous and multi-spot USW methods [48] since their clamping and heat input patterns are different. For this reason, their optimal parameters may differ significantly.

Based on the above, optimization of the joint formation conditions is a vector approximation task due to the variety of possible USW modes and their numerous parameters (for example, the process duration, clamping pressure [49–51], amplitude and frequency of US vibrations, etc.) and deployed equipment. Therefore, among the most effective approaches to its solution are approaches based on the methods of ANN.

In the previous study [52], the authors developed a methodology for determining the optimal USW parameters. The methodology included the sequential solution of a set of tasks. Among them were (i) planning an experiment; (ii) determining and analysis of the experimental data, including the selection of the most suitable functional properties; (iii) the USW process simulation; (iv) searching for the optimal parameters; (v) the experimental verification of the adopted model. The Taguchi method was applied to plan the experiment and analyze its results. At the same time, the simulation algorithms were based on ANN with varying parameters (the number of neurons and the type of activation functions, as well as training methods). As part of the experimental verification of the adopted model, the samples of the USW lap joints were fabricated, containing a PEEK-based prepreg with the unidirectional continuous CFs as a central layer. It was found that damage to the unidirectional CF tapes might occur when the prepreg is melted. The results revealed the problem of the USW lap joint formation with the prepreg from the CFF impregnated via a thermoplastic solution. It has stimulated further research on improving the ANN approach for the USW process simulation.

Thereby, the aim of this study is to find out the optimal USW parameters for the lap joints with the neat PEEK adherends and the PEI-impregnated CFF prepreg as the central layer via a simulation process carried out with ANNs. The paper is structured as follows. Section 2 describes both the implemented techniques and deployed equipment. Section 3 briefly characterizes the prepreg structure. Section 4 presents the results of the mechanical testing and analysis of the lap joints' structure. Section 5 includes the simulation results aimed at determining the optimal USW parameters. Finally, Sections 6 and 7 discuss the results and their prospects.

2. Materials and Methods

The "770PF" PEEK powder (Zeepeek, Changchun, Jilin Province, China) was used as a feedstock for manufacturing plates (adherends) with dimensions of 100 mm × 20 mm × 2.2 mm using the "RR/TSMP" injection molding machine (Ray-Ran Test Equipment Ltd., Nuneaton, UK). The mold heating temperature range was 200–205 °C; the powder feeder (hopper) was heated to 395 °C.

The USW lap joints were formed with the ED from a commercially available PEEK film Aptiv 2000 250 μm thick (Victrex, Lancashire, UK) cut into 22 × 22 mm square pieces. Before the USW process, the EDs were placed between the PEEK adherends while the prepreg was located in the center of the joints (Figure 1). A description of the prepreg fabrication technique is described in Section 3. "UZPS-7" ultrasonic welding machine was explored (SpetsmashSonic LLC, Voronezh, Russia). The plates to be welded were placed in a fixing clamp to prevent their mutual movement. A sonotrode with square geometry and 20 × 20 mm dimension was used. The USW lap joints included five layers; two outer neat PEEK adherends, the central CFF prepreg layer, and two intermediate ED (Figure 1). The USW lap joint thinning was measured with a micrometer.

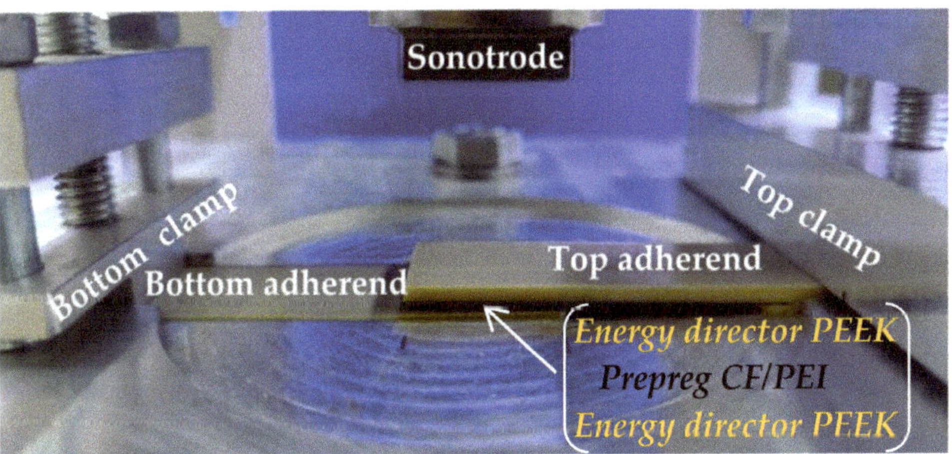

Figure 1. The USW facility.

It should be noted that the USW machine used for this study provided the possibility of varying the USW duration and the clamping pressure. However, the US oscillation amplitude was constant at 10 μm. The US vibration frequency of 20 kHz was also not varied.

In the experimental design, the following levels of the USW parameters (the simulation factors) and their possible variation ranges were preset (Table 1):

1. The USW duration (t) range was set as 600, 850, and 1150 ms since it was not possible to join the PEEK plates at lower t values (since the heat input was low to initiate the melting of the PEEK). However, the prepreg could be seriously damaged at higher t levels, resulting in highly faulty USW lap joints.
2. Ranges of the clamping pressure (P) and its (holding) duration after applying US vibrations (t) were determined based on the technical characteristics of the USW machine and by visual control of the USW lap joints.

Table 1. The combination of the USW parameters and their levels (according to the Taguchi table in L9 format for a three-factor experiment).

Mode/Experiment Number	Level/Factor		
	USW Duration (*t*), ms	**Clamping Duration after US Vibrations (*τ*), ms**	**Clamping Pressure (*P*), atm**
1	1/600	1/2000	1/1.5
2	1/600	2/5000	2/2.0
3	1/600	3/8000	3/2.5
4	2/850	1/5000	2/1.5
5	2/850	2/8000	3/2.0
6	2/850	3/2000	1/2.5
7	3/1100	1/8000	3/1.5
8	3/1100	2/2000	1/2.0
9	3/1100	3/5000	2/2.5

The tensile tests of the USW lap joints were performed according to ASTM D5868. The tests were carried out with an "Instron 5582" electromechanical tensile testing machine (Instron, Norwood, MA, USA). The cross-head speed was 13 mm/min. To minimize misalignment during the tests, gaskets made of identical-thickness PEEK plates were installed in the wedge grips.

The structure of cross-sections of the USW lap joints was analyzed with "Neophot 2" optical microscope (Carl Zeiss, Jena, Germany) after the tensile tests.

As mentioned above, both continuous and multi-spot USW modes could be applied. For a qualitative assessment of suitability of the optimal USW parameters, the samples for the fatigue tests were fabricated by using the PEEK plates, the prepreg from the PEI-impregnated CFF, and the PEEK film EDs (like the data described below in Section 4). All welded elements had the same width and length. The USW process was carried out by joining the details at three spots. Then, the "dog-bone" samples (Figure 2) were cut with a milling machine. The fatigue tests were carried out in the quasi-static axial tension mode using "Biss Nano" universal servo-hydraulic machine with "Biss Bi-06-103 15 kN" load cell (ITW India Private Limited (BISS), Bangalore, India) at a cross-head speed of 1 mm/min and a load sampling rate of 100 Hz. To characterize the strain development quantitatively, the non-contact digital image correlation (DIC) method was applied to calculate the parameters of the mechanical hysteresis loops.

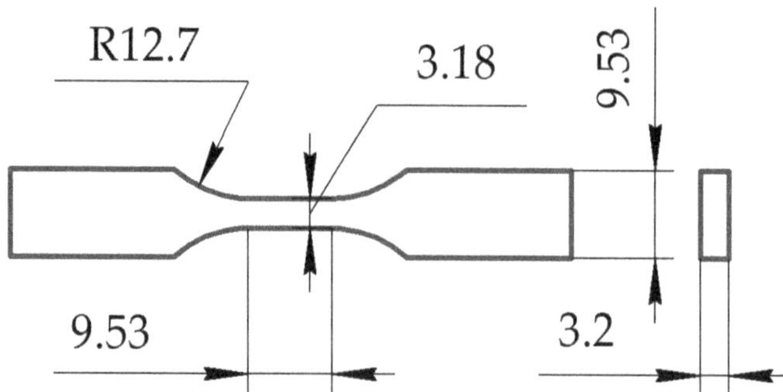

Figure 2. A scheme of the specimens for the static tension and fatigue tests (all dimensions in mm).

The strain values were sampled at a frequency of 5 Hz and then evaluated by the DIC method. For drawing the strain fields, a speckle pattern was deposited on the sample surfaces, which was captured with "Point Gray Grasshopper 50S5M" digital camera (Point Gray Research® Inc., Vancouver, BC, Canada) and "Sony® ICX625 2/3" CCD matrix

2448 × 2048 (a resolution of 5 megapixels, matrix dimensions of 8.4 × 7.0 mm, and pixel sizes of 3.45 × 3.45 μm). The required surface illumination level was provided with "Jinbei EF-100 LEDSun Light" LED illuminator (Shanghai Jinbei Photographic Equipments Co., Ltd., Shanghai, China). The analysis of the sample surface images and a drawing of the strain fields were conducted using VIC 2D 2009 software (Correlated Solutions Inc., Columbia, SC, USA).

To assess the fatigue properties of the USW lap joints, cyclic loading of the samples was carried out in the "tensile–tensile" mode with the load control, a sinusoidal shape of the loading signal, and a zero-asymmetry cycle (the R ratio of 0). The loading frequency was 1 Hz with a periodic deceleration down to 0.05 Hz (measuring cycle) for photographing the sample surfaces with the DIC. For implementing the high-cycle fatigue (HCF) mode, the maximum stress per cycle σ_{max} was chosen, ~50 MPa, which was below the yield point ($0.8 \cdot \sigma_{0.2}$). For the low-cycle fatigue (LCF) mode, a σ_{max} value of 55 MPa was preset, which was 0.85 of the ultimate tensile strength level.

3. Fabrication of the Impregnated Prepregs

Our previous study was conducted on the variation of the PEEK binder contents in a CFF-based prepreg [53]. It was shown that the molten thermoplastic in the prepreg intensively extruded from the fusion zone during the USW process, tearing the CFF if the thickness of the latter was significantly lower than that of the prepreg (at PEEK/CF weight ratios of 50/50 and above). In doing so, the prepreg thickness was to be minimized, and the entire volume of the CFF had to be impregnated with a thermoplastic binder.

Since the prepreg was fabricated under laboratory conditions, additional studies were carried out on impregnating the CFF with thermoplastic solutions without using strong acids. For this reason, the option of dissolving PEEK was not considered. However, assuming that a chemical bond was formed between the adherends and the PEEK film ED from the USW process, polyetherimide (PEI) and polyethersulfone (PES) were taken as options. First, the PEI and PES were dissolved in N,N-Dimethylformamide (C_3H_7NO). Then, the fragments of the CFF were soaked in the solution for one hour, followed by evaporation of the organic solvent on the "IKA C-MAG HP 4" heater (IKA®-Werke GmbH & Co. KG, Staufen im Breisgau, Germany). The latter was conducted under ventilation conditions to guarantee the removal of volatiles.

The Prepregs Impregnated with PES and PEI

The prepreg was fabricated as follows. In order to remove the technological (epoxy) sizing agent from the surfaces, the CFF was annealed at 500 °C for 30 min. Then, it was impregnated with the dissolved PES. Finally, the solvent was evaporated in an oven at 120 °C.

As a result, the impregnated CFF thickness was ~500 μm. After that, it was subjected to compression molding in order to reduce the thickness to that of the initial CFF of 230–250 μm. In the prepreg, the ratio of the components was 66 wt.% CFF/34 wt.% PES. Finally, the thickness of the prepreg (the CFF impregnated with the polymer) was ~250 μm (Figure 3a,b).

The prepreg from the PEI-impregnated CFF was made in a similar way. As a result, the component ratio and the final prepreg thickness were also identical (66 wt.% CFF/34 wt.% PEI and ~250 μm, respectively). Its general view from the surface and SEM micrographs of the cross-section is shown in Figure 3c,d.

Next, the prepregs were compared by their effect on the strength properties of the USW lap joints, obtained with the following parameters preset based on our previous results [52]: a USW duration of 1000 ms, a clamping pressure 1.5 atm, and a clamping duration after US vibrations of 5000 ms. The results of the tensile tests are presented in Table 2.

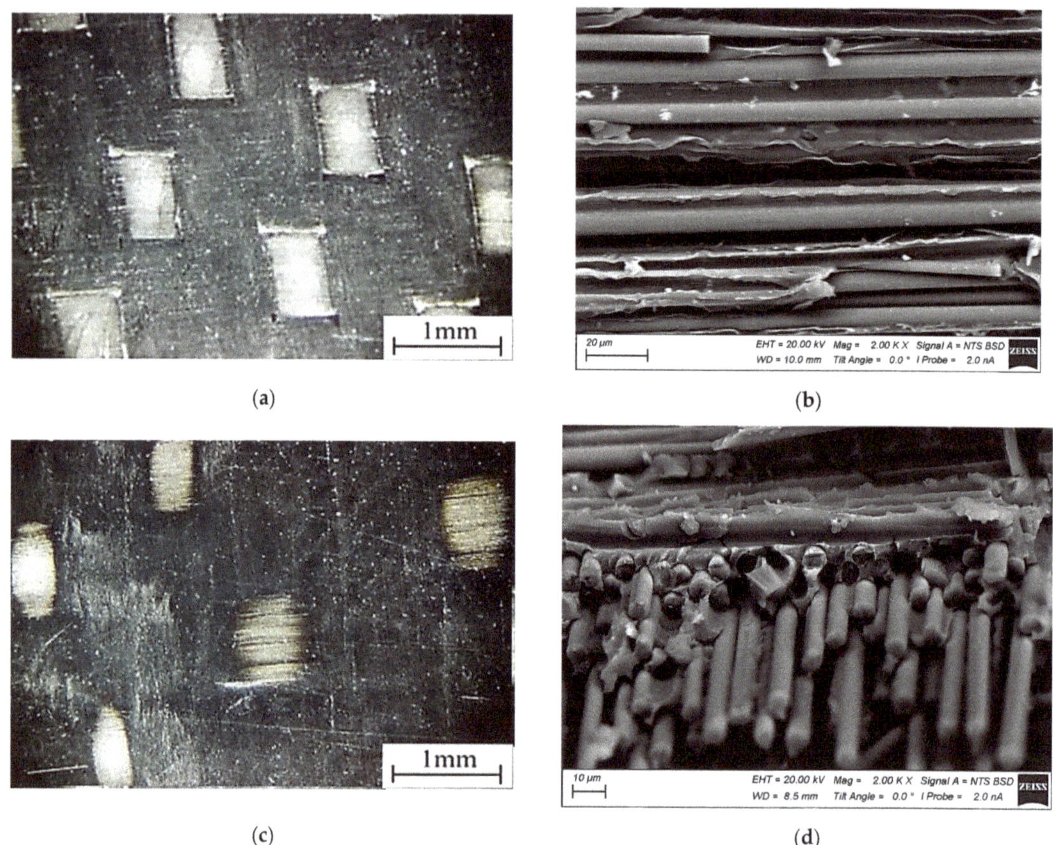

Figure 3. The general views from the surfaces (**a,c**), as well as the SEM micrographs of the cross-sections (**b,d**) of the prepregs from the annealed CFF impregnated with the PES (**a,b**) and the PEI (**c,d**).

Table 2. The mechanical and dimensional properties of the USW lap joints obtained at t = 1000 ms, P = 1.5 atm, and τ = 5000 ms.

Binder	Ultimate Tensile Strength (σ_{UTS}), MPa	Elongation at Break (ε), %	Elastic Modulus (E), MPa	USW Joint Thinning (Δh), mm
PES	40.5 ± 1.5	1.8 ± 0.2	2034 ± 110	410 ± 20
PEI	50.1 ± 1.8	2.1 ± 0.2	2175 ± 90	290 ± 15

Additionally, the USW joint thinning was evaluated with a micrometer. In the case of the PES-impregnated prepreg, it was ~410 μm, while it was reduced to ~290 μm for the PEI one (Table 2). According to the visual inspection results for the cross-sections of the USW lap joints, the polymer melting occurred both in the prepreg and partially in the EDs during the USW process. However, the PEI and PES melting points were lower than that of the PEEK. In addition, the presence of 66 % CFF increased the prepreg's integral thermal conductivity and heat capacity. In the PES case, its more intensive extrusion occurred due to the higher melt flow index, which could be accompanied by damage to the CFF. Based on these data, only PEI-impregnated prepreg was used further.

4. Mechanical Test Results

The strain–stress diagrams for the USW lap joints are shown in Figure 4, and the results of their mechanical tests are presented in Table 3.

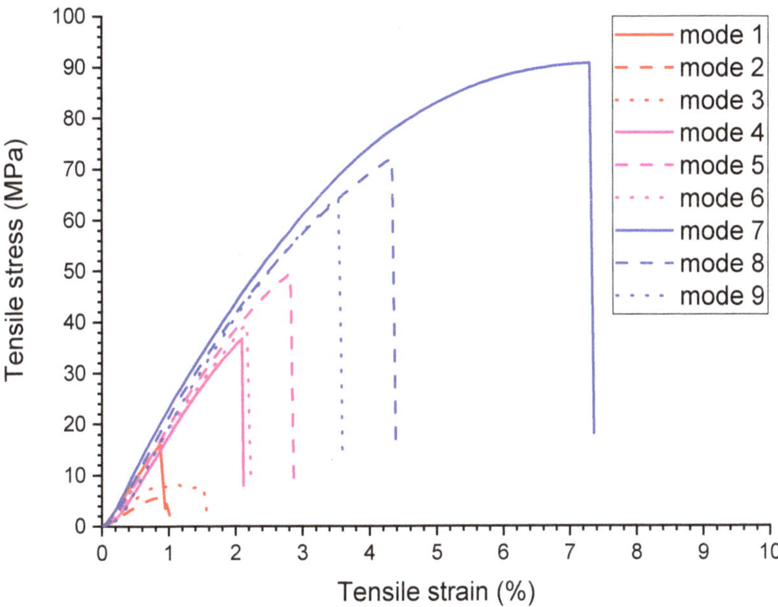

Figure 4. The strain–stress diagrams for the USW lap joints obtained using modes 1–9.

Table 3. The mechanical and dimensional characteristics of the USW lap joints for various USW parameters.

Mode Number	Ultimate Tensile Strength (σ_{UTS}), MPa	Elongation at Break (ε), %	Work of Strain (A), N·m	Elastic Modulus (E), MPa	USW Joint Thinning (Δh), mm
1	16.1 ± 1.5	0.9 ± 0.1	7.6 ± 0.5	2014 ± 110	0.09 ± 0.03
2	5.6 ± 0.4	1.0 ± 0.1	3.4 ± 0.2	740 ± 68	0.16 ± 0.05
3	8.2 ± 0.6	1.7 ± 0.2	9.1 ± 0.7	1180 ± 102	0.12 ± 0.04
4	36.7 ± 1.8	2.1 ± 0.2	38.7 ± 2.5	2017 ± 90	0.30 ± 0.03
5	49.6 ± 3.4	2.9 ± 0.2	76.1 ± 4.4	2314 ± 115	0.40 ± 0.04
6	39.8 ± 1.9	2.2 ± 0.2	45.0 ± 3.2	2230 ± 121	0.40 ± 0.03
7	90.8 ± 4.1	7.4 ± 0.4	449.8 ± 15.7	2576 ± 130	0.70 ± 0.06
8	72.1 ± 3.5	4.4 ± 0.3	180.8 ± 10.4	2511 ± 123	0.70 ± 0.06
9	64.7 ± 3.2	3.6 ± 0.3	123.5 ± 9.3	2471 ± 119	0.55 ± 0.03

Additionally, the USW joint thinning values, measured with the micrometer, are reported. The primary data processing was carried out by the Taguchi method. The results are presented in the Supplementary Materials (Tables S1–S5, Figure S1).

After the tensile tests, the failure patterns of the USW lap joints were analyzed (Figures 5 and 6). Table 4 presents distances between the joined PEEK plates and the CFF thicknesses, determined by the analysis of the optical images shown in Figure 6. These data were used to interpret the revealed patterns.

Note that the authors failed to solve the macrobending problem caused by the double thickness of the central part of the specimens [54] in the tensile tests of the USW lap joints. This issue was investigated numerically in the previous study [55]. Macrobending greatly determined the resistance of the specimens to failure and contributed to minimizing the influence of the interlayer shear mechanism. For the same reason, the ultimate tensile strength σ_{UTS} parameter was used as a strength characteristic of the lap joints rather than the lap shear strength (LSS) one. Its impact was minimized by ANN simulation and optimization of the USW parameters (Section 5).

When forming the USW lap joint using mode 1 (t = 600 ms, P = 1.5 atm; Figure 5a), the EDs did not (or partly) melt locally. This fact reflected a non-uniform distribution of the temperature field (and, accordingly, the lap joint structure). In the prepreg, the CFF was impregnated with the PEI, while the EDs were the PEEK film. Thus, their thickness decreased slightly to 180–240 μm relative to the initial thickness of 250 μm (Figure 6a). The EDs' thicknesses differed to the left and right of the prepreg, but they did not exceed 90 μm (Table 3) at σ_{UTS} = 16.1 MPa.

Figure 5. General view photos of the fractured USW lap joints after the tensile tests ((**a**–**i**), for modes 1–9, respectively).

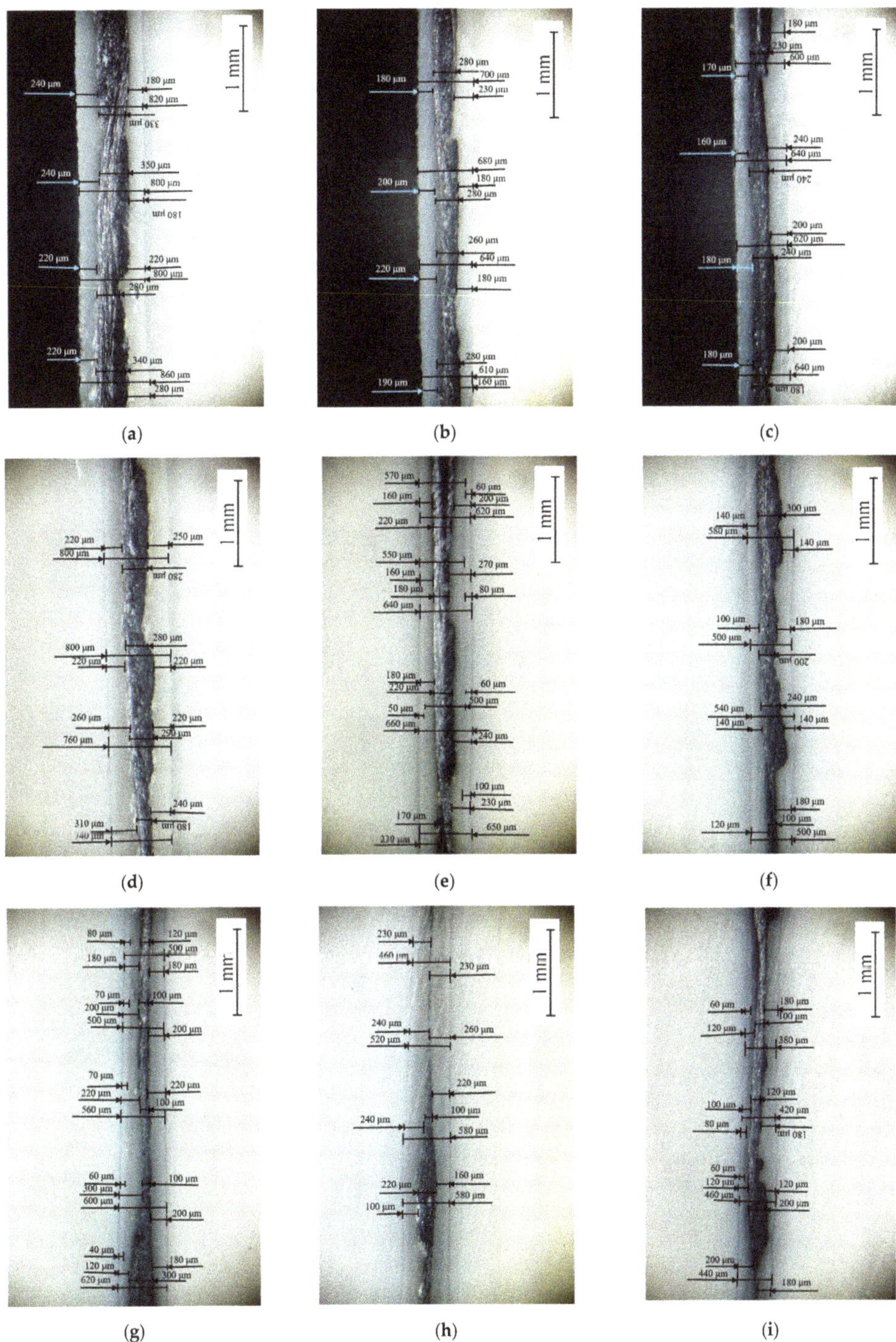

Figure 6. General-view photos of the fractured USW lap joints after the tensile tests ((**a–i**), for modes 1–9, respectively).

Table 4. The dimensional characteristics of the USW lap joints.

Mode Number	Initial Total Thickness of Lap Joined Parts, mm	USW Lap Joint Thickness, mm	Distance between PEEK Plates after USW, µm	CFF Thickness after USW, d_{CF}, µm
	Contact		Optical (Photo)	
1	5.09	5.00	820 ± 40	350 ± 70
2	5.16	5.00	660 ± 80	270 ± 90
3	5.07	4.95	650 ± 50	260 ± 80
4	5.00	4.70	740 ± 60	250 ± 70
5	5.13	4.73	680 ± 120	190 ± 90
6	5.17	4.77	560 ± 60	200 ± 100
7	5.04	4.34	560 ± 60	160 ± 140
8	5.19	4.49	530 ± 70	110 ± 110
9	5.19	4.64	440 ± 60	160 ± 160

The subsequent increase in clamping pressure to 2.0 and 2.5 atm only prevented the frictional heat release between the joined parts since higher applied pressure hindered the possibility of upper adherend expansion and their mutual displacements under US vibrations (at a low USW duration of $t = 600$ ms). As a result, the USW lap joint thinning did not exceed 160 and 120 µm, respectively, (Table 3) after the USW processes when mode 2 was in use ($t = 600$ ms, $P = 2.0$ atm) and mode 3 ($t = 600$ ms, $P = 2.5$ atm). At the same time, the CFF thickness changed slightly relative to its initial value of ~250 µm (Table 4). Since the ED almost did not melt at such a short duration of US vibrations (Figure 6b,c), the formed USW lap joints were unreliable. They fractured at $\sigma_{UTS} = 6$–8 MPa (Figure 4) due to an adhesive delamination of one of the PEEK adherends (Figure 5b,c). In both cases, the EDs melted to a lesser extent than at the clamping pressure of 1.5 atm (mode 1) since a higher clamping pressure suppressed mutual displacements between the adherends, EDs, and the prepreg.

With increasing the USW duration to 850 ms, melting of the PEI binder in the prepreg actively developed due to frictional heating intensification. This resulted in USW lap joint thinning of 300 µm (Table 3) for mode 4 ($t = 850$ ms, $P = 1.5$ atm). The "excessive" polymer melted in the US-welding zone and was squeezed out beyond its boundaries (Figure 5d). At the same time, the thickness of the CFF remained about its initial value of ~250 µm (Table 4) with a variation in its thickness of ~80 µm (Figure 6d), maintaining the prepreg structural integrity. The reliable lap joint formation was confirmed by its high ultimate tensile strength of 37 MPa (Table 3).

The subsequent increase in clamping pressure to 2.0 and 2.5 atm was accompanied by greater squeezing out of the molten polymer from the prepreg and partially from the EDs. The USW lap joint thinning was about ~400 µm in both cases (Table 3). At the same time, the CFF thickness decreased at ~50 µm (Table 4). Visually, the CFF retained its structural integrity (Figure 6e for mode 5 at $t = 850$ ms and $P = 2.0$ atm; Figure 6f for mode 6 at $t = 850$ ms and $P = 2.5$ atm). However, its thickness variation slightly increased to 90–100 µm (Table 4). The ultimate tensile strength values were ~50 and ~40 MPa, respectively (Figure 4). Nevertheless, lateral tearing of the CFF was observed for mode 5, the fragments of which were squeezed out by the molten polymer flow from the welding zone (Figure 5e). The delamination of one PEEK adherend was accompanied by cohesive tearing of the CFF fragment. On the one hand, this increase in adhesion provided the maximum ultimate tensile strength of ~50 MPa. On the other hand, even the PEI-impregnated prepreg (some of which were extruded in the USW process) failed during the tensile test.

Because of the generally similar macrostructure (Figure 5d,f, respectively), the USW lap joint formed using mode 6 ($t = 850$ ms, $P = 2.5$ atm) and mode 4 failed similarly. Therefore, it is highly likely, the maximum implemented clamping pressure hindered the frictional heating development, which was also facilitated by the minimum clamping duration after US vibrations of 2000 ms. In this case, the distance between the joined PEEK adherends was ~190 μm, and the CFF thickness was ~200 μm at the center of the specimen after the USW process, according to the optical observation (Table 4).

Increasing the USW duration up to 1100 ms resulted in further squeezing out of the EDs and the prepreg due to enhanced frictional heating. For modes 7 and 8 (Figure 5g,h), the USW lap joint thinning was 700 μm, while it was 550 μm for mode 9 (Table 3). At the same time, the CFF thickness was significantly reduced to 110–160 μm (Table 4), and the variation in its value reached the initial level (Figure 5i). In other words, the CFF was locally completely damaged (Figure 6h), while locally, it was significantly "over-impregnated" with the molten polymer (Figure 6g–i). Thus, the CFF failure did not enable us to recommend modes 7–9 (at t $\geq$ 1100 ms) for the USW procedures developed for the PEEK plates with the PEI-impregnated CFF prepreg placed between them, despite the high ultimate tensile strength of 65–91 MPa.

Note that significant polymer remelting and mixing of the components, including the fragments of the fractured CFF, prevented the failure of the USW lap joints due to macrobending in all three cases (modes 7–9). Probably, the USW lap joint thinning down to 700 μm contributed to this phenomenon for modes 7 and 8.

By way of summarizing, the following remark on frictional heating is to be added. The USW duration was shown to mostly affect the frictional heating since mechanical vibrations are transformed into a mutual displacement of the US-weld components. The clamping pressure initially suppresses the mutual displacement and frictional heating. However, after melting the EDs and the prepreg, the former stimulates mass transfer and mutual mixing. Thus, the effect of the USW parameters is complex and cannot be characterized in an isolated way.

5. Neural Network Simulation

Determining the optimal USW conditions (combinations of the process parameters) from a limited number of experiments might be reduced to the problem of vector approximation of the mechanical properties of the USW lap joints in the multidimensional space of the USW modes [52]. Among the numerous mathematical methods designed to solve approximation problems, ANN simulation is considered one of the most appropriate.

Many papers reported some ANN applications and their advantages clearly [56,57]. They are most widely applied for solving problems of classification and recognition on large volumes of training samples (big data). However, in the case of a small amount of initial information (for instance, within nine experiments in the present study), a few issues arose [58,59], causing a high error in the simulation results in the approximation region and, especially outside the determination areas of the experimental parameters (forecast zones).

Previously, we proposed a solution based on increasing the sample size by adding a random noise and synthesizing several models [52]. However, there is another way to minimize simulation errors. It is based on adding a priori known data for the boundary values of the parameters to the experimental data. For example, it was known that any joint would not form (since zero pressure eliminates frictional heating) under conditions of zero-clamping pressure and arbitrary values of clamping duration after US vibrations [52]. In this case, the mechanical properties would have zero values, while the original structural characteristics of the joined parts would remain unchanged. Similar results would be obtained for any clamping pressure and negligible USW durations ($t \leq 1000$ ms). Figure 7 presents the analysis results of the distribution of the experimental parameters and selected priors for which the USW results were known (the tabulated data values, shown in Figure 7, are presented in the Supplementary section, Table S6).

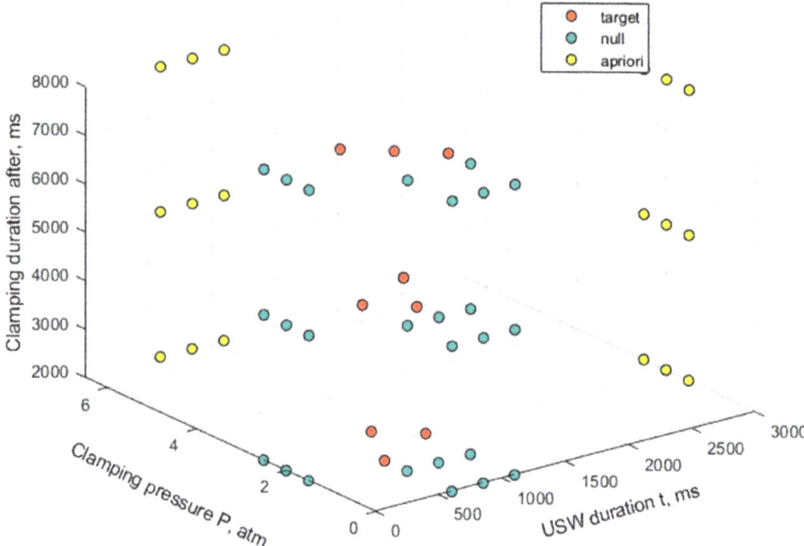

Figure 7. The distribution of the experimental parameters and selected priors for the USW results. The USW parameters used in the training sample: ● the experimental parameters; ● the parameters under which lap joints are not formed; ○ the parameters when the prepreg is fractured.

ANN simulation was carried out with the MathLab software package, which has the standard tools for the synthesis, training, and analysis of ANN. In addition, feedforward neural networks (FFNs) and radial basis function networks (RBFs) were implemented. The number of neurons in the hidden layer varied from four to eight, and the activation functions (linear, hyperbolic tangent, and logistic) were used.

By analogy with the technique described in [52], two-sided restrictions were applied, which determined the region of the optimal USW lap joint characteristics (Table 5) and the USW parameters studied in Section 4.

Table 5. Threshold values of the optimal USW lap joint characteristics.

Ultimate tensile strength (σ_{UTS}), MPa	$30 < \sigma_{UTS} < 60$
Elongation at break (ε), mm	$2.0 < \varepsilon < 3.5$
USW lap joint thinning (Δd), mm	$0.15 < \Delta d < 0.50$
Distance between PEEK adherends after USW (d_{adh}), µm	$550 < d_{adh} < 750$
CFF thickness after USW (d_{CF}), µm	$170 < d_{CF} \leq 350$

Among many developed models based on experimental and a priori data, the two were chosen. They provided the minimum acceptable root mean square approximation error (the bottom limit) and excluded excessive nonlinearity of approximation (due to overtraining of networks). The latter condition was the top limit.

Model 1. The FNNs were implemented with six neurons in the hidden layer and the activation functions: linear in the hidden layer and hyperbolic tangent in the output one. The training was carried out on a sample of the normalized data (with zero mean and unit variance). The sample was added with a synthesized data set obtained by a random spread of the normalized experimental data with the uniform distribution law [−0.05, 0.05].

Model 2. The RBN parameters were determined based on the analysis of the initial data: a spread of radial basis functions of 0.35 and the mean squared error goal of 0.0001. Then, training was done on a sample of the normalized data (with zero mean and unit variance).

The areas with the determined optimal parameters for both models are shown in Figure 8a,b. The models revealed the complex shape pattern and the ambivalence of the solution to the problem. However, the approach applied in this research with the use of a priori data enabled us to localize the areas.

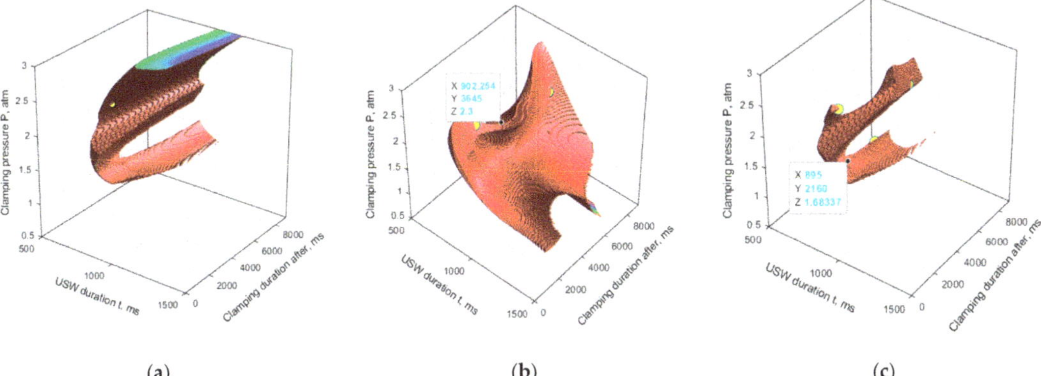

(a)   (b)   (c)

Figure 8. The areas of the optimal parameters gained due to the ANN simulation results: ◯ the experimental parameters corresponding to an optimal condition; (**a**) model 1, (**b**) model 2, (**c**) intersection of the areas.

To identify general patterns of the simulation results, crossing the areas of the optimal parameters was performed for both models (Figure 8c). The following parameters were chosen t = 900 ms, P = 1.7 atm, τ = 2.000 ms (marked in Figure 8c and designated as mode 10) for the model verification due to the following considerations:

1. The point was in the center of the bottom segment of a predicted area for the optimal parameters.
2. The point did not fall in the top segment. Instead, it referred to lower clamping pressures P. According to the authors, this had to minimize the damage of the melted prepreg during the USW.
3. The point was equidistant from two neighboring experimental ones.

However, the top segment of the "region of interest" in this setting remained uncovered. For this reason, we chose a point from this segment for subsequent verification, shown in Figure 8b (t = 900 ms, P = 2.3 atm, τ = 3600 ms; designated as mode 11). The point in the bottom segment (mode 10) was between modes 4 and 5, while another (mode 11) in the top segment was between modes 5 and 6.

6. Verification of the Neural Network Model

Tables 6 and 7 show the results of the mechanical tests of the USW lap joints obtained under optimal USW conditions. For mode 10, both ultimate tensile strength and elongation at break values were significantly higher than those for mode 4. However, they were almost identical to those for mode 5 (which did not fall into the optimal region). In this case, there was a noticeable USW joint thinning to 430 μm. This meant that the USW process was accompanied by high frictional heating and the plastic strains of both the prepreg and the ED (Table 6). According to Table 7, the CFF thickness remained almost at the initial level of ~270 μm, although the distance between the adherends decreased to 560 μm (according to the optically measured data, Figure 9a).

Table 6. The mechanical and dimensional characteristics of the USW lap joints obtained with the parameters optimized by ANN simulation.

Mode Number	Ultimate Tensile Strength (σ_{UTS}), MPa	Elongation at Break (ε), %	Elastic Modulus (E), MPa	USW Joint Thinning (Δh), mm
10 (900/1.7/2000)	47.1 ± 4.1	2.7 ± 0.2	2197 ± 109	0.43 ± 0.03
11 (900/2.3/3600)	39.9 ± 3.2	2.1 ± 0.2	2357 ± 121	0.45 ± 0.03

Table 7. The dimensional characteristics of the USW lap joints obtained with the parameters optimized by ANN simulation.

Mode Number	Initial Total Thickness of Lap Joined Parts, mm	USW Lap Joint Thickness, mm	Distance between PEEK Plates after USW, µm	CFF Thickness after USW, d_{CF}, µm
	Contact		Optical (Photo)	
10	5.11	4.68	560 ± 60	270 ± 30
11	4.98	4.53	610 ± 130	240 ± 100

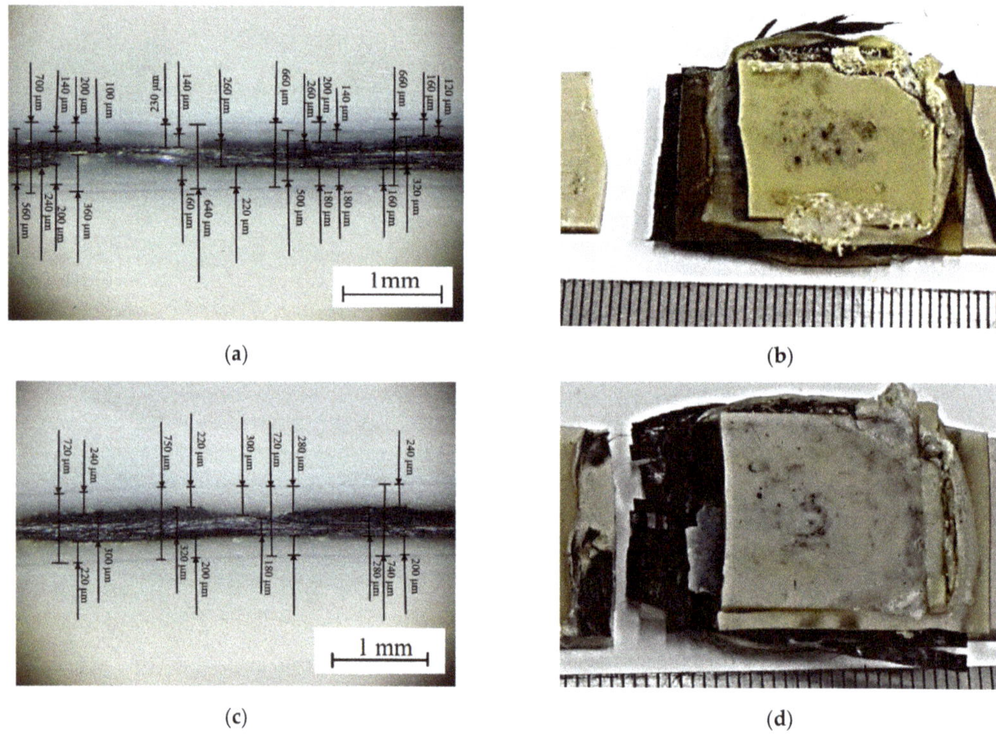

Figure 9. Optical images of the cross sections (**a**,**c**) and general views of the failed USW lap joints after the tensile tests (**b**,**d**) obtained by using modes 10 (**a**,**b**) and 11 (**c**,**d**).

After the implementation of mode 11, both ultimate tensile strength and elongation at break values were lower, while the USW joint thinning was slightly greater (450 µm, according to Table 6). In general, the parameters of mode 11, given in Table 6, were almost identical to those for mode 6 (Table 3). The same could be stated about the dimensional characteristics given in Table 7 and evaluated optically (Figure 9c). The fracture pattern of the USW lap joints, obtained by using modes 6 and 11, could also be considered similar (Figure 9b,d).

Based on the results of the verification of ANN simulation data, mode 10 was chosen as the optimal mode, using which the USW lap joints were obtained for both tension and fatigue tests. Their typical stress–strain diagram (Figure S2) indicated that they exhibited a near-elastic strain behavior almost up to the fracture point ($\varepsilon \sim 1.9\%$), characterized by the brittle type. The latter was manifested through an abrupt stress drop exactly at reaching the ultimate strength. Table 8 presents the key mechanical properties.

Table 8. The mechanical properties of the USW lap joint obtained with mode 10.

Ultimate Tensile Strength (σ_{UTS}), MPa	Elastic Modulus (E), GPa	Elongation at Break (ε), %	Yield Point 0.2% ($\sigma_{0.2}$), MPa
66.7 ± 4.3	3.9 ± 0.2	1.9 ± 0.1	64.4 ± 4.3

Under quasi-static loads, the ability to be plastically strained for the PEEK plates was suppressed by the presence of the prepreg (the yield point of the USW lap joints was comparable to its tensile strength), causing the brittle fracture pattern. In addition, a significant difference in the elastic modulus values of the CFF and the PEEK plates (adherends) contributed to the development of both shear stresses and strains at the interlayer boundary. Thus, the strength properties of the USW lap joints depended on a level of adhesion between the prepreg and the PEEK adherends. Respectively, the fracture was induced by discontinuities at the interlayer boundaries.

Photographs of the USW lap joint surfaces, captured in the tensile tests, with superimposed longitudinal ε_{YY} strain fields under analysis (Figure S3). Up to the pre-fracture stage, the CFF retained its structural integrity, and its fragments did not protrude beyond the specimen contours (Figure S3a–c). Before the failure, on the contrary, numerous fractured fibers were observed, bordering the contour of the specimen gauge length, while the CFF was torn together with the PEEK adherends in the fracture zone (Figure S3d). This fact indicated a satisfactory level of their adhesion to the prepreg.

Figure 10 shows the fields of ε_{YY} strain calculated from the HCF test results ($\sigma_{max} = 50$ MPa). Up to 1000 cycles (Figure 10a), ε_{YY} strains were uniformly distributed, and their maximum values did not exceed ~1.5%. At ~10,000 cycles (Figure 10b), a strain localization zone was formed at the bottom of the sample gauge length. Simultaneously, some fractured fibers of the CFF were observed on the side surfaces of the sample gauge length (mainly on the right). At the pre-fracture stage (Figure 10c), the CFF was damaged along the entire gauge length, which corresponded to an increase in ε_{YY} values up to ~2.0% over its entire surface.

It should be noted that the testing time of 82,000 cycles was achieved after 1.5 days at loading per cycle of 50 MPa. However, local discontinuities of the CFF were observed already after 6000 cycles. In order to reduce the testing duration, the load per cycle was increased to 60 MPa. After the load enhancement, the sample withstood about 638 cycles. This point is indicated on the fatigue curve (Figure 11b) as a separate test mode. Thus, the durability was assumed to be conditionally equal to 82,638 cycles for the load per cycle of 50 MPa. The fatigue curve was linear, which might indicate that the sample would have a durability comparable to 638 cycles at the maximum load per cycle of 60 MPa.

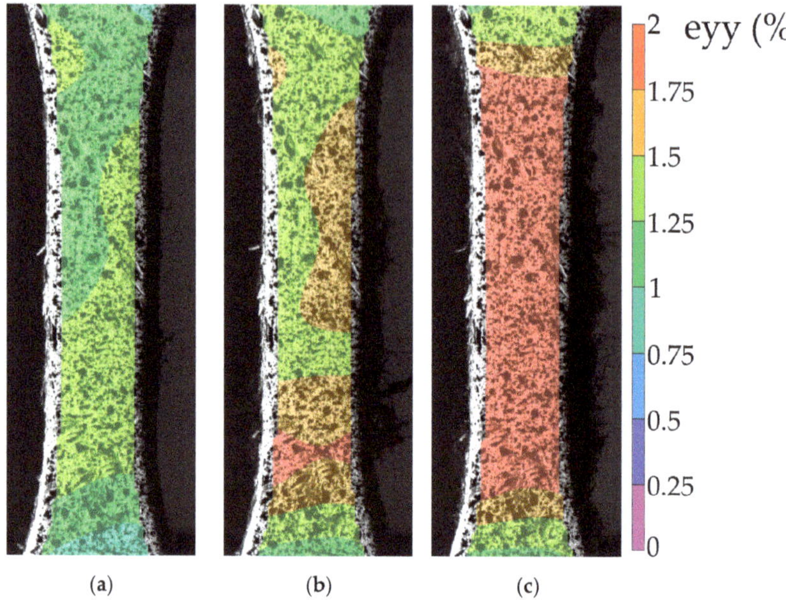

Figure 10. The strain development in the HCF tests: 1000 (**a**); 3000 (**b**); 7000 cycles (**c**).

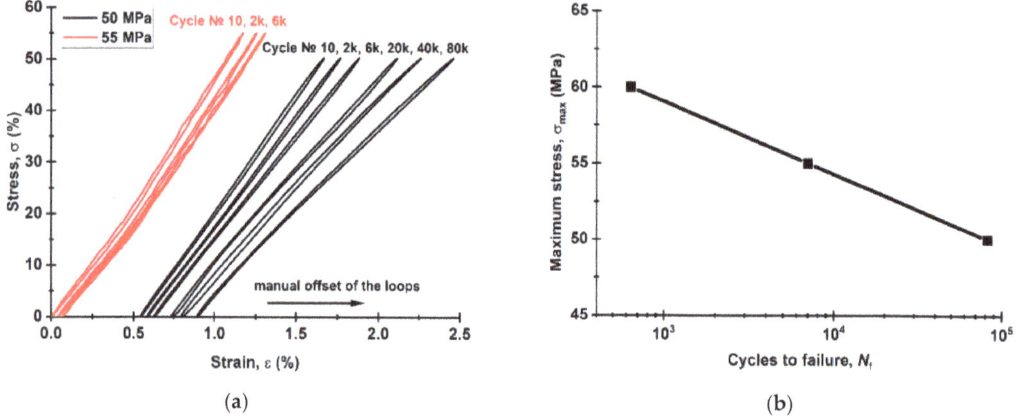

Figure 11. The fatigue test results for the USW lap joints: the mechanical hysteresis loops (**a**) and the fatigue curve (**b**).

The mechanical hysteresis loops, drawn with the DIC method, are shown in Figure 11a (black curves). The small area of the loops testifies to the low ductility of the samples. As the fatigue tests progressed, both loop shapes and their inclination angles slightly changed. This fact indicated that the proportion of the inelastic strains was minimal at the maximum load per cycle. At the same time, a fatigue failure was mainly due to damaging the CFF and lowering its adhesion to the PEEK binder.

For the LCF mode (σ_{max} = 55 MPa), a change in the hysteresis loop slope was observed when the operating time reached 2000 and then 6000 cycles, reflecting a decrease in the sample stiffness (Figure 11a). In contrast to the HCF mode, the upper half of the loop was tilted to the left (typically, it was tilted to the right, see black loops). The authors attributed the loss of the sample stiffness to the damage development mainly in the region of the interlayer boundary of the CFF and the PEEK adherends. This agrees well with the sample's appearance during the fatigue tests (Figure 12). However, this stress level, despite the occurrence of local adhesive damage, did not result in a fast failure. The fatigue (Wöller) curve, shown in the semi-logarithmic coordinates in Figure 11b, possessed a linear pattern in the studied load range. The LCF-to-HCF transition corresponded to the range from 10^3 to 10^4 cycles. The endurance limit could not be identified in the framework of the tests carried out in this research. The graphs in Figure 11 indicate the reliable US-welding of the PEEK adherends and might be further used for establishing a correlation between the structure of the joints and their loss to resist cyclic loading.

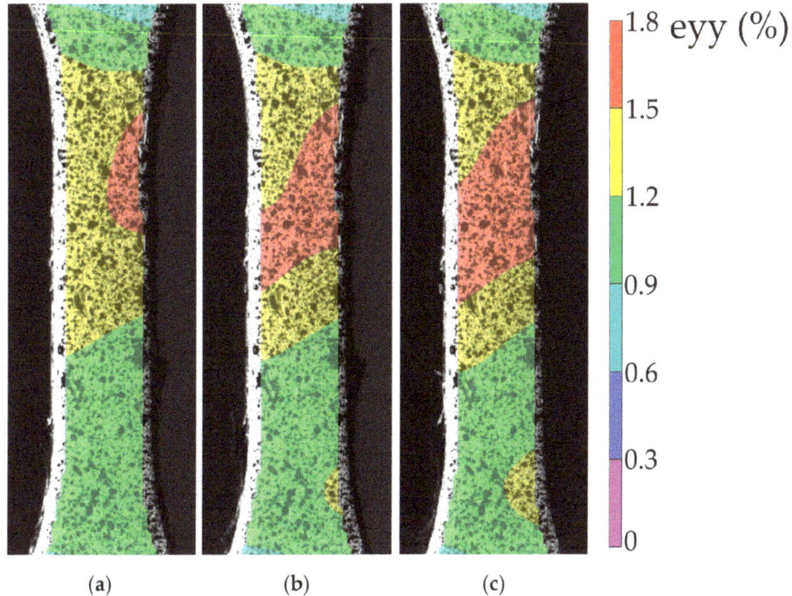

Figure 12. The strain development in the LCF tests: 1000 (**a**); 3000 (**b**); 7000 cycles (**c**).

For the LCF mode (σ_{max} = 55 MPa), the strain fields are shown in Figure 12. At the stress per cycle of 55 MPa, the strains were localized in the upper part of the sample already at 1000 cycles (Figure 12a). A gradual increase in the size of the localized strain area up to ~1.8% indicated the delamination development under the PEEK adherends (Figure 12b). Just before the fracture (Figure 12c), the strain localization extended to the entire cross-section but not to the entire area of the gauge length.

Using the DIC method, the parameters of the mechanical hysteresis loops were calculated for both LCF and HCF modes, namely, the dynamic and secant moduli, as well as the loop areas and the residual strain values [60]. Figure 13 shows that both secant and dynamic moduli were higher at the load per cycle of 55 MPa. For the HCF mode, the dynamic modulus decreased smoothly at ~6000 cycles (Figure 13a), while a downward trend appeared for the secant modulus already after ~1000 cycles (Figure 13b; black curve). In the HCF mode at the load per cycle of 55 MPa, both moduli reduced insignificantly at ~1000 cycles, after which the lowering rate increased, possessing an almost constant value starting from ~4000 cycles until a failure. The graphs in Figure 13 describe the evident indicators of losing the bearing capacity as the structure of the welds degrades.

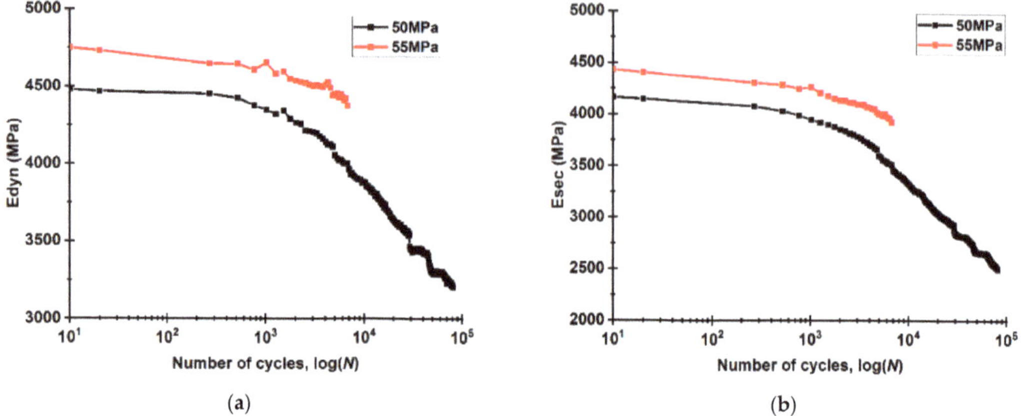

(a)

(b)

Figure 13. The changes in the dynamic characteristics of the USW lap joints in the fatigue tests: dynamic (**a**) and secant (**b**) moduli.

It was assumed that the failure of the CFF and the USW lap joint was almost simultaneous in the HCF case. However, the fracture of the CFF occurred while maintaining the integrity of the PEEK adherends when reaching 6000 cycles. An accelerated decrease in the elastic modulus indicated a loss of stiffness and stretching, which was only possible if the CFF reinforcement was absent.

Further, changes in both areas of the hysteresis loops, reflecting the mechanical energy loss (mainly due to dissipation), and the residual strains were drawn for both LCF and HCF modes. According to Figure 14a, there was a gradual decrease in the loop areas up to 6000 cycles for both modes. At the same time, the loop areas were almost equal at a level of about 8 kJ/m^3. For the HCF case, the loop areas further increased up to 11 kJ/m^3 by the failure point (Figure 14a). The scatter of the hysteresis loop area measurements was related to the specific features of the measuring technique [60]; however, it is not critical in terms of the revealed regularities.

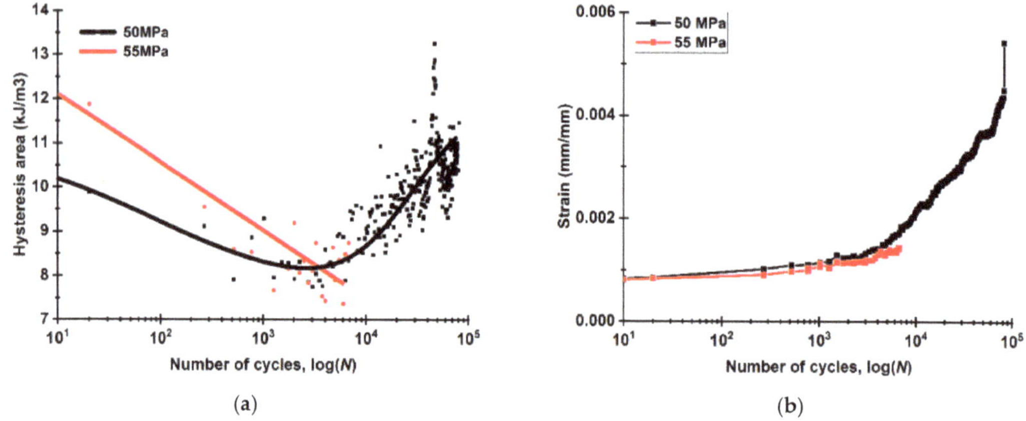

(a)

(b)

Figure 14. The strain behavior of the USW lap joints in the fatigue tests in terms of the mechanical hysteresis loop areas (**a**) and the residual strains (**b**).

Figure 14b shows the residual strain kinetics. At the initial loading stages (up to 6000 cycles), the residual strains enhanced insignificantly. However, at the next stage, there was an inflection and a subsequent non-linear increase in the residual strains (especially for the HCF case). By the failure point, their values reached 0.0055 mm/mm (0.55%), indicating damage to the CFF, which was not able to bear the load at these loading stages. This caused stretching of the USW lap joints.

7. Prospects (USW of Particulate and Laminated PEEK-CF Lap Joints)

This study was a further development of our previous research [52], both in terms of the formation of the USW lap joints from the neat PEEK adherends with the CFF interlayer. The current study has shown that it is necessary to impregnate the CFF with the PEI solution, at which both thermal and mechanical properties were comparable with those of the PEEK adherends in order to increase the strength characteristics. In addition, ANN simulation was further developed. This enabled us to determine the optimal USW parameters. As a result, the high strength characteristics were achieved, while the structural integrity of the CFF was maintained. The ANN was trained based on the results of the limited experimental sample with the additional expert characteristics of the materials considered by the authors.

Further, it was important to estimate the applicability of the results of ANN simulation for the development of the USW procedures for particulate or laminated composites. In response, the particulate composite samples were fabricated containing 30 wt.% CFs (MCFs; Tenax®-A, Teijin Carbon Europe GMBH) in the "PEEK770 MCF" matrix (Tenax®-A, Teijin Carbon Europe GMBH). They possessed a characteristic length of 200 μm at an aspect ratio of ~30. In addition, the "CF/PEEK770" laminated composites were made, consisting of 10 layers of bidirectional "ACM C285S" CFF at a ratio of 40 wt.% PEEK/60 wt.% CF. They were used as the adherends, while the ED films and the PEI-impregnated prepreg with the CFF were inserted in the same way as in the previous sections.

It should be noted that all attempts to form a USW lap joint at the "optimal" parameters established for the neat PEEK adherends (modes 10 and 11) failed for the CF-enforced PEEK ones (the melting of the weld components did not initiate at all). Nevertheless, experimentally, by increasing the USW duration (at constant both P and τ values), the USW lap joints were obtained for the particulate (mode 12) and laminate PEEK/CF (mode 13) composites. Their mechanical properties are presented in Table 9, and the dimensional characteristics, calculated from the optical images (Figure 15a,c), are reported in Table 10. Their strain–stress diagrams are shown in Figure S4. The strength properties were greater for the USW lap joint of the laminates. In this case, both elastic modulus and ultimate tensile strength values were maximal (Table 9), while their failure was developed by the interlayer shear mechanism (Figure 15d). So, the LSS value was estimated as LSS = 11.4 MPa. At the same time, the failure of the USW lap joint of the particulate composite occurred due to its macrobending (Figure 15b).

Table 9. The mechanical and dimensional properties of the USW lap joint obtained with modes 12, 13.

Mode Number	Ultimate Tensile Strength (σ_{UTS}), MPa	Elongation at Break (ε), %	Elastic Modulus (E), MPa	USW Joint Thinning (Δh), mm
12 (1200/1.7/2000)	57.7 ± 4.5	1.5 ± 0.1	5613 ± 405	0.460 ± 0.03
13 (1600/1.7/2000)	84.5 ± 6.2	1.4 ± 0.1	9926 ± 524	0.350 ± 0.02

Although the USW lap joint of the particulate composite was formed at the longer USW duration of t = 1200 ms, the USW joint thinning was 460 μm, comparable to that for the neat adherends using mode 10. For the laminates, the Δh value was lower (350 μm) at t = 1600 ms. At the same time, the CFF was noticeably thinned (Figure 15a) to 140 μm (Table 10). In the laminate case, the CFF retained its structural integrity, while the ED did completely melt (Figure 15c).

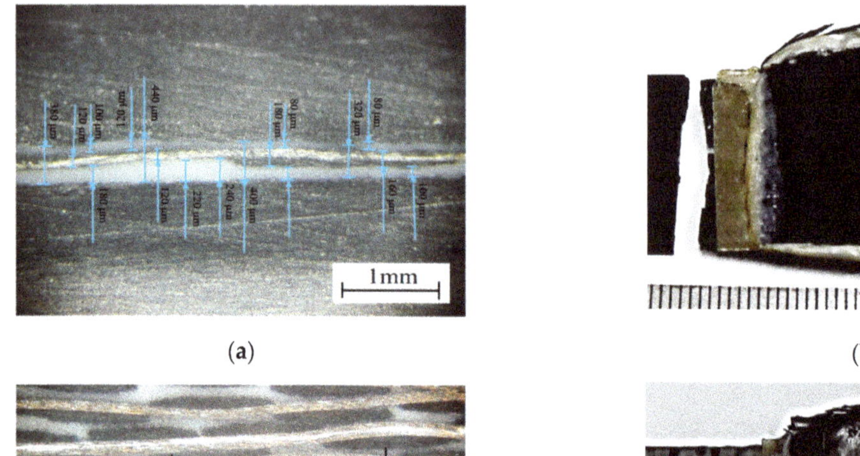

(a)

(b)

(c)

(d)

Figure 15. Optical images of the cross sections (**a,c**) and general views of the failed USW lap joints after the tensile tests (**b,d**), obtained with modes 12 (particulate composite adherends, (**a,b**)) and 13 (laminate composite adherends, (**c,d**)).

Table 10. The dimensional characteristics of the USW lap joints with the CFF prepreg between the adherends.

Mode Number	Initial Total Thickness of Lap Joined Parts, mm	USW Lap Joint Thickness, mm	Distance between PEEK Plates after USW, μm	CFF Thickness after USW, d_{CF}, μm
	Contact		Optical (Photo)	
12 (1200/1.7/2000)	4.56	4.1	380 ± 60	140 ± 40
13 (1600/1.7/2000)	6.14	5.79	460 ± 20	185 ± 95

8. Conclusions and Outlook

The obtained results allowed us to draw the following conclusions.

- The proposed method of impregnating the CFF with the PEI solution was shown to be efficient for implementing the USW process for the neat PEEK adherends and allowed us to maintain the integrity of the CFF under US vibrations exposure (at the prepreg thickness of ~250 μm, which was comparable to that of the initial CFF). However, when USW the "PEEK–ED (PEEK)–prepreg (PEI impregnated CFF)–ED (PEEK)–PEEK" composites, the lower melting point and the higher melting flow index caused an extrusion of the molten PEI, which could be accompanied by damage to the CFF.

- By ANN simulation, based on the sample of the experimental data expanded with the expert data set, the optimal modes were determined for the USW lap joint of the "PEEK–ED (PEEK)–prepreg (PEI impregnated CFF)–ED (PEEK)–PEEK" composites. Furthermore, the experimental verification of the simulation results was carried out, which showed that mode 10 ($t = 900$ ms, $P = 1.7$ atm, $\tau = 2000$ ms) ensured the high strength properties and preservation of the structural integrity of the CFF.
- It was shown that the "PEEK– CFF prepreg–PEEK" the USW lap joint could be fabricated by the "multi-spot" USW method using the optimal mode 10. It was capable of resisting the load per cycle of 50 MPa (the bottom HCF level).
- The USW mode, determined by ANN simulation for the neat adherends, did not provide joining both the particulate and laminated composite adherends with the CFF prepreg reinforcement. As a result, the USW lap joints were formed only by significantly increasing the USW durations t up to 1200 and 1600 ms, respectively. This occurred since the elastic energy transferred to the welding zone through the upper adherend.

The proposed approach for ANN simulation of the USW process can be implemented to determine the optimal USW modes for the particulate and PEEK/CF laminates, including those containing the CFF (prepreg) layer. However, a new training sample should be formed for this purpose, based on the experimental data and augmented with a priori results.

Supplementary Materials: The following supporting information can be downloaded at: https://www.mdpi.com/article/10.3390/ma16052115/s1, Table S1: The influence levels of the studied factors on the ultimate tensile strength values.; Table S2: The influence levels of the studied factors on the elongation at break values; Table S3: The influence levels of the studied factors on the work of fracture values; Table S4: The influence levels of the studied factors on the USW joint thinning values; Figure S1: The dependences of the physical and mechanical characteristics of the USW lap joints on the parameter levels: ultimate tensile strength (a), elongation at break (b), work of strain (c), USW joint thinning (d); Table S5: The general influence levels of the studied factors on the USW lap joint characteristics; Figure S2. The strain–stress diagrams for the spot USW joints for fatigue tests; Figure S3: Photographs of the USW lap joint surfaces in the tensile tests: at the beginning stage (a); the strain localization stage (b); the pre-fracture stage (c); the fracture stage (d): Table S6. The USW parameters and a priori data values for the neural network simulation.

Author Contributions: Conceptualization, S.V.P.; methodology, S.V.P., V.O.A., and D.Y.S.; software, D.Y.S.; formal analysis, S.V.P., and A.V.B.; investigation, A.V.B., D.T., A.A.B., and D.G.B.; data curation, A.V.B.; writing—original draft preparation, V.O.A., S.V.P., K.S.P., and D.Y.S.; writing—review and editing, S.V.P.; visualization, A.V.B., V.O.A., D.Y.S.; supervision, S.V.P.; project administration, S.V.P. All authors have read and agreed to the published version of the manuscript.

Funding: The research on neural network design and simulation (Section 5) was funded by the ISPMS SB RAS, project FWRW-2021-0010. The research on USW tests and structural analysis (Section 3, Section 4, and Section 6) was funded by the Russian Science Foundation (grant no. 21-19-00741).

Institutional Review Board Statement: Not applicable.

Informed Consent Statement: Not applicable.

Data Availability Statement: The data presented in this study are available on request from the corresponding author. The data are not publicly available due to confidential disclosure reasons.

Conflicts of Interest: The authors declare no conflict of interest.

References

1. Zhang, J.; Lin, G.; Vaidya, U.; Wang, H. Past, Present and Future Prospective of Global Carbon Fibre Composite Developments and Applications. *Compos. Part B Eng.* **2023**, *250*, 110463. [CrossRef]
2. Sam-Daliri, O.; Faller, L.-M.; Farahani, M.; Roshanghias, A.; Oberlercher, H.; Mitterer, T.; Araee, A.; Zangl, H. MWCNT–Epoxy Nanocomposite Sensors for Structural Health Monitoring. *Electronics* **2018**, *7*, 143. [CrossRef]
3. Ghabezi, P.; Flanagan, T.; Harrison, N. Short Basalt Fibre Reinforced Recycled Polypropylene Filaments for 3D Printing. *Mater. Lett.* **2022**, *326*, 132942. [CrossRef]

4. Centea, T.; Grunenfelder, L.K.; Nutt, S.R. A Review of Out-of-Autoclave Prepregs—Material Properties, Process Phenomena, and Manufacturing Considerations. *Compos. Part A Appl. Sci. Manuf.* **2015**, *70*, 132–154. [CrossRef]

5. Liu, G.; Xiong, Y.; Zhou, L. Additive Manufacturing of Continuous Fiber Reinforced Polymer Composites: Design Opportunities and Novel Applications. *Compos. Commun.* **2021**, *27*, 100907. [CrossRef]

6. Karaş, B.; Smith, P.J.; Fairclough, J.P.A.; Mumtaz, K. Additive Manufacturing of High Density Carbon Fibre Reinforced Polymer Composites. *Addit. Manuf.* **2022**, *58*, 103044. [CrossRef]

7. Dickson, A.N.; Barry, J.N.; McDonnell, K.A.; Dowling, D.P. Fabrication of Continuous Carbon, Glass and Kevlar Fibre Reinforced Polymer Composites Using Additive Manufacturing. *Addit. Manuf.* **2017**, *16*, 146–152. [CrossRef]

8. Frketic, J.; Dickens, T.; Ramakrishnan, S. Automated Manufacturing and Processing of Fiber-Reinforced Polymer (FRP) Composites: An Additive Review of Contemporary and Modern Techniques for Advanced Materials Manufacturing. *Addit. Manuf.* **2017**, *14*, 69–86. [CrossRef]

9. Yassin, K.; Hojjati, M. Processing of Thermoplastic Matrix Composites through Automated Fiber Placement and Tape Laying Methods. *J. Thermoplast. Compos. Mater.* **2018**, *31*, 1676–1725. [CrossRef]

10. Benatar, A. Ultrasonic Welding of Plastics and Polymeric Composites. In *Power Ultrasonics*; Elsevier: Amsterdam, The Netherlands, 2015; Volume 21, pp. 295–312. ISBN 9789896540821.

11. Brasington, A.; Sacco, C.; Halbritter, J.; Wehbe, R.; Harik, R. Automated fiber placement: A review of history, current technologies, and future paths forward. *Compos. Part C* **2021**, *6*, 100182. [CrossRef]

12. Bhudolia, S.K.; Gohel, G.; Leong, K.F.; Islam, A. Advances in Ultrasonic Welding of Thermoplastic Composites: A Review. *Materials* **2020**, *13*, 1284. [CrossRef] [PubMed]

13. Benatar, A.; Gutowski, T.G. Ultrasonic Welding of PEEK Graphite APC-2 Composites. *Polym. Eng. Sci.* **1989**, *29*, 1705–1721. [CrossRef]

14. Harras, B.; Cole, K.C.; Vu-Khanh, T. Optimization of the Ultrasonic Welding of PEEK-Carbon Composites. *J. Reinf. Plast. Compos.* **1996**, *15*, 174–182. [CrossRef]

15. Villegas, I.F.; Bersee, H.E.N. Ultrasonic Welding of Advanced Thermoplastic Composites: An Investigation on Energy-Directing Surfaces. *Adv. Polym. Technol.* **2010**, *29*, 112–121. [CrossRef]

16. Villegas, I.F. Ultrasonic Welding of Thermoplastic Composites. *Front. Mater.* **2019**, *6*, 291. [CrossRef]

17. Köhler, F.; Villegas, I.; Dransfeld, C.; Herrmann, A. Static Ultrasonic Welding of Carbon Fibre Unidirectional Thermoplastic Materials and the Influence of Heat Generation and Heat Transfer. *J. Compos. Mater.* **2021**, *55*, 2087–2102. [CrossRef]

18. Khmelev, V.N.; Slivin, A.N.; Abramov, A.D. Model of Process and Calculation of Energy for a Heat Generation of a Welded Joint at Ultrasonic Welding Polymeric Thermoplastic Materials. In Proceedings of the 2007 8th Siberian Russian Workshop and Tutorial on Electron Devices and Materials, Novosibirsk, Russia, 1–5 July 2007; pp. 316–322.

19. Tian, Z.; Zhi, Q.; Feng, X.; Zhang, G.; Li, Y.; Liu, Z. Effect of Preload on the Weld Quality of Ultrasonic Welded Carbon-Fiber-Reinforced Nylon 6 Composite. *Polymers* **2022**, *14*, 2650. [CrossRef]

20. Palardy, G.; Shi, H.; Levy, A.; Le Corre, S.; Fernandez Villegas, I. A Study on Amplitude Transmission in Ultrasonic Welding of Thermoplastic Composites. *Compos. Part A Appl. Sci. Manuf.* **2018**, *113*, 339–349. [CrossRef]

21. Nguyen, T.; Thanh, L.Q.; Loc, N.H.; Huu, M.N.; Nguyen Van, A. Effects of Different Roller Profiles on the Microstructure and Peel Strength of the Ultrasonic Welding Joints of Nonwoven Fabrics. *Appl. Sci.* **2020**, *10*, 4101. [CrossRef]

22. Villegas, I.F. In Situ Monitoring of Ultrasonic Welding of Thermoplastic Composites through Power and Displacement Data. *J. Thermoplast. Compos. Mater.* **2015**, *28*, 66–85. [CrossRef]

23. Jongbloed, B.; Vinod, R.; Teuwen, J.; Benedictus, R.; Villegas, I.F. Improving the Quality of Continuous Ultrasonically Welded Thermoplastic Composite Joints by Adding a Consolidator to the Welding Setup. *Compos. Part A Appl. Sci. Manuf.* **2022**, *155*, 106808. [CrossRef]

24. Levy, A.; Le Corre, S.; Fernandez Villegas, I. Modeling of the Heating Phenomena in Ultrasonic Welding of Thermoplastic Composites with Flat Energy Directors. *J. Mater. Process. Technol.* **2014**, *214*, 1361–1371. [CrossRef]

25. Li, W.; Frederick, H.; Palardy, G. Multifunctional Films for Thermoplastic Composite Joints: Ultrasonic Welding and Damage Detection under Tension Loading. *Compos. Part A Appl. Sci. Manuf.* **2021**, *141*, 106221. [CrossRef]

26. Bhudolia, S.K.; Gohel, G.; Kantipudi, J.; Leong, K.F.; Barsotti, R.J. Ultrasonic Welding of Novel Carbon/Elium® Thermoplastic Composites with Flat and Integrated Energy Directors: Lap Shear Characterisation and Fractographic Investigation. *Materials* **2020**, *13*, 1634. [CrossRef]

27. Tsiangou, E.; Teixeira de Freitas, S.; Fernandez Villegas, I.; Benedictus, R. Investigation on Energy Director-Less Ultrasonic Welding of Polyetherimide (PEI)- to Epoxy-Based Composites. *Compos. Part B Eng.* **2019**, *173*, 107014. [CrossRef]

28. Zhao, T.; Broek, C.; Palardy, G.; Villegas, I.F.; Benedictus, R. Towards Robust Sequential Ultrasonic Spot Welding of Thermoplastic Composites: Welding Process Control Strategy for Consistent Weld Quality. *Compos. Part A Appl. Sci. Manuf.* **2018**, *109*, 355–367. [CrossRef]

29. Koutras, N.; Amirdine, J.; Boyard, N.; Fernandez Villegas, I.; Benedictus, R. Characterisation of Crystallinity at the Interface of Ultrasonically Welded Carbon Fibre PPS Joints. *Compos. Part A Appl. Sci. Manuf.* **2019**, *125*, 105574. [CrossRef]

30. Yang, Y.; Liu, Z.; Wang, Y.; Li, Y. Numerical Study of Contact Behavior and Temperature Characterization in Ultrasonic Welding of CF/PA66. *Polymers* **2022**, *14*, 683. [CrossRef] [PubMed]

31. Goto, K.; Imai, K.; Arai, M.; Ishikawa, T. Shear and Tensile Joint Strengths of Carbon Fiber-Reinforced Thermoplastics Using Ultrasonic Welding. *Compos. Part A Appl. Sci. Manuf.* **2019**, *116*, 126–137. [CrossRef]
32. Eveno, E.C.; Gillespie, J.J.W. Experimental Investigation of Ultrasonic Welding of Graphite Reinforced Polyetheretherketone Composites. In Proceedings of the National SAMPE Technical Conference, Atlantic City, NJ, USA, 25–28 September 1989; pp. 923–934.
33. Rizzolo, R.H.; Walczyk, D.F. Ultrasonic Consolidation of Thermoplastic Composite Prepreg for Automated Fiber Placement. *J. Thermoplast. Compos. Mater.* **2016**, *29*, 1480–1497. [CrossRef]
34. Gomer, A.; Zou, W.; Grigat, N.; Sackmann, J.; Schomburg, W. Fabrication of Fiber Reinforced Plastics by Ultrasonic Welding. *J. Compos. Sci.* **2018**, *2*, 56. [CrossRef]
35. Wang, K.; Shriver, D.; Li, Y.; Banu, M.; Hu, S.J.; Xiao, G.; Arinez, J.; Fan, H.-T. Characterization of Weld Attributes in Ultrasonic Welding of Short Carbon Fiber Reinforced Thermoplastic Composites. *J. Manuf. Process.* **2017**, *29*, 124–132. [CrossRef]
36. Kiss, Z.; Temesi, T.; Bitay, E.; Bárány, T.; Czigány, T. Ultrasonic Welding of All-polypropylene Composites. *J. Appl. Polym. Sci.* **2020**, *137*, 48799. [CrossRef]
37. Lee, T.H.; Fan, H.-T.; Li, Y.; Shriver, D.; Arinez, J.; Xiao, G.; Banu, M. Enhanced Performance of Ultrasonic Welding of Short Carbon Fiber Polymer Composites Through Control of Morphological Parameters. *J. Manuf. Sci. Eng.* **2020**, *142*, 011009. [CrossRef]
38. Zhi, Q.; Li, Y.; Shu, P.; Tan, X.; Tan, C.; Liu, Z. Double-Pulse Ultrasonic Welding of Carbon-Fiber-Reinforced Polyamide 66 Composite. *Polymers* **2022**, *14*, 714. [CrossRef]
39. Zhou, W.; Ji, X.; Yang, S.; Liu, J.; Ma, L. Review on the Performance Improvements and Non-Destructive Testing of Patches Repaired Composites. *Compos. Struct.* **2021**, *263*, 113659. [CrossRef]
40. Qiu, J.; Zhang, G.; Sakai, E.; Liu, W.; Zang, L. Thermal Welding by the Third Phase Between Polymers: A Review for Ultrasonic Weld Technology Developments. *Polymers* **2020**, *12*, 759. [CrossRef]
41. Reis, J.P.; de Moura, M.; Samborski, S. Thermoplastic Composites and Their Promising Applications in Joining and Repair Composites Structures: A Review. *Materials* **2020**, *13*, 5832. [CrossRef]
42. Chu, Q.; Li, Y.; Xiao, J.; Huan, D.; Zhang, X.; Chen, X. Processing and Characterization of the Thermoplastic Composites Manufactured by Ultrasonic Vibration–Assisted Automated Fiber Placement. *J. Thermoplast. Compos. Mater.* **2018**, *31*, 339–358. [CrossRef]
43. Lionetto, F.; Dell'Anna, R.; Montagna, F.; Maffezzoli, A. Modeling of Continuous Ultrasonic Impregnation and Consolidation of Thermoplastic Matrix Composites. *Compos. Part A Appl. Sci. Manuf.* **2016**, *82*, 119–129. [CrossRef]
44. Dell'Anna, R.; Lionetto, F.; Montagna, F.; Maffezzoli, A. Lay-Up and Consolidation of a Composite Pipe by In Situ Ultrasonic Welding of a Thermoplastic Matrix Composite Tape. *Materials* **2018**, *11*, 786. [CrossRef] [PubMed]
45. Kim, J.; Lee, J. The Effect of the Melt Viscosity and Impregnation of a Film on the Mechanical Properties of Thermoplastic Composites. *Materials* **2016**, *9*, 448. [CrossRef] [PubMed]
46. Bonmatin, M.; Chabert, F.; Bernhart, G.; Cutard, T.; Djilali, T. Ultrasonic Welding of CF/PEEK Composites: Influence of Welding Parameters on Interfacial Temperature Profiles and Mechanical Properties. *Compos. Part A Appl. Sci. Manuf.* **2022**, *162*, 107074. [CrossRef]
47. Khatri, B.; Roth, M.F.; Balle, F. Ultrasonic Welding of Additively Manufactured PEEK and Carbon-Fiber-Reinforced PEEK with Integrated Energy Directors. *J. Manuf. Mater. Process.* **2022**, *7*, 2. [CrossRef]
48. Zhao, Q.; Wu, H.; Chen, X.; Ni, Y.; An, X.; Wu, W.; Zhao, T. Insights into the Structural Design Strategies of Multi-Spot Ultrasonic Welded Joints in Thermoplastic Composites: A Finite Element Analysis. *Compos. Struct.* **2022**, *299*, 115996. [CrossRef]
49. Purtonen, T.; Laakso, P.; Salminen, A. The Effect of Clamping Force When Using Ultrasonic Welding Ridge on Ultra High Speed Fiber Laser Welding of Polycarbonate. In Proceedings of the International Congress on Applications of Lasers & Electro-Optics, Temecula, CA, USA, 20–23 October 2008; p. 602.
50. Lee, D.; Kannatey-Asibu, E.; Cai, W. Ultrasonic Welding Simulations for Multiple Layers of Lithium-Ion Battery Tabs. *J. Manuf. Sci. Eng.* **2013**, *135*, 061011. [CrossRef]
51. Al-Sarraf, Z.; Lucas, M. A Study of Weld Quality in Ultrasonic Spot Welding of Similar and Dissimilar Metals. *J. Phys. Conf. Ser.* **2012**, *382*, 012013. [CrossRef]
52. Panin, S.V.; Stepanov, D.Y.; Byakov, A.V. Optimizing Ultrasonic Welding Parameters for Multilayer Lap Joints of PEEK and Carbon Fibers by Neural Network Simulation. *Materials* **2022**, *15*, 6939. [CrossRef]
53. Byakov, A.V.; Alexenko, V.O.; Panin, S.V. Ultrasonic Welding Assisted Formation of PEEK-CF Layered Composites. *AIP Conf. Proc.* **2022**, *2509*, 020037.
54. Tutunjian, S.; Dannemann, M.; Fischer, F.; Eroğlu, O.; Modler, N. A Control Method for the Ultrasonic Spot Welding of Fiber-Reinforced Thermoplastic Laminates through the Weld-Power Time Derivative. *J. Manuf. Mater. Process.* **2018**, *3*, 1. [CrossRef]
55. Bochkareva, S.A.; Panin, S.V. Investigation of Fracture of 'PEEK-CF-Prepreg' US-Consolidated Lap Joints. *Procedia Struct. Integr.* **2022**, *40*, 61–69. [CrossRef]
56. Haykin, S.S. *Neural Networks and Learning Machines*, 3rd ed.; Pearson Education: Upper Saddle River, NJ, USA, 2009.
57. Serrano, L.G. *Grokking Machine Learning*; Manning: Shelter Island, NY, USA, 2021.
58. Shu, J.; Xu, Z.; Meng, D. Small Sample Learning in Big Data Era. *arXiv* **2018**. [CrossRef]

59. Bornschein, J.; Visin, F.; Osindero, S. Small Data, Big Decisions: Model Selection in the Small-Data Regime. In Proceedings of the 37th International Conference on Machine Learning (ICML'20), London, UK, 13–18 July 2020; pp. 1035–1044.
60. Bogdanov, A.A.; Panin, S.V.; Lyubutin, P.S.; Eremin, A.V.; Buslovich, D.G.; Byakov, A.V. An Automated Optical Strain Measurement System for Estimating Polymer Degradation under Fatigue Testing. *Sensors* **2022**, *22*, 6034. [CrossRef] [PubMed]

Review

Friction Stir Welding of Aluminum in the Aerospace Industry: The Current Progress and State-of-the-Art Review

Mohamed M. Z. Ahmed [1,*], Mohamed M. El-Sayed Seleman [2], Dariusz Fydrych [3] and Gürel Çam [4]

[1] Department of Mechanical Engineering, College of Engineering at Al Kharj, Prince Sattam Bin Abdulaziz University, Al Kharj 11942, Saudi Arabia
[2] Department of Metallurgical and Materials Engineering, Faculty of Petroleum and Mining Engineering, Suez University, Suez 43512, Egypt
[3] Institute of Machines and Materials Technology, Faculty of Mechanical Engineering and Ship Technology, Gdańsk University of Technology, Gabriela Narutowicza Street 11/12, 80-233 Gdańsk, Poland
[4] Department of Mechanical Engineering, Iskenderun Technical University, Iskenderun 31200, Hatay, Türkiye
* Correspondence: moh.ahmed@psau.edu.sa; Tel.: +966-011-588-8273

Abstract: The use of the friction stir welding (FSW) process as a relatively new solid-state welding technology in the aerospace industry has pushed forward several developments in different related aspects of this strategic industry. In terms of the FSW process itself, due to the geometric limitations involved in the conventional FSW process, many variants have been required over time to suit the different types of geometries and structures, which has resulted in the development of numerous variants such as refill friction stir spot welding (RFSSW), stationary shoulder friction stir welding (SSFSW), and bobbin tool friction stir welding (BTFSW). In terms of FSW machines, significant development has occurred in the new design and adaptation of the existing machining equipment through the use of their structures or the new and specially designed FSW heads. In terms of the most used materials in the aerospace industry, there has been development of new high strength-to-weight ratios such as the 3rd generation aluminum–lithium alloys that have become successfully weldable by FSW with fewer welding defects and a significant improvement in the weld quality and geometric accuracy. The purpose of this article is to summarize the state of knowledge regarding the application of the FSW process to join materials used in the aerospace industry and to identify gaps in the state of the art. This work describes the fundamental techniques and tools necessary to make soundly welded joints. Typical applications of FSW processes are surveyed, including friction stir spot welding, RFSSW, SSFSW, BTFSW, and underwater FSW. Conclusions and suggestions for future development are proposed.

Keywords: friction stir welding; aerospace industry; SSFSW; BTFSW; RFSSW; aluminum–lithium alloys

Citation: Ahmed, M.M.Z.; El-Sayed Seleman, M.M.; Fydrych, D.; Çam, G. Friction Stir Welding of Aluminum in the Aerospace Industry: The Current Progress and State-of-the-Art Review. *Materials* **2023**, *16*, 2971. https://doi.org/10.3390/ma16082971

Academic Editor: Raul D. S. G. Campilho

Received: 21 February 2023
Revised: 30 March 2023
Accepted: 5 April 2023
Published: 8 April 2023

1. Introduction

The materials used in aerospace applications are numerous, starting from metallic, ceramic, polymeric, and composite materials [1]. The primary metallic materials used in the aerospace industry include but are not limited to aluminum alloys [2], magnesium alloys [3–9], titanium alloys [10–14], steel alloys, and Ni-based superalloys. These materials can be classified into low-softening-temperature materials (aluminum and magnesium alloys) and high-softening-temperature materials (nickel-based superalloys, steel alloys, and titanium alloys). During the manufacturing processes of aerospace structures, all types of materials require welding and joining at the highest quality possible. Friction stir welding (FSW) has been proven to satisfy the required quality of welding different types of materials, especially those with low-softening-temperature alloys [15–20], and is relatively applied in high-softening-temperature materials such as titanium alloys [21–25] and steel alloys [26–35]. This is mainly because the available tool materials can be used satisfactorily to produce very long-distance joints without any significant degradation [36–38].

FSW's invention was driven by the need to join the high-strength aluminum alloy series 7xxx and 2xxx, known as non-weldable aluminum alloys, using conventional welding techniques [39]. FSW has progressed and is used in many industrial applications such as marine, railway, automotive, and aerospace [2,40–47]. FSW has been progressively adopted in aerospace applications for welding structures made from high-strength aluminum alloys such as large-volume fuel tanks [45]. Fuel tanks for Delta II and Delta IV rockets were the first significant aerospace applications to use the FSW process to replace the fusion welding techniques [2]. Boeing (the manufacturer) has reported high-cost savings over the previous variable polarity plasma arc (VPPA) process and almost zero defect incidence [2]. In the replacement of the existing rivets in many structures, major airframe manufacturers are investigating the use of FSW [48,49]. The Eclipse 500 business jet was one of the first aircraft to adapt FSW technology in its upper and lower wing skins, cabin skins, side cockpit skins, engine beam, and aft fuselage skins [48]. FSW technology enables faster joining with speeds up to sixty times faster than manual riveting or six times faster than automated riveting with improved quality, resulting in a significant cost reduction. The assembly of the Eclipse 500 using FSW required the design, development, and fabrication of a custom, high-performance FSW system with manipulation and process control capabilities beyond what had previously been produced by the FSW system [48]. NASA's Space Launch System (SLS) used FSW to manufacture a 39 m-long liquid hydrogen tank using a giant 52 m-tall friction stir welding facility specially built for the SLS [50–53]. The SLS main stage also comprises a liquid oxygen tank, an aft engine section, an intertank section, and a forward skirt. The thickest aluminum structures ever assembled used friction stir welding in the SLS core stage [50–53]. Recently, Indian Space Research Organization (ISRO) launched a rocket in 2018, the first to fly with propellant tanks constructed using FSW, and claimed that FSW is a more efficient manufacturing method to improve the productivity and payload capability of the vehicle [18]. In terms of high-softening-temperature materials, although the tool materials still limit the wide industrial applications of the FSW process, it has also progressed in some applications, such as in the use of the cast Ti-6Al-4V in the manufacture of the spacecraft propellant tank, aimed at reducing lead time and costs compared to the routes of conventional manufacturing [42]. FSW has numerous advantages over other solid-state methods of severe plastic deformation with the purpose of bonding or joining, such as accumulative roll bonding [54–56]. FSW can be used to produce different configuration joints such as butt [57], lap [58], T [37] and corner joints [59]. Recently, FSW principles have been adopted for additive manufacturing in the solid state as well as in different configurations [60].

The fuel tanks of space shuttles and spaceships have been manufactured from welded structures of high-strength aluminum alloys. These welded structures usually experience a complex internal/external pressure and structure torque during the service, which requires high-standard welds [45]. The use of conventional fusion welding has commonly resulted in porosities and hot cracking in the joined structures. On the other hand, FSW has attracted extensive interest from the aerospace industry owing to its exceptional advantages including fewer defects [45]. Boeing has reported virtually zero defect incidence and significant cost savings over the previous variable polarity plasma arc (VPPA) process [2], as well as low distortion and excellent joint performance [45]. Thus, FSW has been accepted as an ideal technique for joining large aerospace structures made of high-strength aluminum alloys [2,45], and has been investigated and optimized for the welding of titanium alloys and stainless steel alloys [61].

As described above, there has been significant progress on the FSW process, but only limited literature can be found regarding a more comprehensive review of FSW in the aerospace industry. With this background, we tried to provide a review of the historical development of FSW technology followed by the state of the art of FSW for aerospace applications. A literature survey was conducted in the Web of Science and Scopus databases and the Google Scholar internet search engine based on the terms: "FSW + aerospace" and "friction stir welding + aerospace". Due to the novelty of the FSW subject (not longer than 30 years), the search was not limited by time. This resulted in a collection of over 300 articles. After removing duplicate, substantively distant articles and those of dubious quality (e.g., unpublished and unreviewed research reports), almost 200 papers were left for further analysis. Figure 1 shows a schematic flowchart of the strategy used to prepare the current review. This review reports the principals of FSW, its advantages, and limitations concerning aerospace applications in Section 2. The main FSW variants applied in the aerospace industry, such as friction stir spot welding, stationary shoulder friction stir welding, and bobbin tool friction stir welding, are outlined in Section 3. The FSW production machines in the aerospace industry are described in Section 4, mainly the Eclipse FSW machine and fuel tank FSW machines. The up-to-date research in the FSW of aerospace materials is summarized in Section 4 with a focus on aluminum alloys. Figure 2 shows the structure of this review.

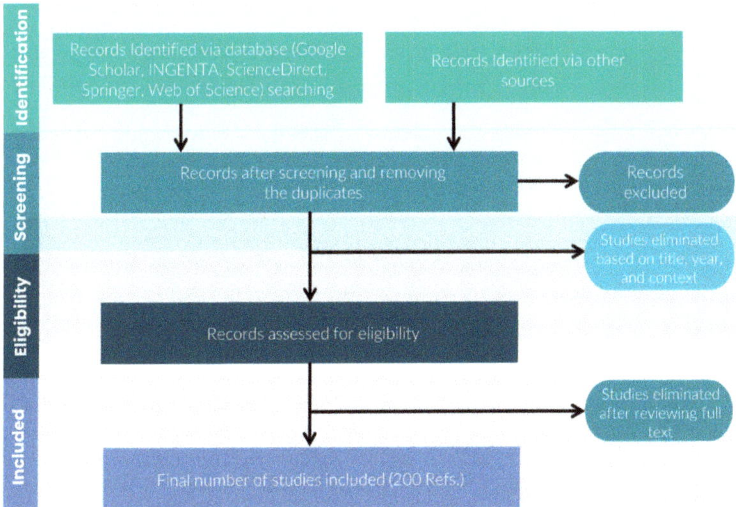

Figure 1. A schematic flowchart of the strategy used to prepare the current review.

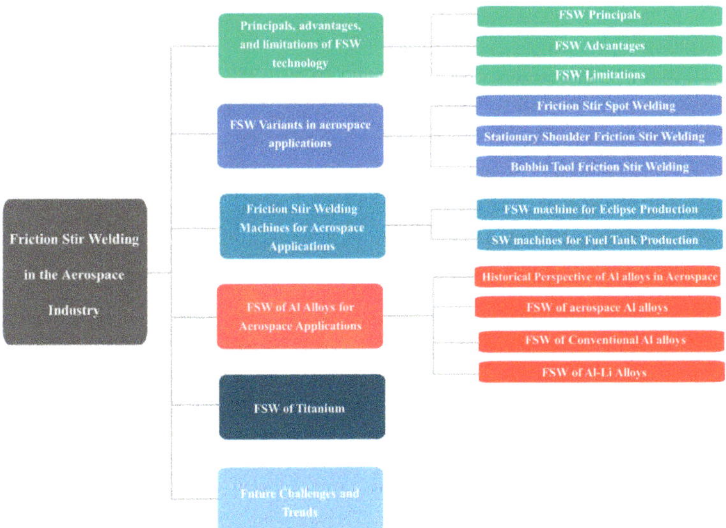

Figure 2. The structure of the current review.

2. FSW Principals, Advantages, and Limitations

2.1. FSW Principals

Friction stir welding (FSW) is currently considered a well-developed solid-state joining process that was invented by The Welding Institute (TWI) in 1991 [62–64], mainly for the purpose of joining the aerospace aluminum alloys 2xxx and 7xxx series of relatively high strength, which at the time were known to be non-weldable due to both porosity formation in the fusion zone and the poor solidification microstructure, and thus poor mechanical properties, as is the case for other high strength aluminum alloys [65–73]. Since then, FSW has been rapidly developed into a doable joining technology for a range of metals and alloys, and is used in applications from microelectronics to space shuttles [74].

The FSW innovation was in the use of an external non-consumable rotating tool to accomplish the welding in the solid state. The tool as the key player in the FSW process consists mainly of a shoulder and a probe (pin). The ratio between the shoulder (larger diameter) and the probe (smaller diameter) depends mainly on the type and thickness of the welded material [75] and sometimes on the type of tool material [76]. To conduct FSW, this tool, while rotating at a predetermined rotation rate (rpm), is plunged into two abutting or overlapped plates or sheets until achieving full penetration of the probe with enough pressure from the shoulder on the top surface of the plates. This will heat and cause the softening of the materials around the tool in this area and make its plastic deformation possible and steady. At this stage, the tool can traverse with a predetermined speed (mm/min) along the joint line to produce the joint in a solid state. During FSW, the constrained soft material around the tool is moved from the advancing side (in which the tool traversal direction and tool rotation direction are similar) to the retreating side (in which the tool traversal direction and tool rotation direction are opposite). This sequence of actions will build the joint area behind the tool at a rate that depends on the ratio between the tool rotation rate and traversal speed. At the end of the predetermined joint length, the tool exists while rotating, leaving the keyhole behind, which is known to be one of the characteristic features of FSW. Figure 3 shows a schematic of the FSW process in which all the FSW-related terms are indicated. This schematic shows the FSW tool after exit, and the tilt angle is exaggerated for clarity. Figure 4 shows images of the FSW stages of a steel alloy using a WC tool. The tool, while rotating, can be seen plunging between the abutting plates in (a), and then after the complete plunging of the tool and just before traversing along the joint line in (b), and at the end of the welding pass and just before extraction in (c).

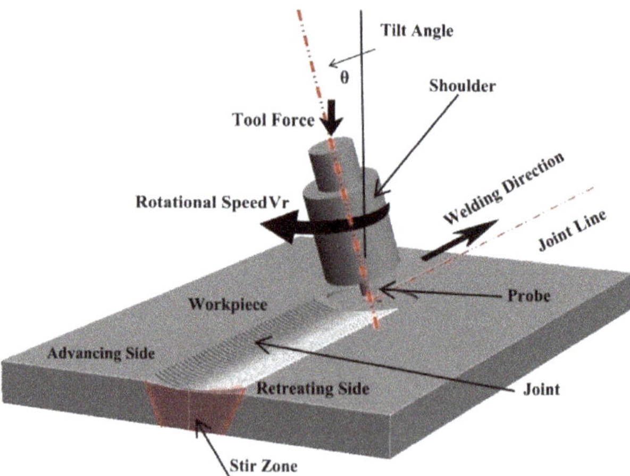

Figure 3. A schematic of the FSW process, indicates the process's main characteristic features. The FSW tool is shown with the exit hole just below the tool after extraction.

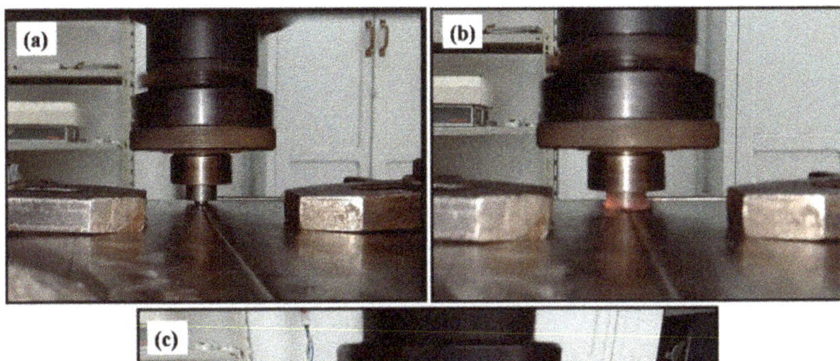

Figure 4. Images showing the stages of FSW steel alloy using the WC tool. (**a**) Initial stage of a tool plunging at the abutting edge of the two tightly clamped plates on the table of the FSW machine, (**b**) after plunging and traversing, and (**c**) at the end of the FSW and just before extracting the tool.

The weld area of the friction-stir-welded materials has some characteristic features that distinguish the FSW weld area from the weld areas of other joining techniques. It consists of four main zones: (1) The base material (BM) represents the part of the material that neither experiences any plastic deformation nor enough heat, so all the microstructural features and properties of the BM are preserved. (2) The heat-affected zone (HAZ) represents the second zone towards the center of the weld area that does not experience any plastic deformation, and only enough heat to affect some microstructural features and their dependent properties. The width of the HAZ and the effect of the thermal cycle on its microstructural constituents mainly depend on the heat input experienced during the welding process and the type of welded material. (3) The thermo-mechanically affected zone (TMAZ) represents the third zone towards the center of the weld area that experiences both plastic deformation and enough heat to distort the grain structure due to the passage and the shear effect of the tool. The TMAZ is highly affected in terms of heat but slightly in terms of deformation, which is why it represents the weakest area in terms of the hardness of the FSWed heat-treatable aluminum alloys, and failure always occurs at the TMAZ. (4) The stir zone or the nugget zone (NG) represents the central zone of the weld area where the highest heat and plastic deformation occurs due to the continuous stirring of the softened material around the FSW tool. The NG, due to this severe thermo-mechanical process, undergoes significant microstructural changes represented by the formation of a completely new microstructure, either due to the recrystallization processes and/or phase transformation processes that take place at the high strain rate, temperature, and strain based on the type of welded material. Figure 5 shows the transverse cross-section optical macrograph of friction-stir-welded 75 mm-thick AA6082, on which the HAZ, TMAZ, and NG zones can clearly be observed. It can be noted that the interface between the NG and the TMAZ at the AS is quite sharp, while at the RS it is diffusive.

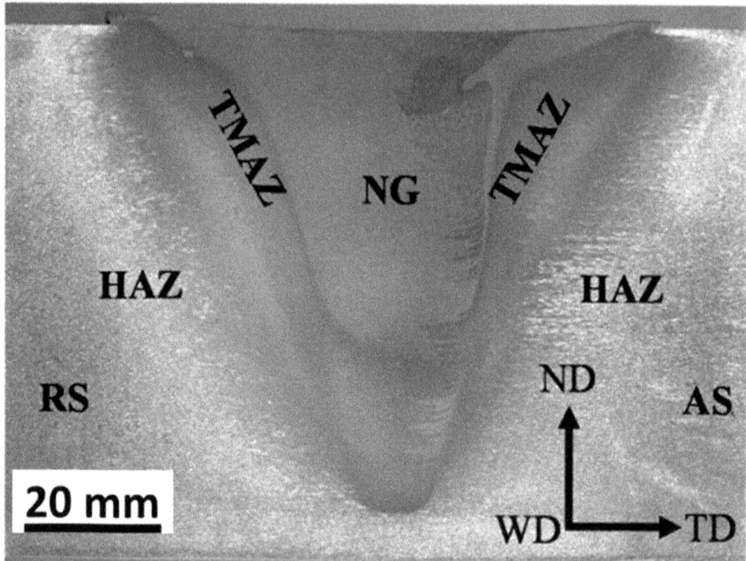

Figure 5. Transverse cross-section optical macrograph of friction-stir-welded 75 mm-thick AA6082 on which the different zones are labeled. TD, WD, and ND stand for transverse direction, welding direction, and normal direction, respectively.

The optical microstructure across the different zones of the weld area in FSWed 20 mm-thick AA7075 is presented in Figure 6a–f. The microstructure sequence is shown according to the arrow indicated on the optical macrograph of the joint above the figure. Figure 6a,b clearly shows the diffusive interface and the rotated large grain structure in the TMAZ at the interface between the TMAZ and the NG at the RS. Figure 6c,d, inside the NG zone, indicates the recrystallized grain structure, and it can be observed that the grain size is slightly more prominent at the AS (d) than at the RS (c) due to the height of the AS. Figure 6e,f clearly shows the sharp transition and the rotated large grain structure at the TMAZ at the interface between the TMAZ and the NG at the AS. These macro- and microstructural features of the FSWed materials have a strong implication for the enhancement of properties and the integrity of joints.

2.2. Advantages of FSW

FSW is a solid-state process that has many advantages to be used in aerospace applications [2,65]:

1. The weld nugget experiences a high-strain-rate plastic deformation process at a relatively high temperature, resulting in a dynamically recrystallized structure that, in most of the alloys, is a refined grain structure.
2. The weld zone experiences low heat input, resulting in low distortion in the welded plates.
3. It is a fully automated, repeatable process with a limited number of variables involved.
4. Different aerospace materials both in similar and dissimilar configurations can be welded in all kinds of joint configurations.
5. Joints with improved mechanical properties comparable to conventional fusion welding techniques can be fabricated.
6. Significant cost and time savings as the tool is almost non-consumable and the thick sections can be welded in one pass.

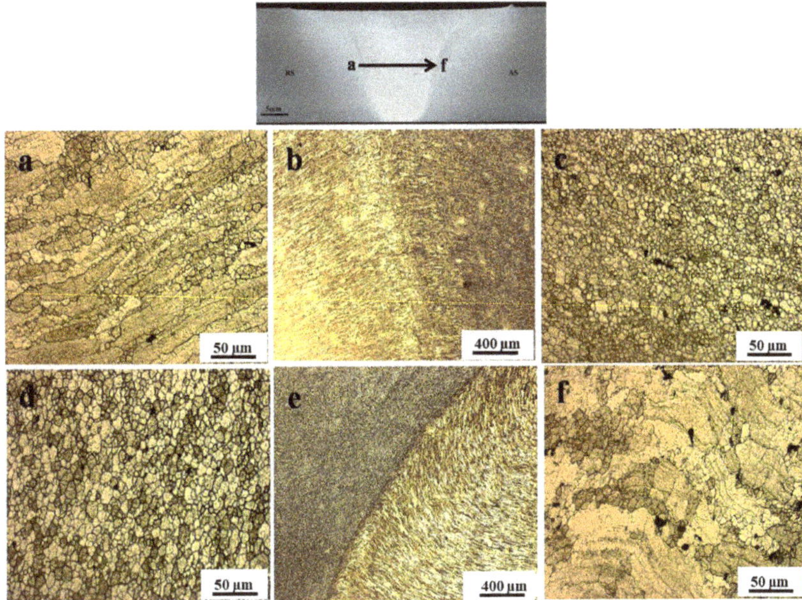

Figure 6. Optical microstructure across the weld area along the indicated arrow in the optical macrograph of FSWed AA7075-T6. (**a,b**) TMAZ–NG RS interface, (**c,d**) NG, and (**e,f**) TMAZ–NG AS interface.

2.3. Limitations of FSW

There are some limitations to the usage of the FSW process in aerospace applications to be summarized below [2,65]:

1. The workpiece to be welded has to be clamped and strained on top of the backing plate to avoid separation and flowing down the material upon tool plunging and traversing.
2. The machines are not flexible in terms of accessibility, and some parts require manual welding. In addition, the FSW machines are specially designed for specific applications that can cause the capital investment to be high.
3. The tool life for the FSW of high-melting-point materials is still one of the challenges that limit the use of FSW in some applications.

3. FSW Variants in Aerospace Applications

3.1. Friction Stir Spot Welding

Friction stir spot welding (FSSW) is one of the FSW variants developed for local welding applications, mainly to replace riveting in some aerospace applications. The basic principle of FSSW is the same as that of FSW; instead, there is no traversing in this case. In this regard, FSSW consists of three stages: (1) FSSW tool plunging while rotating up to a specified plunge depth, (2) dwelling for a specific time after penetration for the specified plunge depth, and (3) retracting the tool, leaving behind the joint with a keyhole as a characteristic feature [77,78]. Figure 7 shows a schematic of the FSSW process stages, and Figure 8 shows a top view of a series of FSSW points with the exit holes apparent in (a) and the transverse section of the joints presented in (b), which shows the reduction in the thickness in the sheets after FSSW. This process is termed conventional FSSW with some limitations such as the keyhole, thickness reduction of the top sheet, and the presence of "hook" bonding features [79]. To overcome these limitations, a refill FSSW (RFSSW) was developed for aerospace applications by Kawasaki Heavy Industries (KHI) as a new derivative of the conventional FSSW process. RFSSW does not leave an exit hole behind in the workpiece after producing the solid-state lap joint between sheet metals [47]. Figure 9 shows a schematic diagram of the stages of the refill FSSW process [47,80]. For conducting RFSSW, a preheating stage is started while the probe and shoulder are aligned at the same level at the top sheet surface. The stirring friction effect in this stage softens the workpiece, allowing the rotating tool to start plunging in alternating movements between the probe and the shoulder. In the second stage, the shoulder begins to plunge, thus causing more softening of the material, so the plasticized material is injected up into the pin slot. During the third stage, the probe starts plunging to re-inject the displaced material. In the final stage, the shoulder and the probe are aligned parallel to each other again on the top surface to induce a spot joint without a keyhole [47,81–84].

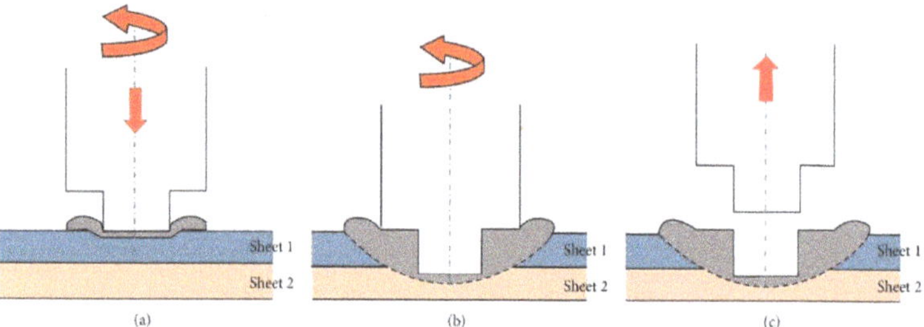

Figure 7. A FSSW process schematic of the tool actions steps. (**a**) Plunging while rotating, (**b**) dwelling after plunging to the specified depth, and (**c**) FSSW tool extraction [85].

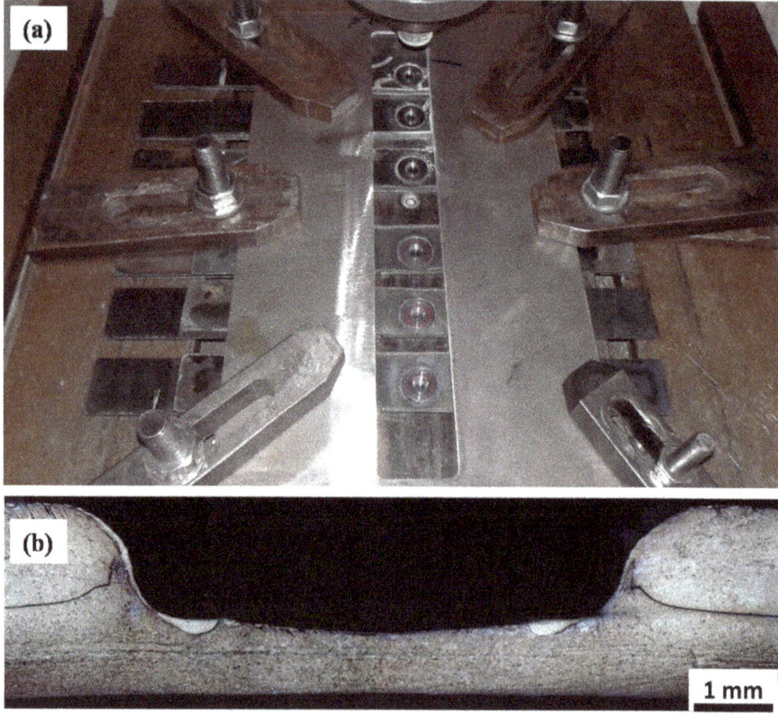

Figure 8. (**a**) Top view of FSSWed TWIP steel sheets showing the exit holes and (**b**) macrograph of the transverse cross-section showing the reduction in the sheet thickness.

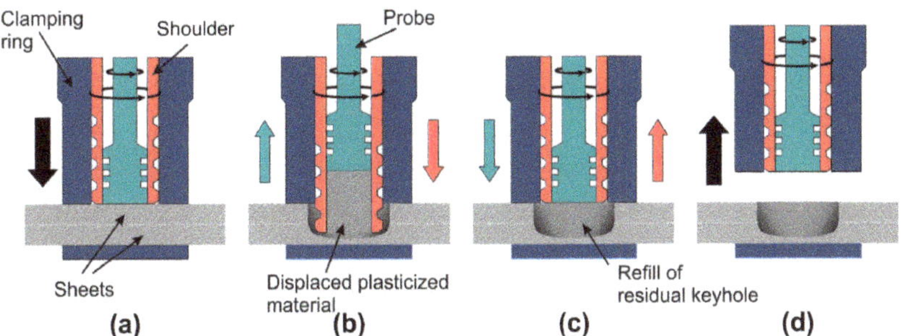

Figure 9. Schematic representation of the shoulder-plunge refill FSSW mode: (**a**) clamping of the sheets, (**b**) shoulder plunging and probe retraction, (**c**) shoulder and probe reaching back to the sheet's surface and refilling the keyhole, and (**d**) releasing of the clamping force and tool set lifting [80].

The RFSSW technique was used successfully by Boldsaikhan et al. [47] for the dissimilar welding of aerospace aluminum alloys AA7075-T6 and AA2024-T3. In this work, AA2024-T3 was used as the lower sheet representing the skin side of the aircraft structure, and AA7075-T6 was used as the top sheet representing the stiffener side of a skin-stiffener structure of the aircraft. Figure 10 shows the top and bottom view after RFSSW application in (a) and (b), respectively. Figure 10c shows the joint transverse cross-section macrograph where the keyhole is wholly eliminated. In their investigation of the fracture mode of the RFSSWed joints, Boldsaikhan et al. [47] reported two failure modes: a nugget pullout failure as shown in Figure 11a and an interfacial failure as shown in Figure 11b. In terms of failure they reported a load of 5.45 kN for the optimized RFSSW parameters that produced the nugget pullout failure. They also reported that this failure load was substantially greater than the shear load of 2.65 kN of a standard rivet with similar size [47]. This implies that the refilling technique enhances the spot joint quality and strength. Recently, Ahmed et al. [86] developed a refill technique based on friction stir deposition [87–92]. They reported that the RFSSW lap joints of AA6082 that were filled with AA2011 showed higher tensile shear loads than those of the FSSW (before refill) lap joints. The RFSSW joint (welded at 600 rpm/3 s and refilled at 400 rpm/1 mm/min) showed a higher tensile shear load of 5400 N ± 100 compared with that recorded by the unfilled joint (4300 N ± 80) [86]. Zu et al. [93,94] investigated the RFSSW of 2.0 mm-thick 2219-O (upper plate) and 2219-C10S (lower plate) with a different thickness. They reported that the lap shear load of the joints fabricated using a lower plate thickness of 4, 10, and 14 mm was 7.4 ± 0.3 kN, 6.7 ± 0.2 kN, and 6.4 ± 0.4 kN, respectively; all of them failed as a plug fracture mode. De Castro et al. [80] investigated the effect of AA2198-T8 RFSSW on tool wear. They performed a total of 2350 welds of AA2198-T8 sheets, and the effect of wear on the probe and shoulder was investigated. While the probe did not suffer any considerable wear after this number of welds, the shoulder underwent wear in different areas, with distinct wear mechanisms. Adhesive wear and plastic deformation were determined as the primary damage mechanisms affecting other shoulder areas. They suggested that the worn shoulder surface reduced the lap shear strength of the joints while all the tested welds surpassed the minimum standard lap shear strength requirements for aeronautical applications [80]. Numerous studies are available in the literature investigating the RFSSW of various aluminum alloys such as AA7075 [95–97], AA7050 [98], AA2024 [99,100], AA2198 [80], AA2014 [101], AA6061 [102,103], AA2219 [93], dissimilar Al alloys [102,104], dissimilar Al/steel [105], Mg alloys [9,106,107], dissimilar Mg, and steel [7].

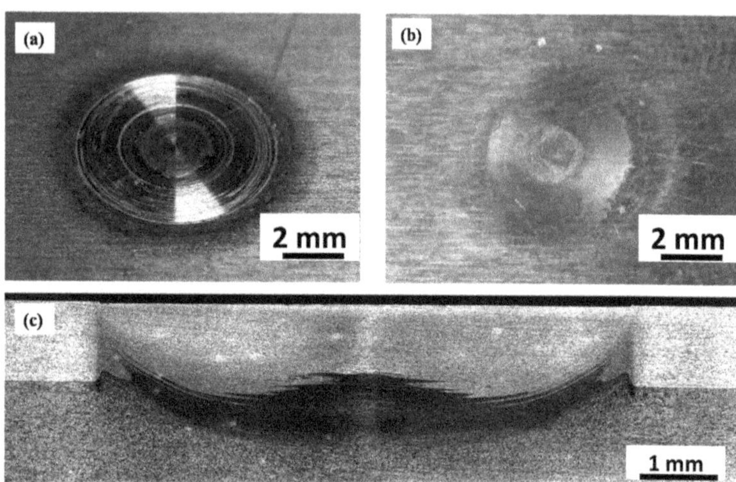

Figure 10. RFSSW joint between AA2024-T3 and AA7075-T6 sheets for aerospace applications. (**a**) Top view, (**b**) bottom view and (**c**) transverse cross-section macrograph [47] (has permission from Elsevier).

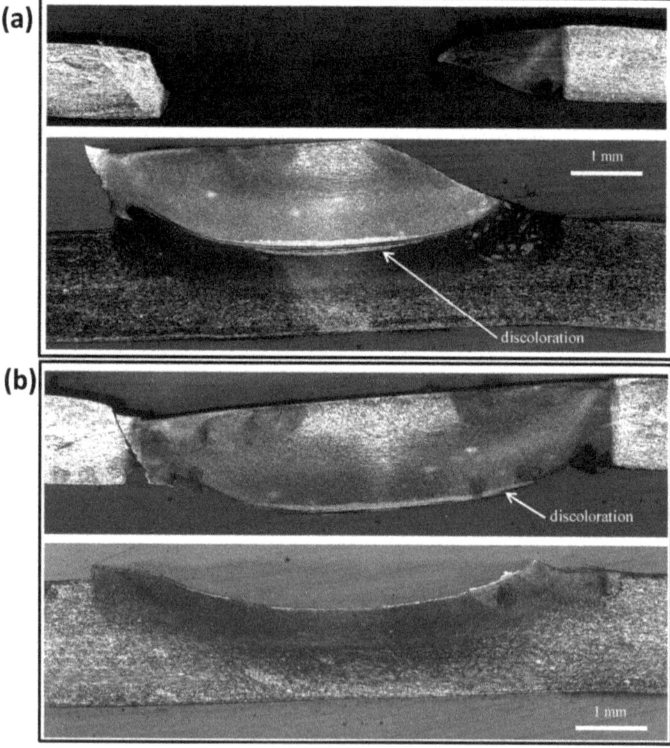

Figure 11. Failure mode of the RFSSWed AA2024-T3 and AA7075-T6. (**a**) Cross-Section of nugget pull-out failure: top sheet cross-section (**top**) and bottom sheet cross-section (**bottom**). (**b**) Cross-Section of interfacial failure: top sheet cross-section (**top**) and bottom sheet cross-section (**bottom**). The cross-section plane is parallel to the pull direction [47] (has permission from Elsevier).

3.2. Stationary Shoulder Friction Stir Welding

The FSW of titanium-based alloys such as Ti-6Al-4V, which are used for aerospace applications, has been limited due to their poor thermal conductivity. The heat is mainly generated at the upper surface when conventional FSW tools are used, resulting in a substantial through-thickness temperature gradient. Combined with the limited but relatively high hot working range of alloys such as Ti-6Al-4V, the conventional FSW of titanium is virtually impossible [108]. The TWI has developed the stationary shoulder friction stir welding (SSFSW) variant to overcome this problem and weld the titanium alloys using FSW [109]. In the SSFSW process, the tool pin only rotates through a non-rotating shoulder that only slides over the joint area. Having the shoulder stationary significantly reduces the shoulder contribution to heat generation and affects its distribution through the joint thickness. SSFSW generates highly focused heat input around the tool pin (probe) and eliminates excessive surface heating [110]. Figure 12 shows (a) an external view of the SSFSW setup upon plunging, (b) an underneath view showing the stationary shoulder and the rotating pin, (c) a top view of the SSFSWed AA7075, and (d) a transverse cross-section macrograph of the SSFSWed AA7075. The SSFSW approach was used by Russell et al. [110] in the welding of 6.35 mm-thick Ti-6Al-4V, they reported that the stationary shoulder allows more uniform heating through the thickness; thus, the microstructure is uniform along the whole cross-section. The SSFSW approach has also been used in welding aluminum to develop a through-thickness uniform microstructure and crystallographic texture [111].

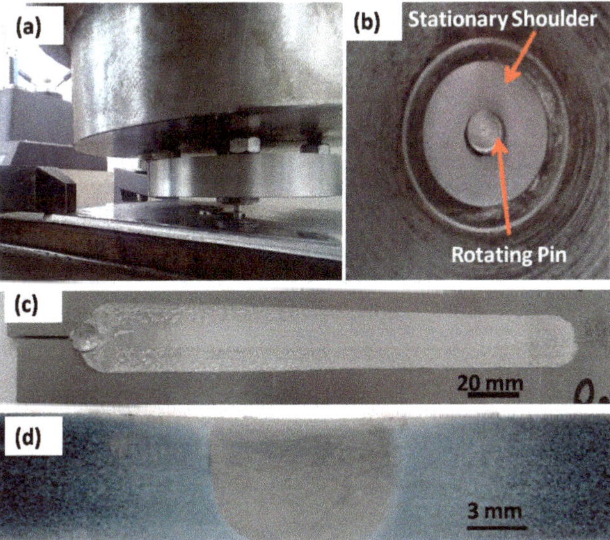

Figure 12. (**a**) External view of SSFSW setup upon plunging. (**b**) Underneath view showing the stationary shoulder and the rotating pin. (**c**) Top view of the SSFSWed AA7075. (**d**) Transverse cross-section macrograph of SSFSWed AA7075 [112].

The SSFSW approach has been extensively used to join several dissimilar and similar aluminum alloys of high strength [113–117]. Wu et al. [118] carried out a detailed investigation to compare the SSFSW and the conventional FSW of high-strength aerospace AA7050-T765. They used the same pin tool geometries schematically shown in Figure 13 in both FSW and SSFSW, aiming to study the FSW parameters' effect on the power consumption in each case. They concluded that the welding using SSFSW requires a lower heat input (30%) than that required for the conventional FSW. In addition, the use of the SSFSW resulted in welds with a number of characteristics: (1) narrower heat-affected zone width and a parallel shape; (2) lower through-thickness variation in terms of microstructure and other properties; (3) better cross-sectional tensile properties than the conventional FSW; (4) improvement in the surface roughness due to the non-rotating tool causing ironing on the top surface, as can be noted in Figure 14. On the other hand, they reported that when using too high a tool rotation traversal speed or too high a rotation rate, an adverse effect might occur, such as "speed cracking" that is observed in hot extrusion [118]. These results suggest SSFSW to be one of the promising FSW variants for aerospace applications.

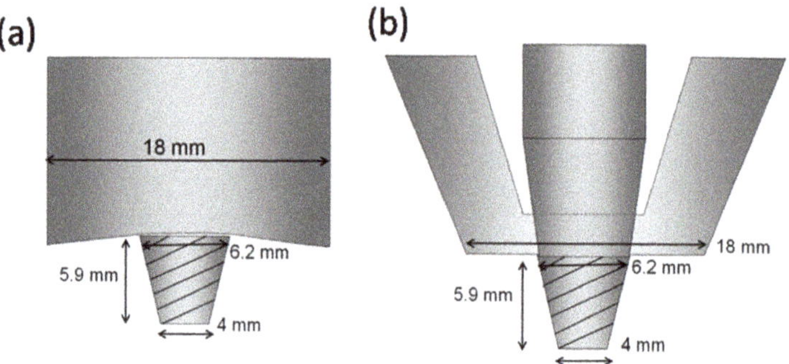

Figure 13. Schematic drawings of the different FSW tools: (**a**) the conventional FSW tool and (**b**) the SSFSW tools used by Wu et al. in their investigation [118] (reprint with permission from Elsevier).

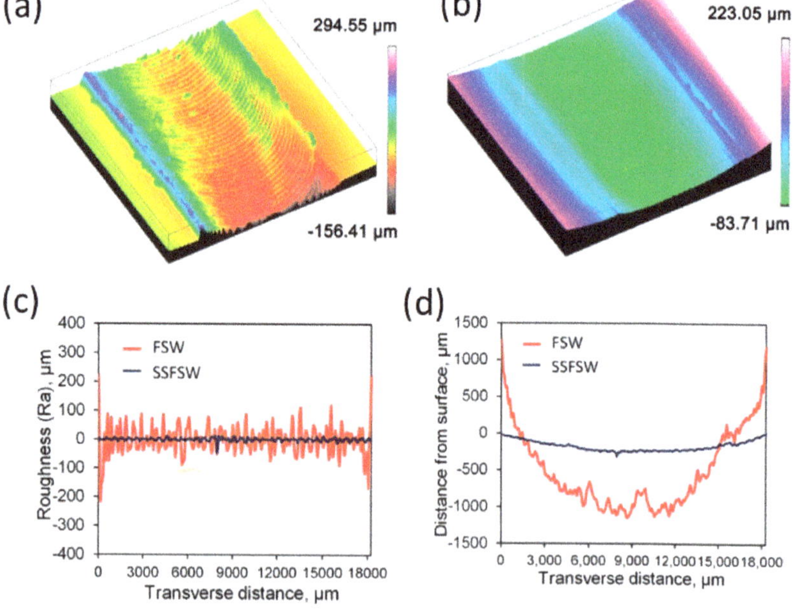

Figure 14. The surface quality characteristics and the surface roughness, respectively, obtained using (**a**,**c**) conventional FSW tool and (**b**,**d**) SSFSW tool [118] (has permission from Elsevier).

Airbus Group has adopted the SSFSW approach as a significant breakthrough for using the FSW technology in aerospace applications for welding low-melting-point alloys and dissimilar metals, which offers the opportunity for high-quality welds and improvements in production. Airbus Group Innovations, the research and technology arm of Airbus Group, has innovated the DeltaN FS friction stir welding technology, and a stationary shoulder FSW tool system. Mazak machining centers have implemented this tool system to be combined into a range of production processes [119]. The FSW machine's capital investment is always a significant challenge to using the method in many sectors. Mazak has recently developed and incorporated a combined machine tool and friction stir welding solution into a Vertical Centre Smart 430, a machining center [119] to overcome the high cost of the FSW capital investment. Additionally, Mazak has incorporated the DeltaN FS technology into their range of products, enabling welding functions and machining to be undertaken in the same platform. In the same trend, a new French company has developed an FSW head to be used with the existing machine centers or CN machine tools, drastically reducing the technology cost [119].

Marie et al. [120] proposed and investigated the use of DeltaN FS technology in joining dissimilar metals for satellite feedthrough. The main function of the feedthrough as a small part of a satellite box is energy supply and transport. Their main specification is vacuum tightness. Feedthroughs are made of dissimilar materials (aluminum alloy/steel or aluminum alloy/titanium) to save weight and facilitate their integration into an aluminum box. To prevent sealant from degassing in space, Marie et al. [120] investigated the use of the stationary shoulder FSW (DeltaN FS) as an alternative to mechanical fastening to join the outer aluminum part around the hard metal connector (steel or titanium) [120]. Additionally, in collaboration with Airbus Defence and Space for the manufacturing of Titanium, TWI has investigated propellant tanks using SSFSW technology for low-cost space applications, and the work was supported by The European Space Agency (ESA) [42]. These tanks used to be manufactured using forging with a combination of fusion welding techniques such as electron beam welding and tungsten inert gas welding (TIG), which require extensive machining that can reach up to a 90% reduction in mass. This makes the propellant tanks one of the most time-consuming and costly items to manufacture for the spacecraft motor system. Thus, the use of FSW in manufacturing these tanks will result in a significant reduction in the time and cost of production in addition to the high-quality joints, which will make FSW a smart option for the future manufacturing of the spacecraft tank. Figure 15a,b shows the propellant tank prototype and the hemicylinder manufactured using FSW.

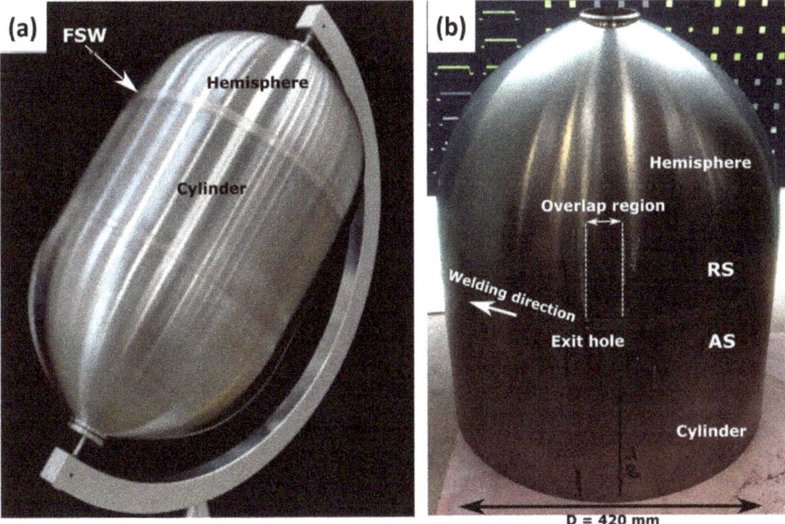

Figure 15. Images of (**a**) prototype of the propellant tank, (**b**) friction stir welded hemicylinder showing overlap region [42] (has permission from Elsevier).

3.3. Bobbin Tool Friction Stir Welding

The use of double-shoulder (upper and lower shoulders) FSW tools, known as bobbin tools, represents one of the new developments of FSW technology that has many advantages in terms of joint quality and machine capabilities, which can be summarized as follows [121–123]:

- The full penetration joint eliminates weld root flaws and leads to a lack of penetration defects.
- Low Z forces on fixture and machine.
- Due to the use of the lower shoulder, no backing plate is required.
- Low distortion due to low Z force applied.
- The ability for thickness variation tolerance.
- Capable of joining closed profiles such as hollow extrusions.
- More uniform mechanical properties through the thickness.

Figure 16 shows a schematic for the bobbin tool FSW (BTFSW) tool with the two shoulders (upper and lower) in (a), as well as an example of the fixture setup used for BTFSW presented in (b). It can be mentioned that BTFSW does not need a backing plate, which makes it ideal for welding hollow sections and reduces the applied vertical force that can result in reduced torsion.

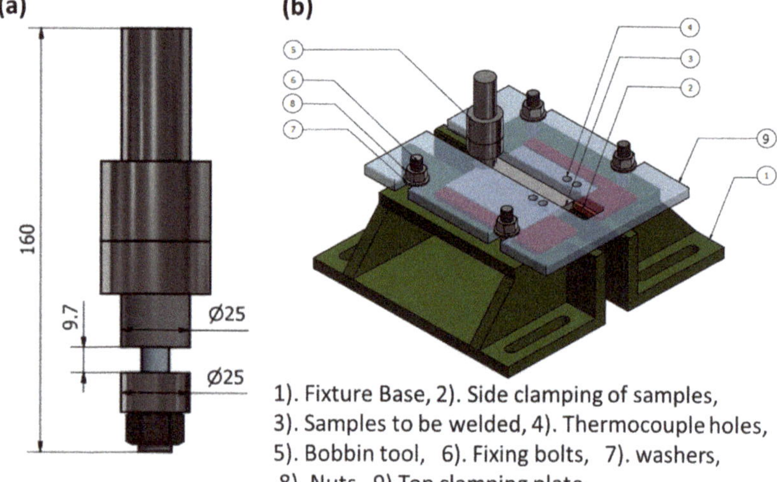

Figure 16. Schematic for (**a**) the FSW bobbin tool for welding 10 mm-thick aluminum, and (**b**) an example of the fixture setup for BTFSW.

The performance of the high-strength aluminum alloys was investigated after welding using BTFSW and compared with that welded by the conventional FSW (CFSW) [121–126]. Threadgill et al. [121] investigated the welding of thick sections of AA6082-T6 using both a bobbin tool and a conventional tool and reported that both tools produced sound joints with the difference that the net axial force on the workpiece was almost zero in the case of BTFSW, which has significant beneficial implications in machine design and cost. Xu et al. [124,125] investigated the aluminum alloy AA7085-T7452 after welding using both BTFSW and CFSW. They obtained sound joints in 12 mm-thick sections with lower joint efficiency after BTFSW due to the presence of a Lazy S defect produced by a larger extent of heat input during BTFSW [125]. Yang et al. [126] conducted a comparative study on the use of BTFSW and CFSW in the welding of AA6061-T4. They reported that the strength of the joints produced with BTFSW reached the same level, i.e., about 93%, as that of the CFSW [126]. Wang et al. [122] investigated the BTFSW of aerospace high-strength aluminum alloy AA2198, and they successfully produced sound joints of 3.2 mm thickness using different FSW parameters. The maximum joint efficiency obtained was 80% [122]. Ahmed et al. [58,127–130] conducted several studies investigating the effect of the tool pin profile and traversal speed mechanical properties of aluminum alloys. Their finding confirmed that the mechanical properties of the base material can be preserved in the weld zone through the optimization of the BTFSW parameters.

The high forces generated during the conventional FSW process that requires backing support underneath the parts to be joined make the FSW system too costly for the aerospace industry due to the very large and varying geometries that require unique large fixtures and support for each. The bobbin tool FSW technology can be the best solution to overcome this limitation. Fraunhofer IWS engineers have adopted the bobbin tool FSW technology to develop an FSW system using flexible fixtures that do not require fixed counter points [44], mainly for welding fuselage structures. This system design does not require additional support structures underneath the joining point, substantially reducing the forces that the machine and the parts must handle. The system developed is a welding robot that autonomously moves on a three-dimensional rail using an internal drive with intelligent clamping that can be suitable for clamping the curved parts. They built a demonstration section size of up to 2.5 m, which was successfully carried out and passed testing [44].

4. Friction Stir Welding Machines for Aerospace Applications

4.1. FSW Machine for Eclipse Production

First flight of the Eclipse 500 was on 26 August 2002 in which FSW was used to weld the cabin skins, aft fuselage skins, upper and lower wing skins, side cockpit skins and the engine beam [100]. This required the Eclipse Aviation Corporation in collaboration with MTS System Corporation to develop a special FSW system for thin-gauge aerospace aluminum structures with complex contours. It was a gantry type machine with a seven-axis system that was completely instrumented to control and monitor the FSW process. The machine was incorporated with additional sensing systems to ensure the quality of joints. The gantry was selected by MTS from the ones that were available in order to save time in designing and developing a new type. The one that was selected based on their investigation was the U5 gantry from Cincinnati Machine with sufficient stiffness to react the FSW process loads. For this gantry, a welding head was specially designed and built by MTS to allow for the accurate control of the FSW process of complex-contour, thin-gauge applications. The welding head included two rotational axes in addition to the X-, Y- and Z-axes of the gantry motion. One of the welding head axes provided rotation and the other provided pitch. The weld head consisted of a hydraulic spindle motor, a three-degrees-of-freedom load cell, an actuated sensing ring and a patent pending independent spindle axis [100]. To control the pin penetration to ·0.025 mm, a spindle actuator was incorporated to meet one of the main requirements for Eclipse structure manufacturing. The seventh controllable axis of the Eclipse FSW system was the sensing ring. This sensing ring encircled the spindle and was used as a surface for mounting the process monitoring sensors. The controller incorporated within the system was able to monitor the position and load feedback from the redundant sensors and was programmed to abort the process if the signals did not match within an acceptable error band. The communication within Eclipse FSW system does allow the operator to communicate through either a user interface personal computer (PC) or through a remote pendant [100].

Figure 17 shows two images for MMZA left and MMES right with the world's first production aircraft to use friction stir welding. It was one of the four prototypes used by Eclipse Aerospace during the development and Federal Aviation Administration (FAA) approval of the Eclipse 500. The plane was donated to TWI by Eclipse Aerospace and professionally restored for display by Marshall Aviation Services. The plane was photographed during the 11th International Symposium on Friction Stir Welding, which took place 17–19 May 2016 at TWI in Cambridge.

Figure 17. Images for MMZA left and MMES right with the experimental Eclipse 500 business jet presented at TWI–Cambridge as the first aircraft to be manufactured using FSW technology.

4.2. FSW Machines for Fuel Tank Production

For a launching vehicle, spaceship and space shuttle, the fuel tank represents an important structure that mainly consists of a number of cylindrical parts in the middle, top and bottom domes, and one short section at each end as shown schematically in Figure 18 [45]. The requirements for the friction stir welding of fuel tanks are: (1) FSW joints of 3~4 pieces of arc plate sections to form the longitudinal barrel; (2) a variable-curvature longitudinal joint of gores to form the gird; and (3) closed circumferential FSW joints from the gird to the cap, and from the dome to the barrel section, and from section to section [45]. The joint types that are required for the manufacturing of the complete fuel tank include butt joints, variable-curvature joints, closed circumferential joints and lock joints. To manufacture these different types of joints, the FSW machines have to be able to work in two control modes: constant-distance mode and constant-force mode. The constant-force mode is used in the case of the variable-curvature joints. In the constant-force mode, the plunge distance is controlled by keeping the force constant along the joint. This will require a constant-force control unit to be integrated with the welding spindle for the real-time detection of the vertical force. Accordingly, the real-time force is compared with the preset value, which results in the lifting and dropping of the spindle to keep the force constant along the joint profile. The constant-force control mode has been used in FSW for welding the variable-curvature dome [45]. The constant-distance control mode means keeping the plunge distance constant along the joint line, and this is suitable for the straight butt joints such as in welding of the sections and from section to section.

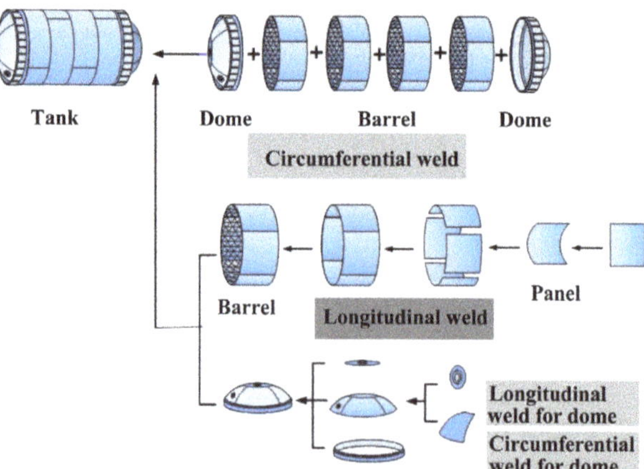

Figure 18. Schematic diagram for structure and main welds of launch vehicle tank [45] (has permission from Elsevier).

Based on that, the FSW machines used in the manufacturing of fuel tanks are classified into three categories: (1) equipment for tank barrel welding, which is a vertical-frame-type longitudinal FSW machine (Figure 19a); (2) equipment for ellipsoid dome welding, which is a large enclosed-frame-type FSW (Figure 19b); (3) circumferential FSW equipment with the integration of outside positioning and main driving (Figure 19c).

Figure 19. Fuel tank FSW machines. (**a**) Longitudinal FSW equipment of tank section, (**b**) FSW equipment of tank dome, (**c**) circumferential FSW equipment for tank, and (**d**) automatic online PAUT for tank welding with non-planar path [45] (has permission from Elsevier).

The use of FSW in aerospace applications has resulted in high weld quality and geometry accuracy. In addition, the number of defects that need to be repaired has been reduced from around ~45 to below 3, with a significant high first pass rate. In comparison to fusion welding techniques, the tensile strength of FSW joints has been improved by over 15% from 270~300 MPa to 320~350 MPa. Figure 20 shows some manufactured fuel tank sections of different sizes and geometries.

5. FSW of Al Alloys for Aerospace Applications

5.1. Historical Perspective of Al Alloys in Aerospace

Al and its alloys possess unique combinations of properties. Thus, Al can be considered one of the most attractive metallic materials for a wide range of applications, from soft, highly ductile foil to highly demanding materials in severe conditions. Al alloys have a long history in aircraft applications. Al was used in aviation before inventing airplanes. In the early nineteenth century, Ferdinand Zeppelin (8 July 1838–8 March 1917) manufactured the frames for his Zeppelin airships from aluminum sheet Al (LZ1–LZ5). On 2 July 1900, Zeppelin made the first flight with the Al-frame LZ1 airship over Lake Constance near Friedrichshafen, southern Germany. The LZ1 airship was developed until reaching the LZ5 airship version, in order to overcome all the accidents that occurred from 1900 to 1910, while retaining the presence of the Al frame as a main component [131,132]. Since then, Al alloys have been of interest in the aerospace industry. During the early nineteenth century (17 December 1903), Al was also chosen as a lightweight material by the Wright brothers for their airplane's cylinder block and engine components during their first attempt at human flight. This event was accompanied by the first attempt to thermally treat an Al alloy [131,133]. As a result of this discovery, aluminum alloys are preferred in the aerospace industry. A German aircraft designer Hugo Junkers developed the world's first full-metal aircraft (the Junkers J 1 monoplane) in 1915. Its fuselage was constructed entirely of an aluminum alloyed with magnesium, manganese, and copper [133,134].

Figure 20. Application of FSW in longitudinal weld of tank section: (**a**) Φ 3350 section; (**b**) Φ 5000 section. (**c**) Application of FSW on Φ 3350 dome. (**d**) Application of FSW on Φ 3350 tank [45] (has permission from Elsevier).

Al alloys are widely acceptable for aerospace applications as they possess a light weight, relatively high strength, workability, and corrosion resistance. Besides these advantages, Al base alloys have high availability. Compared to steel, Al is approximately one-third the weight of steel, which enables aircraft to be more fuel efficient and carry greater weight. Steel is utilized in aircraft only when great strength is required, such as in extremely high-speed planes [131,134,135]. The wing panes, the fuselage, the rudder, the exhaust pipes, the floor and door, the seats, the cockpit instruments, and the engine turbines of today's planes are all made of aluminum. Additionally, all current spacecraft are composed of a 50–90% aluminum alloy. Al alloys were widely employed in the Apollo spacecraft, space shuttles, Skylab space station, and International Space Station [131].

Over the years, the development of the aerospace industry has led to a growing need for special light materials with high durability and resistance to fatigue. This has led to a focus on Al alloys to achieve specific specifications. Several types of Al alloys are available today [134–137], but some are more suited for aerospace applications than others. The most commonly used Al alloys in aerospace applications are outlined in Table 1 with their applications [131,138].

Table 1. The most commonly used Al alloys in aerospace applications.

Al Alloy Series	Representative Alloys	Applications
2xxx	Al clad 2024 AA2014 AA2219	Wing and fuselage sheet structures, fasteners, screws and rivets [41,134,139,140]. Aircraft internal structure, External fuel tank [41,138,140]
3xxx	AA3003, AA3005, AA3105	Air conditional tube, heat exchange Parts for aircraft engines [131,134,135]
5xxx	AA 5052	Engine components, fittings, inner body panels and structural parts [41,137,138,140]
6xxx	AA6061 AA6063	Light aircraft applications (wing and fuselage structures) Finer details of an aircraft (aesthetic and architectural finishes) [131,139,140].
7xxx	AA7050, AA7068 AA7075, AA7475	Military aircraft (wing skins and fuselage) Fuselage bulkheads of larger aircraft, aerospace applications [41,131,134,138–140]
8xxx	AA8009, AA8019, AA8090	Helicopter components [131,138,140]

5.2. Future Perspective of Al in Aerospace Applications

Demand for Al alloys in aerospace applications is anticipated to double over the next decade. By 2025, the worldwide demand for aluminum will reach 80 million tons. As a result, the aerospace sector increasingly relies on recycled aluminum Al to meet its growing demand. In addition, there is a push for development in the materials used, the joining techniques, and the design structure of aircraft [41,141].

Aluminum–lithium (Al–Li) alloys have been developed for use in the aerospace sector to lower aircraft weight and improve their performance. Al–Li alloys are considered new materials due to their high strength-to-weight ratio, excellent fatigue, and high toughness properties [142,143]. As more countries enter the aerospace business, there will be more development in Al–Li alloys in the years to come [41]. Table 2 lists the most popular third generation of Al–Li alloys, and Figure 21 shows the proposed use of Al–Li alloys for various aerospace applications.

Table 2. The proposed Al–Li alloys replace the traditional Al alloys in the aircraft industry [143] (has permission from Elsevier).

	Al–Li Alloys		Required Property		Traditional Al Alloy		Aircraft Parts
				Sheets			
1.	2199T8E74 & 2060-T8E30	1.	Medium strength	1.	2524 T351	1.	Cabin skins
2.	2098-T851 & 2198-T8	2.	Damage tolerant	2.	2024-T3	2.	Fuselage
				Plates			
1.	2098-T82P (sheet/plate)			1.	2024-T62	1.	F-16 fuselage panels
2.	2050-T84, 2055-T8X, 2195-T82	1.	Medium strength	2.	7050-T7451	2.	Upper wing covers
		2.	Medium strength	3.	2124-T851	3.	Spars, ribs, other internal structures
3.	2050-T84	3.	Medium strength	4.	7050-T7451, 7X75-T7XXX		
4.	2195-T82/T84	4.	High strength	5.	7150-T7751, 7055-T7751, 7055-T7951,	4.	Launch vehicle cryogenic tanks
5.	2297-T87, 2397-T87	5.	Medium strength			5.	F-16 fuselage bulkheads
6.	2099-T86	6.	High strength	6.	7255-T7951	6.	Internal fuselage structures
7.	2199-T86, 2050-T84, 2060-T8E86	7.	Damage tolerant	7.	2024-T351, 2324-T39, 2624-T351, 2624-T39	7.	Lower wings covers
				Forging			
1.	2060-T8E50 & 2050-T852	1.	High strength	1.	7050-T7452 & 7175-T7351,	1.	Wings/fuselage attachments & window and crown

Table 2. Cont.

	Al–Li Alloys		Required Property		Traditional Al Alloy		Aircraft Parts
				Extrusions			
1. 2.	2099-T81, 2076-T8511 2099-T83, 2099-T81, 2196-T8511, 2055-T8E83, 2065-T8511		1. 2.	Damage tolerant Medium/ High strength	1. 2.	2024-T3511, 2026-T3511, 2024-T4312 & 6110-T6511 7075-T73511, 7075-T79511, 7150-T6511,	1. Lower wings stringers Fuselage/Pressure cabin 2. Fuselage/Pressure cabin Stringer and framers, upper

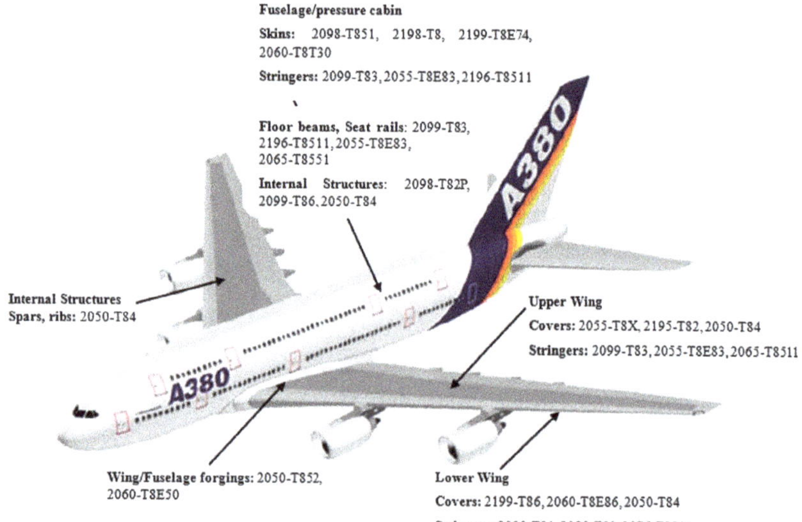

Figure 21. The used third generation of Al–Li alloys in the aircraft [143] (has permission from Elsevier).

Al–Cu–Li and Al–Mg–Li alloys are two of the most common types of Al–Li alloys that are used in the industry [144]. Al–Cu–Li alloys have high strength compared to the 7xxx series Al alloys and are therefore aimed to be used in high-strength engineering applications [145,146]. Al–Mg–Li alloys are extremely lightweight (density = 2.54 g/cm^3) and exhibit a moderate strength equivalent to that of 2xxx Al alloys (except Al–Cu–Li alloys) [147,148] and Mg–Li alloys [149,150]. The AA1424 (Al–Mg–Li–Zr) Al alloy is a heat-treatable alloy that was developed out of the 1420 and 1421 alloys [144,147,148]. It gains strength from both the precipitation of intermetallic Al$_3$Li and the solid-solution strengthening attainable by Mg [144,151]. Despite its welding difficulties, Al–Mg–Li alloys have garnered interest in aerospace. In addition, mechanical joinings (fasteners and riveting) are still used in the aerospace industry. FSW can be a good choice to solve most problems when combined with the fusion welding of Al–Li alloys [65,151], and to replace mechanical fasteners.

5.3. FSW of Conventional Al alloys

Since the invention of FSW at TWI [62–64], FSW has gained extensive interest from research centers, universities, and industries to identify its applicability for welding similar and dissimilar joints of different materials in various configurations. FSW possesses exceptional advantages over the traditional fusion welding method, including fewer defects, low distortion, low residual stresses, environmental friendliness, and usually excellent joint performance. Recently, it has been recognized as an ideal technology for the solid-state welding of aerospace parts made of high-strength similar [65,122,152] and dissimilar Al alloys with different thicknesses [153–156]. Additionally, weight is one of the most significant difficulties facing aircraft manufacturers. By joining Al-alloy stringers to skins for aircraft wings and fuselage components using FSW, thousands of rivets and any overlapping Al materials are eliminated. According to one renowned aircraft manufacturer, weight savings of around 1 kg/m from FSW might be realized. Aerospace producers cannot afford to ignore FSW since this welding technology can join practically any alloys—including some previously non-weldable precipitation-reinforced 2xxx and 7xxx series Al alloys [154–156].

During fusion welding, copper as an alloying element in the 2xxx series of Al alloys results in hot cracking, a poor solidification microstructure, and porosity in the fusion zone, which makes joining difficult. Because of this, fusion welding is not a suitable method for joining these alloy series together. Benavides et al. [157] studied the microstructural evolution during the friction stir welding of AA2024 and found that the welding process was beneficial in joining the Al 2xxx alloy type with improved mechanical properties. AA2024-T3 is a high-strength Al alloy often used in the aerospace industry. It exhibits high tensile strength, fatigue strength, a sleek surface, and low fracture spreading. It is commonly used in the exteriors of the fuselage, longitudinal beams, structures underneath wings, and sometimes in reinforcing structures and the maintenance and repair of aircraft. Sutton et al. [158] studied the variations in microstructure within a 2024-T3 Al alloy FSWed at 360 rpm and 3.3 mm/s. They concluded that the FSW could create two types of segregated, banded microstructures: hard particle-rich and particle-poor bands. The spacing of the bands was directly correlated with the welding parameters. These banded microstructures affected the macroscopic fracture process in the welds. Furthermore, by manipulating the FSW process parameters, the inhomogeneity microstructure could be avoided by increasing the fracture resistance.

The FSW approach has several potential applications in aircraft structures, especially dissimilar joints. Compared to similar welds, the dissimilar welds display a microstructure equivalent to that of similar welds but with a single lamellae flow pattern of the base materials (BM) in the SZ. This is explained by the different viscosities of the alloys during welding. Many studies are still needed to completely homogenize the mechanical properties and microstructures of the friction stir welds (FSWs) and their surrounding affected regions: the thermo-mechanical heat-affected zone (TMAZ) and the heat-affected zone (HAZ). Amancio-Filho et al. [159] investigated the effect of different rotation speeds (500–1200 rpm) and travel speeds (150–400 mm/mon) on the mechanical properties and microstructures of the dissimilar aircraft Al-alloy FSWs AA2024-T351 and AA6056-T4. The results showed that sound butt joints were obtained at the FSW parameters of 800 rpm rotation speed and 150 mm/min travel speed. This study established that in a dissimilar FSW, the weaker component determines the joint's performance, with failure occurring in the region of strength loss due to annealing processes. Da Silva et al. [154] reported that the boundary between both BMs at the SZ for the dissimilar 2024/7075 FSWs was clearly delineated. The microstructural analysis demonstrated the formation of a recrystallized fine-grained SZ with two distinct grain sizes due to the two distinct BMs. Additionally, the threaded pin geometry also had an impact on the material flow and mixing pattern during the FSW process of the dissimilar AA2024-T3 and AA7075-T6 Al alloys. Lee et al. [153] related the strength of FSWed lap joints of AA6061 and AA5052 alloys mainly to the interface morphology and the vertical transport of each alloy material with FSW parameters. Avinash et al. [160] produced a defect-free, AA2024-T3/AA7075-T6 friction-stir-welded dissimilar butt joint at the welding parameters of 80 mm/min and 1000 rpm for the welded plate thickness ratio of 1.3. The joint strength was lower than the BMs, possibly because the dissimilar joint thickness ratio was higher than uniform. RaviKumar et al. [161] examined the dissimilar FSW of 7075-T651 and 6061-T651 Al alloys under various welding conditions: rotation speed, traversal speed, and pin geometry. They concluded that the ultimate tensile strength of 205.23 MPa was achieved at the welding parameters of 900 rpm and 10 mm/min using a taper cylindrical threaded profile. Moreover, the two material alloys were unevenly distributed in the SZ.

AA5052 provides the highest strength and ductility, making it ideal for manufacturing engine components and fittings. Additionally, it is very corrosion-resistant. AA5052-H32 is now finding applications in the aerospace industry for fabricating lightweight and low-cost TV screen frames on the back of passengers' airplane seats. Shanavas et al. [162] studied the influences of rotational speed and travel speed on the UTS of underwater and normal FSW of 6 mm-thick AA5052-H32 aluminum alloy. It was noted that the UTS gained by underwater FSW was about 2% higher than that of the conventional FSW process. A microstructural examination revealed that the heat-affected region was not found in underwater welding [14,163–165]. A fractography investigation showed that all the FSWs displaying higher joint efficiency failed through ductile mode fracture.

The 6xxx series Al alloys (medium-strength aerospace alloys) are used for fuselage structures and wing skins. FSW is a viable approach for modifying the AA 6xxx alloy microstructure, resulting in a refined and homogeneous grain structure with good weld efficiency. Kumbhar and Bhanumurthy [166] investigated the effect of FSW variables on the microstructural changes and the associated mechanical performance of the AA6061-O Al alloy welded in butt joints at different rotational speeds from 710 to 1400 rpm and welding speeds from 63 to 100 mm/min. The post-weld heat treatment (PWHT) of the joints was also examined. They recommended that it is beneficial to the weld joints at lower rotational speeds and at a higher welding speed, thus improving productivity. FSW of AA6061-O increases the UTS of the welds compared to that of the BM in the O-condition for all welded joints. Moreover, PWHT for up to 8 h restores the ductility and strength while improving the microstructure homogeneities of the welds compared to that of the BM in T6. Scialpi et al. [167] studied the influence of FSW tool shoulder profiles on the microstructural and mechanical properties of FSWed AA6082-T6 Al alloy joints (1.5 mm thickness) at 1810 rpm and 460 mm/min. They found that a shoulder with a fillet and cavity worked well for thin sheets to achieve the best joints compared to the other shoulder geometries. Sato et al. [168] reported that the hardness profile across the SZ of the FSWed 6 mm-thick AA6063-T5 Al depended on the precipitate distribution and the grain size microstructure.

The 7xxx series Al alloys are one of the strongest Al alloys currently in use in the industry. A variety of aircraft structural applications benefit from its high strength-to-weight ratio and natural aging properties. Kimura et al. [169] investigated the AA 7xxx FSW joining phenomena. They concluded that the AA7075 FSW mechanism was comparable to that of low-carbon steel and attributed this phenomenon to the similarities in their strength properties. Fu et al. [170] studied the role of three ambient conditions (hot water, cold water and air) on the mechanical properties and joint efficiency of submerged 5.5 mm-thick FSW AA7050 alloys at 100 mm/min and 800 rpm in butt joints. They found that the hot-water-welded joints had the best mechanical properties of all the welded joints. Moreover, the ratio of the elongation and UTS of the joint to the BM in hot water achieved 150% and 92%, respectively. Venugopal et al. [171] investigated the microstructure and corrosion resistance of the FSW 12 mm-thick AA7075-T6 alloy welded at 350 rpm and 60 mm/min. The results showed that the pitting corrosion resistance of the welded metal (high grain refining) was better than that of TMAZ and the BM.

The available data in the published research indicate that FSW is the most suitable welding method for joining dissimilar and similar Al alloys. Friction stir welding out-performs fusion welding in strength, ductility, fatigue resistance, and fracture toughness. Ahmed et al. [172] studied the effect of varying FSW travel speeds from 50 to 200 mm/min at a constant rotation speed of 300 rpm on the mechanical properties and microstructure of similar and dissimilar AA7075-T6 and AA5083-H111 butt joints. The results showed that the applied welding parameters succeeded in producing defect-free joints. A marked grain refining was achieved in the stir zones of all the similar and dissimilar joints. The hardness profile of the similar AA7075 welds revealed typical behavior for age-hardened

Al alloys with a hardness loss in the SZ, and in the case of similar AA5083 welds, typical behavior for work-hardened Al alloys with a detected enhancement in the SZ hardness. In contrast, the dissimilar welds revealed a smooth transition in the hardness profile between the two hardness values of the AA7075 and AA5083 alloys. Furthermore, the dissimilar joints showed that the UTS ranged from 245 to 267 MPa with a weld joint efficiency ranging from 77 to 87% of the strength of AA5083 BM.

5.4. FSW of Aluminum–Lithium Alloys

Al–Li alloys have become of great interest in the aerospace industry with the aim of reducing the structural weight of the aircraft while reducing fuel consumption. The fusion welding of Al–Li alloys results in common fusion welding defects such as hot cracks, pores, element loss and joint softening, which result in low joint strength and limit the further application of Al–Li alloys in the aerospace field [173]. Thus, FSW is one of the best methods known to eliminate these fusion welding defects as a solid-state welding process. Therefore, extensive research has been conducted to investigate the effect of FSW parameters on the microstructure and mechanical properties of different Al–Li alloys. This has been recently summarized in comprehensive reviews by Yang et al. [173] and by Mishra and Sidhar [174]. This section will summarize some of the FSW research to investigate the effect of FSW parameters on the microstructural features and mechanical properties. Wei et al. [175] investigated the effect of FSW variables in terms of pin rotational speed, travel speed, and downward force on the mechanical properties and microstructure of the FSWed AA1420 (Al–Mg–Li) alloy. They related the SZ grain size increase to the increased heat input. Furthermore, the UTS of the joints reached 86% of the BM with a 180° maximum bending angle. Sidhar et al. [151] examined the influence of post-welding treatment on the FSWs of AA1424 (Al–Mg–Li) alloy welded at 800 rpm and 305 mm/min. The results showed that the HAZ and the SZ showed a full recovery of strength. A joint efficiency of around 97% of the BM was obtained. Moreover, they ascribed the high strength of joints to high density and homogenous dispersion of the fine Al_3Li precipitation phase. Altenkirch et al. [176] investigated the effect of FSW parameters on the residual stresses and hardness of Al–Li AA2199 during friction stir welds. They reported that the low hardness region widened with increasing downforce and tool rotation and decreased as the traversal speed increased. In terms of distortion, they reported that the conditions that reduced the heat input led to lower distortion levels.

Shukla and Baeslack [177] studied the microstructure of a friction-stir-welded thin-sheet Al–Cu–Li alloy using transmission electron microscopy. They interrelated the micro-hardness reduction to the dissolution and coarsening of T1 and θ' precipitates. Additionally, Cavaliere et al. [178,179] investigated the microstructure of FSWed Al–Li 2198 alloys using TEM and obtained the same results. Steuwer et al. [180] studied the microstructure Al–Li AA2199 friction stir welds. They attributed the W-shaped hardness profile across FSW in third-generation Al–Li–Cu–Mg alloys to the dissolution of the age-hardening phases in different regions. Ma et al. [181] investigated the mechanical properties of the friction-stir-welded nugget of 2198-T8 Al–Li alloy joints. They reported that yield and tensile strength had a "U" shape through the weld zone and a lower value in the weld zone, while the elongation was reversed. De Geuser et al. [182] investigated the microstructure of a friction-stir-welded AA2050 Al–Li–Cu in the T8 state. They reported a strict correlation between the volume fraction of the T1 precipitates and the hardness of the material. Gao et al. [183] investigated the correlation between microstructure and mechanical properties in a friction-stir-welded 2198-T8 Al–Li alloy. They reported a reduction in the hardness of the weld zone and strength due to the dissolution of the T1 phase that existed in the base material. Qin et al. [184] investigated the evolution of precipitation in a friction-stir-welded 2195-T8 Al–Li alloy. Their results showed that precipitations in the base metal primarily consisted of T1 (Al_2CuLi) platelets and small amounts of the θ' (Al_2Cu) and $\tau2$ (Al_7Cu_2Fe) phases. In the heat-affected zone (HAZ), these precipitations dissolved during welding, allowing the re-precipitation of δ' (Al_3Li) and β' (Al_3Zr) during cooling. The δ' and β'

phases were the primary strengthening phases in the weld nugget zone (WNZ), which resulted in the observed lower microhardness of the nugget region. Table 3 summarizes the FSW conditions and the resulting mechanical properties for Al–Li alloys.

Table 3. Al–Li alloys FSW conditions and the resulting mechanical properties.

Alloy, Thickness (mm)	Tool Shape	Rotation Rate, (rpm)	Traverse Speed, (mm/min)	UTS(FSW)/UTS(BM) (%)	Hardness Profile Shape	Refs.
AA2195-T87, 5	Taper threaded pin	200–1000	100–300	390/573 (68%)–425/573 (74%)	W	[185]
AA2060-T8, 2	Cylindrical straight pin	300–1400	100	375/530 (71%)–443/530 (83%)	W	[186]
AA2198-T851 3.2	Bobbin with cylindrical pin	800	42	380/473 (80%)	W	[187]
AA2099 T8, 5	Threaded Cylindrical pin	700–1100	45	275/540 (51%)–340/540 (64%)	W	[188]
AA2099-T83, 5	threaded, tapered, triangular pin	400–1200	75–550	343/558 (61%)–390/558 (70%)	Not available	[189]
AA2050-T8, 15	Threaded pin with 3 flats	400	200	Not available	W	[182]
AA2198-T8, 2	Tapered pin	600	200	300/491 (60%)	W	[181]
AA2198-T8, 1.8	Tapered pin	800	300	386/518 (70%)	W	[183]
AA2198-T8, 3.2	Bobbin tool	400–1000	42	270/473 (57%)–380/473 (80%)	W	[190]

5.5. ISO Standard for Aluminum FSW

The ISO standards available for aluminum and aluminum alloys FSW are as the following:

1. ISO 25239-1:2020 Friction stir welding—Aluminum—Part 1: Vocabulary
2. ISO 25239-2:2020 Friction stir welding—Aluminum—Part 2: Design of weld joints
3. ISO 25239-3:2020 Friction stir welding—Aluminum—Part 3: Qualification of welding operators
4. ISO 25239-4:2020 Friction stir welding—Aluminum—Part 4: Specification and qualification of welding procedures
5. ISO 25239-5:2020 Friction stir welding—Aluminum—Part 5: Quality and inspection requirements

6. FSW of Titanium

Due to the poor thermal conductivity of titanium alloys, using conventional FSW tools results in excessive heat generation at the surface and, consequently, a significant temperature gradient across the thickness [108]. In addition, alloys such as Ti–6Al–4V have a relatively high working temperature. These two reasons have motivated the TWI to innovate a unique FSW tool known as the stationary shoulder FSW (SSFSW) tool (introduced in Section 3.2) [108,110]. The early publications in this regard were published by Wynne et al. [108], where they investigated the microstructure and texture of FSWed 6 mm-thick Ti–6Al–4V alloy. They reported that the microstructure was uniform across the thickness and significantly refined compared to the base material [108]. Zhang et al. [191] investigated the FSW of commercially pure titanium using the PCBN tool for 2 mm thickness. They reported that the SZ consisted of fine lath-shaped grains that contained PCBN debris and Ti borides. This implies the difficulty in using FSW titanium with the conventional tool even for small thicknesses. Ramulu et al. [192] evaluated the tensile properties of FSWed Ti–6Al–4V of 2 and 2.5 mm thicknesses. They reported that FSWs in the Ti–6Al–4V alloy can possess yields and ultimate tensile strengths superior to that of the parent material due to grain refinement in the weld nugget, while the decreased elongations were associated with root defects leading to premature failure in addition to the refined grain structure. Farias et al. [193] investigated WC tool wear during the FSP of Ti–6Al–4V of 2 mm thickness. They reported that the severe tool wear caused a loss of surface quality and the inclusion of fragments inside the joining, and recommended the replacement of cemented carbide with tungsten alloys. Additionally, Wang et al. [193] investigated the different FSW tool (W–1.1%La$_2$O$_3$ and two different grades of WC–Co-based materials) wear during the FSW of Ti–6Al–4V, and they reported that tool degradation occurred due to plastic deformation in the W–1.1%La$_2$O$_3$ tool. Additionally, shear-stress-induced cracks were observed at the pin tip and tool debris was left in the processed material. The mechanical fracture along with the diffusion was responsible for tool weight loss [193]. Yoon et al. [194–196] successfully obtained FSW joints with 5 mm-thick Ti–6Al–4V plates using a Co-based alloy tool. Due to the difficulty of obtaining FSW joints using the available tool materials, some researchers have used heat-assisting systems to obtain successful joints. For example, Ji et al. [197,198] investigated the joint formation and mechanical properties of back-heating-assisted friction-stir-welded Ti–6Al–4V. They reported that the back-heating method reduces the temperature gradient along the thickness, which is beneficial for eliminating the tearing defect; therefore, defect-free joints can be attained using a wider parameter range. Li et al. [199] designed a new FSW tool where the pin was made from W-Re25% alloy because of its excellent high-temperature wear resistance and the shoulder was made from nickel-based superalloy (GH4043) due to its good high-temperature impact property and low cost. They produced a number of defect-free welds using 2 mm-thick Ti–6Al–4V. Recently, Amirov et al. [200] obtained FSW joints with titanium ($\alpha + \beta$) alloys using a nickel superalloy tool.

7. Future Challenges and Trends in FSW for Aerospace Industries

FSW as an innovative solid-state welding technology is clearly replacing the conventional welding techniques in the aerospace industry. However, the development of launch vehicle models and other aerospace applications requires the FSW community to work hard towards the development of the technology to overcome a number of limitations that hinder the extension of applying FSW to high-softening-temperature materials, composite materials, and polymeric materials. One of the key challenges that limit the use of the FSW process in high-softening-temperature materials is the tool materials that need to be cost-effective relative to conventional welding techniques. In addition, the high cost of the FSW machines needs to be continuously solved by adopting the existing CNC machines and the development of welding heads to be used within the workshop machines.

The aerospace industry needs to ensure high accuracy and high-quality manufacturing; thus, the exclusive defects of FSW joints need to be detected using advanced NDT techniques such as the phased array ultrasonic that was used during the manufacturing of the fuel tanks. Additionally, the ability to repair those defects at the highest standard and feasibility needs to be continuously developed. Provided that these shortcomings are overcome, FSW technology in conjunction with laser beam welding (LBW) will enable significant weight savings in the aerospace industry as well as in other transport systems. It is also worth pointing out that FSW technology should not be considered a competing joining process to fusion welding techniques such as LBW, but a supporting one.

Author Contributions: Conceptualization, M.M.Z.A., D.F. and M.M.E.-S.S.; methodology, M.M.Z.A., G.Ç. and M.M.E.-S.S.; software, M.M.Z.A., D.F. and M.M.E.-S.S.; validation, M.M.Z.A., G.Ç. and M.M.E.-S.S.; formal analysis, M.M.Z.A. and M.M.E.-S.S.; investigation, M.M.Z.A., D.F. and M.M.E.-S.S.; resources, M.M.Z.A., D.F. and G.Ç.; data curation, M.M.Z.A. and G.Ç.; writing—original draft preparation, M.M.Z.A. and M.M.E.-S.S.; writing—review and editing, M.M.Z.A., G.Ç., D.F. and M.M.E.-S.S.; visualization, G.Ç. and D.F.; supervision, M.M.Z.A.; project administration, M.M.Z.A. and D.F.; funding acquisition, M.M.Z.A. All authors have read and agreed to the published version of the manuscript.

Funding: This study was sponsored by the Prince Sattam bin Abdulaziz University via project number 2023/RV/018.

Institutional Review Board Statement: Not applicable.

Informed Consent Statement: Not applicable.

Data Availability Statement: Will be available through corresponding author.

Acknowledgments: The authors extend their appreciation for the Prince Sattam bin Abdulaziz University for sponsor this study via project number 2023/RV/018.

Conflicts of Interest: The authors declare no conflict of interest.

References

1. Prasad, N.E.; Wanhill, R.J.H. *Aerospace Materials and Material Technologies*; Springer: Singapore, 2017. [CrossRef]
2. Threadgill, P.L.; Leonard, A.J.; Shercliff, H.R.; Withers, P.J. Friction stir welding of aluminium alloys. *Int. Mater. Rev.* **2009**, *54*, 49–93. [CrossRef]
3. Rakshith, M.; Seenuvasaperumal, P. Review on the effect of different processing techniques on the microstructure and mechanical behaviour of AZ31 Magnesium alloy. *J. Magnes. Alloy.* **2021**, *9*, 1692–1714. [CrossRef]
4. Kuai-she, W.; Xun-hong, W. Evaluation of Microstructure and Mechanical Property of FSW Welded MB3 Magnesium Alloy. *J. Iron Steel Res. Int.* **2006**, *13*, 75–78.
5. Hsu, H.H.; Hwang, Y.M. A study on friction stir process of magnesium alloy AZ31 sheet. *Key Eng. Mater.* **2007**, *340–341*, 1449–1454. [CrossRef]
6. Yang, Z.; Li, J.; Zhang, J.; Lorimer, G.; Robson, J. Review on Research and Development of Magnesium Alloys. *Acta Metall. Sin. (Engl. Lett.)* **2008**, *21*, 313–328. [CrossRef]
7. Fu, B.; Shen, J.; Suhuddin, U.F.; Pereira, A.A.; Maawad, E.; dos Santos, J.F.; Klusemann, B.; Rethmeier, M. Revealing joining mechanism in refill friction stir spot welding of AZ31 magnesium alloy to galvanized DP600 steel. *Mater. Des.* **2021**, *209*, 109997. [CrossRef]
8. Ahmed, M.M.Z.; El-Sayed Seleman, M.M.; Sobih, A.M.E.-S.; Bakkar, A.; Albaijan, I.; Touileb, K.; Abd El-Aty, A. Friction Stir-Spot Welding of AA5052-H32 Alloy Sheets: Effects of Dwell Time on Mechanical Properties and Microstructural Evolution. *Materials* **2023**, *16*, 2818. [CrossRef]
9. Liu, Z.; Fan, Z.; Liu, L.; Miao, S.; Lin, Z.; Wang, C.; Zhao, Y.; Xin, R.; Dong, C. Refill friction stir spot welding of AZ31 magnesium alloy sheets: Metallurgical features, microstructure, texture and mechanical properties. *J. Mater. Res. Technol.* **2022**, 105242. [CrossRef]
10. Robelou, A.; Bellarosa, R.; Norman, A.; Andrews, D.; Martin, J.; Nor, K. *Friction Stir Welding of Low Cost Space Hardware-Titanium Propellant Tank (2)*; Airbus Defence and Space: Stevenage, UK, 2016; pp. 3–4.
11. Chen, Y.C.; Nakata, K. Microstructural characterization and mechanical properties in friction stir welding of aluminum and titanium dissimilar alloys. *Mater. Des.* **2009**, *30*, 469–474. [CrossRef]
12. Fujii, H.; Sun, Y.; Kato, H.; Nakata, K. Investigation of welding parameter dependent microstructure and mechanical properties in friction stir welded pure Ti joints. *Mater. Sci. Eng. A* **2010**, *527*, 3386–3391. [CrossRef]
13. Nirmal, K.; Jagadesh, T. Numerical simulations of friction stir welding of dual phase titanium alloy for aerospace applications. *Mater. Today Proc.* **2020**, *46*, 4702–4708. [CrossRef]
14. Meikeerthy, S.; Ethiraj, N.; Neme, I.; Masi, C. Evaluation of Pure Titanium Welded Joints Produced by Underwater Friction Stir Welding. *Adv. Mater. Sci. Eng.* **2023**, *2023*, 2092339. [CrossRef]
15. Mabuwa, S.; Msomi, V.; Muribwathoho, O.; Motshwanedi, S.S. The microstructure and mechanical properties of the friction stir processed TIG-welded aerospace dissimilar aluminium alloys. *Mater. Today Proc.* **2021**, *46*, 658–664. [CrossRef]
16. Ahmed, M.M.Z.; Essa, A.R.S.; Ataya, S.; El-Sayed Seleman, M.M.; El-Aty, A.A.; Alzahrani, B.; Touileb, K.; Bakkar, A.; Ponnore, J.J.; Mohamed, A.Y.A. Friction Stir Welding of AA5754-H24: Impact of Tool Pin Eccentricity and Welding Speed on Grain Structure, Crystallographic Texture, and Mechanical Properties. *Materials* **2023**, *16*, 2031. [CrossRef] [PubMed]
17. Pandian, V.; Kannan, S. Numerical prediction and experimental investigation of aerospace-grade dissimilar aluminium alloy by friction stir welding. *J. Manuf. Process.* **2020**, *54*, 99–108. [CrossRef]
18. Rajan, D.; Prasad, V.S. Evaluation of NCMRWF numerical weather prediction models for SHAR region Space-Launch programme of India. *Adv. Space Res.* **2022**. [CrossRef]

19. Ipekoglu, G.; Erim, S.; Kiral, B.G.; Çam, G. Investigation into the effect of temper condition on friction stir weldability of AA6061 Al-alloy plates. *Met. Mater.* **2021**, *51*, 155–163. [CrossRef]

20. Çam, G.; Ipekoğlu, G.; Serindağ, H.T. Effects of use of higher strength interlayer and external cooling on properties of friction stir welded AA6061-T6 joints. *Sci. Technol. Weld. Join.* **2014**, *19*, 715–720. [CrossRef]

21. Su, Y.; Li, W.; Shen, J.; Bergmann, L.; dos Santos, J.F.; Klusemann, B.; Vairis, A. Comparing the fatigue performance of Ti-4Al-0.005B titanium alloy T-joints, welded via different friction stir welding sequences. *Mater. Sci. Eng. A* **2022**, *859*, 144227. [CrossRef]

22. Du, S.; Liu, H.; Jiang, M.; Hu, Y.; Zhou, L. Eliminating the cavity defect and improving mechanical properties of TA5 alloy joint by titanium alloy supporting friction stir welding. *J. Manuf. Process.* **2021**, *69*, 215–222. [CrossRef]

23. Campanella, D.; Buffa, G.; Lamia, D.; Fratini, L. Residual stress and material flow prediction in Friction Stir Welding of Gr2 Titanium T-joints. *Manuf. Lett.* **2022**, *33*, 249–258. [CrossRef]

24. Zhang, Z.; Tan, Z.J.; Wang, Y.F.; Ren, D.X.; Li, J.Y. The relationship between microstructures and mechanical properties in friction stir lap welding of titanium alloy. *Mater. Chem. Phys.* **2023**, *296*, 127251. [CrossRef]

25. Çam, G. Friction stir welded structural materials: Beyond Al-alloys. *Int. Mater. Rev.* **2011**, *56*, 1–48. [CrossRef]

26. Liu, J.; Wu, B.; Wang, Z.; Li, C.; Chen, G.; Miao, Y. Microstructure and mechanical properties of aluminum-steel dissimilar metal welded using arc and friction stir hybrid welding. *Mater. Des.* **2023**, *225*, 111520. [CrossRef]

27. Han, S.-C.; Chaudry, U.M.; Yoon, J.-Y.; Jun, T.-S. Investigating local strain rate sensitivity of the individual weld zone in the friction stir welded DP 780 steel. *J. Mater. Res. Technol.* **2022**, *20*, 508–515. [CrossRef]

28. Duan, R.H.; Wang, Y.Q.; Luo, Z.A.; Wang, G.D.; Xie, G.M. Achievement of excellent strength and plasticity in the nugget zone of friction stir welded bainitic steel and its deformation behavior. *J. Mater. Res. Technol.* **2022**, *20*, 3381–3390. [CrossRef]

29. Gain, S.; Das, S.K.; Acharyya, S.K.; Sanyal, D. Friction stir welding of industrial grade AISI 316L and P91 steel pipes: A comparative investigation based on mechanical and metallurgical properties. *Int. J. Press. Vessel. Pip.* **2023**, *201*, 104865. [CrossRef]

30. Varghese, J.; Rajulapati, K.V.; Rao, K.B.S.; Meshram, S.D.; Reddy, G.M. Ambient, elevated temperature tensile properties and origin of strengthening in friction stir welded 6 mm thick reduced activation ferritic-martensitic steel plates in as-welded and post-weld normalised conditions. *Mater. Sci. Eng. A* **2022**, *857*, 144019. [CrossRef]

31. Ahmed, M.M.Z.; Abdelazem, K.A.; El-Sayed Seleman, M.M.; Alzahrani, B.; Touileb, K.; Jouini, N.; El-Batanony, I.G.; El-Aziz, H.M.A. Friction stir welding of 2205 duplex stainless steel: Feasibility of butt joint groove filling in comparison to gas tungsten arc welding. *Materials* **2021**, *14*, 4597. [CrossRef] [PubMed]

32. Chaudry, U.; Han, S.-C.; Alkelae, F.; Jun, T.-S. Efffect of PWHT on the Microstructure and Mechanical Properties of Friction Stir Welded DP780 Steel. *Metals* **2021**, *11*, 1097. [CrossRef]

33. Ahmed, M.M.Z.; El-Sayed Seleman, M.M.; Touileb, K.; Albaijan, I.; Habba, M.I.A. Microstructure, crystallographic texture, and mechanical properties of friction stir welded mild steel for shipbuilding applications. *Materials* **2022**, *15*, 2905. [CrossRef] [PubMed]

34. Küçükömero, T.; Aktarer, S.M.; Çam, G. Investigation of mechanical and microstructural properties of friction stir welded dual phase (DP) steel. *IOP Conf. Ser. Mater. Sci. Eng.* **2019**, *629*, 012010. [CrossRef]

35. Ipeko, G.; Küçükömero, T.; Aktarer, S.M.; Sekban, D.; Çam, G. Investigation of microstructure and mechanical properties of friction stir welded dissimilar St37/St52 joints. *Mater. Res Express* **2019**, *6*, 046537. [CrossRef]

36. Hirata, T.; Tanaka, T.; Chung, S.W.; Takigawa, Y.; Higashi, K. Relationship between deformation behavior and microstructural evolution of friction stir processed Zn–22wt.% Al alloy. *Scr. Mater.* **2007**, *56*, 477–480. [CrossRef]

37. Ahmed, M.M.Z.; El-Sayed Seleman, M.M.; Zidan, Z.A.; Ramadan, R.M.; Ataya, S.; Alsaleh, N.A. Microstructure and mechanical properties of dissimilar friction stir welded AA2024-T4/AA7075-T6 T-butt joints. *Metals* **2021**, *11*, 128. [CrossRef]

38. Ahmed, M.M.Z.; Ataya, S.; El-Sayed Seleman, M.M.; Mahdy, A.M.A.; Alsaleh, N.A.; Ahmed, E. Heat input and mechanical properties investigation of friction stir welded aa5083/aa5754 and aa5083/aa7020. *Metals* **2021**, *11*, 68. [CrossRef]

39. Charit, I.; Mishra, R.S. Low temperature superplasticity in a friction-stir-processed ultrafine grained Al–Zn–Mg–Sc alloy. *Acta Mater.* **2005**, *53*, 4211–4223. [CrossRef]

40. Gomez, A.; Smith, H. Liquid hydrogen fuel tanks for commercial aviation: Structural sizing and stress analysis. *Aerosp. Sci. Technol.* **2019**, *95*, 105438. [CrossRef]

41. Dursun, T.; Soutis, C. Recent developments in advanced aircraft aluminium alloys. *Mater. Des.* **2014**, *56*, 862–871. [CrossRef]

42. Meisnar, M.; Bennett, J.M.; Andrews, D.; Dodds, S.; Freeman, R.; Bellarosa, R.; Adams, D.; Norman, A.F.; Rohr, T.; Ghidini, T. Microstructure characterisation of a friction stir welded hemi-cylinder structure using Ti-6Al-4V castings. *Mater. Charact.* **2018**, *147*, 286–294. [CrossRef]

43. Wang, G.-Q.; Zhao, Y.-H.; Tang, Y.-Y. Research Progress of Bobbin Tool Friction Stir Welding of Aluminum Alloys: A Review. *Acta Met. Sin.* **2019**, *33*, 13–29. [CrossRef]

44. Grimm, A. *New Joining Technologies for Future Fuselage Metal Structures*; Fraunhofer IWS Annual Report 2014; Fraunhofer-Gesellschaft: München, Germany, 2014.

45. Wang, G.; Zhao, Y.; Hao, Y. Friction stir welding of high-strength aerospace aluminum alloy and application in rocket tank manufacturing. *J. Mater. Sci. Technol.* **2018**, *34*, 73–91. [CrossRef]

46. Tavares, S.; dos Santos, J.; de Castro, P. Friction stir welded joints of Al-Li Alloys for aeronautical applications: Butt-joints and tailor welded blanks. *Theor. Appl. Fract. Mech.* **2013**, *65*, 8–13. [CrossRef]

47. Boldsaikhan, E.; Fukada, S.; Fujimoto, M.; Kamimuki, K.; Okada, H. Refill friction stir spot welding of surface-treated aerospace aluminum alloys with faying-surface sealant. *J. Manuf. Process.* **2019**, *42*, 113–120. [CrossRef]

48. Christner, B.; Hansen, M.; Skinner, M.; Sylva, G. Friction Stir Welding System Development for Thin-Gauge Aersopace Structures. In Proceedings of the Fourth International Symposium on Friction Stir Welding, Park City, UT, USA, 14–16 May 2003; Volume 332, pp. 1–6. [CrossRef]

49. Shepherd, G.E. The Evaluation of Friction Stir Welded Joints on Airbus Aircraft Wing Structure. In Proceedings of the Fourth International Symposium on Friction Stir Welding, Park City, UT, USA, 14–16 May 2003; pp. 1–5.

50. Amini, A.; Asadi, P.; Zolghadr, P. Friction stir welding applications in industry. In *Woodhead Publishing Series in Welding and Other Joining Technologies*; Givi, M.K.B., Asadi, P., Eds.; Woodhead Publishing: Sawston, UK, 2014; pp. 671–722. [CrossRef]

51. Kallee, S.W. Industrial applications of friction stir welding. In *Woodhead Publishing Series in Welding and Other Joining Technologies*; Lohwasser, D., Chen, W., Eds.; Woodhead Publishing: Sawston, UK, 2010; pp. 118–163. [CrossRef]

52. Meng, X.; Huang, Y.; Cao, J.; Shen, J.; dos Santos, J.F. Recent progress on control strategies for inherent issues in friction stir welding. *Prog. Mater. Sci.* **2020**, *115*, 100706. [CrossRef]

53. Mishra, R.S.; Sidhar, H. *Chapter 1—Friction Stir Welding*; Mishra, R.S., Sidhar, A., Eds.; Butterworth-Heinemann: Oxford, UK, 2017; pp. 1–13. [CrossRef]

54. Rahmatabadi, D.; Pahlavani, M.; Gholami, M.D.; Marzbanrad, J.; Hashemi, R. Production of Al/Mg-Li composite by the accumulative roll bonding process. *J. Mater. Res. Technol.* **2020**, *9*, 7880–7886. [CrossRef]

55. Rahmatabadi, D.; Tayyebi, M.; Najafizadeh, N.; Hashemi, R.; Rajabi, M. The influence of post-annealing and ultrasonic vibration on the formability of multilayered Al5052/MgAZ31B composite. *Mater. Sci. Technol.* **2021**, *37*, 78–85. [CrossRef]

56. Rahmatabadi, D.; Pahlavani, M.; Marzbanrad, J.; Hashemi, R.; Bayati, A. Manufacturing of three-layered sandwich composite of AA1050/LZ91/AA1050 using cold roll bonding process. *Proc. Inst. Mech. Eng. Part B J. Eng. Manuf.* **2021**, *235*, 1363–1372. [CrossRef]

57. Ahmed, M.M.Z.; Jouini, N.; Alzahrani, B.; El-Sayed Seleman, M.M.; Jhaheen, M. Dissimilar friction stir welding of AA2024 and AISI 1018: Microstructure and mechanical properties. *Metals* **2021**, *11*, 330. [CrossRef]

58. Ahmed, M.M.Z.; Habba, M.I.A.; El-Sayed Seleman, M.M.; Hajlaoui, K.; Ataya, S.; Latief, F.H.; El-Nikhaily, A.E. Bobbin Tool Friction Stir Welding of Aluminum Thick Lap Joints: Effect of Process Parameters on Temperature Distribution and Joints' Properties. *Materials* **2021**, *14*, 4585. [CrossRef]

59. Lohwasser, D.; Chen, Z. *Friction Stir Welding from Basics to Applications*; Woodhead Publishing Limited: Sawston, UK, 2010.

60. Yu, H.Z.; Jones, M.E.; Brady, G.W.; Griffiths, R.J.; Garcia, D.; Rauch, H.A.; Cox, C.D.; Hardwick, N. Non-beam-based metal additive manufacturing enabled by additive friction stir deposition. *Scr. Mater.* **2018**, *153*, 122–130. [CrossRef]

61. Boitsov, A.G.; Kuritsyn, D.N.; Siluyanova, M.V.; Kuritsyna, V.V. Friction Stir Welding in the Aerospace Industry. *Russ. Eng. Res.* **2018**, *38*, 1029–1033. [CrossRef]

62. Thomas, W.M.; Nicholas, E.D.; Needham, J.C.; Murch, M.G.; Templesmith, P.; Dawes, C.J. Friction Stir Welding. G.B. Patent Application No. 9125978, 6 December 1991.

63. Dawes, C.; Thomas, W. *Friction Stir Joining of Aluminium Alloys*; TWI Bulletin 6; TWI: Cambridge, UK, 1995.

64. Thomas, W.M.; Nicholas, E.D.; Needham, J.C.; Murch, M.G.; Templesmith, P.; Dawes, C.J. Friction welding. U.S. Patent No. 5,460,317, 24 October 1995.

65. Mishra, R.S.; Ma, Z.Y. Friction stir welding and processing. *Mater. Sci. Eng. R Rep.* **2005**, *50*, 1–78. [CrossRef]

66. Çam, G.; Javaheri, V.; Heidarzadeh, A. Advances in FSW and FSSW of dissimilar Al-alloy plates. *J. Adhes. Sci. Technol.* **2022**, *37*, 162–194. [CrossRef]

67. Kashaev, N.; Ventzke, V.; Çam, G. Prospects of laser beam welding and friction stir welding processes for aluminum airframe structural applications. *J. Manuf. Process.* **2018**, *36*, 571–600. [CrossRef]

68. Heidarzadeh, A.; Mironov, S.; Kaibyshev, R.; Çam, G.; Simar, A.; Gerlich, A.; Khodabakhshi, F.; Mostafaei, A.; Field, D.; Robson, J.; et al. Friction stir welding/processing of metals and alloys: A comprehensive review on microstructural evolution. *Prog. Mater. Sci.* **2021**, *117*, 100752. [CrossRef]

69. Çam, G.; İpekoğlu, G. Recent developments in joining of aluminum alloys. *Int. J. Adv. Manuf. Technol.* **2016**, *91*, 1851–1866. [CrossRef]

70. Çam, G.; Mistikoglu, S. Recent developments in friction stir welding of al-Alloys. *J. Mater. Eng. Perform.* **2014**, *23*, 1936–1953. [CrossRef]

71. Ipeko, G.; Çam, G. Formation of weld defects in cold metal transfer arc welded 7075-T6 plates and its effect on joint performance. *IOP Conf. Ser. Mater. Sci. Eng.* **2019**, *629*, 012007. [CrossRef]

72. Çam, G. Prospects of producing aluminum parts by wire arc additive manufacturing (WAAM). *Mater. Today Proc.* **2022**, *62*, 77–85. [CrossRef]

73. Çam, G.; Koçak, M. Microstructural and mechanical characterization of electron beam welded Al-alloy 7020. *J. Mater. Sci.* **2007**, *42*, 7154–7161. [CrossRef]

74. Ahmed, M.M.Z. The Development of Thick Section Welds and Ultra-Fine Grain Aluminium Using Friction Stir Welding and Processing. Ph.D. Thesis, The University of Sheffield, Sheffield, UK, 2009.

75. Ahmed, M.M.Z.Z.; Wynne, B.P.; Rainforth, W.M.; Addison, A.; Martin, J.P.; Threadgill, P.L. Effect of Tool Geometry and Heat Input on the Hardness, Grain Structure, and Crystallographic Texture of Thick-Section Friction Stir-Welded Aluminium. *Met. Mater. Trans. A* **2018**, *50*, 271–284. [CrossRef]

76. Ahmed, M.M.Z.; Wynne, B.P.; Martin, J.P. Effect of friction stir welding speed on mechanical properties and microstructure of nickel based super alloy Inconel 718. *Sci. Technol. Weld. Join.* **2013**, *18*, 680–687. [CrossRef]

77. Li, G.; Zhou, L.; Luo, L.; Wu, X.; Guo, N. Microstructural evolution and mechanical properties of refill friction stir spot welded alclad 2A12-T4 aluminum alloy. *J. Mater. Res. Technol.* **2019**, *8*, 4115–4129. [CrossRef]

78. Ahmed, M.M.Z.; Seleman, M.M.E.S.; Ahmed, E.; Reyad, H.A.; Touileb, K.; Albaijan, I. Friction Stir Spot Welding of Different Thickness Sheets of Aluminum Alloy AA6082-T6. *Materials* **2022**, *15*, 2971. [CrossRef]

79. Sun, Y.; Fujii, H.; Zhu, S.; Guan, S. Flat friction stir spot welding of three 6061-T6 aluminum sheets. *J. Mater. Process. Technol.* **2018**, *264*, 414–421. [CrossRef]

80. de Castro, C.C.; Shen, J.; Plaine, A.H.; Suhuddin, U.F.; de Alcântara, N.G.; dos Santos, J.F.; Klusemann, B. Tool wear mechanisms and effects on refill friction stir spot welding of AA2198-T8 sheets. *J. Mater. Res. Technol.* **2022**, *20*, 857–866. [CrossRef]

81. Kubit, A.; Kluz, R.; Trzepieciński, T.; Wydrzyński, D.; Bochnowski, W. Analysis of the mechanical properties and of micrographs of refill friction stir spot welded 7075-T6 aluminium sheets. *Arch. Civ. Mech. Eng.* **2018**, *18*, 235–244. [CrossRef]

82. Kubit, A.; Bucior, M.; Wydrzyński, D.; Trzepieciński, T.; Pytel, M. Failure mechanisms of refill friction stir spot welded 7075-T6 aluminium alloy single-lap joints. *Int. J. Adv. Manuf. Technol.* **2017**, *94*, 4479–4491. [CrossRef]

83. Kubit, A.; Wydrzynski, D.; Trzepiecinski, T. Refill friction stir spot welding of 7075-T6 aluminium alloy single-lap joints with polymer sealant interlayer. *Compos. Struct.* **2018**, *201*, 389–397. [CrossRef]

84. Kubit, A.; Trzepiecinski, T.; Faes, K.; Drabczyk, M.; Bochnowski, W.; Korzeniowski, M. Analysis of the effect of structural defects on the fatigue strength of RFSSW joints using C-scan scanning acoustic microscopy and SEM. *Fatigue Fract. Eng. Mater. Struct.* **2018**, *42*, 1308–1321. [CrossRef]

85. Yang, X.W.; Fu, T.; Li, W.Y. Friction Stir Spot Welding: A Review on Joint Macro- and Microstructure, Property, and Process Modelling. *Adv. Mater. Sci. Eng.* **2014**, *2014*, 697170. [CrossRef]

86. Ahmed, M.M.Z.; El-Sayed Seleman, M.M.; Ahmed, E.; Reyad, H.A.; Alsaleh, N.A.; Albaijan, I. A Novel Friction Stir Deposition Technique to Refill Keyhole of Friction Stir Spot Welded AA6082-T6 Dissimilar Joints of Different Sheet Thicknesses. *Materials* **2022**, *15*, 6799. [CrossRef] [PubMed]

87. Rutherford, B.A.; Avery, D.Z.; Phillips, B.J.; Rao, H.M.; Doherty, K.J.; Allison, P.G.; Brewer, L.N.; Jordon, J.B. Effect of thermomechanical processing on fatigue behavior in solid-state additive manufacturing of Al-Mg-Si alloy. *Metals* **2020**, *10*, 947. [CrossRef]

88. Ahmed, M.M.Z.; El-Sayed Seleman, M.M.; Elfishawy, E.; Alzahrani, B.; Touileb, K.; Habba, M.I.A. The Effect of Temper Condition and Feeding Speed on the Additive Manufacturing of AA2011 Parts Using Friction Stir Deposition. *Materials* **2021**, *14*, 6396. [CrossRef] [PubMed]

89. El-Sayed Seleman, M.M.; Ataya, S.; Ahmed, M.M.Z.; Hassan, A.M.M.; Latief, F.H.; Hajlaoui, K.; El-Nikhaily, A.E.; Habba, M.I.A. The Additive Manufacturing of Aluminum Matrix Nano Al2O3 Composites Produced via Friction Stir Deposition Using Different Initial Material Conditions. *Materials* **2022**, *15*, 2926. [CrossRef] [PubMed]

90. Jahangir, M.N.; Mamun, M.A.H.; Sealy, M.P. A review of additive manufacturing of magnesium alloys. In *AIP Conference Proceedings*; AIP Publishing LLC: Melville, NY, USA, 2018; Volume 1980, p. 030026. [CrossRef]

91. Elfishawy, E., Ahmed, M.M.Z.; El-Sayed Seleman, M.M. Additive Manufacturing of Aluminum Using Friction Stir Deposition. In Proceedings of the TMS 2020 149th Annual Meeting & Exhibition Supplemental Proceedings, San Diego, CA, USA, 23–27 February 2020; Minerals, Metals & Materials Society, Ed.; The Minerals, Metals & Materials Series; Springer: Cham, Switzerland, 2020. [CrossRef]

92. Alzahrani, B.; El-Sayed Seleman, M.M.; Ahmed, M.M.Z.; Elfishawy, E.; Ahmed, A.M.Z.; Touileb, K.; Jouini, N.; Habba, M.I.A. The Applicability of Die Cast A356 Alloy to Additive Friction Stir Deposition at Various Feeding Speeds. *Materials* **2021**, *14*, 6018. [CrossRef] [PubMed]

93. Zou, Y.; Li, W.; Yang, X.; Patel, V.; Shen, Z.; Chu, Q.; Wang, F.; Tang, H.; Cui, F.; Chi, M. Characterizations of dissimilar refill friction stir spot welding 2219 aluminum alloy joints of unequal thickness. *J. Manuf. Process.* **2022**, *79*, 91–101. [CrossRef]

94. Zou, Y.; Li, W.; Xu, Y.; Yang, X.; Chu, Q.; Shen, Z. Detailed characterizations of microstructure evolution, corrosion behavior and mechanical properties of refill friction stir spot welded 2219 aluminum alloy. *Mater. Charact.* **2021**, *183*, 111594. [CrossRef]

95. Janga, V.S.R.; Awang, M.; Yamin, M.F.; Suhuddin, U.F.H.; Klusemann, B.; Dos Santos, J.F. Experimental and numerical analysis of refill friction stir spot welding of thin AA7075-T6 sheets. *Materials* **2021**, *14*, 7485. [CrossRef]

96. Janga, V.S.R.; Awang, M. Influence of Plunge Depth on Temperatures and Material Flow Behavior in Refill Friction Stir Spot Welding of thin AA7075-T6 Sheets: A Numerical Study. *Metals* **2022**, *12*, 927. [CrossRef]

97. Fritsche, S.; Draper, J.; Toumpis, A.; Galloway, A.; Amancio-Filho, S.T. Refill friction stir spot welding of AlSi10Mg alloy produced by laser powder bed fusion to wrought AA7075-T6 alloy. *Manuf. Lett.* **2022**, *34*, 78–81. [CrossRef]

98. Zhang, D.; Dong, J.; Xiong, J.; Jiang, N.; Li, J.; Guo, W. Microstructure characteristics and corrosion behavior of refill friction stir spot welded 7050 aluminum alloy. *J. Mater. Res. Technol.* **2022**, *20*, 1302–1314. [CrossRef]

99. Deng, L.; Li, S.; Ke, L.; Liu, J.; Kang, J. Microstructure and fracture behavior of refill friction stir spot welded joints of AA2024 using a novel refill technique. *Metals* **2019**, *9*, 286. [CrossRef]

100. Gera, D.; Fu, B.; Suhuddin, U.F.; Plaine, A.; Alcantara, N.; dos Santos, J.F.; Klusemann, B. Microstructure, mechanical and functional properties of refill friction stir spot welds on multilayered aluminum foils for battery application. *J. Mater. Res. Technol.* **2021**, *13*, 2272–2286. [CrossRef]

101. Wang, S.; Wei, X.; Xu, J.; Hong, J.; Song, X.; Yu, C.; Chen, J.; Chen, X.; Lu, H. Strengthening and toughening mechanisms in refilled friction stir spot welding of AA2014 aluminum alloy reinforced by graphene nanosheets. *Mater. Des.* **2019**, *186*, 108212. [CrossRef]
102. Yousefi, A.; Serjouei, A.; Hedayati, R.; Bodaghi, M. Fatigue modeling and numerical analysis of re-filling probe hole of friction stir spot welded joints in aluminum alloys. *Materials* **2021**, *14*, 2171. [CrossRef]
103. Zou, Y.; Li, W.; Yang, X.; Su, Y.; Chu, Q.; Shen, Z. Microstructure and mechanical properties of refill friction stir spot welded joints: Effects of tool size and welding parameters. *J. Mater. Res. Technol.* **2022**, *21*, 5066–5080. [CrossRef]
104. Chen, D.; Li, J.; Xiong, J.; Shi, J.; Dou, J.; Zhao, H. Enhance mechanical properties of refill friction stir spot welding joint of alclad 7050/2524 aluminum via suspension rotating process. *J. Mater. Res. Technol.* **2021**, *12*, 1243–1251. [CrossRef]
105. Li, P.; Chen, S.; Dong, H.; Ji, H.; Li, Y.; Guo, X.; Yang, G.; Zhang, X.; Han, X. Interfacial microstructure and mechanical properties of dissimilar aluminum/steel joint fabricated via refilled friction stir spot welding. *J. Manuf. Process.* **2019**, *49*, 385–396. [CrossRef]
106. Fu, B.; Shen, J.; Suhuddin, U.F.; Chen, T.; dos Santos, J.F.; Klusemann, B.; Rethmeier, M. Improved mechanical properties of cast Mg alloy welds via texture weakening by differential rotation refill friction stir spot welding. *Scr. Mater.* **2021**, *203*, 114113. [CrossRef]
107. Sarila, V.; Koneru, H.P.; Cheepu, M.; Chigilipalli, B.K.; Kantumuchu, V.C.; Shanmugam, M. Microstructural and Mechanical Properties of AZ31B to AA6061 Dissimilar Joints Fabricated by Refill Friction Stir Spot Welding. *J. Manuf. Mater. Process.* **2022**, *6*, 95. [CrossRef]
108. Wynne, B.P.; Threadgill, P.L.; Davies, P.S.; Thomas, M.J.; Ng, B.S. Microstructure and Texture in Static Shoulder Friction Stir Welds of Ti–6Al–4V. In Proceedings of the 7th International Friction Stir Welding Symposium, Awaji, Japan, 20–22 May 2008; pp. 1–8.
109. Russell, M.J.; Blignault, C. Recent developments in friction stir welding of Ti alloys. In Proceedings of the 6th International Symposium on Friction Stir Welding, Saint Sauveur, QC, Canada, 10–13 October 2006.
110. Russell, M.J.; Threadgill, P.L.; Thomas, M.J.; Wynne, B.P. Static shoulder friction stir welding of Ti-6Al-4V; process and evaluation. In Proceedings of the 11th World Conference on titanium (Ti-2007), (JIMIC-5), Kyoto, Japan, 3–7 June 2007.
111. Ahmed, M.M.Z.; Wynne, B.P.; Rainforth, W.M.; Threadgill, P.L. Through-thickness crystallographic texture of stationary shoulder friction stir welded aluminium. *Scr. Mater.* **2011**, *64*, 45–48. [CrossRef]
112. Hammad, A.S.; Ahmed, M.M.; Lu, H.; El-Shabasy, A.B.; Alzahrani, B.; El-Sayed Seleman, M.M.; Zhang, Y.; El Megharbel, A. An investigation on mechanical and microstructural evolution of stationary shoulder friction stir welded aluminum alloy AA7075-T651. *Proc. Inst. Mech. Eng. Part C J. Mech. Eng. Sci.* **2022**, *236*, 6665–6676. [CrossRef]
113. Barbini, A.; Carstensen, J.; Santos, J.F. Influence of a non-rotating shoulder on heat generation, microstructure and mechanical properties of dissimilar AA2024 / AA7050 FSW joints. *J. Mater. Sci. Technol.* **2018**, *34*, 119–127. [CrossRef]
114. Li, D.; Yang, X.; Cui, L.; He, F.; Zhang, X. Journal of Materials Processing Technology Investigation of stationary shoulder friction stir welding of aluminum alloy 7075-T651. *J. Mater. Process. Technol.* **2015**, *222*, 391–398. [CrossRef]
115. Sun, T.; Roy, M.J.; Strong, D.; Simpson, C.; Withers, P.J.; Prangnell, P.B. Weld zone and residual stress development in AA7050 stationary shoulder friction stir T-joint weld. *J. Mater. Process. Technol.* **2019**, *263*, 256–265. [CrossRef]
116. Sun, Z.; Yang, X.; Li, D.; Cui, L. The local strength and toughness for stationary shoulder friction stir weld on AA6061-T6 alloy. *Mater. Charact.* **2016**, *111*, 114–121. [CrossRef]
117. Ji, S.D.; Meng, X.C.; Liu, J.G.; Zhang, L.G.; Gao, S.S. Formation and mechanical properties of stationary shoulder friction stir welded 6005A-T6 aluminum alloy. *Mater. Des.* **2014**, *62*, 113–117. [CrossRef]
118. Wu, H.; Chen, Y.; Strong, D.; Prangnell, P. Stationary shoulder FSW for joining high strength aluminum alloys. *J. Mater. Process. Technol.* **2015**, *221*, 187–196. [CrossRef]
119. Richardson, M. A Stirring Work of Friction. *Aerospace Manufacturing Magazine*. 2014. Available online: https://www.aero-mag.com/a-stirring-work-of-friction/ (accessed on 11 March 2020).
120. Marie, F.; Silvanous, J.; Hahan, S. Joining of dissimilar metals for satellite feedtroughs using DeltaN FS. In Proceedings of the Tenth FSW Symposium, Beijing, China, 20–22 May 2014.
121. Threadgill, P.L.; Ahmed, M.M.Z.; JMartin, J.P.; Perrett, J.G.; Wynne, B.P. The use of bobbin tools for friction stir welding of aluminium alloys. *Mater. Sci. Forum* **2010**, *638–642*, 1179–1184. [CrossRef]
122. Ahmed, M.M.Z.; Touileb, K.; El-Sayed Seleman, M.M.; Albaijan, I.; Habba, M.I.A. Bobbin Tool Friction Stir Welding of Aluminum: Parameters Optimization Using Taguchi Experimental Design. *Materials* **2022**, *15*, 2771. [CrossRef]
123. Ataya, S.; Ahmed, M.M.Z.; El-Sayed Seleman, M.M.; Hajlaoui, K.; Latief, F.H.; Soliman, A.M.; Elshaghoul, Y.G.Y.; Habba, M.I.A. Effective Range of FSSW Parameters for High Load-Carrying Capacity of Dissimilar Steel A283M-C/Brass CuZn40 Joints. *Materials* **2022**, *15*, 1394. [CrossRef] [PubMed]
124. Xu, W.F.; Luo, Y.X.; Fu, M.W. Microstructure evolution in the conventional single side and bobbin tool friction stir welding of thick rolled 7085-T7452 aluminum alloy. *Mater. Charact.* **2018**, *138*, 48–55. [CrossRef]
125. Xu, W.; Luo, Y.; Zhang, W.; Fu, M. Comparative study on local and global mechanical properties of bobbin tool and conventional friction stir welded 7085-T7452 aluminum thick plate. *J. Mater. Sci. Technol.* **2018**, *34*, 173–184. [CrossRef]
126. Yang, C.; Ni, D.; Xue, P.; Xiao, B.; Wang, W.; Wang, K.; Ma, Z. A comparative research on bobbin tool and conventional friction stir welding of Al-Mg-Si alloy plates. *Mater. Charact.* **2018**, *145*, 20–28. [CrossRef]
127. Ahmed, M.M.Z.; El-Sayed Seleman, M.M.; Eid, R.G.; Zawrah, M.F. Production of AA1050/silica fume composite by bobbin tool-friction stir processing: Microstructure, composition and mechanical properties. *CIRP J. Manuf. Sci. Technol.* **2022**, *38*, 801–812. [CrossRef]

128. Ahmed, M.M.Z.; Habba, M.I.A.; Jouini, N.; Alzahrani, B.; El-Sayed Seleman, M.M.; El-Nikhaily, A. Bobbin tool friction stir welding of aluminum using different tool pin geometries: Mathematical models for the heat generation. *Metals* **2021**, *11*, 438. [CrossRef]

129. Ahmed, M.M.Z.; Seleman, M.M.E.S.; Eid, R.G.; Albaijan, I.; Touileb, K. The Influence of Tool Pin Geometry and Speed on the Mechanical Properties of the Bobbin Tool Friction Stir Processed AA1050. *Materials* **2022**, *15*, 4684. [CrossRef]

130. Albaijan, I.; Ahmed, M.M.Z.; El-Sayed Seleman, M.M.; Touileb, K.; Habba, M.I.A.; Fouad, R.A. Optimization of Bobbin Tool Friction Stir Processing Parameters of AA1050 Using Response Surface Methodology. *Materials* **2022**, *15*, 6886. [CrossRef]

131. Mouritz, A.P. *Introduction to Aerospace Materials*; Woodhead Publishing Limited: Sawston, UK, 2012.

132. Kaempf and Harris. A Quick History of Metal Fabrication in the Aerospace Industry. Industry Articles from Kaempf and Harris. 2019. Available online: https://www.kaempfandharris.com/industry-news/a-quick-history-of-metal-fabrication-in-the-aerospace-industry (accessed on 4 December 2019).

133. Smye, B. Aluminum Alloys for Aerospace. Aerospace Manufacturing and Design. 2018. Available online: https://www.aerospacemanufacturinganddesign.com/article/aluminum-alloys-for-aerospace/ (accessed on 11 March 2020).

134. Polmear, I.J. *Light Alloys: Metallurgy of the Light Metals*, 3rd ed.; Edward Arnold: London, UK, 1995.

135. Hatch, J.E. *Aluminum: Properties and Physical Metallurgy*, 1st ed.; ASM: Metals Park, OH, USA, 1984.

136. Campbell, F.C. *Elements of Metallurgy and Engineering Alloys*; ASM International: Materials Park, OH, USA, 2008.

137. Polmear, I. *Light Alloys from Traditional Alloys to Nanocrystals*, 4th ed.; Elsevier: London, UK, 2005.

138. Davis, J.R. *Light Metals and Alloys, Alloying: Understanding the Basics*; ASM International: Novelty, OH, USA, 2001.

139. Seleman, M.M.E.S.; Ahmed, M.M.Z.; Ramadan, R.M.; Zaki, B.A. Effect of FSW Parameters on The Microstructure and Mechanical Properties of T-joints between Dissimilar Al-Alloys. *Int. J. Integr. Eng.* **2022**, *14*, 1–12. [CrossRef]

140. Polmear, I.J. Recent Developments in light alloys. *Mater. Trans. JIM* **1996**, *37*, 12–31. [CrossRef]

141. Ahmed, M.M.Z.; Hajlaoui, K.; El-Sayed Seleman, M.M.; Elkady, M.F.; Ataya, S.; Latief, F.H.; Habba, M.I.A. Microstructure and Mechanical Properties of Friction Stir Welded 2205 Duplex Stainless Steel Butt Joints. *Materials* **2021**, *14*, 6640. [CrossRef] [PubMed]

142. Wanhill, R.J.H.; Bray, G.H. Aerostructural design and its application to aluminum lithium alloys. In *Aluminum-Lithium Alloys: Processing, Properties, and Applications*; Prasad, N.E., Gokhale, A.A., Wanhill, R.J.H., Eds.; Butterworth-Heinemann–Elsevier: Amsterdam, The Netherlands, 2014; pp. 27–57.

143. El-Aty, A.A.; Xu, Y.; Guo, X.; Zhang, S.H.; Ma, Y.; Chen, D. Strengthening mechanisms, deformation behavior, and anisotropic mechanical properties of Al-Li alloys: A review. *J. Adv. Res.* **2018**, *10*, 49–67. [CrossRef]

144. Prasad, N.E.; Gokhale, A.; Wanhill, R. *Aluminum-Lithium Alloys: Processing, Properties, and Applications*; Butterworth-Heinemann–Elsevier: Amsterdam, The Netherlands, 2013.

145. Rioja, R.J.; Liu, J. The evolution of Al-Li base products for aerospace and space applications. *Metall. Mater. Trans. A Phys. Metall. Mater. Sci.* **2012**, *43*, 3325–3337. [CrossRef]

146. Lequeu, P.; Smith, K.P.; Daniélou, A. Aluminum-copper-lithium alloy 2050 developed for medium to thick plate. *J. Mater. Eng. Perform.* **2009**, *19*, 841–847. [CrossRef]

147. Fridlyander, I.N.; Khokhlatova, L.B.; Kolobnev, N.I.; Rendiks, K.; Tempus, G. Thermally stable aluminum-lithium alloy 1424 for application in welded fuselage. *Met. Sci. Heat Treat.* **2002**, *44*, 3–8. [CrossRef]

148. Khokhlatova, L.B.; Kolobnev, N.I.; Oglodkov, M.S.; Mikhaylov, E.D. Aluminum-lithium alloys for aircraft building. *Metallurgist* **2012**, *56*, 336–341. [CrossRef]

149. Rahmatabadi, D.; Hashemi, R.; Tayyebi, M.; Bayati, A. Investigation of mechanical properties, formability, and anisotropy of dual phase Mg-7Li-1Zn. *Mater. Res. Express* **2019**, *6*, 096543. [CrossRef]

150. Pahlavani, M.; Marzbanrad, J.; Rahmatabadi, D.; Hashemi, R.; Bayati, A. A comprehensive study on the effect of heat treatment on the fracture behaviors and structural properties of Mg-Li alloys using RSM. *Mater. Res. Express* **2019**, *6*, 076554. [CrossRef]

151. Sidhar, H.; Martinez, N.Y.; Mishra, R.S.; Silvanus, J. Friction stir welding of Al-Mg-Li 1424 alloy. *Mater. Des.* **2016**, *106*, 146–152. [CrossRef]

152. Mishra, A. Friction Stir Welding of Aerospace Alloys. *Int. J. Res. Appl. Sci. Eng. Technol.* **2019**, *7*, 863–870. [CrossRef]

153. Lee, C.Y.; Lee, W.B.; Kim, J.W.; Choi, D.H.; Yeon, Y.M.; Jung, S.B. Lap joint properties of FSWed dissimilar formed 5052 Al and 6061 Al alloys with different thickness. *J. Mater. Sci.* **2008**, *43*, 3296–3304. [CrossRef]

154. da Silva, A.A.M.; Arruti, E.; Janeiro, G.; Aldanondo, E.; Alvarez, P.; Echeverria, A. Material flow and mechanical behaviour of dissimilar AA2024-T3 and AA7075-T6 aluminium alloys friction stir welds. *Mater. Des.* **2011**, *32*, 2021–2027. [CrossRef]

155. Mehdi, H.; Mishra, R.S. Mechanical and microstructure characterization of friction stir welding for dissimilar alloy-A Review. *Int. J. Res. Eng. Innov.* **2017**, *1*, 57–67. Available online: http://www.ijrei.com (accessed on 11 March 2020).

156. Dubourg, L.; Merati, A.; Jahazi, M. Process optimisation and mechanical properties of friction stir lap welds of 7075-T6 stringers on 2024-T3 skin. *Mater. Des.* **2010**, *31*, 3324–3330. [CrossRef]

157. Benavides, S.; Li, Y.; Murr, L.; Brown, D.; McClure, J. Low-temperature friction-stir welding of 2024 aluminum. *Scr. Mater.* **1999**, *41*, 809–815. [CrossRef]

158. Sutton, M.; Yang, B.; Reynolds, A.; Taylor, R. Microstructural studies of friction stir welds in 2024-T3 aluminum. *Mater. Sci. Eng. A* **2002**, *323*, 160–166. [CrossRef]

159. Amancio-Filhoa, S.T.; Sheikhia, S.; Santosa, J.F.D.; Bolfarini, C. Preliminary study on the microstructure and mechanical properties of dissimilar friction stir welds in aircraft aluminium alloys 2024-T351 and 6056-T4. *J. Mater. Process. Technol.* **2008**, *206*, 132–142. [CrossRef]

160. Avinash, P.; Manikandan, M.; Arivazhagan, N.; Ramkumar, K.D.; Narayanan, S. Friction stir welded butt joints of AA2024 T3 and AA7075 T6 aluminum alloys. *Procedia Eng.* **2014**, *75*, 98–102. [CrossRef]

161. RaviKumar, S.; Rao, V.S.; Pranesh, R.V. Effect of Welding Parameters on Macro and Microstructure of Friction Stir Welded Dissimilar Butt Joints between AA7075-T651 and AA6061-T651 Alloys. *Procedia Mater. Sci.* **2014**, *5*, 1726–1735. [CrossRef]

162. Shanavas, S.; Dhas, J.E.R.; Murugan, N. Weldability of marine grade AA 5052 aluminum alloy by underwater friction stir welding. *Int. J. Adv. Manuf. Technol.* **2018**, *95*, 4535–4546. [CrossRef]

163. Khalaf, H.I.; Al-sabur, R.; Abdullah, M.E.; Kubit, A.; Derazkola, H.A. Effects of Underwater Friction Stir Welding Heat Generation on Residual Stress of AA6068-T6 Aluminum Alloy. *Materials* **2022**, *15*, 2223. [CrossRef]

164. Saravanakumar, R.; Rajasekaran, T.; Pandey, C. Underwater Friction Stir Welded Armour Grade AA5083 Aluminum Alloys: Experimental Ballistic Performance and Corrosion Investigation. *J. Mater. Eng. Perform.* **2023**, 1–16. [CrossRef]

165. Heidarzadeh, A.; Javidani, M.; Mofarrehi, M.; Farzaneh, A.; Chen, X.G. Submerged dissimilar friction stir welding of aa6061 and AA7075 aluminum alloys: Microstructure characterization and mechanical property. *Metals* **2021**, *11*, 1592. [CrossRef]

166. Kumbhar, N.T.; Bhanumurthy, K. Friction stir welding of 6061 alloy. *Asian J. Exp. Sci.* **2008**, *22*, 63–74. [CrossRef]

167. Scialpi, A.; De Filippis, L.A.C.; Cavaliere, P. Influence of shoulder geometry on microstructure and mechanical properties of friction stir welded 6082 aluminium alloy. *Mater. Des.* **2007**, *28*, 1124–1129. [CrossRef]

168. Sato, Y.S.; Kokawa, H.; Enomoto, M.; Jogan, S. Microstructural evolution of 6063 aluminum during friction-stir welding. *Met. Mater. Trans. A* **1999**, *30*, 2429–2437. [CrossRef]

169. Kimura, M.; Kusaka, M.; Seo, K.; Fuji, A. Joining phenomena during friction stage of A7075-T6 aluminium alloy friction weld. *Sci. Technol. Weld. Join.* **2005**, *10*, 378–383. [CrossRef]

170. Fu, R.; Sun, Z.; Sun, R.; Li, Y.; Liu, H.; Liu, L. Improvement of weld temperature distribution and mechanical properties of 7050 aluminum alloy butt joints by submerged friction stir welding. *Mater. Des.* **2011**, *32*, 4825–4831. [CrossRef]

171. Venugopal, T.; Rao, K.S.; Rao, K.P. Studies on friction stir welded aa 7075 aluminum alloy. *Trans. Indian Inst. Met.* **2004**, *57*, 659–663.

172. Ahmed, M.M.Z.; Ataya, S.; El-Sayed Seleman, M.M.; Ammar, H.R.; Ahmed, E. Friction stir welding of similar and dissimilar AA7075 and AA5083. *J. Mater. Process. Technol.* **2017**, *242*, 77–91. [CrossRef]

173. Yang, Y.; Bi, J.; Liu, H.; Li, Y.; Li, M.; Ao, S.; Luo, Z. Research progress on the microstructure and mechanical properties of friction stir welded AlLi alloy joints. *J. Manuf. Process.* **2022**, *82*, 230–244. [CrossRef]

174. Mishra, R.S.; Sidhar, H. *Chapter 5—Friction Stir Welding of Al–Li Alloys*; Mishra, R.S., Sidhar, A., Eds.; Butterworth-Heinemann: Oxford, UK, 2017; pp. 79–95. [CrossRef]

175. Wei, S.; Hao, C.; Chen, J. Study of friction stir welding of 01420 aluminum-lithium alloy. *Mater. Sci. Eng. A* **2007**, *452–453*, 170–177. [CrossRef]

176. Altenkirch, J.; Steuwer, A.; Withers, P.J. Process—Microstructure—Property correlations in Al–Li AA2199 friction stir welds. *Sci. Technol. Weld. Join.* **2010**, *15*, 522–527. [CrossRef]

177. Shukla, A.K.; Baeslack, W.A. Study of microstructural evolution in friction-stir welded thin-sheet Al–Cu–Li alloy using transmission-electron microscopy. *Scr. Mater.* **2007**, *56*, 513–516. [CrossRef]

178. Cavaliere, P.; Cabibbo, M.; Panella, F.; Squillace, A. 2198 Al–Li plates joined by Friction Stir Welding: Mechanical and microstructural behavior. *Mater. Des.* **2009**, *30*, 3622–3631. [CrossRef]

179. Cavaliere, P.; De Santis, A.; Panella, F.; Squillace, A. Effect of anisotropy on fatigue properties of 2198 Al–Li plates joined by friction stir welding. *Eng. Fail. Anal.* **2009**, *16*, 1856–1865. [CrossRef]

180. Steuwer, A.; Dumont, M.; Altenkirch, J.; Birosca, S.; Deschamps, A.; Prangnell, P.B.; Withers, P.J. A combined approach to microstructure mapping of an Al–Li AA2199 friction stir weld. *Acta Mater.* **2011**, *59*, 3002–3011. [CrossRef]

181. Ma, Y.E.; Zhao, Z.; Liu, B.; Li, W. Mechanical properties and fatigue crack growth rates in friction stir welded nugget of 2198-T8 Al–Li alloy joints. *Mater. Sci. Eng. A* **2013**, *569*, 41–47. [CrossRef]

182. De Geuser, F.; Malard, B.; Deschamps, A. Microstructure mapping of a friction stir welded AA2050 Al–Li–Cu in the T8 state. *Philos. Mag.* **2014**, *94*, 1451–1462. [CrossRef]

183. Gao, C.; Zhu, Z.; Han, J.; Li, H. Correlation of microstructure and mechanical properties in friction stir welded 2198-T8 Al–Li alloy. *Mater. Sci. Eng. A* **2015**, *639*, 489–499. [CrossRef]

184. Qin, H.; Zhang, H.; Wu, H. The evolution of precipitation and microstructure in friction stir welded 2195-T8 Al–Li alloy. *Mater. Sci. Eng. A* **2015**, *626*, 322–329. [CrossRef]

185. Muthumanickam, A. Effect of Friction Stir Welding Parameters on Mechanical Properties and Microstructure of AA2195 Al—Li Alloy Welds. *Trans. Indian Inst. Met.* **2019**, *72*, 1557–1561. [CrossRef]

186. Yan, K.; Wang, T.; Liang, H.; Zhao, Y. Effects of Rotation Speed on Microstructure and Mechanical Properties of 2060 Al-Cu-Li Alloy in Friction Stir Welding. *J. Mater. Eng. Perform.* **2018**, *27*, 5803–5814. [CrossRef]

187. Wang, F.F.; Li, W.Y.; Shen, J.; Zhang, Z.H.; Li, J.L.; dos Santos, J.F. Global and local mechanical properties and microstructure of Bobbin tool friction-stir-welded Al–Li alloy. *Sci. Technol. Weld. Join.* **2016**, *21*, 479–483. [CrossRef]

188. Alam, M.P.; Sinha, A.N. Effect of heat assisting backing plate in friction stir welding of high strength Al-Li alloy. *Energy Sources Part A Recover. Util. Environ. Eff.* **2019**, *44*, 2851–2862. [CrossRef]

189. Cisko, A.R.; Jordon, J.B.; Amaro, R.L.; Allison, P.G.; Wlodarski, J.S.; McClelland, Z.B.; Garcia, L.; Rushing, T.W. A parametric investigation on friction stir welding of Al-Li 2099. *Mater. Manuf. Process.* **2020**, *35*, 1069–1076. [CrossRef]

190. Wang, F.F.; Li, W.Y.; Shen, J.; Hu, S.Y.; dos Santos, J.F. Effect of tool rotational speed on the microstructure and mechanical properties of bobbin tool friction stir welding of Al–Li alloy. *Mater. Des.* **2015**, *86*, 933–940. [CrossRef]

191. Zhang, Y.; Sato, Y.S.; Kokawa, H.; Hwan, S.; Park, C.; Hirano, S. Stir zone microstructure of commercial purity titanium friction stir welded using pcBN tool. *Mater. Sci. Eng. A* **2008**, *488*, 25–30. [CrossRef]

192. Ramulu, M.; Edwards, P.D.; Sanders, D.G.; Reynolds, A.P.; Trapp, T. Tensile properties of friction stir welded and friction stir welded-superplastically formed Ti-6Al-4V butt joints. *Mater. Des.* **2010**, *31*, 3056–3061. [CrossRef]

193. Farias, A.; Batalha, G.F.; Prados, E.F.; Magnabosco, R.; Delijaicov, S. Tool wear evaluations in friction stir processing of commercial titanium Ti-6Al-4V. *Wear* **2013**, *302*, 1327–1333. [CrossRef]

194. Yoon, S.; Ueji, R.; Fujii, H. Microstructure and texture distribution of Ti-6Al-4V alloy joints friction stir welded below β-transus temperature. *J. Mater. Process. Technol.* **2016**, *229*, 390–397. [CrossRef]

195. Yoon, S.; Ueji, R.; Fujii, H. Effect of initial microstructure on Ti-6Al-4V joint by friction stir welding. *Mater. Des.* **2015**, *88*, 1269–1276. [CrossRef]

196. Yoon, S.; Ueji, R.; Fujii, H. Effect of rotation rate on microstructure and texture evolution during friction stir welding of Ti-6Al-4V plates. *Mater. Charact.* **2015**, *106*, 352–358. [CrossRef]

197. Ji, S.; Li, Z.; Wang, Y.; Ma, L. Joint formation and mechanical properties of back heating assisted friction stir welded Ti–6Al–4V alloy. *Mater. Des.* **2017**, *113*, 37–46. [CrossRef]

198. Ji, S.; Li, Z.; Zhang, L.; Wang, Y. Eliminating the tearing defect in Ti-6Al-4V alloy joint by back heating assisted friction stir welding. *Mater. Lett.* **2016**, *188*, 21–24. [CrossRef]

199. Li, J.; Shen, Y.; Hou, W.; Qi, Y. Friction stir welding of Ti-6Al-4V alloy: Friction tool, microstructure, and mechanical properties. *J. Manuf. Process.* **2020**, *58*, 344–354. [CrossRef]

200. Amirov, A.; Chumaevskii, A.; Savchenko, N.; Gurianov, D.; Nikolaeva, A.; Krasnoveykin, V.; Ivanov, A.; Rubtsov, V.; Kolubaev, E. Features of Permanent Joints of Titanium (α+β)-Alloys Obtained by Friction Stir Welding Using a Nickel Superalloy Tool. *Metals* **2023**, *13*, 222. [CrossRef]

materials

Article

Microstructure and Mechanical Properties Analysis of Al/Cu Dissimilar Alloys Joining by Using Conventional and Bobbin Tool Friction Stir Welding

Kishan Fuse [1], Vishvesh Badheka [1], Ankit D. Oza [2], Chander Prakash [3,4,*], Dharam Buddhi [5], Saurav Dixit [5,6,*] and N. I. Vatin [5]

1   Department of Mechanical Engineering, School of Technology, Pandit Deendayal Energy University, Raysan, Gandhinagar 382007, India; kishan.fuse@sot.pdpu.ac.in (K.F.); vishvesh.badheka@spt.pdpu.ac.in (V.B.)
2   Department of Computer Sciences and Engineering, Institute of Advanced Research, Gandhinagar 382426, India; ankit.oza@iar.ac.in
3   School of Mechanical Engineering, Lovely Professional University, Phagwara 144411, India
4   Division of Research and Development, Lovely Professional University, Phagwara 144411, India
5   Peter the Great St. Petersburg Polytechnic University, 195251 Saint Petersburg, Russia; dbuddhi@gmail.com (D.B.); vatin@mail.ru (N.I.V.)
6   Division of Research & Innovation, Uttaranchal University, Dehradun 248007, India
*   Correspondence: chander.mechengg@gmail.com (C.P.); sauravarambol@gmail.com (S.D.)

Abstract: The feasibility of producing welding joints between 6061-T6 aluminum and pure copper sheets of 6 mm thickness by conventional friction stir welding (CFSW) and bobbin tool friction stir welding (BTFSW) by using a slot-groove configuration at the joining surface was investigated. The microstructure of the welded samples was examined by using an optical microscope and X-ray diffraction. Furthermore, the mechanical properties of the weld samples·are compared based on the results of the tensile test, hardness measurement, and fractography test. The slot-groove configuration resulted in the presence of a bulk-sized Al block on the Cu side. The microscopic observations revealed the dispersion of fine Cu particles in the stir zone. The presence of intermetallic compounds (IMCs) $CuAl_2$, which are hard and brittle, lowered the strength of the weld joints. The strength of the weld joints produced with BTFSW was superior to that of the C-FSW. The maximum hardness values of 214 HV and 211 HV are reported at the stir zone for BTFSW and CFSW, respectively. The fracture location of all the joints was at the intersection of the stir zone and the thermomechanically affected zone was on the Cu side.

Keywords: bobbin tool; friction stir welding; dissimilar; intermetallic; microhardness; microstructure

Citation: Fuse, K.; Badheka, V.; Oza, A.D.; Prakash, C.; Buddhi, D.; Dixit, S.; Vatin, N.I. Microstructure and Mechanical Properties Analysis of Al/Cu Dissimilar Alloys Joining by Using Conventional and Bobbin Tool Friction Stir Welding. *Materials* **2022**, *15*, 5159. https://doi.org/10.3390/ma15155159

Academic Editor: Raul D.S.G. Campilho

Received: 4 July 2022
Accepted: 22 July 2022
Published: 25 July 2022

Publisher's Note: MDPI stays neutral with regard to jurisdictional claims in published maps and institutional affiliations.

1. Introduction

The dissimilar material joining in one of the recent challenges faced by manufacturing industries. The amalgamation of Al and Cu is of extreme interest in electricity, cryogenic and refrigeration fields pertaining to benefits such as weight reduction and cost saving [1,2]. This necessitates a reliable joining technique for fabricating Al/Cu components. The challenges in joining of Al/Cu combination is due to large dissimilarities in their physical, thermal, and chemical properties, such as differences in melting temperatures and the strong affinity between Al and Cu [3]. The formation of brittle intermetallic compounds (IMCs) produces joints with less mechanical strength. Thus, the conventional fusion welding techniques are not fit for such joining as it requires melting of material followed by cooling.

Solid-state welding has emerged as a solution for joining dissimilar material systems. One novel type of solid-state welding is conventional friction stir welding (CFSW). The process can join Al and Cu at low heat input as it utilizes friction heat and mechanical deformation to produce joints [4–6]. The low heat input suppresses the formation of excessive IMCs and results in better joint characteristics [7]. Comprehensive research has been reported to investigate and improve the joint strength of Al/Cu joining by using the conventional FSW technique.

Several process parameters affect the weld joint quality during CFSW of Al/Cu joining. Shankar et al. [8] investigated the effect of welding speed and tool offset on FSW of dissimilar Al 1050 and oxygen-free copper joining. They reported the highest joint strength, 91% AA 1050 alloy strength. Khajeh et al. [9] pointed out that welding speed (W) and traverse speed (V) ratio (W/V) significantly affect joint properties as they stand for heat input. They reported IMCs and voids as prime reasons for lower strength and ductility in non-optimum joints. Hou et al. [10] investigated tool offset during FSW of 6061 aluminium alloy and commercially pure copper via FSW. They reported an increase in tensile strength with an increase in tool offset from 0 mm to 1.4 mm. They also found peak temperature dependency on tool offset. Zhang et al. [11] produced a sound weld between 1060 aluminium alloy and annealed a pure copper sheet by using CFSW at a tool rotating speed of 1050 rpm and a weld travel speed of 30 mm/min. They inferred metallurgical bonding between base materials to form an intercalation structured at the intersection of Cu and weld nugget area. The hardness increment was mainly due to higher dislocation density and developed dislocation loops.

Optimization of process parameters is essential to recognize the best process parameter window during CFSW of Al/Cu joining. Mahdianikhotbesara et al. [12] optimized micro FSW for producing dissimilar material joining between 1050 aluminium and pure copper alloys. They performed experiments by using Taguchi L9 arrays. They found the rotation speed of the tool and welding speed as the most significant parameters affecting tensile strength. Eslami et al. [13] adopted the Taguchi technique for experimental design to optimize process variables during FSW of aluminum 1050A and copper EN CW004A. They concluded that the lowest offset of 1.4 mm led to the best electrical and mechanical properties. Sahu et al. [14] presented the fuzzy-grey Taguchi technique for optimizing Al/Cu dissimilar FSWed joints. They varied rotational tool speed, traverse speed, plunge depth, and offset distance of tool. They measured the effect of selected parameters in terms of compressive strength, percentage of elongation, angle of bend, weld bead thickness, and average hardness at the stir zone (SZ).

Enhancing joint properties by using different welding strategies remained an active area of research in CFSW of Al/Cu joining. Sahu et al. [15] conducted CFSW of Al/Cu with Ti, Ni, and Zn foil as an interlayer to enhance joint properties. They reported enhanced metallurgical and mechanical characteristics with Zn interlayer due to the formation of uniform, thin, and continuous IMCs. Hou et al. [16] cold sprayed the Cu plate with a Ni coating of 90 μm and further FSWed to a 6061 aluminium alloy in a butt configuration. They found significant improvement in mechanical strength. They reported average peak stress and percentage elongation of 190 MPa and 14%, respectively. Argesi et al. [17] used SiC nano-composites to decline the adverse effect of IMCs during FSW of pure copper and Al 5754 alloy. They observed a reduction in Al and Cu grain size in the SZ due to SiC particles. They also found a significant increase in microhardness of the weld zone from 160 HV to 320 HV. Zhang et al. [18] replaced the surface-to-surface joint configuration at the joining area with a tooth-shaped joint configuration for bonding Al/Cu dissimilar metals by using CFSW. This proved a better design to tailor the base materials content into the weld. They reported a failure load of 9.6 KN and 7.9 KN with tool-shaped joint configuration and routine butt configuration, respectively.

Bobbin tool friction stir welding (BTFSW) is a recently developed variant of the CFSW technique. The exception in BTFSW is that the tool of BTFSW is attached with an additional shoulder at the lower side of the pin, known as the lower shoulder [19,20]. The lower shoulder helps avoid use of a backing plate, which is an integral part of CFSW. The BTFSW offers many advantages over CFSW, such as eliminating root flaw defects like lack of penetration, full penetration, simplified fixture, uniform through thickness grain structure, effective utilization of generated frictional heat etc. [21–23]. The investigation using BTFSW includes studying the effect of rotation speed [24,25], welding speed [26,27], optimization of process parameters [28], comparative study of CFSW and BTFSW [29], and numerical modelling [30,31].

The literature study revealed that conventional FSW had been extensively investigated for joining of dissimilar Al/Cu materials considering welding process variables such as tool rotating speed, tool traverse speed, axial force, tool pin offset, positioning of base plates, the effect of geometrical features of the tool such as tool diameter, pin diameter, pin length etc. However, the current research needs more attention on the other welding strategies, such as the alignment of base plates at the joining surface (slot-groove configuration). On the other hand, despite the proven potential of BTFSW over CFSW, limited research is present in the literature. The research on BTFSW was focused on similar material joining only. The use of BTFSW for dissimilar Al/Cu joining needs to be addressed to explore the technique. Thus, the present work addresses the research questions about joining Al/Cu dissimilar materials by using conventional FSW and BTFSW. The slot-groove configuration fixes the base plates at the joining surface. Weld samples of Al/Cu joints will be put through mechanical and microstructural characterization to ensure their quality.

2. Experimental Methodology

This study aimed to make dissimilar material joining by using slot-groove joint configuration by using BTFSW and further compared with the CFSW joint. The base plates used for the study were 6061-T6 aluminium and pure copper. The chemical composition (% weight) of the Al plate is Si-0.457%, Fe-0.552%, Zn-0.010%, Cu-0.173%, Mg-0.937%, Mn-0.106%, and Al-balanced. The copper plate used contained 99.9% Cu. The schematic of the joint configuration and experimental setup is shown in Figure 1. During the experiment, copper was placed on the advancing side (AS), and 6061-T6 aluminium was placed on the retreating side (RS) for proper material mixing. In this work, both the plates used were 65 mm × 65 mm × 6 mm dimensions. The butt joint configuration is attempted in this study.

(a)　　　　　　　　　　　　　　　　　**(b)**

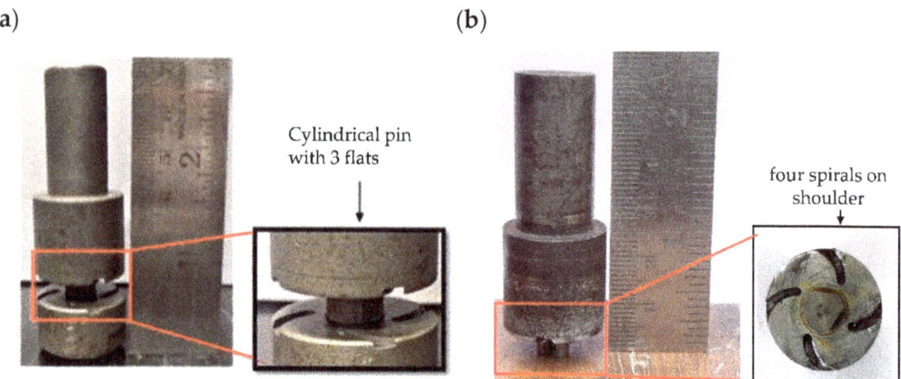

Figure 1. Stirring tool used for experiments (**a**) BTFSW tool (**b**) CFSW tool.

The welding experiments were worked on the vertical milling machine. The tool rotation speed of 1500 rpm and travel speed of 31.5 mm/min was used in both the welding conditions. The tilt angle of 2° was used in CFSW. The H13 tool steel was used to make the non-consumable tools for CFSW and BTFSW. The BTFSW tool design comprised of top and bottom shoulder having 24-mm diameter. The shoulders were provided with four spiral grooves each, which assisted in flowing the material from the outer shoulder edges to the tool's pin. The cylindrically shaped tool pin of 8-mm diameter was designed with a three flats feature. The length of the tool pin was 6 mm. The CFSW tool used during experiments had all the dimensions and features similar to the BTFSW tool except for the lower shoulder. The Figure 1 shows the CFSW and BTFSW tools used for the experimentation. The edge of the joining plates was modified to ensure proper mixing of material from AS to RS. The slot-groove configuration was prepared in such that copper plate had a groove which worked as the female part, and aluminium had a slot which worked as the male part. A longitudinal groove of size 130 mm × 10 mm × 3 mm was prepared (as a female part) along the thickness of the copper plate. Further, the male counterpart in 6061-T6 aluminium was machined to a thickness of 3.5 mm and width of 10 mm. It was inserted into the female copper groove as a press-fit before welding, as shown in Figure 2.

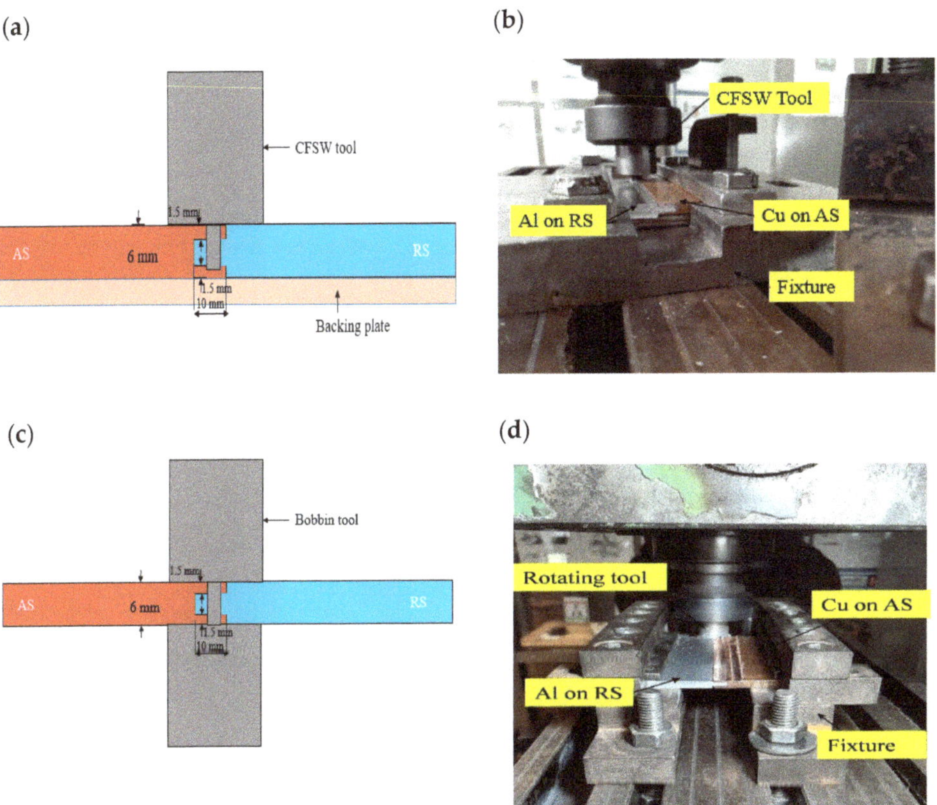

Figure 2. Joint configuration for dissimilar Al/Cu joining using CFSW. (**a**) Schematic (**b**) experimental set-up and BTFSW. (**c**) Schematic (**d**) experimental set-up.

After dissimilar Al/Cu welding, the weld samples were sectioned transversely to the welding route. The cut samples were prepared for the metallographic study by grinding, polishing, and further etching. The modified Keller's reagent was used on the Al side, and the Cu side was etched with 5 mL $FeCl_3$, 10 mL ethanol, 10 mL HCl and remaining distilled water. Metallographic inspections were carried out by Olympus-GX51 optical microscopy (Shinjuku, Japan). The Vickers microhardness indentation machine (NEXUS 4302) was used to assess microhardness. It is measured along the centerline of polished cross-sections at 1-mm intervals. During the measurement, a load of 300 g was used with a dwell time of 10 s. The tensile testing was executed at ambient temperature conditions by using computer-controlled UTM (FSA/M-100) with a speed of 1 mm/min. X-ray diffraction (XRD) studies can be used to confirm the production of distinct IMCs at the NZ.

3. Results and Discussion

3.1. Surface Appearance of the Joints

Figure 3 presents morphologies of the top surface of Al/Cu butt joints welded under CFSW and BTFSW conditions. It can be observed that the weld surface is present with uniform ripples without any surface defects such as cracks or tunnels. The CFSW joint is present with a keyhole at the exit, whereas the BTFSW tool comes out of the joining zone by cutting the plates at the exit. The flash can be observed only on AS in the CFSW joint, whereas the flash is present on both sides in the BTFSW joint. The flash in CFSW is mainly comprised of aluminium material. The specific movement of plasticized material during FSW instigated the formation of flash. In the present work, 6061-T6 aluminium is less viscous compared to copper. Hence, during the stirring process, less-viscous aluminium smoothly flowed from the advancing side to the retreating side, resulting in a flash. The flash on both sides in BTFSW can be attributed to more plunge depth at the top surface. In BTFSW, plunge is decided from pin length. When pin length in BTFSW is less than the thickness of the base plates, the shoulders penetrate more on either top or bottom surface. In the present work, it can be concluded that more penetration of the upper shoulder of the BTFSW tool on the top surface of the work material resulted in a flash on both sides.

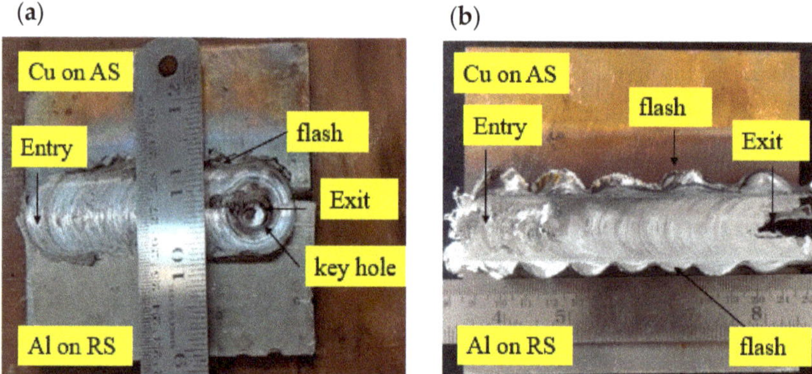

Figure 3. The top surface images of the Al/Cu joints welded with (**a**) CFSW, (**b**) BTFSW.

3.2. Macrostructure Observations

The macrostructure observed for dissimilar Al/Cu joints using CFSW and BTFSW is shown in Figure 4. It can be observed that the macrostructure for the CFSW is similar to BTFSW. The various sizes and shapes of the Cu particles can be seen distributed in a wide area of the joints. The hook formation phenomenon can also be observed in both joints. The SZ exhibits a heterogeneous structure. The interface of Al and Cu can be identified as unsmooth, which can be reasoned to scratching copper particles from the copper base material with the rotating tool. The big Cu particle can be seen at the intersection of SZ and the thermomechanically affected zone (TMAZ) on RS. The big Cu particle in CFSW is located at the top surface while in the mid and bottom of SZ for BTFSW. This can be attributed to the single shoulder in CFSW and the double shoulder arrangement in BTFSW. Figure 4b reveals that the Al has formed a bonding with Cu in the form of a hook. The hook formation can be attributed to a special slot-groove joint configuration. It is observed that some big-sized copper particles have dispersed at the intersection of SZ and TMAZ on RS. The stir zone was observed to be defect free. The red circle shows the fine copper particles dispersed at the top surface in the aluminium. Voids can be seen at the intersection of SZ and TMAZ on RS, as shown by the yellow circle. One of the reasons for forming the void is the presence of big Cu particles in the zone, which restricts the material flow in the region.

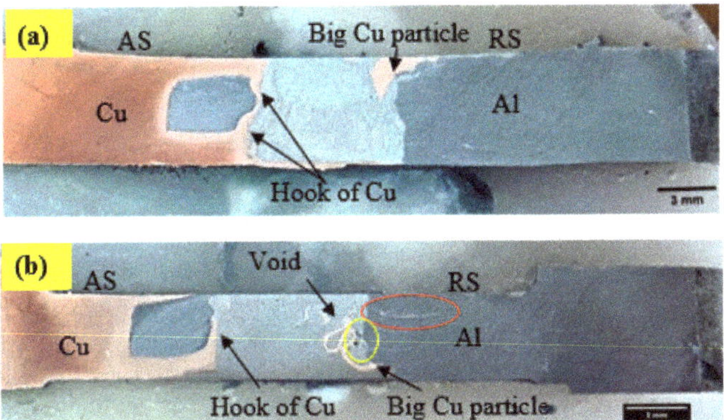

Figure 4. Macrostructure of Al/Cu joint by (**a**) CFSW (**b**) BTFSW.

3.3. Microstructure Observations

Microstructural features of the welds produced under CFSW and BTFSW are shown in Figures 5 and 6, respectively. The SZ in CFSW presents excellent bonding of Al and Cu, as shown in Figure 5a. No defect was observed in SZ. The SZ can be identified with the number of Cu particles in the Al matrix, which is obviously different from similar welding material. More dense dispersion of fine Cu particles was observed at the top than at the bottom, as shown in Figure 5b,f. It can be attributed to the combined effect of the pin and shoulder in the dispersion of Cu at the top surface. However, dispersion at the bottom was only due to pin effect. No bonding between the Al and Cu can be seen at some locations, as shown in Figure 5a. The hook intersection is surrounded by a small amount of Al/Cu mixed material on one side than the other (Figure 5c). The big Cu particle can be seen at the intersection of SZ and TMAZ on RS in CFSW, as shown in Figure 5d. The strong mechanical stirring produced by the rotating tool results in the scratching of Cu particles from the Cu base and further deposited in the aluminium matrix on RS.

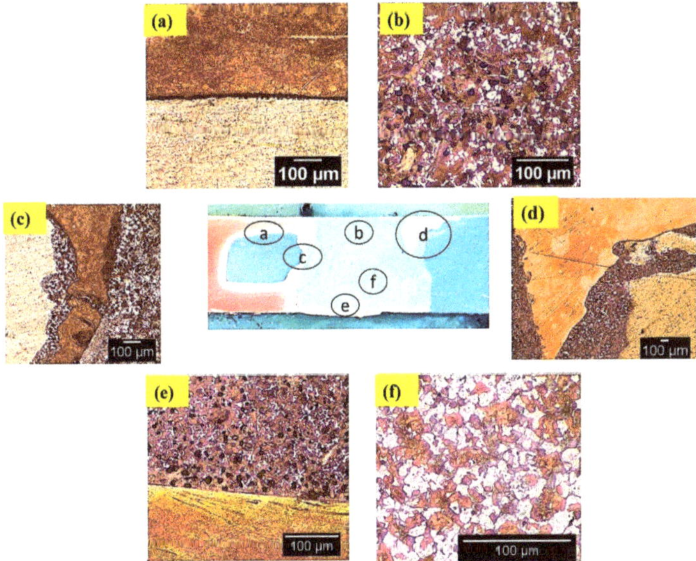

Figure 5. Microstructure in Al/Cu joint formed by CFSW at (**a,b,d**) top, (**c**) middle, (**e,f**) bottom.

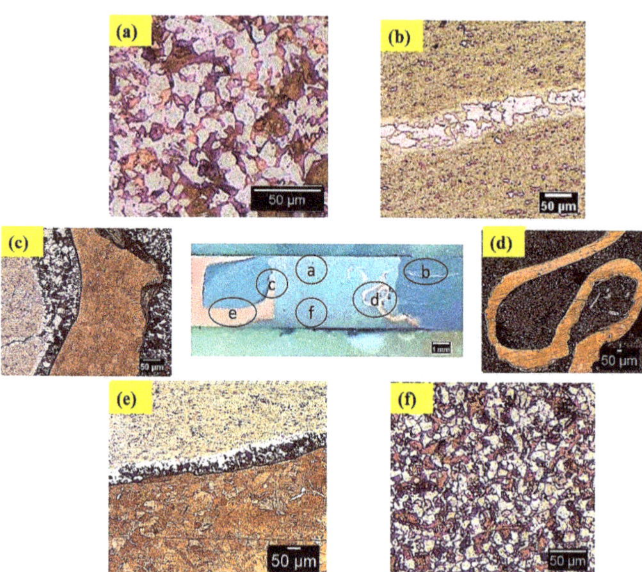

Figure 6. Microstructure in Al/Cu joint formed by BTFSW at (**a,b**) top, (**c,d**) middle, (**e,f**) bottom.

Figure 6a shows how dissimilar joints produced by BTFSW revealed that the SZ consists of a composite structure of Al and Cu materials. In SZ, superior bonding between Al and Cu was highlighted. The size of Cu particles is larger in SZ at the top compared to SZ at the bottom as shown in Figure 6a,f. This can be attributed to more net heat at the bottom surface than at the top surface in BTFSW. The net heat is the difference between the heat generated and heat loss at respective locations. In presented study, during BTFSW, the friction heat generated is the same at top and bottom surfaces due to the same size of the shoulder diameters. However, the amount of heat loss is more at top than on the bottom side. This is due to heat dissipation take place at the top surface via heat conduction into spindle and fixture and heat convention into the air. However, in the case of the bottom shoulder, only heat convention into the air is the mode of the heat dissipation, thus resulting in more net heat at the bottom side. Thus, more net heat at bottom might have caused the copper to stir more intensely and resulted in the higher deformation, reasoned in the dispersion of smaller copper particles in the Al matrix at bottom of the SZ. The dispersion of Cu base material particles in the Al matrix at RS can be observed in Figure 6b. This may be due to severe stirring of the base material and movement of Cu from AS to RS. The hook of Cu is surrounded by excellently boned Al and Cu, as seen in Figure 6c. The interface mixing of Al and Cu at the bottom is presented in Figure 6e.

3.4. Tensile Properties

The UTS and fracture to elongation for the welds produced by CFSW and BTFSW are shown in Figure 7. The UTS of the welds produced under CFSW and BTFSW is reported as 29.29 ± 2.19 MPa and 58.70 ± 9.41 MPa, respectively. The lower UTS for CFSW can be attributed to less strength at hook intersections. The hook intersection acted as a crack initiation location in CFSW. On the other hand, the hook of Cu formed in the BTFSW joint is surrounded by intermixed Al and Cu on both sides. It restricted easy crack formation at hook location resulting in better strength than CFSW. However, the achieved strength in both the welds is not acceptable to the strength of base materials.

The low strength in both the joints may be the presence of intermetallic (IMCs). It is known that the IMC layer is the most important for metallurgical bonding between Al and Cu at the interface. But the excessive thickness of the IMC layer is drastically detrimental to the joint properties [3]. Cu particle-mixing in the Al matrix is typically problematic due to incompatibilities in the chemical and physical properties of the joining metals, resulting in the development of IMCs. The enormous number of IMCs has an unfavourable impact on the weld joint strength and hardness as IMCs are hard and brittle. In CFSW and BTFSW joints, the end of extruded part of Al has not jointed properly in the slot of Cu, as presented in Figures 5a and 6e. The joining between the base materials was due to a formed Cu hook during the process, which resulted in reported strength in the joint. The fracture to elongation is observed as very low as 2.08% and 0.28% for the welds produced by BTFSW and CFSW, respectively. Low elongation is caused by the presence of a large number of hard and brittle IMCs as well as void defects.

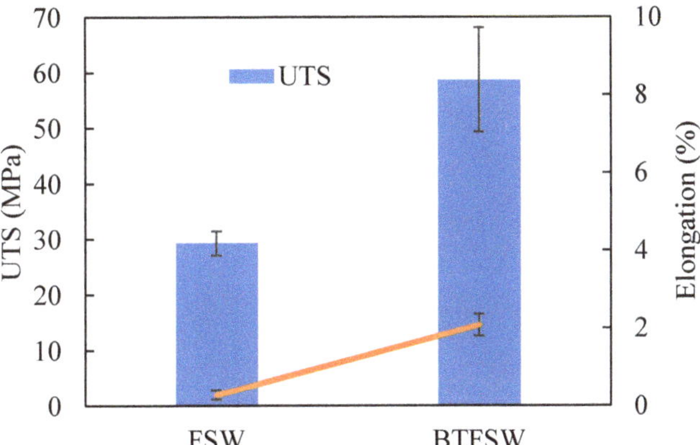

Figure 7. Tensile properties of dissimilar Al/Cu joint produced by FSW and BTFSW.

3.5. Microhardness Observations

Figure 8 shows the microhardness distribution of the joints produced by CFSW and BTFSW. The peak hardness at the SZ for both the welds can be seen in Figure 7. Maximum hardness values of 211 HV and 214 HV are reported at the stir zone for CFSW and BTFSW, respectively. The hard IMCs presence and fine-grain microstructure are attributed to the higher hardness value in SZ compared to base materials. The fine grains are induced by dynamic recrystallization caused by the combined role of heat and stirring force. The presence of fine grains and extremely thin IMCs layer in the Cu side share area resulted in marginally higher hardness (See Figures 5 and 6). Lower hardness was recorded at the Al side shear zone. The substantial hardness difference between Cu and Al sides in the weld nugget zone (WNZ) is not observed for both samples. This is due to slot and groove configuration instead of regular face-to-face butt configuration because in the slot and groove configuration, the copper is only in contact with aluminium from top and bottom.

3.6. XRD Analysis

XRD analysis was performed on the typical dissimilar FSW and BTFSW Al/Cu joints cross-section to reveal the existent phases, especially IMCs. The results of the XRD analysis are presented in Figure 9. The Al–Cu binary equilibrium phase diagram indicates some of the commonly found Al−Cu IMCs as Al_3Cu_4 ($\zeta2$), Al_4Cu_9 ($\gamma2$), Al_2Cu_3 (δ), AlCu ($\eta2$) and Al_2Cu (θ). It is reported that the Al_2Cu (Al-rich phase) and Al_4Cu_9 (Cu-rich phase) are the initially formed IMCs near the Al side and Cu side, respectively. In this study, only Al_2Cu were found in both samples. However, other compounds, such as AlCu, Al_3Cu_4, and Al_2Cu_3, should also be formed per the Al–Cu phase diagram. Despite the tool pin's high stirring activity, stronger $CuAl_2$ peaks in the SZ suggest an insufficient contact time [32]. The formation of $CuAl_2$ IMCs is mainly due to small Cu particles [33].

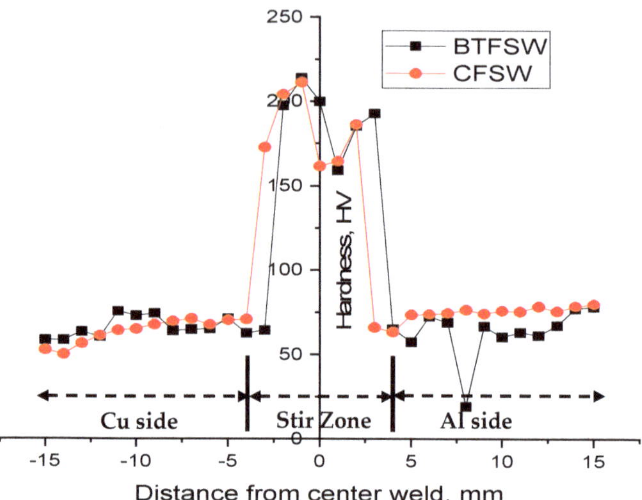

Figure 8. Microhardness of dissimilar Al/Cu joint produced by FSW and BTFSW.

(**a**)

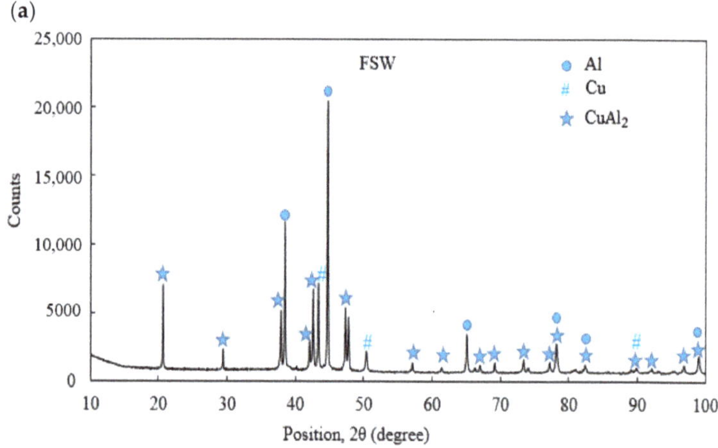

(**b**)

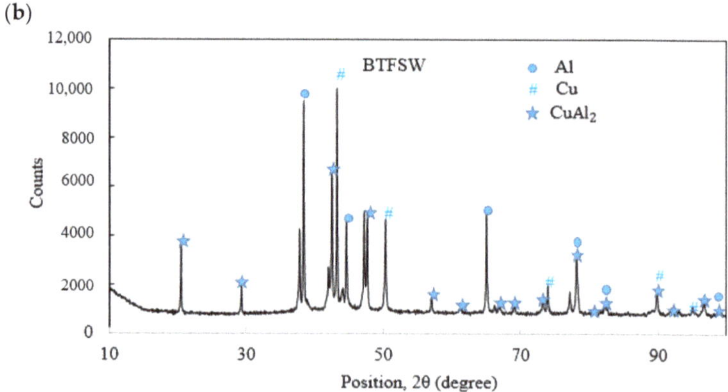

Figure 9. XRD results (**a**) CFSW (**b**) BTFSW.

3.7. Fractography

Table 1 presents fracture features of the welds produced by BTFSW and CFSW. All the joints failed at the intersection of TMAZ and SZ on the Cu side. The flat fracture surfaces of all the joints indicate failure is predominantly by brittle fracture. As load increases in the tension test, the tiny cavities in SZ expand rapidly and becomes cracks from which the sample failure starts. Also, inhomogeneous structure in the weld zone may have affected the failure behaviour of the samples [12]. Higher heat input must have promoted the formation of lamellar structures with substantial IMCs, which would harm the samples mechanical properties.

Table 1. Fracture features of the tensile specimens under BTFSW and CFSW.

Welding	Fracture Sorphologies	Fracture Location	Remarks
BTFSW		The intersection of SZ and TMAZ on the Cu side	Brittle fracture
		The intersection of SZ and TMAZ on the Cu side	Fracture initiated from the end of extruded Al
CFSW		The intersection of SZ and TMAZ on the Cu side	Brittle fracture
		The intersection of SZ and TMAZ on the Cu side	Brittle fracture

4. Scope for Future Work

The research and development work in dissimilar materials joining by using CFSW has been studied to a large extent with a wealth of process knowledge. But, BTFSW which is an emerging daughter technology of CFSW has been investigated to a lesser extent in joining dissimilar materials even though earlier research on BTFSW showed remarkable advantages of it over CFSW. There is a clear need to understand the feasibility of the BTFSW in the dissimilar material joining. The effect of process parameters such as rotation speed and welding speed can be researched in 6061-T6 aluminium-pure copper joining by using BTFSW. Furthermore, the slot-groove configuration can also been researched by changing male and female combination to know the effect of it. The research can be extended to know the effect of placement of joining materials on AS and RS. The optimization of process parameters by using optimization techniques, which gives correlative model between input parameters and the desired quality output, needs to be explored.

5. Conclusions

This study applied the novel implementation of the bobbin tool technique for joining dissimilar 6061-T6 aluminium and pure copper. The slot-groove configuration's feasibility for joining dissimilar materials was also attempted. Furthermore, the Al/Cu dissimilar welded joints by using the BTFSW technique were compared with conventional FSW. The concluding remarks of this study can be summarized as follows.

- The research on dissimilar Al/Cu joining by using slot-groove configuration demonstrated that some of the fine copper particles dispersed at the top surface of the aluminium. The voids were present at the intersection of SZ and TMAZ on RS for both the joints.
- Maximum hardness values of 214 HV and 211 HV were recorded at the stir zone for BTFSW and CFSW, respectively. The increased hardness value in SZ against the base metals was credited to the newly formed hard IMCs and fine-grain microstructure.
- XRD analysis of both the samples revealed the presence of Cu2Al IMC in the stir zone of CFSW and BTFSW joints.
- All specimens had a brittle fracture, and the specimens fractured mostly at the intersection of SZ and TMAZ on the Cu side.
- The results of the presented investigation will help explore the wide advantages of the bobbin tool technique in joining dissimilar materials.

Author Contributions: Conceptualization, K.F., V.B., D.B., S.D. and N.I.V.; methodology, K.F., A.D.O., S.D. and N.I.V.; software, C.P.; validation, K.F. and C.P.; formal analysis, K.F., S.D.; investigation, K.F., V.B. and A.D.O.; resources, V.B., K.F., S.D.; data curation, C.P.; writing—original draft preparation, K.F.; writing—review and editing, C.P., V.B., A.D.O., S.D. and N.I.V.; visualization, K.F.; supervision, C.P. and D.B. All authors have read and agreed to the published version of the manuscript.

Funding: The research is partially funded by the Ministry of Science and Higher Education of the Russian Federation as part of the World-class Research Center program: Advanced Digital Technologies: contract No. 075-15-2022-311 dated 20 April 2022.

Institutional Review Board Statement: Not applicable.

Informed Consent Statement: Not applicable.

Data Availability Statement: Not applicable.

Conflicts of Interest: The authors declare no conflict of interest.

References

1. Honarpisheh, M.; Asemabadi, M.; Sedighi, M. Investigation of annealing treatment on the interfacial properties of explosive-welded Al/Cu/Al multilayer. *Mater. Des.* **2012**, *37*, 122–127. [CrossRef]
2. Sahu, P.K.; Pal, S.; Pal, S.K.; Jain, R. Influence of plate position, tool offset and tool rotational speed on mechanical properties and microstructures of dissimilar Al/Cu friction stir welding joints. *J. Mater. Process. Technol.* **2016**, *235*, 55–67. [CrossRef]
3. Mao, Y.; Ni, Y.; Xiao, X.; Qin, D.; Fu, L. Microstructural characterization and mechanical properties of micro friction stir welded dissimilar Al/Cu ultra-thin sheets. *J. Manuf. Process.* **2020**, *60*, 356–365. [CrossRef]
4. Fuse, K.; Badheka, V. Hybrid self-reacting friction stir welding of AA 6061-T6 aluminium alloy with cooling assisted approach. *Metals* **2020**, *11*, 16. [CrossRef]
5. Liu, X.C.; Sun, Y.F.; Nagira, T.; Ushioda, K.; Fujii, H. Evaluation of dynamic development of grain structure during friction stir welding of pure copper using a quasi in situ method. *J. Mater. Sci. Technol.* **2019**, *35*, 1412–1421. [CrossRef]
6. Shen, Z.; Li, W.Y.; Ding, Y.; Hou, W.; Liu, X.C.; Guo, W.; Chen, H.Y.; Liu, X.; Yang, J.; Gerlich, A.P. Material flow during refill friction stir spot welded dissimilar Al alloys using a grooved tool. *J. Manuf. Process.* **2020**, *49*, 260–270. [CrossRef]
7. Muthu, M.F.X.; Jayabalan, V. Tool travel speed effects on the microstructure of friction stir welded aluminum–copper joints. *J. Mater. Process. Technol.* **2020**, *217*, 105–113. [CrossRef]
8. Shankar, S.; Vilaça, P.; Dash, P.; Chattopadhyaya, S.; Hloch, S. Joint strength evaluation of friction stir welded Al-Cu dissimilar alloys. *Measurement* **2019**, *146*, 892–902. [CrossRef]
9. Khajeh, R.; Jafarian, H.R.; Seyedein, S.H.; Jabraeili, R.; Eivani, A.R.; Park, N.; Kim, Y.; Heidarzadeh, A. Microstructure, mechanical and electrical properties of dissimilar friction stir welded 2024 aluminum alloy and copper joints. *J. Mater. Res. Technol.* **2021**, *14*, 1945–1957. [CrossRef]

10. Hou, W.; Shah, L.H.A.; Huang, G.; Shen, Y.; Gerlich, A. The role of tool offset on the microstructure and mechanical properties of Al/Cu friction stir welded joints. *J. Alloys Compd.* **2020**, *825*, 154045. [CrossRef]

11. Zhang, Q.Z.; Gong, W.B.; Wei, L.I.U. Microstructure and mechanical properties of dissimilar Al–Cu joints by friction stir welding. *Trans. Nonferrous Met. Soc. China* **2015**, *25*, 1779–1786. [CrossRef]

12. Mahdianikhotbesara, A.; Sehhat, M.H.; Hadad, M. Experimental study on micro-friction stir welding of dissimilar butt joints between Al 1050 and pure copper. *Metallogr. Microstruct. Anal.* **2021**, *10*, 458–473. [CrossRef]

13. Eslami, N.; Hischer, Y.; Harms, A.; Lauterbach, D.; Böhm, S. Optimization of process parameters for friction stir welding of aluminum and copper using the taguchi method. *Metals* **2019**, *9*, 63. [CrossRef]

14. Sahu, P.K.; Kumari, K.; Pal, S.; Pal, S.K. Hybrid fuzzy-grey-Taguchi based multi weld quality optimization of Al/Cu dissimilar friction stir welded joints. *Adv. Manuf.* **2016**, *4*, 237–247. [CrossRef]

15. Sahu, P.K.; Pal, S.; Pal, S.K. Al/Cu dissimilar friction stir welding with Ni, Ti, and Zn foil as the interlayer for flow control, enhancing mechanical and metallurgical properties. *Metall. Mater. Trans. A* **2017**, *48*, 3300–3317. [CrossRef]

16. Hou, W.; Oheil, M.; Shen, Z.; Shen, Y.; Jahed, H.; Gerlich, A. Enhanced strength and ductility in dissimilar friction stir butt welded Al/Cu joints by addition of a cold-spray Ni interlayer. *J. Manuf. Process.* **2020**, *60*, 573–577. [CrossRef]

17. Argesi, F.B.; Shamsipur, A.; Mirsalehi, S.E. Preparation of bimetallic nano-composite by dissimilar friction stir welding of copper to aluminum alloy. *Trans. Nonferrous Met. Soc. China* **2021**, *31*, 1363–1380. [CrossRef]

18. Zhang, W.; Shen, Y.; Yan, Y.; Guo, R.; Guan, W.; Guo, G. Microstructure characterization and mechanical behavior of dissimilar friction stir welded Al/Cu couple with different joint configurations. *Int. J. Adv. Manuf. Technol.* **2018**, *94*, 1021–1030. [CrossRef]

19. Fuse, K.; Badheka, V. Bobbin tool friction stir welding: A review. *Sci. Technol. Weld. Join.* **2019**, *24*, 277–304. [CrossRef]

20. Fuse, K.; Badheka, V.; Patel, V.; Andersson, J. Dual sided composite formation in Al 6061/B4C using novel bobbin tool friction stir processing. *J. Mater. Res. Technol.* **2021**, *13*, 1709–1721. [CrossRef]

21. Wang, G.Q.; Zhao, Y.H.; Tang, Y.Y. Research progress of bobbin tool friction stir welding of aluminum alloys: A review. *Acta Metall. Sin.* **2020**, *33*, 13–29. [CrossRef]

22. Li, W.Y.; Fu, T.; Hütsch, L.; Hilgert, J.; Wang, F.F.; Dos Santos, J.F.; Huber, N. Effects of tool rotational and welding speed on microstructure and mechanical properties of bobbin-tool friction-stir welded Mg AZ31. *Mater. Des.* **2014**, *64*, 714–720. [CrossRef]

23. Li, G.; Zhou, L.; Luo, S.; Dong, F.; Guo, N. Microstructure and mechanical properties of bobbin tool friction stir welded ZK60 magnesium alloy. *Mater. Sci. Eng. A* **2020**, *776*, 138953. [CrossRef]

24. Wang, F.F.; Li, W.Y.; Shen, J.; Hu, S.Y.; Dos Santos, J.F. Effect of tool rotational speed on the microstructure and mechanical properties of bobbin tool friction stir welding of Al–Li alloy. *Mater. Des.* **2015**, *86*, 933–940. [CrossRef]

25. Zhou, L.; Li, G.H.; Zha, G.D.; Shu, F.Y.; Liu, H.J.; Feng, J.C. Effect of rotation speed on microstructure and mechanical properties of bobbin tool friction stir welded AZ61 magnesium alloy. *Sci. Technol. Weld. Join.* **2018**, *23*, 596–605. [CrossRef]

26. Zhang, H.; Wang, M.; Zhang, X.; Yang, G. Microstructural characteristics and mechanical properties of bobbin tool friction stir welded 2A14-T6 aluminum alloy. *Mater. Des.* **2015**, *65*, 559–566. [CrossRef]

27. Wen, Q.; Li, W.; Patel, V.; Gao, Y.; Vairis, A. Investigation on the effects of welding speed on bobbin tool friction stir welding of 2219 aluminum alloy. *Met. Mater. Int.* **2020**, *26*, 1830–1840. [CrossRef]

28. Zhao, S.; Bi, Q.; Wang, Y.; Shi, J. Empirical modeling for the effects of welding factors on tensile properties of bobbin tool friction stir-welded 2219-T87 aluminum alloy. *Int. J. Adv. Manuf. Technol.* **2017**, *90*, 1105–1118. [CrossRef]

29. Xu, W.; Luo, Y.; Zhang, W.; Fu, M. Comparative study on local and global mechanical properties of bobbin tool and conventional friction stir welded 7085-T7452 aluminum thick plate. *J. Mater. Sci. Technol.* **2018**, *34*, 173–184. [CrossRef]

30. Hilgert, J.; Schmidt, H.N.B.; Dos Santos, J.F.; Huber, N. Thermal models for bobbin tool friction stir welding. *J. Mater. Process. Technol.* **2011**, *211*, 197–204. [CrossRef]

31. Wen, Q.; Li, W.Y.; Gao, Y.J.; Yang, J.; Wang, F.F. Numerical simulation and experimental investigation of band patterns in bobbin tool friction stir welding of aluminum alloy. *Int. J. Adv. Manuf. Technol.* **2019**, *100*, 2679–2687. [CrossRef]

32. Ouyang, J.; Yarrapareddy, E.; Kovacevic, R. Microstructural evolution in the friction stir welded 6061 aluminum alloy (T6-temper condition) to copper. *J. Mater. Process. Technol.* **2006**, *172*, 110–122. [CrossRef]

33. Xue, P.; Ni, D.R.; Wang, D.; Xiao, B.L.; Ma, Z.Y. Effect of friction stir welding parameters on the microstructure and mechanical properties of the dissimilar Al–Cu joints. *Mater. Sci. Eng. A* **2011**, *528*, 4683–4689. [CrossRef]

materials

Article

Effects of Partial-Contact Tool Tilt Angle on Friction Stir Welded AA1050 Aluminum Joint Properties

Mahmoud E. Abdullah [1,*], **M. Nafea M. Rohim** [1], **M. M. Mohammed** [1] **and Hamed Aghajani Derazkola** [2,*]

1 Mechanical Department, Faculty of Technology and Education, Beni-Suef University, Beni-Suef 62511, Egypt; mohamed.rahim@techedu.bsu.edu.eg (M.N.M.R.); moustafa.mahmoud@techedu.bsu.edu.eg (M.M.M.)
2 Department of Mechanics, Design and Industrial Management, University of Deusto, Avda Universidades 24, 48007 Bilbao, Spain
* Correspondence: iec.mahmoud@gmail.com (M.E.A.); h.aghajani@deusto.es (H.A.D.)

Abstract: This study aims to investigate the impact of partial-contact tool tilt angle (TTA) on the mechanical and microstructure properties of the AA1050 alloy friction stir weld (FSW). Three levels of partial-contact TTA were tested, 0°, 1.5°, and 3°, compared to previous studies on total-contact TTA. The weldments were evaluated using surface roughness, tensile tests, microhardness, microstructure, and fracture analysis. The results show that in partial-contact conditions, increasing TTA decreases the generated heat in the joint line and increases the possibility of FSW tool wear. This trend was the opposite of joints that were friction stir welded via total-contact TTA. The microstructure of the FSW sample was finer at higher partial-contact TTA, while the possibility of defect formation at the root of the stir zone in higher TTA was more than in lower TTA. The robust sample prepared at 0° TTA had 45% of AA1050 alloy strength. The maximum recorded heat in 0° TTA was 336 °C and the ultimate tensile strength of this sample was 33 MPa. The elongation of the 0° TTA welded sample was 75% base metal, and the average hardness of the stir zone was 25 Hv. The fracture surface analysis of the 0° TTA welded sample consisted of a small dimple, indicating the brittle fracture mode.

Keywords: tool tilt angle; friction stir welding; AA 1050; mechanical properties; welding parameters; microstructure

Citation: Abdullah, M.E.; M. Rohim, M.N.; Mohammed, M.M.; Derazkola, H.A. Effects of Partial-Contact Tool Tilt Angle on Friction Stir Welded AA1050 Aluminum Joint Properties. *Materials* **2023**, *16*, 4091. https://doi.org/10.3390/ma16114091

Academic Editor: Raul D. S. G. Campilho

Received: 1 May 2023
Revised: 27 May 2023
Accepted: 29 May 2023
Published: 31 May 2023

1. Introduction

The friction stir welding (FSW) process is a solid-state welding technology used in many industries recently for several reasons, such as high productivity, low defects related to melting, and high welding strength [1]. A rotational non-consumable tool penetrates inside the weld seam in the FSW process to join workpieces [2]. The friction between the tool and workpiece increases the weld line temperature and converts the base materials to a pasty form. After that, the FSW tool starts to move forward along the joint line [3]. During the forward movement, the raw material from the tool's leading edge (LE) extrudes inside the joint area and fills the tool's trailing edge (TE) [4].

The main FSW process parameters that directly affect the quality of joint material flow and mechanical properties are the FSW tool geometry, rotational velocity, traverse velocity, offset, plunge depth, and tilt angle [5]. One of the most critical FSW parameters that can affect the surface and internal flow is the tool tilt angle (TTA) [6]. TTA is defined as the difference angle between the tool's normal axis and the normal axis of the workpiece during welding [7]. The TTA can be implemented in two types. The first type is when the whole tool is inside the workpiece (total-contact TTA), and the other type is when part of the tool is inside the workpiece (partial-contact TTA) [8].

In total-contact TTA, the whole shoulder of the tool is in contact with the top surface of the workpiece, and in partial-contact TTA, a part of the tool is not in contact with the workpiece [9]. TTA is a crucial factor that can affect the final properties of the joint. This factor is a leading technical factor during FSW of similar and dissimilar materials [10]. Research shows that TTA significantly impacts the quality of aluminum alloy FSW joints. Due to the wide range of applications of aluminum alloys in various industries, an investigation of the effects of TTA on the final quality of friction stir welded samples is necessary [11]. The TTA was evaluated by researchers in two different categories. The first is related to experimental studies, and the second to simulation studies.

Shah and Badheka [12] studied the effect of total-contact TTA on the tensile strength of AA7075 aluminum alloy welded specimens. It was found that the TTA can be classified as a vital welding parameter due to its ability to control the defects and fracture behavior of welded joints. Acharya et al. [13] evaluated the effect of total-contact TTA on microstructure and microhardness for joining different aluminum series. They reported a difference in the heat-affected and stirred zones with the different tool tilt angles of the welding tool. They also found the microhardness of the weld center slightly higher than the base metal. Birol and Kasman [14] studied the effect of total-contact TTA and tool rotational speed. They found an improvement when increasing the rotational tool speed and reducing the tool tilt angle. Zhai et al. [15] investigated the effect of TTA on material flow and evaluated the defects resulting from the friction stir welding process of AA2219-T6. It was concluded that the best TTA was between 1° and 2°, which made joints defect-free. Gupta et al. [16] implemented 1.5° as a TTA for joining thick AA 7017-T651 aluminum alloy. Several mechanical tests and microstructure analyses were performed to evaluate the efficiency of the welded specimen. It was found that the welding pitch (tool rotational speed/traveling speed) is one of the most significant parameters in controlling the quality of welded specimens at the same TTA.

Acharya et al. [13] investigated the effect of TTA on the mechanical properties and fracture behavior of AA 6092. They found the best TTA in the range 1° to 2°; after that, the strength of joints decreased because of grain growth in the stir zone. It was noted that on increasing the TTA, both the flash formed and the ripple spacing increased. Dialami et al. [17] investigated the effect of total-contact TTA on heat generation and material flow. They found that high heat generation and friction force on the advancing side improved material flow on the retreating side. Zhai et al. [18] investigated the thermo-mechanical modeling of partial-contact TTA during FSW of AA6061 aluminum alloy. During forward movement, they determined an incomplete contact region between the tool shoulder and the workpiece. They measured the incomplete contact region between the shoulder and workpiece and compared it with calculation results.

Among various aluminum alloys, AA1050 aluminum alloy is widely used in various industries such as automotive, aerospace, and railway structures. FSW of AA1050 aluminum can help to produce light structures for various industries. For this reason, investigation of the effects of TTA during FSW of AA1050 alloy seems necessary. Tsarkov et al. [2] investigated the effects of total-contact TTA on the FSW joint of AA1050 alloy. They studied joint configuration with pin-less and regular FSW tools. On the other hand, Barlas [1] investigated the total-contact TTA on FSW of AA1050 aluminum alloy in joint lap configuration.

As mentioned, the difference between total-contact TTA and non-contact TTA is not clearly understood. Due to the available literature, the effects of partial-contact TTA on the AA1050 aluminum alloy have not been considered. In a previous study, the total-contact TTA effects on FSW of AA1050 aluminum alloy were evaluated. These results give base information to compare with other contact conditions of TTA. This study attempts to evaluate the effect of partial-contact TTA on the welding properties and compare the results with total-contact TTA joints. However, it notes a lack of research on the effects of partial-contact TTA on AA1050 aluminum alloy. Therefore, this study aims to fill this gap by evaluating the impact of partial-contact TTA on welding properties and comparing the results with total-contact TTA joints.

2. Materials and Methods

Commercial pure aluminum 1050 sheets of 5.3 mm thick material were used in this study as weldment. The weldment was provided from a local market, and the chemical composition of the provided aluminum was evaluated in a laboratory. The chemical composition after being tested three times is shown in Table 1. The raw aluminum sheets were cut and made a certain size via a hydraulic shearing machine. The final aluminum pieces were 160×100 mm rectangular shapes. Due to the selected parameters, 18 plates were cut and prepared for testing. This number was selected because all FSW conditions were implemented three times and the average values are presented in this study. The welding tool is made from W302 steel with a flat shoulder and cylindrical pin. The tool has a 15 mm shoulder diameter and a 5 mm pin diameter and 4.5 mm length. It has been suggested that the optimum pin length should be approximately 85% of the base metal thickness [19]. Based on previous research, 4.5 mm is the optimum length. A schematic image and a picture of the tool are depicted in Figures 1a and 1b, respectively.

Table 1. Average chemical composition of pure aluminum 1050 (wt.%).

Element	Fe	Cu	Zn	Ni	Ca	Ga	Pb	B	Al
Wt.%	0.335	0.016	0.007	0.007	0.007	0.007	0.005	0.003	Balance
StdDev.	0.0000000	0.0005774	0.0017321	0.0005774	0.001	0.0005774	0.000	0.0005774	

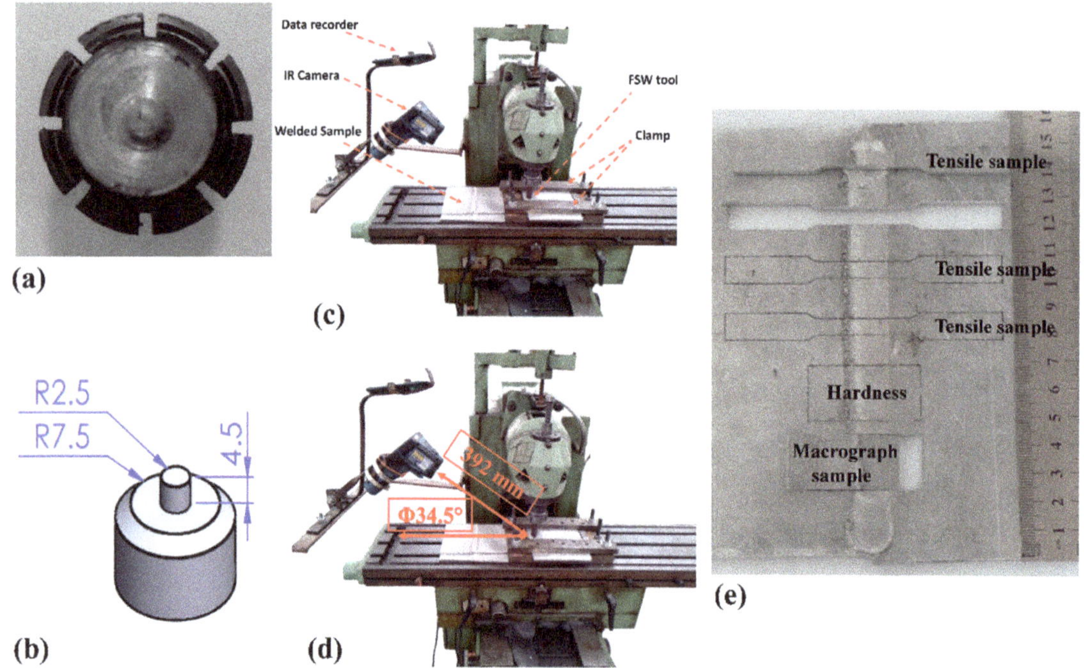

Figure 1. (**a**) Picture of FSW tool and (**b**) Schematic view of FSW tool dimensions. (**c**) Experimental setup of FSW process; (**d**) IR camera angle. (**e**) Mechanical and microstructure test samples.

In this study the tool rotational and traverse velocities were selected as 910 rpm and 44 mm/min, respectively. These criteria were selected after undergoing a series of trials. For the assessment of the effects of tool tilt angle, $0°$, $1.5°$, and $3°$ tilts were selected. During the experimental procedure, the tilt angle was selected as a variable and other parameters were kept constant. The thermal change during the FSW process was recorded via an infrared thermometer (IR camera). The thermometer was able to record 1000 °C with 0.1 °C resolution and a measuring error of ± 1 °C. This type of camera is an easy-to-use comparison with thermocouples due to its applicability in wireless operations and its high responsiveness with respect to up- or down-measured temperature values. During the tests, the IR camera fixed on the machine and focused at a certain point in the advancing side of the joint line. The IR camera was on during the process and recorded the temperature changes of the focused point from the beginning to the end. The thermal changes at the point during the passing of the tool were recorded by the IR camera every 5 s and the data collected in an Excel file. At the end, the recorded numbers were plotted in an Excel file. The total distance between the IR camera and the selected point at the surface of the workpiece was 392 mm with a 34.5° angle. Pictures of the welding setup and the installed IR camera are shown in Figure 1c and 1d, respectively. After the FSW process, each specimen's average surface roughness (Ra) was measured via the Mitutoyo surface roughness tester at No. of cycles 5.

For evaluation of the microstructure, an optical microscope (OM) was employed. In this study, two types of optical microscopes were used. The first one was an optical stereo microscope (LEICA M165 c) used to evaluate the area of the stir zone, the thermo-mechanical affected zone, and the heat-affected zone in the weld line (micrographic analysis). The second was a light optical microscope (LECO LX 31) connected with a camera Pax-cam for image analysis with software (Pax-it, 1.1).

The hardness of the stir zone was measured according to the Vickers hardness by applying 500 gf for 15 s. The tensile tests of the welded specimen and base metal were performed at room temperature according to the ASTM-E8M standard. The sub-size tensile test specimen dimensions were selected and the test carried out with 2 mm/min crosshead speed. The fracture surface of the tensile tests after testing was evaluated via scanning electron microscope (SEM). Figure 1e presents an image of mechanical and microstructural samples cut from the weld line. After the FSW, the mechanical test samples and metallurgical test samples were cut via a laser cutting machine.

3. Results and Discussion

3.1. Thermal Cycles

The TTA in FSW significantly impact the material's surface flow during welding [20]. The tool tilt angle refers to the angle between the rotating tool axis and the normal axis on the workpiece surface [21]. This angle determines the direction and amount of material flow during the FSW process, which affects the weld quality and microstructure [22]. As discussed before, positive TTA, which refers to a tool tilted toward the forward movement, promotes material flow away from the center of the weld, resulting in a wider and shallower weld. On the other hand, a negative tool tilt angle, which refers to a tool tilted in the direction opposite the forward movement, promotes material flow toward the center of the weld, resulting in a narrower and deeper weld. In fact, the contact area between the tool and the workpiece can affect the final product's quality [23]. The contact area between the tool and the workpiece changes the total heat generation. In this case, the frictional generated heat can change the metallurgical and mechanical properties of the final joint.

Geometrical analysis of TTA shows that the highest contact area between the welding tool and the workpiece was at 1.5° TTA (~263.22 mm^2), while the lowest value of the contact area was at 3° TTA (~225.1 mm^2) with neglected surface tension or flash formation. This number is calculated according to the contact surface of the tool and aluminum alloy. A schematic view of the calculated contact area is depicted in Figure 2.

In this study, the thermal history of all samples was recorded during the welding procedure. By recording thermal history, we intended to find the relation between partial-contact TTA with frictional heat generation during FSW of the AA1050 aluminum alloy. The IR camera was fixed on the advancing side and recorded surface temperature changes from the starting phase of the joining until the ending, and the results are presented in Figure 3a. The results revealed that the temperature rose from room temperature and reached the highest level after that, remaining at the maximum until the end of the welding process. The results show the maximum temperature was recorded in 0° TTA and the minimum temperature was recorded in 3° TTA. In all cases, the recorded temperature peaked and stayed steady after a slight decrease. The steady-state phase in thermal history is related to the traverse length of the tool during the welding procedure. At 0° TTA, the temperature reached 336 °C and steadied between 327 °C and 320 °C along the welding line. The results show that the difference between the maximum and minimum recorded temperature was 16 °C. The temperature difference is related to the contact area between the tool and the workpiece during the traverse movement of the tool. The comparison between the results of this study with previous studies with total-contact TTA is presented in Figure 3b. The recorded maximum temperature in this study was compared with Barlas [1] and Tsarkov et al. [2]. Barlas [1] tested the effects of TTA on the FSW of AA1050 aluminum alloy in joint lap configuration.

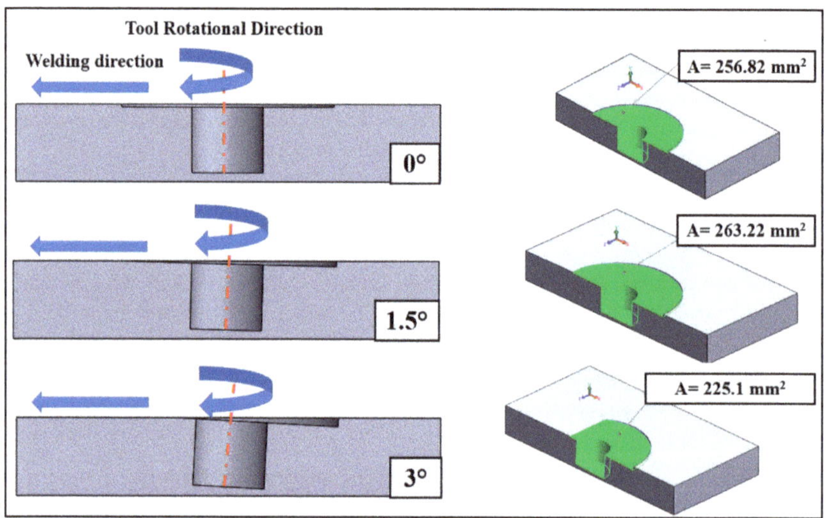

Figure 2. A schematic of the relation between TTA and approximate current contact area at 0.2 mm tool plunging depth.

As mentioned before, the TTA range in Barlas' [1] study was 0°, 1.5°, 2.5°, and 5°, while the tool rotational and traverse velocities were 1200 rpm and 30 mm/min, respectively. In this study, only the recorded temperatures in 0° and 5° TTA were reported, which indicates the minimum and maximum temperatures during testing. The test of Tsarkov et al. [2] was almost similar to this study. They used 0°, 1°, and 2° TTA, while the tool rotational and traverse velocities were 900 rpm and 50 mm/min. They reported the maximum and minimum temperatures in their study that were recorded in 0° and 2° TTA. The TTA was in total contact condition and this is helpful in comparing with our study results. The results show that with increasing TTA in partial-contact conditions, the generated heat decreases, while in total-contact conditions with increasing TTA, the total amount of heat increases. As discussed earlier, the difference between partial-contact TTA and total-contact TTA is related to the contact area of the tool and workpieces. A schematic view of total-contact TTA and partial-contact TTA is presented in Figure 3c. As can be seen, the tool and the top of the pin are not in contact with the workpiece during the forward moving of the FSW tool.

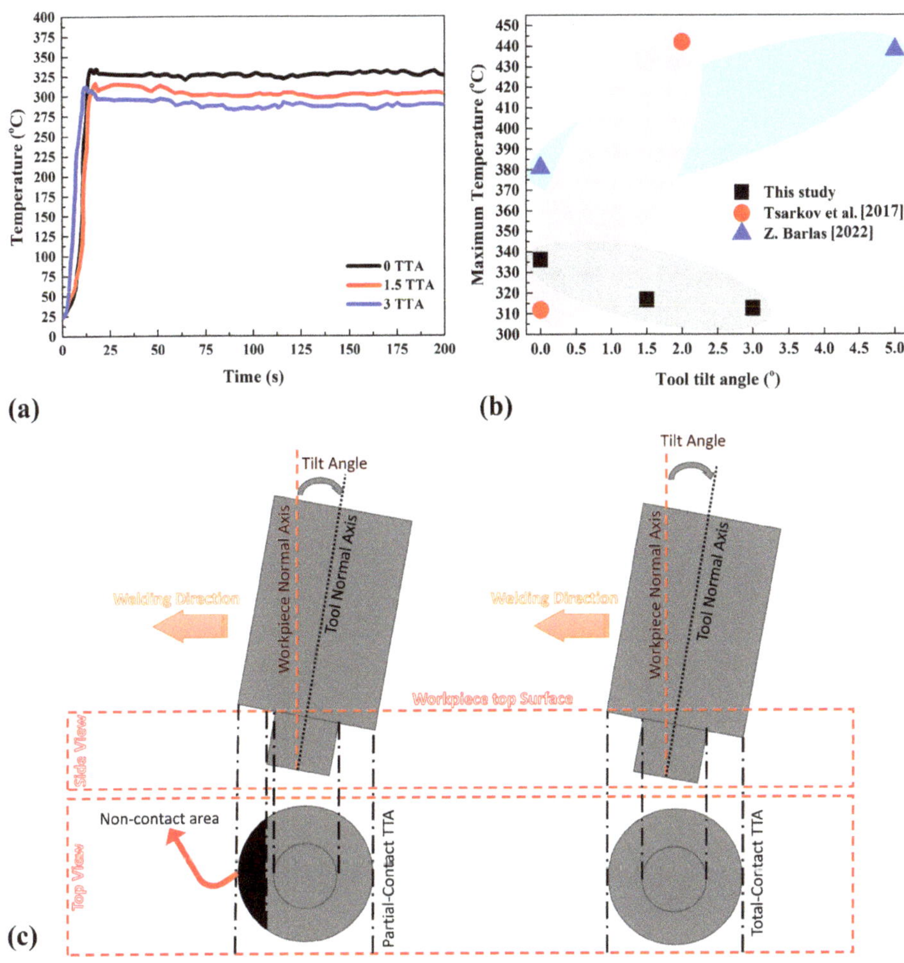

Figure 3. (**a**) Thermal cycles at different TTA. (**b**) Comparison between recorded temperature in this study and two other studies [1,2]; (**c**) schematic view of total-contact TTA and partial-contact TTA.

The exit holes of 0°, 1.5°, and 3° TTA conditions are presented in Figure 4a. This picture is from the top view of the end of the joint line. The obtained result can help to understand the contact condition between the FSW tool and workpiece at various TTA conditions.

As can be seen from the obtained results, the tool in the sample that welded at 0° was in complete contact with the raw sheets. On increasing the TTA, the contact area at the top of the tool in the leading edge decreased. Due to the partial-contact TTA condition in this study, increasing the TTA decreases the generated frictional heat during welding. The contact length from the shoulder with the leading edge of the FSW tool at various TTAs is presented in Figure 4b. The results show that the distance between the pin and leading edge of the FSW tool at 0°, 1.5°, and 3° TTA is 7.5 mm, 5 mm, and 3 mm, respectively. The results reveal that in partial-contact TTA, the distance between the pin and the leading edge decreases on increasing TTA. In fact, on increasing TTA in the partial-contact condition, the friction surface between the tool and AA1050 aluminum alloy decreases, and generation heat decreases. The frictional heat relates to the contact surface between the tool and the workpiece in the FSW process. Lower surface contact leads to lower heat generation and affects the internal and surface flow.

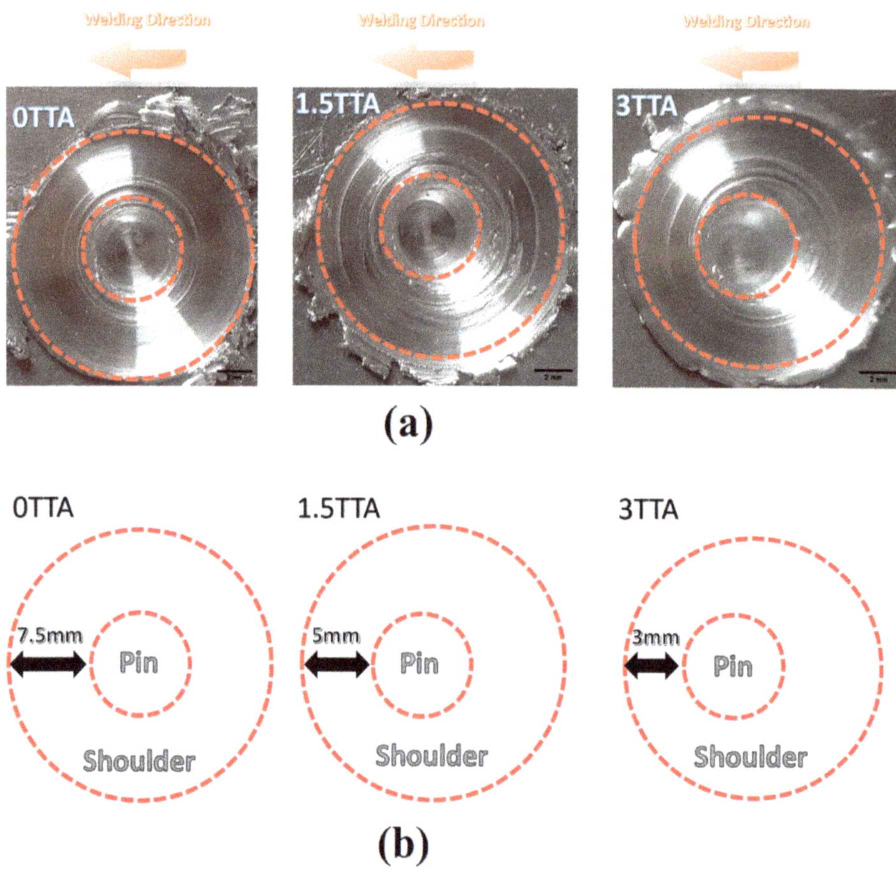

Figure 4. (**a**) The contact surface of the tool and workpiece at 0°, 1.5°, and 3° TTA conditions. (**b**) Distance between leading edge of FSW tool with pin at various TTAs.

3.2. Surface Flow Analysis

The surface flow of welded samples was evaluated via visual examination and Mitutoyo surface roughness tester. The visual examination included the effect of TTA on flash forming and surface flow ring. With this test, the effects of partial-contact TTA on the quality of the joint line surface were evaluated. Figure 5a shows the surface top view of the welded specimen at 0°, 1.5°, and 3° TTA. The first look shows that the surface flash formed at all joint surfaces. Surface flash in the FSW refers to a thin layer of material that is extruded from the surface of the welded joint. This material layer forms due to the high temperature and pressure during the FSW process. The surface flash is often visible as a thin, ribbon-like protrusion on the surface of the welded joint, and it can significantly impact the quality and properties of the welded joint. Surface flash can affect the weld quality in several ways, including welded geometry, weld strength, and weld appearance. It was found that at 0° TTA, a flash formed on the advancing side (AS) because the flow of material on the AS was greater than on the retreating side (RS). This flash resulted from mechanical action and heat generation during the welding process. At 1.5° TTA, a flash also formed on the advancing side, but it was not concentrated or not strong compared with 0° TTA. In the case of 3° TTA, a few flashes formed on both sides of the welding line;

i.e., they look semi-scattered. This indicated that the effect of TTA at 3° led to a reduction in tool contact area. Based on tilt angle, it was established that the tilt angle has a significant effect with respect to proving the material flow around the welding tool. In partial-contact TTA conditions, the flash formation is related to the contact area between the FSW tool and AA1050 aluminum alloy. In other words, at 0° TTA, the material extruding from the AS is insufficient to make a smooth joint line. In 1.5° TTA, the extrusion of AA1050 from the AS to the RS improved, but in 3° TTA, the big gap between tool and workpiece during forward moving increases the flash surface at both the AS and the RS.

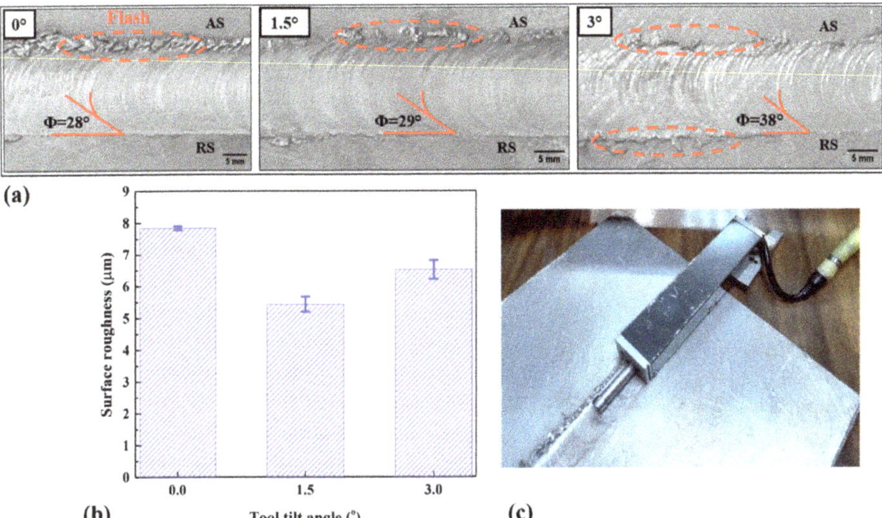

Figure 5. (a) Surface flow of joints welded at 0°, 1.5°, and 3° TTA. (b) Surface roughness at different TTAs and (c) sample of surface roughness measurement.

Flow ring is a phenomenon that occurs via surface ripple in the welding area that is in contact with the tool shoulder [24]. The ripple spacing of this study is measured to indicate the visual surface quality of the welded joint as well as the mechanical action in the stir zone. All welded joints have less distinct flow ring deformation. Moreover, it was found that flow rings were discontinuous at 1.5° and 3° due to the TTAs. As can be seen from the obtained results, the angle of the flow rings at 0°, 1.5°, and 3° TTA were 28°, 29°, and 38°, respectively. This shows that on increasing TTA, the surface shear stress on AA1050 aluminum alloy increases, and more shear force is exerted to extrude plasticized metal from the AS to the RS [25]. The results reveal that the flow ring angle from 0° TTA to 1.5° and 3° TTA increased by 3.5% and 35%, respectively. It seems the increase in surface shear stress at higher TTA leads to the formation of surface flash on both the AS and the RS.

The surface roughness was measured via a Mitutoyo surface roughness tester to evaluate the roughness average (Ra) at different TTAs. Figure 5b,c show surface roughness results and a sample of surface roughness measurements. At TTA 0°, the average Ra was 7.84 ± 0.06 μm. In addition, the Ra at 1.5° and 3° TTA was 5.44 ± 0.24 μm and 6.53 ± 0.3 μm, respectively. It was found that the value of Ra at 0° TTA is higher in comparison with other TTAs. This result indicated that more mechanical action occurred at TTA 0°. Tsarkov et al. [2] did not consider surface flow analysis, while Barlas [1] reported that flashes formed in the vicinity of the joint lines welded with low and very high TTAs.

3.3. Macrostructure Observation

Macroscopic images of the cross-sections of the welds at 0°, 1.5°, and 3° are shown in Figure 6a and 6b, respectively. Figure 6a presents cross-section views of FSW joints welded at 0°, 1.5°, and 3° TTA and Figure 6b indicates the different joint areas. Visual inspection revealed that no defects formed in the joint cross-section at a macroscale. Root defects, tunnel voids, or kissing bond defects are common in FSWs of aluminum alloys formed at a macroscale and usually can be detected via visual inspection. As can be seen, in the produced joint, these defects are not detected in cross-sections. As can be seen from the results, the joint line consists of a stir zone (SZ) that is common between the advancing side (AS) and the retreating side (RS). A thermo-mechanical-affected zone (TMAZ) and a heat-affected zone (HAZ) formed at both the AS and the RS. Figure 6b shows the different parts of joint lines. The joint without TTA indicates a symmetrical SZ from the weld center and narrow HAZ and TMAZ areas. The results show the welded area of 0° TTA formed by cup shape. The SZ of joints welded at 1.5° and 3° TTA has similar morphology.

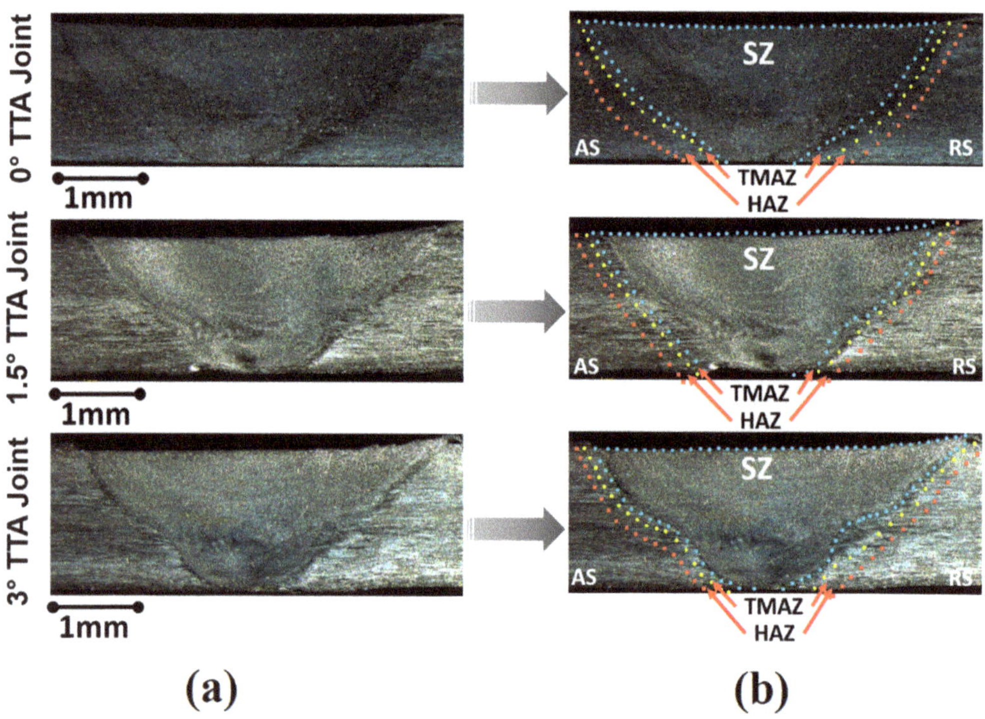

Figure 6. (a) Optical macrograph of friction stir joints at various TTAs and (b) different areas of joint lines.

Figure 7a compares defective and defect-free joints in this study with other studies.

The TTAs used in this research are compared with previous studies, without considering total-contact or partial-contact TTAs. With this comparison, we aimed to understand the effects of TTA quantity and quality on the quality of the joints. The results indicated that defective joints form in total-contact TTA at a low tilt angle, and defect-free joints can be achieved at high TTAs. On the other hand, at least on the macroscale, defect-free joints are achievable in partial-contact TTA joints. This study revealed that partial-contact TTA could achieve a defect-free AA1050 aluminum joint at the macroscale, but in total-contact TTA, sound joints at the macroscale are achievable with a TTA of more than 0°. Figure 7b–d present the thicknesses of HAZ, TMAZ, and SZ of welded samples. The comparison between various joints indicated that the areas of SZ welded at 0°, 1.5°, and 3° TTA were 6.104 mm^2, 5.46 mm^2, and 5.059 mm^2, respectively. The results revealed that the size of the SZ decreased by 17% from 0° to 3° TTA. The TMAZ and HAZ areas were different from the SZ trend. The results show that the TMAZ and HAZ areas decreased from 0° to 1.5°

TTA and, after that, increased from 1.5° to 3° TTA. The TMAZ areas of the joints welded at 0°, 1.5°, and 3° TTA were 0.817 mm^2, 0.742 mm^2, and 0.78 mm^2, respectively. On the other hand, the HAZ areas of the joints welded at 0°, 1.5°, and 3° TTA were 1.165 mm^2, 0.799 mm^2, and 0.975 mm^2, respectively.

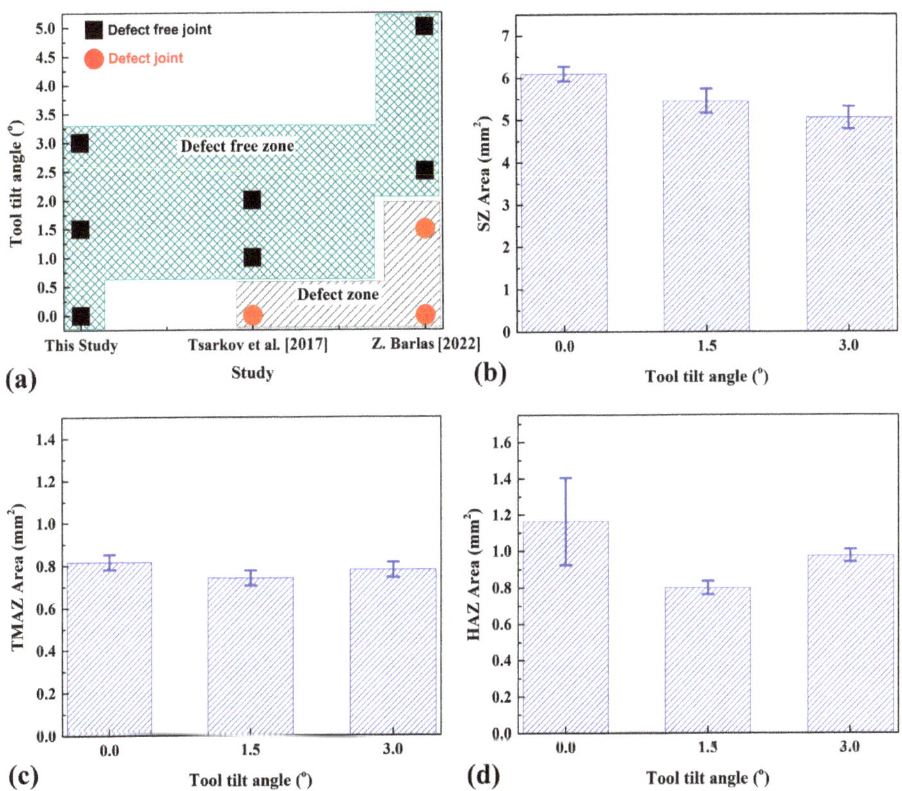

Figure 7. (**a**) Comparison between total-contact and partial-contact TTA joint quality [1,2]. Comparison between total area of (**b**) SZ, (**c**) TMAZ, and (**d**) HAZ at different welded joints.

3.4. Internal Flow Analysis

The internal flow pattern of welded samples on a microscale gives useful information about the defect formation and voids [26]. For this reason, the internal flow pattern of welded samples was assessed via scanning electron microscope (SEM). A sample SEM image of the welded sample with 3° TTA is presented in Figure 8a. As can be seen, the area of the stir zone and other parts are distinguished clearly. The interfaces of the stir zone and other parts of welded samples with 0°, 1.5°, and 3° TTA are presented in Figure 8b, 8c and 8d, respectively. The depicted SEM image was recorded from the advancing side of the welded sample. The results presented for the advancing side and the retreating side were similar. The results revealed that small voids are formed at the interface of the stir zone and TMAZ for welding at 0° TTA joints. Micro-voids were formed in the stir zone of the joint friction stir welded at 1.5° TTA, and micro-cracks were formed inside the stir zone of the joint friction stir welded at 3° TTA.

High-magnification images of internal flow of friction stir welded samples welded at 0°, 1.5°, and 3° TTA are presented in Figure 9a–c, respectively. The results show that small voids formed near the stir zone of the sample friction stir welded at 0° TTA. This void seems to result from chemical changes in the AA1050 aluminum alloy after severe plastic deformation with the FSW tool. This trend can be seen in other samples. The results reveal that the size and distribution of these voids decrease at higher TTA. The results show that high non-contact TTA increases the possibility of FSW tool wear. For this reason, shiny particles can be detected, as in Figure 8b,c. On the other hand, the small shiny point can be detected at 1.5° and 3° TTA, which indicates that a small particle of welding tool remained in the joint area.

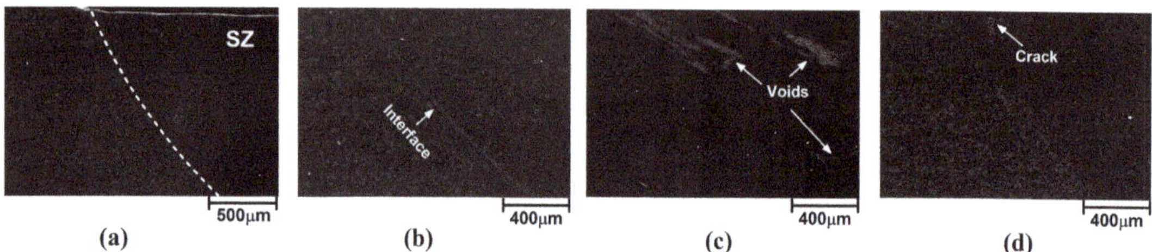

Figure 8. (**a**) Macroscale SEM image of internal flow of friction stir welded sample at 3° TTA. SEM image of SZ interface of friction stir welded sample welded at (**b**) 0° TTA, (**c**) 1.5° TTA, and (**d**) 3° TTA.

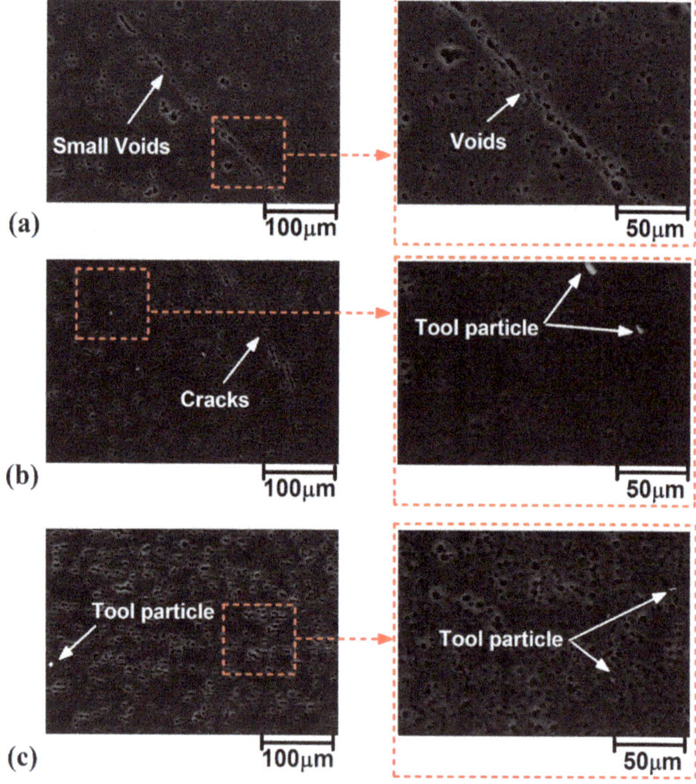

Figure 9. High magnification SEM image of SZ interface of friction stir welded sample welded at (**a**) 0° TTA, (**b**) 1.5° TTA, and (**c**) 3° TTA.

3.5. Microstructure Analysis

The microstructure of the weld zone in the FSW process is related to the stirring action of the tool [7]. Severe plastic deformation can change the grain size and microstructure of AA1050 aluminum alloy after FSW. Figure 10a presents a high-magnification optical microscopy image of the various areas of the joint lines. The SZ, TMAZ, and HAZ areas for 0°, 1.5°, and 3° TTA are presented. The microstructures of friction stir welded samples welded at 0°, 1.5°, and 3° TTA are presented in Figure 10b–d, respectively. The presented results are high-magnification images captured via an optical microscope from the centre of the SZ. The average grain size of the SZ was found via image processing software (Version 1.1). As the results show, the average grain sizes of the friction stir welded sample at 0°, 1.5°, and 3° TTA are 51 μm, 58 μm, and 62 μm, respectively. The microstructure of the 0° TTA joint was 17% smaller than the 3° TTA joint. The total contact of the FSW tool with AA1050 aluminum alloy seems to form fine grains in SZ. On increasing TTA, the contact area of the tool shoulder in the leading edge decreased, and the main parts of the tool for stirring action were the pin and back side of the FSW tool in the trailing edge. A lower contact area leads to lower thermo-mechanical action and the formation of bigger grain sizes in the SZ.

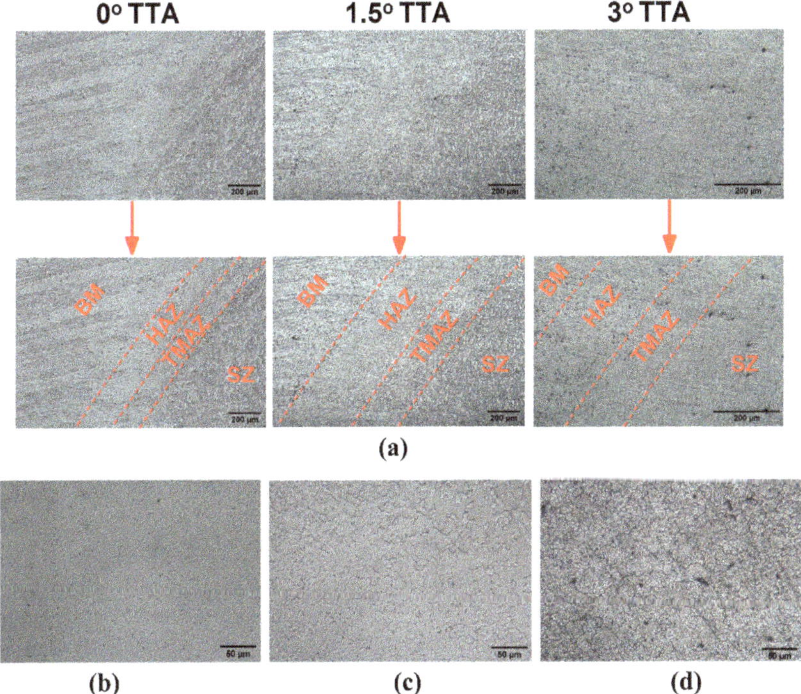

Figure 10. (**a**) Optical image from welding areas of friction stir welded joint at 0°, 1.5°, and 3° TTA. High magnification optical microscope image of SZ microstructure in (**b**) 0°, (**c**) 1.5°, and (**d**) 3° TTA samples.

3.6. Microhardness

The hardness of the welding area gives information about the mechanical properties of the final FSW joint [27]. Figure 11a–c indicate the hardness of the welded area that was welded three times. The vertical line from the centre of the SZ was selected for hardness evaluations. The results revealed that the hardness of the HAZ area was slightly lower than other parts in all joints. On the other hand, the SZ is the hardest area in all cases. The results show that the average hardness of the friction stir welded sample with 0° TTA is more than other cases, and the average hardness of 1.5° TTA is lower than other samples. The average hardnesses of the SZ sample welded at 0°, 1.5°, and 3° TTA were 28 Hv, 26 Hv, and 25 Hv. The average hardnesses of the TMAZ of samples friction stir welded at 0°, 1.5°, and 3° TTA were 26 Hv, 24 Hv, and 23 Hv, respectively. On the other hand, the average hardnesses of the HAZ area of friction stir welded samples were 25 Hv, 23 Hv, and 24 Hv at 0°, 1.5°, and 3° TTA, respectively. The hardness difference between 1.5° and 3° TTA is related to the tool particles that remained in the SZ. In partial-contact TTA, the applied force in the FSW tool increases, and the wear rate of the tool is high. In this case, particles remaining in the SZ increase the hardness of the SZ locally. SEM images of tool particles in the SZ are presented in Figure 11d. As can be seen, the shiny particles are remainders of the tool material in the SZ. The used aluminum alloy in this study is categorized as 1XXX aluminum alloy, a low-element aluminum alloy. This means that this aluminum category's chemical interaction and precipitate formation are very low. On the other hand, in SEM images, the high-weight elements can be seen as shinier than light elements. The tool used was steel, and after FSW, a worn surface can be detected in the tool's pin. In this case, it can be concluded that the shiny particles in the SZ detected in the SEM image would be small particles of the FSW tool that remained in the SZ. Barlas [1] considered the hardness of the optimum joint in his study. As mentioned earlier, the optimum TTA in Barlas' [1] study was 2.5° TTA, and the hardnesses of the SZ, HAZ, and AA1050 aluminum alloy were 31.5 Hv, 28.3 Hv, and 40.7 Hv, respectively. The hardness of the SZ in the Barlas [1] study was near this investigation's result (28 Hv hardness in the stir zone of 3° TTA joint).

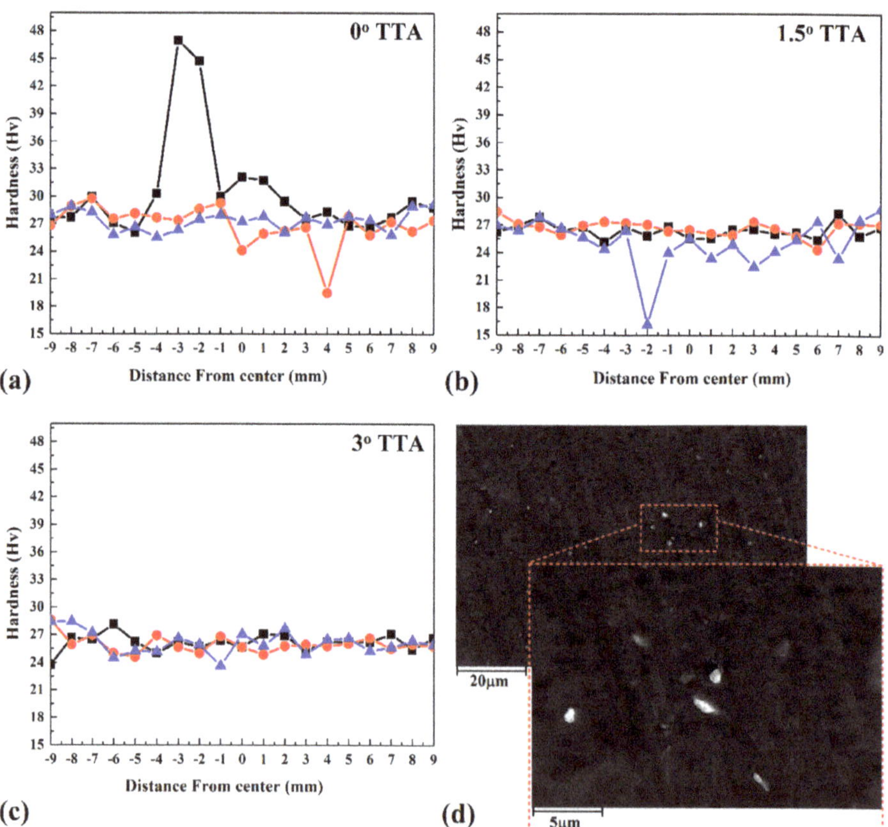

Figure 11. Microhardness profile of specimens friction stir welded at (**a**) 0°, (**b**) 1.5°, and (**c**) 3° TTA. Particle remainder of the FSW tool in the SZ of the 3° TTA joint. (**d**) SEM image of the FSW particles that remained in the SZ.

3.7. Tensile Strength

Figure 12a shows engineering stress–strain curves of welded specimens at different TTAs and AA1050 aluminum alloys. The obtained results from tests such as ultimate tensile strength (UTS) and elongation percentage (E%) are presented in Figure 12b,c, respectively. As can be seen from the obtained results, the UTS of AA1050 aluminum alloy and the friction stir welded samples welded at 0°, 1.5°, and 3° TTA were 26 MPa, 33 MPa, 30 MPa, and 22 MPa, respectively. The results show that the UTS of friction stir welded samples welded at 0° and 1.5° TTA increased more than base metal, while the UTS of the 3° TTA joint was lower than base metal. It is found that the differences between AA1050 aluminum alloy and friction stir welded samples were +7 MPa, +4 MPa, and −4 MPa for friction stir welded samples at 0°, 1.5°, and 3° TTA, respectively. By dividing the UTS of friction stir welded samples with AA1050 aluminum alloy, it can be found that the joint efficacies of 0°, 1.5°, and 3° TTA samples were ~127%, ~115%, and ~85%, respectively.

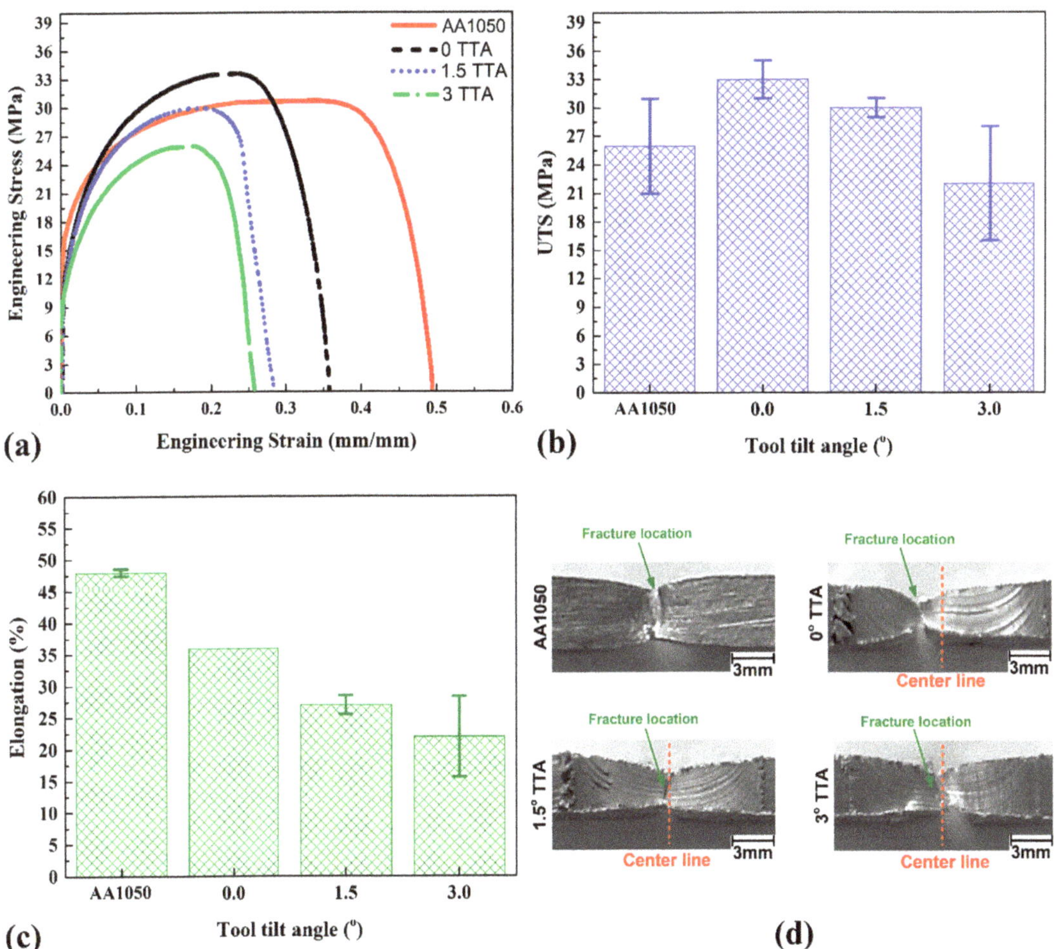

Figure 12. (a) Engineering stress–strain curves of base metal and welded specimens, (b) UTS and (c) elongation of raw metal and welded samples. (d) Fracture locations of base metal and welded samples

The results indicated that the elongations of AA1050 aluminum alloy and the friction stir welded samples welded at 0°, 1.5°, and 3° TTA were 48%, 36%, 27%, and 22% after the tensile test. By comparing AA1050 aluminum alloy with the friction stir welded samples, it can be concluded that the elongations of welded samples at 0°, 1.5°, and 3° TTA were 75%, 56%, and 50% of raw metal. Figure 12d illustrates the fracture location of AA 1050 alloy and friction stir welded samples after the tensile test. The raw metal fracture place was in the middle of the tensile test sample. The results show that the fracture locations of welded samples at 0°, 1.5°, and 3° TTA were TMAZ, SZ, and SZ areas, respectively. Figure 13a–d show the SEM images from the fracture surface of AA1050 aluminum alloy and welded samples at 0°, 1.5°, and 3° TTA, respectively. Small dimples can be seen in all samples, indicating ductile fracture. On the other hand, a small deformation area can be detected in the FSW sample welded at 0°, 1.5°, and 3° TTA.

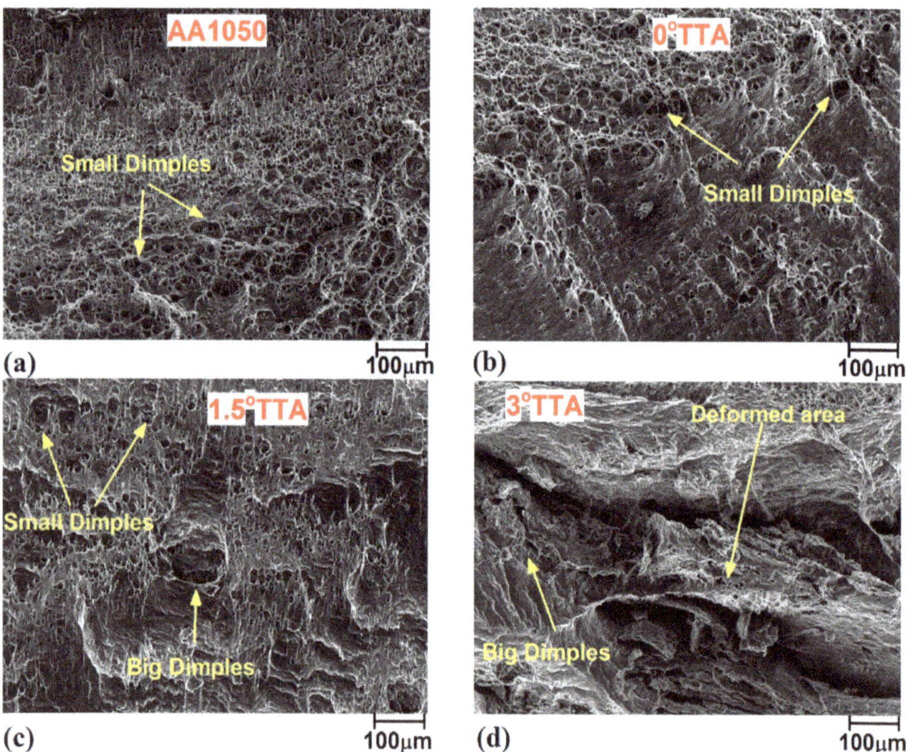

Figure 13. SEM morphology pictures of the fracture surface of tensile tests: (**a**) base metal, (**b**) 0°, (**c**) 1.5°, and (**d**) 3° TTA.

4. Conclusions

This study aimed to investigate the impact of partial-contact TTA on friction stir welding of AA1050 aluminum alloy. The weldments were examined using several tests such as surface roughness, microstructure analysis, thermal cycles, and tensile tests to achieve the objective of this research. Based on the results of this investigation, the following conclusions are presented:

- The heat generated during the FSW process by increasing the TTA in partial-contact conditions decreases. The maximum heat was recorded at 0° TTA (336 °C) and the minimum heat was recorded at 3° TTA (320 °C). The generated heat decreased due to the smaller contact area between the tool and the AA1050 alloy at the higher partial-contact TTA.
- Increasing TTA in partial-contact conditions decreases the size of the SZ, TMAZ, and HAZ areas. The formed joints were defect-free at a macroscale, while internal defects beneath of pin increase at higher TTAs in partial-contact conditions. Microscale voids and small interface gaps were detected at 3° TTA, while root void defects were not formed at 0° TTA.
- The highest strength joint was produced at 0° TTA, and the weakest joint was produced at 3° TTA. The ultimate tensile strength and average hardness of the joint friction stir welded at 0° TTA were 33 MPa and 26 Hv, respectively. The joint efficiencies of 0°, 1.5°, and 3° TTA samples were ~127%, ~115%, and ~85%, respectively.

Author Contributions: Conceptualization, M.E.A., M.N.M.R. and M.M.M.; methodology, M.E.A., M.N.M.R. and M.M.M.; software, M.E.A., M.N.M.R. and M.M.M.; validation, M.E.A., M.N.M.R. and M.M.M.; formal analysis, M.E.A., M.N.M.R., M.M.M. and H.A.D.; investigation, M.E.A., M.N.M.R. and M.M.M.; resources, M.E.A., M.N.M.R. and M.M.M.; data curation, M.E.A., M.N.M.R. and M.M.M.; writing—original draft preparation, M.E.A., M.N.M.R., M.M.M. and H.A.D.; writing—review and editing, H.A.D.; visualization, M.E.A., M.N.M.R. and M.M.M.; supervision, M.E.A., M.N.M.R. and M.M.M.; project administration, M.E.A., M.N.M.R. and M.M.M.; funding acquisition, H.A.D. All authors have read and agreed to the published version of the manuscript.

Funding: This research received no external funding.

Institutional Review Board Statement: Not applicable.

Informed Consent Statement: Not applicable.

Data Availability Statement: Not applicable.

Acknowledgments: The researchers would like to thank and appreciation to the technical staff, Faculty of Technology and Education, Beni-Suef University, Egypt.

Conflicts of Interest: The authors declare no conflict of interest.

References

1. Barlas, Z. The Influence of Tool Tilt Angle on 1050 Aluminum Lap Joint in Friction Stir Welding Process. *Acta Phys. Pol. A* **2017**, *132*, 679–681. [CrossRef]
2. Tsarkov, A.; Trukhanov, K.; Zybin, I.; Vichniakov, A. Tilt Angle Effect on Friction Stir Welding Conditions. *Key Eng. Mater.* **2022**, *910*, 115–122. [CrossRef]
3. Aghajani Derazkola, H.; Simchi, A.; Lambiase, F. Friction stir welding of polycarbonate lap joints: Relationship between processing parameters and mechanical properties. *Polym. Test.* **2019**, *79*, 105999. [CrossRef]
4. Lukács, J.; Meilinger, Á.; Pósalaky, D. High cycle fatigue and fatigue crack propagation design curves for 5754-H22 and 6082-T6 aluminium alloys and their friction stir welded joints. *Weld. World* **2018**, *62*, 737–749. [CrossRef]
5. Dinesh Kumar, R.; Ilhar Ul Hassan, M.S.; Muthukumaran, S.; Venkateswaran, T.; Sivakumar, D. Single and Multi-Response Optimization and Validation of Mechanical Properties in Dissimilar Friction Stir Welded AA2219-T87 and AA7075-T73 Alloys Using T-GRA. *Exp. Tech.* **2019**, *43*, 245–259. [CrossRef]
6. Moreira, P.M.G.P.; de Jesus, A.M.P.; Ribeiro, A.S.; de Castro, P.M.S.T. Fatigue crack growth in friction stir welds of 6082-T6 and 6061-T6 aluminium alloys: A comparison. *Theor. Appl. Fract. Mech.* **2008**, *50*, 81–91. [CrossRef]
7. Yuvaraj, K.P.; Ashoka Varthanan, P.; Rajendran, C. Effect of friction stir welding parameters on mechanical and micro structural behaviour of AA7075-T651 and AA6061 dissimilar alloy joint. *Int. J. Comput. Mater. Sci. Surf. Eng.* **2018**, *7*, 130–149. [CrossRef]
8. Aghajani Derazkola, H.; Simchi, A. Experimental and thermomechanical analysis of friction stir welding of poly(methyl methacrylate) sheets. *Sci. Technol. Weld. Join.* **2018**, *23*, 209–218. [CrossRef]
9. Sadeghian, N.; Besharati Givi, M.K. Experimental optimization of the mechanical properties of friction stir welded Acrylonitrile Butadiene Styrene sheets. *Mater. Des.* **2015**, *67*, 145–153. [CrossRef]
10. Fan, G.; Tomków, J.; Abdullah, M.E.; Derazkola, H.A. Investigation on polypropylene friction stir joint: Effects of tool tilt angle on heat flux, material flow and defect formation. *J. Mater. Res. Technol.* **2023**, *23*, 715–729. [CrossRef]

11. Abd Elnabi, M.M.; Osman, T.A.; El Mokadem, A.; Elshalakany, A.B. Evaluation of the formation of intermetallic compounds at the intermixing lines and in the nugget of dissimilar steel/aluminum friction stir welds. *J. Mater. Res. Technol.* **2020**, *9*, 10209–10222. [CrossRef]
12. Shah, P.H.; Badheka, V.J. An Experimental Insight on the Selection of the Tool Tilt Angle for Friction Stir Welding of 7075 T651 Aluminum Alloys. *Indian J. Sci. Technol.* **2016**, *9*, 1–11. [CrossRef]
13. Acharya, U.; Roy, B.S.; Saha, S.C. On the Role of Tool Tilt Angle on Friction Stir Welding of Aluminum Matrix Composites. *Silicon* **2021**, *13*, 79–89. [CrossRef]
14. Birol, Y.; Kasman, S. Effect of Welding Parameters on the Microstructure and Strength of Friction Stir Weld Joints in Twin Roll Cast EN AW Al-Mn1Cu Plates. *J. Mater. Eng. Perform.* **2013**, *22*, 3024–3033. [CrossRef]
15. Zhai, M.; Wu, C.S.; Su, H. Influence of tool tilt angle on heat transfer and material flow in friction stir welding. *J. Manuf. Process.* **2020**, *59*, 98–112. [CrossRef]
16. Gupta, S.; Haridas, R.S.; Agrawal, P.; Mishra, R.S.; Doherty, K.J. Influence of welding parameters on mechanical, microstructure, and corrosion behavior of friction stir welded Al 7017 alloy. *Mater. Sci. Eng. A* **2022**, *846*, 143303. [CrossRef]
17. Dialami, N.; Cervera, M.; Chiumenti, M. Effect of the tool tilt angle on the heat generation and the material flow in friction stir welding. *Metals* **2019**, *9*, 28. [CrossRef]
18. Zhai, M.; Wu, C.S.; Shi, L. Tool tilt angle induced variation of shoulder-workpiece contact condition in friction stir welding. *Sci. Technol. Weld. Join.* **2022**, *27*, 68–76. [CrossRef]
19. Khan, N.Z.; Siddiquee, A.N.; Khan, Z.A. Proposing a new relation for selecting tool pin length in friction stir welding process. *Meas. J. Int. Meas. Confed.* **2018**, *129*, 112–118. [CrossRef]
20. Elyasi, M.; Taherian, J.; Hosseinzadeh, M.; Kubit, A.; Derazkola, H.A. The effect of pin thread on material flow and mechanical properties in friction stir welding of AA6068 and pure copper. *Heliyon* **2023**, *9*, e14752. [CrossRef]
21. Mirabzadeh, R.; Parvaneh, V.; Ehsani, A. Experimental and numerical investigation of the generated heat in polypropylene sheet joints using friction stir welding (FSW). *Int. J. Mater. Form.* **2021**, *14*, 1067–1083. [CrossRef]
22. Sajed, M.; Guerrero, J.W.G.; Derazkola, H.A. A Literature Survey on Electrical-Current-Assisted Friction Stir Welding. *Appl. Sci.* **2023**, *13*, 1563. [CrossRef]
23. Tamjidy, M.; Hang Tuah Baharudin, B.T.; Paslar, S.; Matori, K.A.; Sulaiman, S.; Fadaeifard, F. Multi-objective optimization of friction stir welding process parameters of AA6061-T6 and AA7075-T6 using a biogeography based optimization algorithm. *Materials* **2017**, *10*, 533. [CrossRef]
24. Khalaf, H.I.; Al-Sabur, R.; Derazkola, H.A. Effect of number of tool shoulders on the quality of steel to magnesium alloy dissimilar friction stir welds. *Arch. Civ. Mech. Eng.* **2023**, *23*, 125. [CrossRef]
25. Chitturi, V.; Pedapati, S.R.; Awang, M. Investigation of weld zone and fracture surface of friction stir lap welded 5052 aluminum alloy and 304 stainless steel joints. *Coatings* **2020**, *10*, 1062. [CrossRef]
26. Ghiasvand, A.; Suksatan, W.; Tomków, J.; Rogalski, G.; Derazkola, H.A. Investigation of the Effects of Tool Positioning Factors on Peak Temperature in Dissimilar Friction Stir Welding of AA6061-T6 and AA7075-T6 Aluminum Alloys. *Materials* **2022**, *15*, 702. [CrossRef]
27. Godasu, A.K.; Kumar, A.; Mula, S. Influence of cryocooling on friction stir processing of Al-5083 alloy. *Mater. Manuf. Process.* **2020**, *35*, 202–213. [CrossRef]

 *materials*

Article

Mechanical Performance and Microstructural Evolution of Rotary Friction Welding of Acrylonitrile Butadiene Styrene and Polycarbonate Rods

Chil-Chyuan Kuo [1,2,3,*], Naruboyana Gurumurthy [1,4], Hong-Wei Chen [1] and Song-Hua Hunag [5]

[1] Department of Mechanical Engineering, Ming Chi University of Technology, No. 84, Gungjuan Road, New Taipei City 243, Taiwan
[2] Research Center for Intelligent Medical Devices, Ming Chi University of Technology, No. 84, Gungjuan Road, New Taipei City 243, Taiwan
[3] Department of Mechanical Engineering, Chang Gung University, No.259, Wenhua 1st Rd., Guishan Dist., Taoyuan City 333, Taiwan
[4] Department of Mechanical Engineering, Presidency University, Rajankunte, Near Yelhanka, Bangalore 700073, India
[5] Li-Yin Technology Co., Ltd., No. 37, Lane 151, Section 1, Zhongxing Road, Wugu District, New Taipei City 241, Taiwan
* Correspondence: jacksonk@mail.mcut.edu.tw

Abstract: Rotary friction welding (RFW) is a green manufacturing technology with environmental pollution in the field of joining methods. In practice, the welding quality of the friction-welded parts was affected by the peak temperature in the weld joint during the RFW of dissimilar plastic rods. In industry, polycarbonate (PC) and acrylonitrile butadiene styrene (ABS) are two commonly used plastics in consumer products. In this study, the COMSOL multiphysics software was employed to estimate the peak temperature in the weld joint during the RFW of PC and ABS rods. After RFW, the mechanical performance and microstructural evolution of friction-welded parts were investigated experimentally. The average Shore A surface hardness, flexural strength, and impact energy are directly proportional to the rotation speed of the RFW. The quality of RFW is excellent, since the welding strength in the weld joint is better than that of the ABS base materials. The fracture occurs in the ABS rods since their brittleness is higher than that of the PC rods. The average percentage error of predicting the peak temperature using COMSOL software using a mesh element count of 875,688 for five different rotation speeds is about 16.6%. The differential scanning calorimetry curve for the friction-welded parts welded at a rotation speed of 1350 rpm shows an endothermic peak between 400 to 440 °C and an exothermic peak between 600 to 700 °C, showing that the friction-welded parts have better mechanical properties.

Keywords: solid-state welding; polymer friction welding; polymer welded joints; RFW; dissimilar polymer welds

Citation: Kuo, C.-C.; Gurumurthy, N.; Chen, H.-W.; Hunag, S.-H. Mechanical Performance and Microstructural Evolution of Rotary Friction Welding of Acrylonitrile Butadiene Styrene and Polycarbonate Rods. *Materials* **2023**, *16*, 3295. https://doi.org/10.3390/ma16093295

Academic Editor: Raul D. S. G. Campilho

Received: 31 March 2023
Revised: 20 April 2023
Accepted: 21 April 2023
Published: 22 April 2023

1. Introduction

Rotary friction welding (RFW) [1] is one of the approaches of friction welding that is considered solid-state welding, which has lower energy consumption and environmental impact compared to gas metal arc welding. The process of RFW is that one welded element is rotated at a constant speed while the other welded element remains stationary under an axial force. In general, RFW gives many features based on its practical experience in industry compared with adhesive bonding [2,3]. The RFW process provides a lower peak temperature in the weld joint compared with fusion welding (FW) [4–6]. Therefore, a wide range of similar or dissimilar materials can be joined efficiently and economically. The products obtained by RFW have low defect rates and low distortion. In addition, the manufacturing costs can be reduced significantly compared with the subtractive technique, such as milling machining from buck materials [7,8].

Eslami et al. [9] reviewed the friction stir welding tooling for polymers and analyzed the weld strengths for different polymeric materials. Paoletti et al. [10] investigated the forces and temperatures in the friction spot stir welding of thermoplastic polymers. The results showed that the increase in the tool rotational speed will reduce the processing forces. Lambiase et al. [11] investigated the influence of the plunging force in the friction stir welding of polycarbonate sheets on the mechanical behavior of the welds. The results revealed that the mechanical behavior of the welds can be improved up to 37 % by plunging force. The shear strength of 34.5 MPa that yields the base material can be obtained by optimal conditions. Rehman et al. [12] investigated the effects of preheating on joint quality in the friction stir welding of polyethylene. The results showed that proper welding of this bimodal high-density polyethylene takes place when the material is maintained at high temperatures. Large elongations in the order of 60% and weld efficiencies in excess of 100% were also achieved by optimal welding temperatures. Skowrońska et al. [13] assessed the structural properties of friction-welded joints. The results revealed that a surface hardness above 340 HV was obtained in the weld joint. Dhooge et al. [14] proposed a new variant friction-welding process for the fully automatic joining of pipelines. Optimization of the duration of the friction phase of the friction-welding process was also investigated. Anwar et al. [15] investigated the microstructure, mechanical properties, and grain size of the alloy after RFW. The minimum grain size was successfully met by postweld heat treatment with improved elongation and strength. It was found that the weld metal grain size of the postweld heat treatment joints is about 35 ± 4 µm. The average grain size of the weld metal in the as-welded condition is about 20 ± 2 µm, in contrast to the weld metal grain size of the PWHT joints. Ishraq et al. [16] investigated the weld strength by optimizing the welding process parameters. The results showed that the major reason for the high strength of a welded material used is the optimal level of fiberglass. Hangai et al. [17] studied the effects of the porosity of aluminum foam on the weldability to a polycarbonate plate. A welding strength of polycarbonate plate and Al foam higher than that of the base Al foam with a porosity of 80% can be obtained. Skowrońska et al. [18] investigated the microstructure of a friction-welded joint made of stainless steel with an ultrafine-grained structure made by hydrostatic extrusion. It was found that strength is the criterion for assessing the properties of the joint because of the complexity of the microstructure of the friction-welded joint. Zhang et al. [19] investigated a thermal compression bonding process in friction welding. It was found that the frictional flow greatly enhanced the formation of intermetallic compounds along the weld interface. Eliseev et al. [20] investigated the microstructural evolution in the transfer layer of aluminum alloy welds. The results showed that the size of the incoherent intermetallic particles and the volume fraction decreased towards to the center of the layer. Ma et al. [21] investigated the effects of temperature on mechanical performances of friction-stir-welded aluminum alloy joints. It was found that the reduction in the gradient along the thickness due to the pinhole increased heat input and material flow at the bottom.

Polymers are frequently used in some structures, because the major difference between metal and polymers is that polymers are lighter than metal. Acrylonitrile butadiene styrene (ABS) [22] and polycarbonate (PC) [23] are two commonly used plastics in consumer products. ABS is an thermoplastic engineering material that has high tensile strength and high resistance to chemical corrosion and physical impacts. In addition, ABS is easy to use in the injection-molding process since it has low melting point. Therefore, ABS plastic is suitable for making consumer products to withstand heavy use. PC plastic is considered as an engineering plastic since it has very good heat resistance, and is widely employed for more robust materials. However, hitherto little is known about the domain knowledge of the RFW of ABS and PC polymer rods. For this reason, the objective of this study is to establish domain knowledge of the RFW of ABS and PC rods. The RFW experiment was performed using a turning machine. During RFW, the peak temperature in the weld joint was determined using an infrared thermal imager. The COMSOL multiphysics software [24–26] was also employed to predict the peak temperature in the weld joint and to compare the results obtained by the experiment. After RFW, the mechanical properties and microstructural evolution of the friction-welded parts were characterized using Shore A hardness, three-point bending, and impact tests. Finally, domain knowledge of the RFW of ABS and PC rods was proposed. The melting behavior, solidification characteristics, and glass transition temperature of friction-welded joints was also conducted with differential scanning calorimetry (DSC).

2. Experimental Details

Figure 1 shows the flowchart of experimental details. The entire process includes designing weld specimen, fabrication of weld specimen, rotary friction welding, determining peak temperature in the weld joint via simulation by COMSOL, and experiment using infrared thermal imager, determining mechanical properties, Shore hardness tests, bending tests, impact tests, fracture surface analysis, DSC thermal analysis, and finally establishing domain knowledge of RFW of dissimilar polymer rods. In the simulation by COMSOL, the entire process involves thermal pattern analysis and suitable boundary conditions for the rotary friction model. At first, a COMSOL model is established, which involves identifying the components of the system and their physical properties. Then, the next step includes setting the parameters for RFW. Then, suitable boundary conditions, such as heat flux at various locations or temperature, need to be selected based on the system geometry and heat transfer mechanisms. Finally, peak temperature in the weld joint through the heat transfer needs to be identified, using heat transfer mechanisms such as conduction, convection, and radiation, to analyze the thermal behavior. The workpiece is a cylindrical rod with a diameter of 20 mm and a length of 40 mm. The welding workpieces were printed with a three-dimensional printing apparatus named fused deposition modeling (FDM), using two different thermoplastic filaments, i.e., PC (Thunder 3D Inc., New Taipei City, Taiwan) and ABS (Thunder 3D Inc., New Taipei City, Taiwan) [27]. Figure 2 shows the two dissimilar workpieces for RFW in this study. The FDM process parameters for ABS rods include infill percentage of 70%, print bed temperature at 100 °C, print speed of 80 mm/s, print temperature of 230 °C, shell thickness of 0.4 mm, and layer thickness of 0.1 mm. The FDM process parameters for PC rods involve infill percentage of 100%, print bed temperature at 100 °C, print speed of 80 mm/s, print temperature of 245 °C, shell thickness of 0.4 mm, and layer thickness of 0.1 mm. The printing strategy for the cylindrical polymer rod of ABS and PC is that the extruder moves in straight lines back and forth to create the parallel lines of the material.

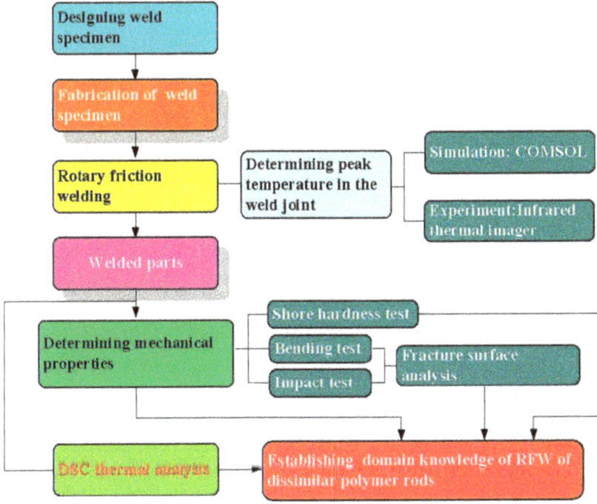

Figure 1. Flowchart of experimental details.

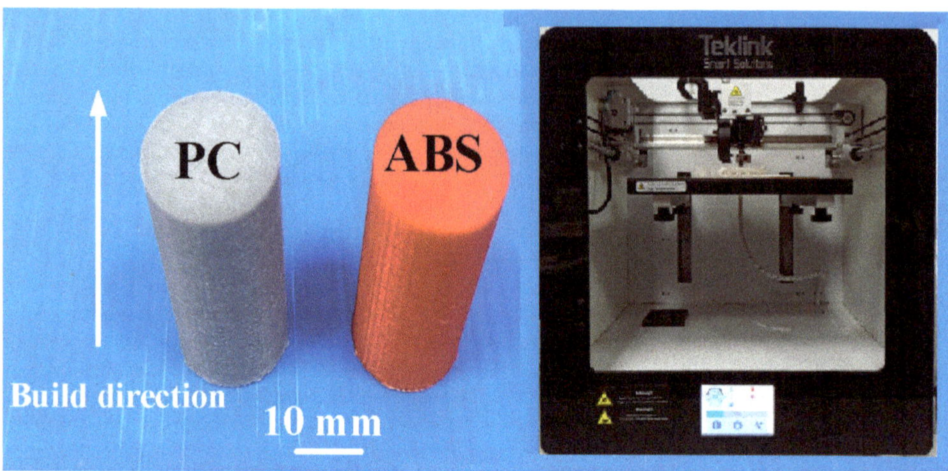

Figure 2. Two dissimilar workpieces for RFW in this study.

Figure 3 shows the situation of RFW. A turning machine was selected as a friction welder to carry out RFW of dissimilar polymer rods. The welding time of RFW was set to 60 s, involving friction time of 30 s, forge time of 20 s, and cooling time of 10 s. To study the effects of rotation speed on the peak temperature in the weld joint, five rotation speeds, i.e., 330 rpm, 490 rpm, 650 rpm, 950 rpm, and 1350 rpm, were performed. During RFW, the peak temperature in the weld joint was recorded using an infrared camera (BI-TM-F01P, Panrico trading Inc., New Taipei City, Taiwan). After RFW, Charpy impact test (780, Instron Inc., Norwood, MA, USA), Shore A surface hardness test (MET-HG-A, SEAT Inc. New Taipei City, Taiwan), and bending test (RH-30, Shimadzu Inc., Kyoto, Japan) were performed to investigate the microstructural evolution and mechanical properties of the friction-welded parts. Figure 4 shows the experimental set-up for impact test. The insert is the close-up view showing the arrangement of the friction-welded parts in the impact test. Figure 5 shows the experimental set-up for bending test, showing the setting of the friction-welded parts during the bending test. The macrostructure of the fracture surfaces after bending and impact tests was performed by a field-emission scanning electron microscope (FE-SEM) (JEC3000-FC, JEOL Inc., Tokyo, Japan) and a stereo optical microscope (Quick Vision, Mitutoyo Inc, Tokyo, Japan). In this work, DSC (STA 409 PC Luxx Simultaneous thermal analyzer, Netzsch-Geratebau GmbH Inc., Waldkraiburg, Germany) analysis was used to calculate the melting and mesomorphic transitions along with their enthalpy and entropy of the friction-welded joints after RFW with five rotation speeds. The DSC experimental setup provides two separate crucibles for heating and cooling. One is for reference and another is for a sample to be investigated. DSC evaluation was carried out under controlled experimental conditions of continuous heating rate of 15 °C/min and continuous cooling rate of 15 °C/min in temperatures ranging from 50 °C to 250 °C. Thermal properties were investigated by two continuous exothermic cycles and an endothermic cycle at 25 mL/min of nitrogen gas supply.

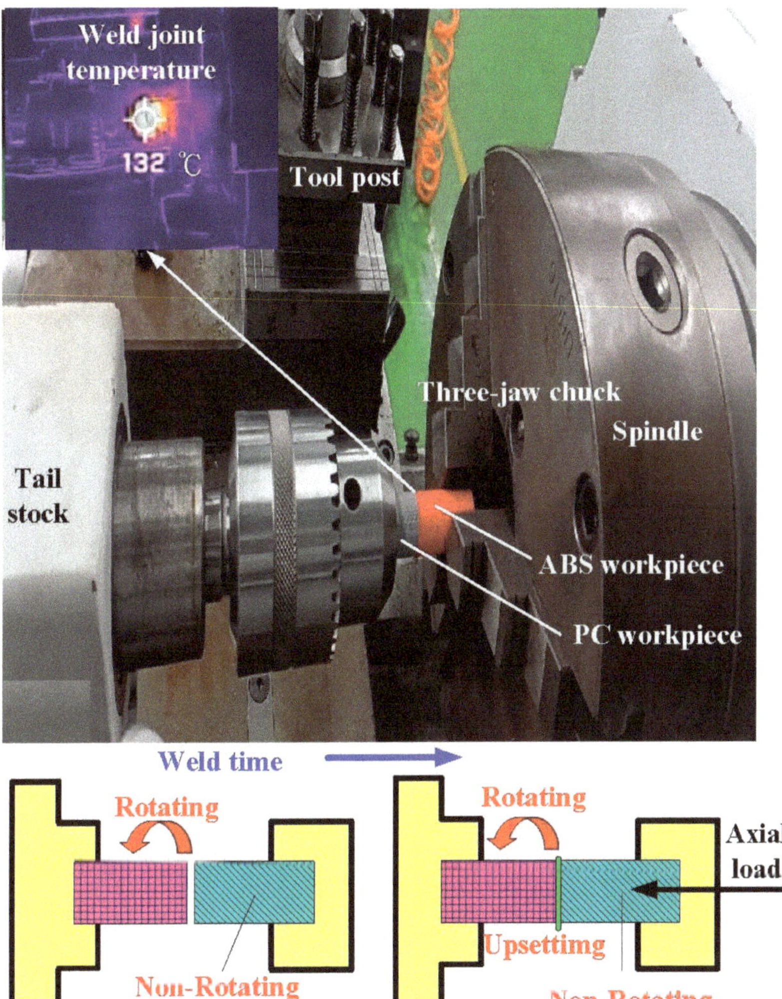

Figure 3. Situation of RFW.

Figure 4. Experimental set-up for impact test.

Figure 5. Experimental set-up for bending test.

3. Results and Discussion

Figure 6 describes the friction-welded PC and ABS rods using RFW. The top of this figure shows panoramic SEM micrographs of the friction-welded joint. This result reveals that the RFW of PC and ABS is acceptable since the bead width is consistent. Figure 7 shows the mechanical properties of surface hardness, impact energy, and flexural strength in the weld joint for RFW of dissimilar polymer rods under five rotation speeds. The maximum joint strength of 132 MPa, Shore A hardness of 80, and impact energy of 156 J can be obtained at a rotation speed of 1350 rpm. As can be seen, the increase in joint strength and surface hardness with increasing rotation speed is observed. This result reveals that the surface hardness, flexural strength, and impact energy are directly proportional to the rotation speed of RFW [28].

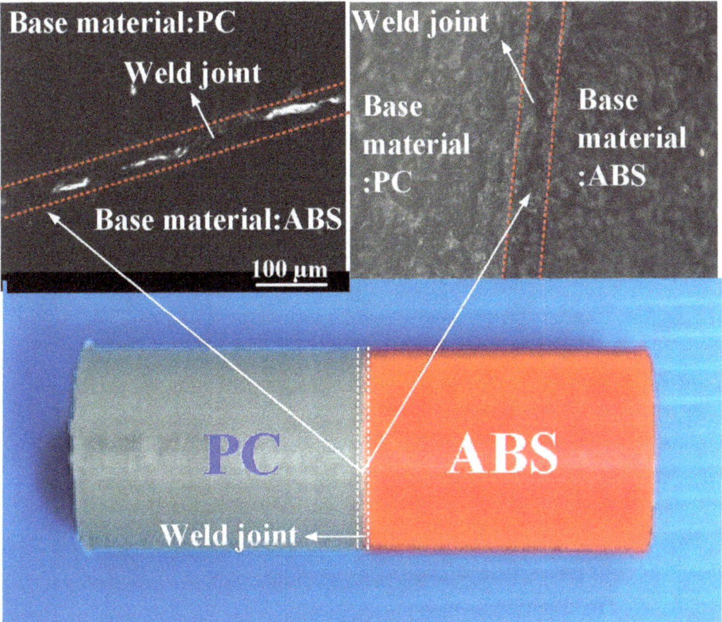

Figure 6. Friction-welded PC and ABS rods using RFW.

Figure 8 shows the number of meshes as a function of peak temperature in the weld joint. Ten different kinds of mesh sizes, i.e., 0.4 mm, 0.5 mm, 0.6 mm, 0.7 mm, 0.8 mm, 0.9 mm, 1.0 mm, 1.1 mm, 1.2 mm, and 1.3 mm, were used to investigate the peak temperature in the weld joint during RFW. The insert shows the thermal model of RFW. It should be noted that a higher number of meshes provides more computation time. As can be seen, the peak temperature predicted by the COMSOL multiphysics software using a mesh element count of 875,688 is very close to that obtained by the experimental result. The mesh size is about 0.7 mm. This shows that a mesh element count of 875,688 is suitable for predicting the peak temperature in the RFW of PC and ABS rods.

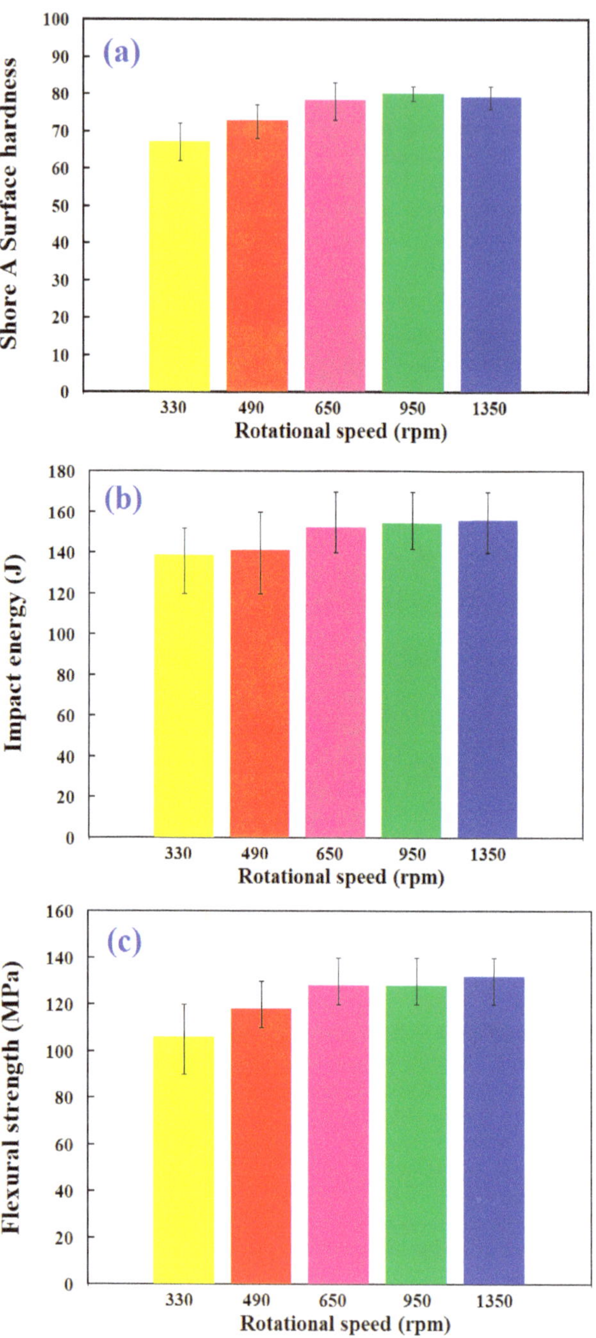

Figure 7. Mechanical properties of (**a**) surface hardness, (**b**) impact energy, and (**c**) flexural strength in the weld joint for RFW of dissimilar polymer rods under five rotation speeds.

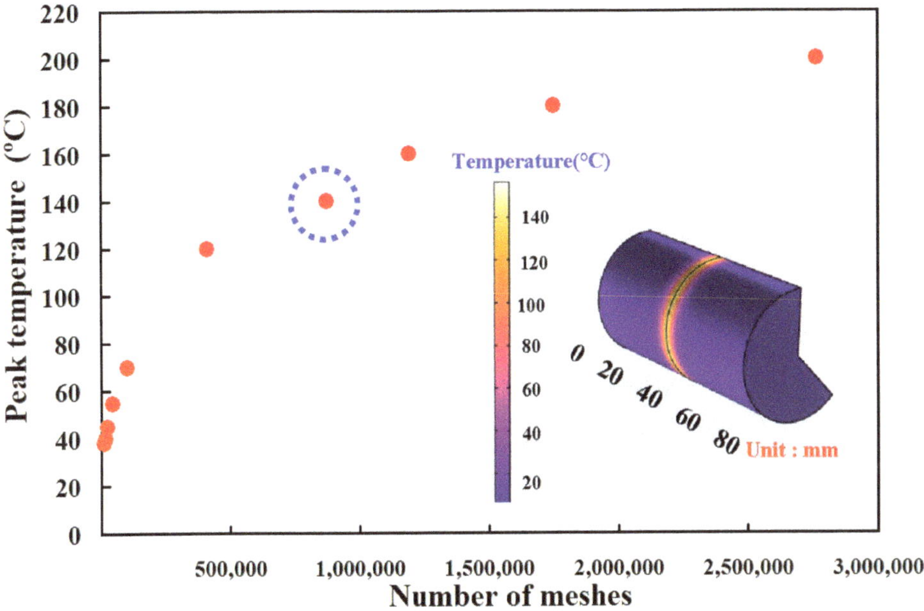

Figure 8. Number of meshes as a function of peak temperature in the weld joint.

Figure 9 shows the friction-welded parts before and after bending tests. It was observed that the joining of ABS and PC by FRW is expected to improve the basic mechanical properties of the single ABS thermoplastic material. It is well known that the flexural strength of the PC material is better than that of the ABS material. According to the experimental results, the average flexural strength of the weld joint and the PC and ABS rods is about 132 MPa, 180 MPa, and 110 MPa, respectively. The fracture initiated in the ABS rods shows that the welding quality of FRW is excellent, since the welding strength in the weld joint is better than that of the ABS base materials [29]. In general, the ABS polymer is a relatively soft and flexible material. The PC polymer has a higher glass transition temperature than the ABS polymer, showing it can withstand higher temperatures before it starts to soften and lose its shape. Therefore, the PC polymer seems to be a better choice for parts that need to operate in high-temperature environments. In particular, the fractured locations in the ABS rods after bending tests are random, as shown in Figure 10.

Figure 11 shows the fracture surfaces of ABS after three-point bending tests for six different rotation speeds. Figure 12 describes the fracture surfaces of ABS after impact tests for six different rotation speeds. A small region of porous surface was also observed and highlighted. The porous surface is caused by insufficient molecular diffusion and crystallinity in FDM printing, and can be reduced by microwave [30]. Figure 13 describes the fracture surfaces of PC after impact tests for six different rotation speeds. The formation of voids and cracks after the impact test can be identified by analyzing the surfaces of fractured parts. As can be seen, jagged and irregular surfaces were found. According to the optical microscopic images of fractured surfaces, two phenomena are found. One is that the ABS material exhibits cracked surface textures. The other is that the PC material exhibits a porous structure.

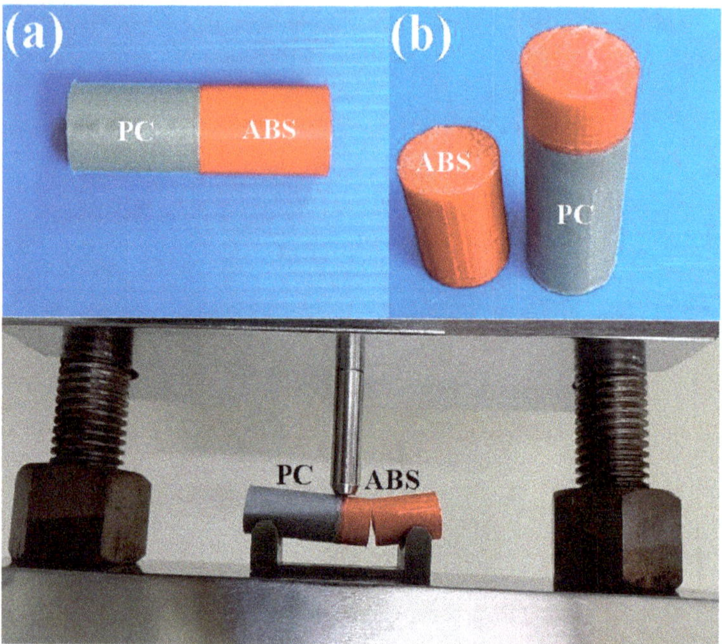

Figure 9. Friction-welded parts (**a**) before and (**b**) after bending tests.

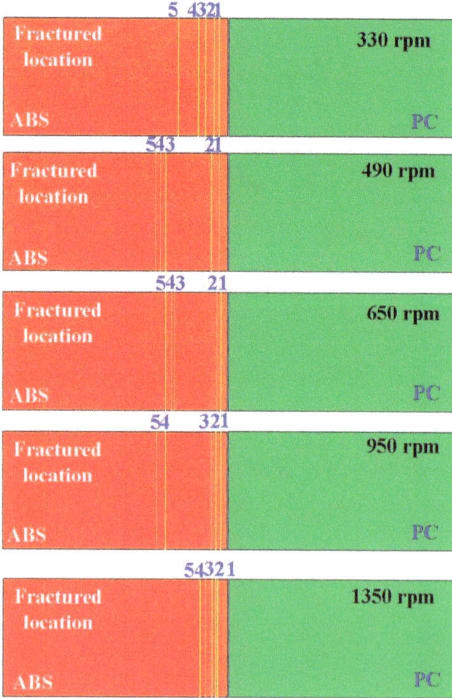

Figure 10. Fractured locations in the ABS rods after bending tests.

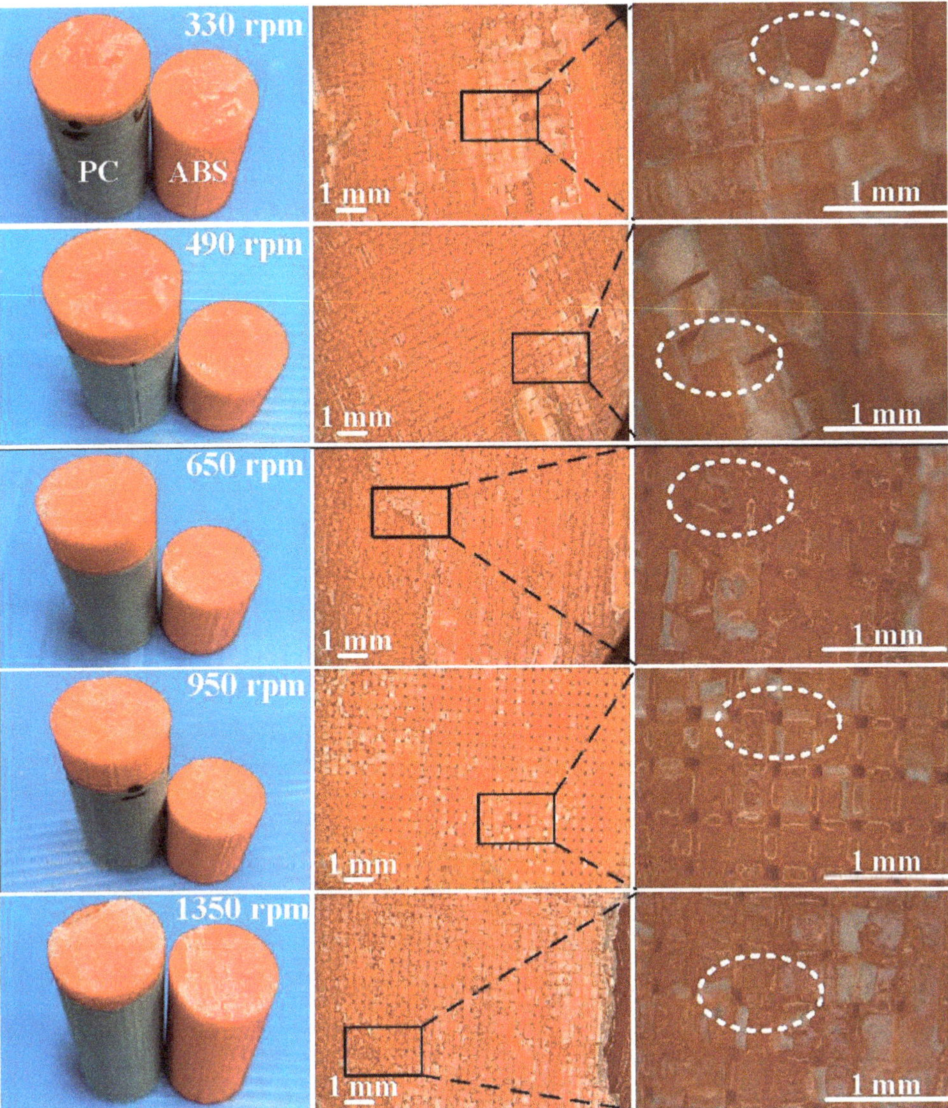

Figure 11. Fracture surfaces of ABS after three-point bending tests for six different rotation speeds.

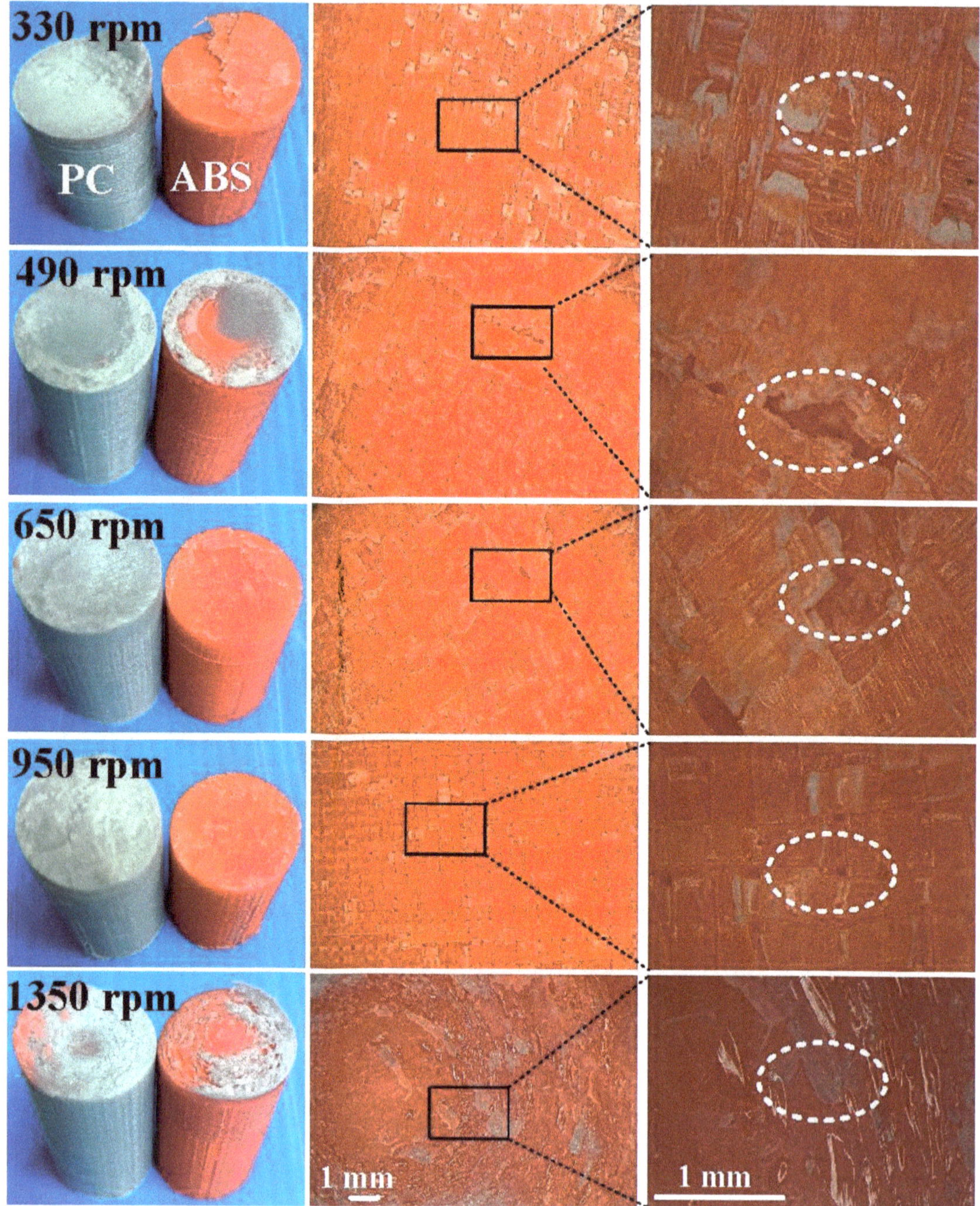

Figure 12. Fracture surfaces of ABS after impact tests for six different rotation speeds.

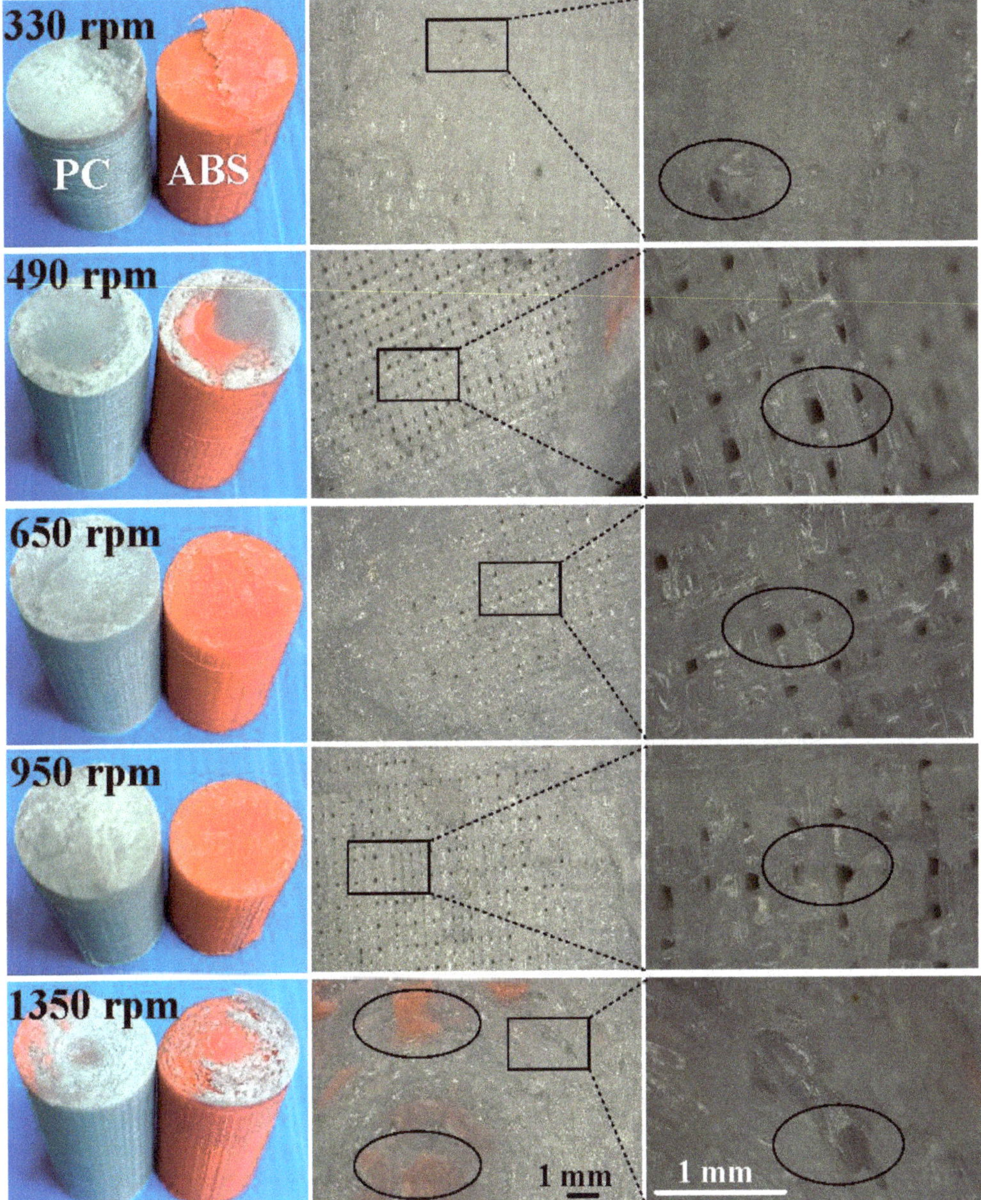

Figure 13. Fracture surfaces of PC after impact tests for six different rotation speeds.

Figure 14 shows the comparison of the experimental and numerical simulation results of the peak temperature for the RFW of PC and ABS rods at five different rotation speeds. As can be seen, the percentage error of the peak temperature between the experimental and numerical simulation results for 330 rpm, 490 rpm, 650 rpm, 950 rpm, and 1350 rpm is about 38.6 %, 29.3%, 18.9%, −1.8%, −2.0%, and 16.6%, respectively. Therefore, the average percentage error of predicting the peak temperature using COMSOL software for five different rotation speeds is about 16.6%. In general, DSC analyzes the melting behavior, solidification characteristics, and glass transition temperature. Figure 15 shows the DSC curve comparisons for the friction-welded part under five rotation speeds. The insert shows the DSC setup for the samples fabricated by five rotation speeds. As can be seen, the DSC peak appears at a temperature of 429 °C, showing that there is a significant thermal event happening in the weld joint. The heat capacities for the friction-welded parts welded by rotation speeds of 330 rpm, 490 rpm, 650 rpm, 950 rpm, and 1350 rpm are −1.002 mW/mg, −0.8127 mW/mg, −0.626 4mW/mg, −1.759 mW/mg, and -2.287 mW/mg, respectively. The DSC curve for the friction-welded parts welded by a rotation speed of 1350 rpm shows an endothermic peak [31] between 400 to 440 °C and an exothermic peak [32] between 600 to 700 °C. This means that higher rotation speed contributes to higher molecular orientation in the weld joints, showing that the friction-welded parts have better mechanical properties. Therefore, the structure of the friction-welded parts welded by a rotation speed of 1350 rpm is stronger.

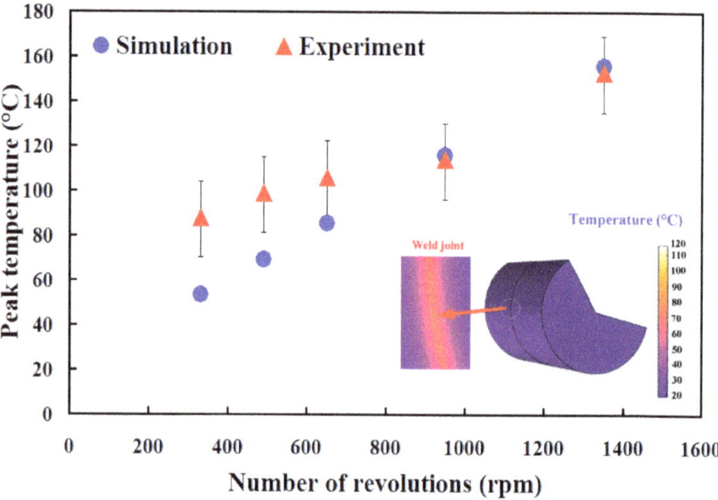

Figure 14. Comparison of the experimental and numerical simulation results of the peak temperature for RFW of PC and ABS rods at five different rotation speeds.

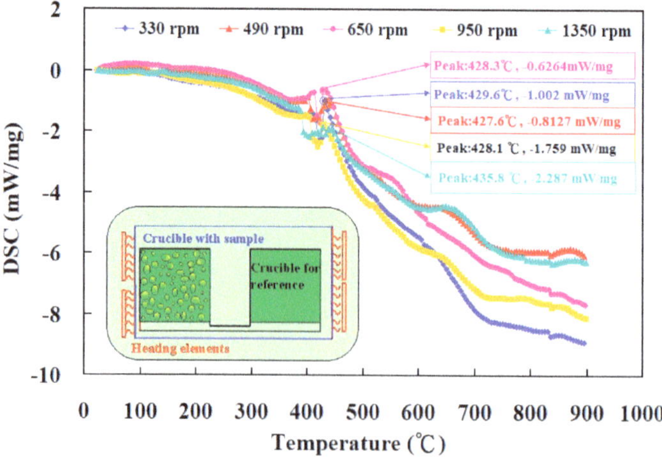

Figure 15. DSC curve comparisons for friction-welded part under five rotation speeds.

According to the research results, the remarkable findings provide potential industrial values in the polymer-welding industry, since the RFW of dissimilar polymer rods is a green manufacturing method based on four sustainable development goals, i.e., SDG$_S$ 7, 8, 9, and 12 [33–37]. In particular, the RFW of polymer rods can be applied for joining the fluid mechanical components, automotive components, axle shafts, or aerospace components [38,39]. In this study, a conventional turning machine was used to carry out the RFW of polymer rods. In future studies, a high-speed lathe or computer numerical control lathe [40,41] are recommended to carry out RFW [42–45]. These topics are currently being investigated, and the results will be presented in future work.

4. Conclusions

The advantages of RFW include efficiency of production, low heat input, and environmental friendliness. PC plastic has very good heat resistance and is widely employed for more robust materials. ABS plastic provides high tensile strength and is very resistant to chemical corrosion and physical impacts. To establish domain knowledge of the RFW of ABS and PC rods, this work reports the joining of FDM-printed dissimilar thermoplastic PC and ABS rods. To establish domain knowledge of the RFW of ABS and PC rods, the RFW experiment was performed using a turning machine. The main conclusions from the experimental work in this study are as follows:

1. The average surface hardness, flexural strength, and impact energy in the weld joint are increased with increasing rotation speed of RFW. The maximum joint strength of 132 MPa, Shore A hardness of 80, and impact energy of 156 J are obtained at a rotation speed of 1350 rpm.
2. The quality of RFW is excellent, since the welding strength in the weld joint is better than that of the ABS base materials. The fracture occurs in the ABS rods since the brittleness of the ABS rods is higher than that of the PC rods.
3. The average percentage error for predicting the peak temperature using COMSOL software using a mesh element count of 875,688 for five different rotation speeds is about 16.6%.
4. The heat capacities for the friction-welded parts welded by rotation speeds of 330 rpm, 490 rpm, 650 rpm, 950 rpm, and 1350 rpm are -1.002 mW/mg, -0.8127 mW/mg, 0.626 4mW/mg, -1.759 mW/mg, and -2.287 mW/mg, respectively. The DSC curve for the friction-welded parts welded by a rotation speed of 1350 rpm shows an endothermic peak between 400 to 440 °C and an exothermic peak between 600 to 700 °C, showing that the friction-welded parts have better mechanical properties.

Author Contributions: C.-C.K. wrote the paper, conceived and designed the analysis, and performed the analysis. N.G., H.-W.C. and S.-H.H. collected the data and contributed data or analysis tools. All authors have read and agreed to the published version of the manuscript.

Funding: This study received financial support by the Ministry of Science and Technology of Taiwan under contract nos. NSTC 111-2221-E-131-015-MY2, MOST 110-2221-E-131-023 and MOST 109-2637-E-131-004.

Institutional Review Board Statement: Not applicable.

Informed Consent Statement: Not applicable.

Data Availability Statement: Data and materials are available.

Conflicts of Interest: The authors declare no conflict of interest.

References

1. Lambiase, F.; Grossi, V.; Paoletti, A. Effect of tilt angle in FSW of polycarbonate sheets in butt configuration. *Int. J. Adv. Manuf. Technol.* **2020**, *107*, 489–501. [CrossRef]
2. Delijaicov, S.; Rodrigues, M.; Farias, A.; Neves, M.; Bortolussi, R.; Miyazaki, M.; Brandão, F. Microhardness and residual stress of dissimilar and thick aluminum plates AA7181-T7651 and AA7475-T7351 using bobbin, top, bottom, and double-sided FSW methods. *Int. J. Adv. Manuf. Technol.* **2020**, *108*, 277–287. [CrossRef]
3. Hassan, A.J.; Boukharouba, T.; Miroud, D. Concept of forge application under effect of friction time for AISI 316 using friction welding process. *Int. J. Adv. Manuf. Technol.* **2021**, *112*, 2223–2231. [CrossRef]
4. Yin, P.; Xu, C.; Pan, Q.; Zhang, W.; Jiang, X. Effect of Different Ultrasonic Power on the Properties of RHA Steel Welded Joints. *Materials* **2022**, *15*, 768. [CrossRef] [PubMed]
5. Li, B.; Liu, Q.; Jia, S.; Ren, Y.; Yang, P. Effect of V Content and Heat Input on HAZ Softening of Deep-Sea Pipeline Steel. *Materials* **2022**, *15*, 794. [CrossRef]
6. Hamedi, H.; Kamyabi-Gol, A. A novel approach to modelling the bond characteristics between CFRP fabrics and steel plate joints under quasi-static tensile loads. *Int. J. Adv. Manuf. Technol.* **2021**, *116*, 3247–3261. [CrossRef]
7. Wang, S.; Liang, W.; Duan, L.; Li, G.; Jinjia, C. Effects of loading rates on mechanical property and failure behavior of single-lap adhesive joints with carbon fiber reinforced plastics and aluminum alloys. *Int. J. Adv. Manuf. Technol.* **2020**, *106*, 2569–2581. [CrossRef]
8. Khedr, M.; Ibrahim, I.R.; Jaskari, M.; Ali, M.; Abdel-Aleem, H.A.; Mahmoud, T.S.; Hamada, A. Microstructural Evolution and Mechanical Performance of Two Joints of Medium-Mn Stainless Steel with Low- and High-Alloyed Steels. *Materials* **2023**, *16*, 1624. [CrossRef]
9. Eslami, S.; Tavares, P.J.; Moreira, P.M.G.P. Friction stir welding tooling for polymers: Review and prospects. *Int. J. Adv. Manuf. Technol.* **2017**, *89*, 1677–1690. [CrossRef]
10. Paoletti, A.; Lambiase, F.; Di Ilio, A. Analysis of forces and temperatures in friction spot stir welding of thermoplastic polymers. *Int. J. Adv. Manuf. Technol.* **2016**, *83*, 1395–1407. [CrossRef]
11. Lambiase, F.; Paoletti, A.; Di Ilio, A. Friction spot stir welding of polymers: Control of plunging force. *Int. J. Adv. Manuf. Technol.* **2017**, *90*, 2827–2837. [CrossRef]
12. Rehman, R.U.; Sheikh-Ahmad, J.; Deveci, S. Effect of preheating on joint quality in the friction stir welding of bimodal high density polyethylene. *Int. J. Adv. Manuf. Technol.* **2021**, *117*, 455–468. [CrossRef]
13. Skowrońska, B.; Chmielewski, T.; Zasada, D. Assessment of Selected Structural Properties of High-Speed Friction Welded Joints Made of Unalloyed Structural Steel. *Materials* **2023**, *16*, 93. [CrossRef]
14. Faes, K.; Dhooge, A.; Jaspart, O.; D'Alvise, L.; De Baets, P. New friction welding process for pipeline girth welds—Welding time optimisation. *Int. J. Adv. Manuf. Technol.* **2009**, *43*, 982–992. [CrossRef]
15. Anwar, S.; Rehman, A.U.; Usmani, Y.; Al-Samhan, A.M. Influence of Post Weld Heat Treatment on the Grain Size, and Mechanical Properties of the Alloy-800H Rotary Friction Weld Joints. *Materials* **2021**, *14*, 4366. [CrossRef] [PubMed]
16. Ishraq, M.Y.; Maqsood, S.; Naeem, K.; Abid, M.; Omair, M. Analysing significant process parameters for friction stir welding of polymer composite. *Int. J. Adv. Manuf. Technol.* **2019**, *105*, 4973–4987. [CrossRef]
17. Hangai, Y.; Omika, K.; Inoue, M.; Kitamura, A.; Mitsugi, H.; Fujii, H.; Kamakoshi, Y. Effect of porosity of aluminum foam on welding between aluminum foam and polycarbonate plate during friction welding. *Int. J. Adv. Manuf. Technol.* **2022**, *120*, 1071–1078. [CrossRef]
18. Skowrońska, B.; Chmielewski, T.; Kulczyk, M.; Skiba, J.; Przybysz, S. Microstructural Investigation of a Friction-Welded 316L Stainless Steel with Ultrafine-Grained Structure Obtained by Hydrostatic Extrusion. *Materials* **2021**, *14*, 1537. [CrossRef]
19. Zhang, D.; Qin, G.; Geng, P.; Ma, H. Study of plastic flow on intermetallic compounds formation in friction welding of aluminum alloy to stainless steel. *J. Manuf. Process.* **2021**, *64*, 20–29. [CrossRef]
20. Eliseev, A.; Osipovich, K.; Fortuna, S. Gradient Structure of the Transfer Layer in Friction Stir Welding Joints. *Materials* **2022**, *15*, 6772. [CrossRef]
21. Ma, X.; Xu, S.; Wang, F.; Zhao, Y.; Meng, X.; Xie, Y.; Wan, L.; Huang, Y. Effect of Temperature and Material Flow Gradients on Mechanical Performances of Friction Stir Welded AA6082-T6 Joints. *Materials* **2022**, *15*, 6579. [CrossRef]
22. Park, S.J.; Park, S.J.; Son, Y.; Ahn, I.H. Influence of warm isostatic press (WIP) process parameters on mechanical properties of additively manufactured acrylonitrile butadiene styrene (ABS) parts. *Int. J. Adv. Manuf. Technol.* **2022**, *122*, 3311–3322. [CrossRef]
23. Ho, H.T.; Nguyen, N.H.; Rollet, M.; Phan, T.N.T.; Gigmes, D. Phosphonate-Functionalized Polycarbonates Synthesis through Ring-Opening Polymerization and Alternative Approaches. *Polymers* **2023**, *15*, 955. [CrossRef]
24. Yang, N.; Gong, Y.; He, P.; Zhou, C.; Zhou, R.; Shao, H.; Chen, G.; Lin, X.; Bie, H. Influence of Circular through Hole in Pt–Rh Bushing on Temperature Propagation at High Temperature. *Materials* **2022**, *15*, 7832. [CrossRef]
25. Deng, X.; Li, J.; Xie, X. Effect of Preheating Temperature on Thermal–Mechanical Properties of Dry Vibrating MgO-Based Material Lining in the Tundish. *Materials* **2022**, *15*, 7699. [CrossRef]
26. Jiang, J.; Chen, Q.; Hu, S.; Shi, Y.; He, Z.; Huang, Y.; Hui, C.; Chen, Y.; Wu, H.; Lu, G. Effect of Electro-Thermo-Mechanical Coupling Stress on Top-Cooled E-Mode AlGaN/GaN HEMT. *Materials* **2023**, *16*, 1484. [CrossRef]
27. Issabayeva, Z.; Shishkovsky, I. Prediction of The Mechanical Behavior of Polylactic Acid Parts with Shape Memory Effect Fabricated by FDM. *Polymers* **2023**, *15*, 1162. [CrossRef]

28. Xie, M.; Shang, X.; Li, Y.; Zhang, Z.; Zhu, M.; Xiong, J. Rotary Friction Welding of Molybdenum without Upset Forging. *Materials* **2020**, *13*, 1957. [CrossRef]

29. Kim, J.K.; Kang, S.-S.; Kim, H.G.; Kwac, L.K. Mechanical Properties and Electromagnetic Interference Shielding of Carbon Composites with Polycarbonate and Acrylonitrile Butadiene Styrene Resins. *Polymers* **2023**, *15*, 863. [CrossRef]

30. Liu, Z.; Wang, Y.; Wu, B.; Cui, C.; Guo, Y.; Yan, C. A critical review of fused deposition modeling 3D printing technology in manufacturing polylactic acid parts. *Int. J. Adv. Manuf. Technol.* **2019**, *102*, 2877–2889. [CrossRef]

31. El-Geassy, A.A.; Abdel Halim, K.S.; Alghamdi, A.S. A Novel Hydro-Thermal Synthesis of Nano-Structured Molybdenum-Iron Intermetallic Alloys at Relatively Low Temperatures. *Materials* **2023**, *16*, 2736. [CrossRef] [PubMed]

32. Assawakawintip, T.; Santiwong, P.; Khantachawana, A.; Sipiyaruk, K.; Chintavalakorn, R. The Effects of Temperature and Time of Heat Treatment on Thermo-Mechanical Properties of Custom-Made NiTi Orthodontic Closed Coil Springs. *Materials* **2022**, *15*, 3121. [CrossRef] [PubMed]

33. Skowrońska, B.; Bober, M.; Kołodziejczak, P.; Baranowski, M.; Kozłowski, M.; Chmielewski, T. Solid-State Rotary Friction-Welded Tungsten and Mild Steel Joints. *Appl. Sci.* **2022**, *12*, 9034. [CrossRef]

34. Li, K.; Zhou, T.; Liu, B.-h. Internet-based intelligent and sustainable manufacturing: Developments and challenges. *Int. J. Adv. Manuf. Technol.* **2020**, *108*, 1767–1791. [CrossRef]

35. Rahman, M.A.; Ali, A.; Rahamathulla, M.; Salam, S.; Hani, U.; Wahab, S.; Warsi, M.H.; Yusuf, M.; Ali, A.; Mittal, V.; et al. Fabrication of Sustained Release Curcumin-Loaded Solid Lipid Nanoparticles (Cur-SLNs) as a Potential Drug Delivery System for the Treatment of Lung Cancer: Optimization of Formulation and In Vitro Biological Evaluation. *Polymers* **2023**, *15*, 542. [CrossRef] [PubMed]

36. Varghese, S.A.; Pulikkalparambil, H.; Promhuad, K.; Srisa, A.; Laorenza, Y.; Jarupan, L.; Nampitch, T.; Chonhenchob, V.; Harnkarnsujarit, N. Renovation of Agro-Waste for Sustainable Food Packaging: A Review. *Polymers* **2023**, *15*, 648. [CrossRef]

37. Morales, M.A.; Maranon, A.; Hernandez, C.; Michaud, V.; Porras, A. Colombian Sustainability Perspective on Fused Deposition Modeling Technology: Opportunity to Develop Recycled and Biobased 3D Printing Filaments. *Polymers* **2023**, *15*, 528. [CrossRef]

38. Kumar, L.; Jain, P.K.; Sharma, A.K. A fuzzy goal programme–based sustainable Greenfield supply network design for tyre retreading industry. *Int. J. Adv. Manuf. Technol.* **2020**, *108*, 2855–2880. [CrossRef]

39. Belkahla, Y.; Mazouzi, A.; Lebouachera, S.E.I.; Hassan, A.J.; Fides, M.; Hvizdoš, P.; Cheniti, B.; Miroud, D. Rotary friction welded C45 to 16NiCr6 steel rods: Statistical optimization coupled to mechanical and microstructure approaches. *Int. J. Adv. Manuf. Technol.* **2021**, *116*, 2285–2298. [CrossRef]

40. Barrionuevo, G.O.; Mullo, J.L.; Ramos-Grez, J.A. Predicting the ultimate tensile strength of AISI 1045 steel and 2017-T4 aluminum alloy joints in a laser-assisted rotary friction welding process using machine learning: A comparison with response surface methodology. *Int. J. Adv. Manuf. Technol.* **2021**, *116*, 1247–1257. [CrossRef]

41. Esangbedo, M.O.; Abifarin, J.K. Cost and Quality Optimization Taguchi Design with Grey Relational Analysis of Halloysite Nanotube Hybrid Composite: CNC Machine Manufacturing. *Materials* **2022**, *15*, 8154. [CrossRef] [PubMed]

42. Bouarroudj, E.; Abdi, S.; Miroud, D. Improved performance of a heterogeneous weld joint of copper-steel AISI 1045 obtained by rotary friction using a metal powder insert. *Int. J. Adv. Manuf. Technol.* **2023**, *124*, 1905–1924. [CrossRef]

43. Szwajka, K.; Zielińska-Szwajka, J.; Trzepieciński, T. Microstructure and Mechanical Properties of Solid-State Rotary Friction Welded Inconel 713C and 32CrMo4 Steel Joints Used in a Turbocharger Rotor. *Materials* **2023**, *16*, 2273. [CrossRef] [PubMed]

44. Insua, P.; Nakkiew, W.; Wisittipanich, W. Post Weld Heat Treatment Optimization of Dissimilar Friction Stir Welded AA2024-T3 and AA7075-T651 Using Machine Learning and Metaheuristics. *Materials* **2023**, *16*, 2081. [CrossRef]

45. Ahmed, M.M.Z.; Essa, A.R.S.; Ataya, S.; El Sayed Seleman, M.M.; El-Aty, A.A.; Alzahrani, B.; Touileb, K.; Bakkar, A.; Ponnore, J.J.; Mohamed, A.Y.A. Friction Stir Welding of AA5754-H24: Impact of Tool Pin Eccentricity and Welding Speed on Grain Structure, Crystallographic Texture, and Mechanical Properties. *Materials* **2023**, *16*, 2031. [CrossRef]

Article

Corrosion Behavior of Friction Stir Welded AA8090-T87 Aluminum Alloy

Chandrasekaran Shyamlal [1], Rajesh Shanmugavel [1], J. T. Winowlin Jappes [1], Anish Nair [1], M. Ravichandran [2], S. Syath Abuthakeer [3], Chander Prakash [4,5,*], Saurav Dixit [6,7,*] and N. I. Vatin [6]

[1] Department of Mechanical Engineering, Kalasalingam Academy of Research and Education, Krishnankoil 626126, India; shyadavidhars@gmail.com (C.S.); s.rajesh@klu.ac.in (R.S.); winowlin@klu.ac.in (J.T.W.J.); anishn@live.com (A.N.)

[2] Department of Mechanical Engineering, K Ramakrishnan College of Engineering, Tiruchirappalli 621112, India; smravichandran@hotmail.com

[3] Department of Mechanical Engineering, PSG College of Technology, Coimbatore 641004, India; ssa.mech@psgtech.ac.in

[4] School of Mechanical Engineering, Lovely Professional University, Phagwara 144411, India

[5] Division of Research and Development, Lovely Professional University, Phagwara 144411, India

[6] Peter the Great St. Petersburg Polytechnic University, 195251 Saint Petersburg, Russia; vatin@mail.ru

[7] Division of Research & Innovation, Uttaranchal University, Dehradun 248007, India

* Correspondence: chander.mechengg@gmail.com (C.P.); sauravarambol@gmail.com (S.D.)

Abstract: Aerospace alloys with reduced wall thickness but possessing higher hardness, good tensile strength and reasonable corrosion resistance are essential in manufacturing of structures such as fuselage. In this work, friction stir welding has been carried out on such an aerospace aluminum alloy AA8090 T87 which contains 2.3% lithium. Tool rotational speed of 900 rpm and traverse speeds of 90 mm/min., 110 mm/min. are the welding parameters. Hardness analysis, surface roughness analysis and corrosion analysis are conducted to analyze the suitability of the joint for the intended application. The samples were corrosion tested in acid alkali solution and they resulted in the formation of pits of varying levels which indicate the extent of surface degradation. Hardness of the samples was measured after corrosion analysis to observe the changes. The analysis suggests that the change in tool traverse speed transformed the corrosion behavior of the joint and affected both the hardness and surface roughness which mitigated the quality of the joint.

Keywords: precipitates; hardness; surface roughness; corrosion; grain boundary; pits

Citation: Shyamlal, C.; Shanmugavel, R.; Jappes, J.T.W.; Nair, A.; Ravichandran, M.; Abuthakeer, S.S.; Prakash, C.; Dixit, S.; Vatin, N.I. Corrosion Behavior of Friction Stir Welded AA8090-T87 Aluminum Alloy. *Materials* **2022**, *15*, 5165. https://doi.org/ma15155165

Academic Editor: Raul D.S.G. Campilho

Received: 4 July 2022
Accepted: 20 July 2022
Published: 26 July 2022

Publisher's Note: MDPI stays neutral with regard to jurisdictional claims in published maps and institutional affiliations.

1. Introduction

The efficacy of Friction Stir Welding (FSW) technology, a solid state joining process [1] founded by TWI (The Welding Institute) [2–4] is used in the aerospace structures due to its production of strong metallurgical joint compared to fusion welding technology. In addition, the resulting weld joints have improved mechanical properties. The finer recrystallized equiaxed grains [5] produced by FSW technique is responsible for adequate strength, toughness and ductility. In FSW, precipitate formation, grain boundary strengthening and dislocation hindrances are the major mechanisms which lead to increased weld performance. The grain boundary strengthening is achieved through increasing the rotation speed and reducing the traverse speed. My adjusting the parameters such as rotational speed and traverse speed the metallurgical characteristics such as precipitation and grain boundary strengthening can be achieved. More the grain boundaries are created during the weld, the smaller the size of the grains become. The grain boundary density and mis-orientation created in the weld zone can be analyzed through EBSD (Electron Back Scattered Diffraction) technique [6].The sub grain boundaries formed due to rotation of the tool [7] in the stir zone (SZ) stops more dislocations by the increased grain boundary area. Hence, the combined effect of precipitate formation and equiaxed grains in the stir zone [8] enhances the joint efficiency. Thus it is understood that the precipitates and dislocations play a major role in increasing the hardness of the structure [9,10]. However, the higher rotational speed will lead to the dissolution [11,12] of precipitates in the stir zone which leads to the decrement in the hardness in the stir zone [13] but enhances the corrosion resistance.

Further, the inclusion of lithium (>2%) as an alloying element improves the elastic modulus of the material. Cu and Mg are added to aluminum lithium alloys to impart strength [14] to the alloy by forming Al-Mg-Cu, Al-Cu-Li precipitates. These precipitates maximize the properties such as hardness and tensile strength of the structure [15]. The precipitate formation may benefit the mechanical properties but it will also affect the corrosion behavior [16]. The corrosion resistance of the joint is considered as an important criterion for the evaluation of joint efficacy. Alloy manufacturing process involves the addition of different alloying elements and as these alloying elements occupy different positions in electrochemical series, they have differing reactivity with the matrix [17]. The formation of intermetallic compounds due to differing electronegativity values will create heterogeneous precipitates such as Al_2Cu [18] in the grain boundaries [19] and in the vicinity of dislocations which may corrode during service. These type of intermetallic compounds forms at the grain boundaries predominantly than inside the grain. These intermetallics may be the reason for the formation of eutectics in the grain boundary. The indication of corrosion occurrence is pit formation [20] in the surface of the material and in grain boundaries which acts as notch and propagate the crack through it. In this work, the major components responsible for pit formation are Al_2Cu (Al-Cu) precipitates. Further, severity in corrosion causes decrease in elongation [21,22] and increase in the surface roughness which finally causes catastrophic failure to the structure.

Limited researchers have investigated mechanical properties of friction stir welded aluminum alloys. However, the reported literatures on the corrosion perspective of AA8090 in the context of precipitate formation are very scanty. The objective of this work is to investigate the hardness, surface roughness, precipitate formation and the effect of precipitate formation on corrosion resistance of AA8090 aluminum alloy.

2. Materials and Methods

The friction stir welding has been executed using CNC machining center (Make: Hurco, Indianapolis, IN, USA). The used sample size is 150 mm × 100 mm × 5 mm which is procured commercially from Bharat Aerospace Alloys, Mumbai, India. The chemical composition of the sample was verified using Optical Emission Spectroscopy (OES) [23,24]. The samples have been cut to specified size by Electronica maxi-cut wire-cut EDM machine. The tool used for welding is H13 tool steel having tool pin height of 4.7 mm, pin root diameter 5 mm and tip diameter 3 mm with shoulder diameter of 23 mm. Figures 1 and 2 shows the machine set-up and the FSW tool used for welding. The joint description is provided in Table 1. The chemical composition of the tool and work is shown in Tables 2 and 3. The hardness testing was performed with 500 gf load, 15 s dwell time using MVH-TS1 make micro-Vickers hardness machine according to ASTM standard E384 using 70 mm × 10 mm × 5 mm strip (polished work piece) [25]. The corrosion test coupons have been extracted from the welded area of size 30 mm × 30 mm × 5 mm, polished and immersion corrosion test was conducted by dipping the sample in 57 g/L NaCl (neither acidic nor alkaline) + 10 mL/L H_2O_2 (a weak acid) solution for 6 h at 32 °C as per ASTM standard G110 [26]. Scanning electron microscopy images were taken using Jeol 6000 plus microscope (Akishima, Japan) to investigate the sub surface after etching the sample using 85 mL distilled water, 15 mL HF, 5 mL H_2SO_4 [27–29]. Electron Back Scatter Diffraction (EBSD) analysis has been carried out to analyze the grain boundary mis-orientation.

Figure 1. Machine set up for welding.

Figure 2. Friction stir welding tool.

Table 1. Terminology used for welded plates.

Sl. No.	Rotating Speed	Traverse Speed	Description of Joint
1.	900 RPM	90 mm/min.	900-90
2.	900 RPM	110 mm/min.	900-110

Table 2. Chemical composition of AA8090 aluminum alloy.

Elements	Al	Li	Cu	Mg	Si	Zr	Cr	Mn	Ti
wt.%	95.2	2.35	1.29	0.88	0.04	0.11	0.0004	0.004	0.0.038

Table 3. Chemical composition of the tool material—H13 Tool steel.

Elements	Fe	Cr	Mo	Si	V	C	Ni	Cu	Mn
wt.%	90	5	1.7	1	1	0.37	0.3	0.25	0.4

3. Investigation of Mechanical Properties

3.1. Analysis on Hardness of the Weld

This section presents the variation of hardness measurements and the variation of hardness across the friction stir welded zone. Generally, the hardness varies based on the variation in the grain size and also with respect to the precipitate formation in the substrate (in this case aluminum) which will come into effect in the weld zones, based on the heat input in the work material.

The source of heat generation is in the stir zone. The heat developed in the stir zone was dispersed throughout the workpiece. The hardness on the base metal is escalated due to the formation of intermetallic compounds Al-Cu-Mg and Al-Cu-Li as confirmed from Table 2. Figures 3 and 4 confirms that recrystallization occurs in and around the stir zone. This zone is also referred to as the Thermo Mechanical Affected Zone (TMAZ) and is characterized by presence of fine grains and coarse precipitates. More precipitates become dissolved as a solid solution in the stir zone and in TMAZ which is observed from the lesser hardness distribution region observed adjacent to the stir zone on both the joints. Secondary reason being, T87 heat treatment was removed in the stir zone due to annealing kind of treatment which took place during welding. The above figures also show the sharp decline in hardness up to point X from the cold worked state (that is from base metal hardness) on both the sides of the joint due to heat escalation. There was an asymmetrical plateau region near the Stir Zone (SZ) in the hardness map of 900-110 joint. This asymmetrical behavior in hardness is attributed to the reduction in the grain boundaries on one side of the joint. However, there is a symmetrical plateau region near the stir zone in the hardness map of 900-90 joint. This indicates the presence of higher grain boundary area and significant precipitation. Hence, the only strengthening mechanism which played crucial role in the stir zone is grain boundary strengthening mechanism along with precipitate formation. The hardness on the advancing side is lesser due to severe plasticization by the tool in the weld region on both 900-90 joint and 900-110 joints. The common observation from both the hardness maps is that the hardness varied with respect to the distance from the stir zone. Hence, on the advancing side and in the retreating side, after point X, there is a sharp increase in the hardness zone which is due to the increase in distance from the stir zone and also due to the partial cold work effect of T87 temper. Table 4 lists the hardness difference between the point marked point X and the adjacent point on the hardness map. The hardness difference of 38.1 VHN is observed for 900-90 joint due to faster cooling rate. This indicates that higher the rotational speed, higher the temperature generation in the advancing side, higher the hardness difference between the neighborhood points. Hence, this confirms the fact that intermetallic compounds (precipitate) formed is more in 900-90 joint than in 900-110 joint. It is also interesting to observe that the lower hardness plateau for 900-110 joint (Row 3 of Table 4), extends from 0 to 5 mm in the advancing side which is also indicated in the hardness map in Figure 4 as a small plateau region. This also confirms scarce precipitate formation in the aforementioned lower hardness plateau region.

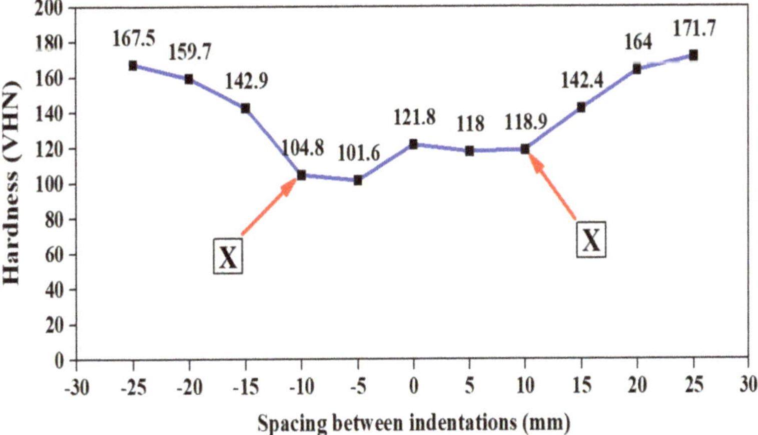

Figure 3. Hardness Map for 900-90 joint. X–Hardness transition point

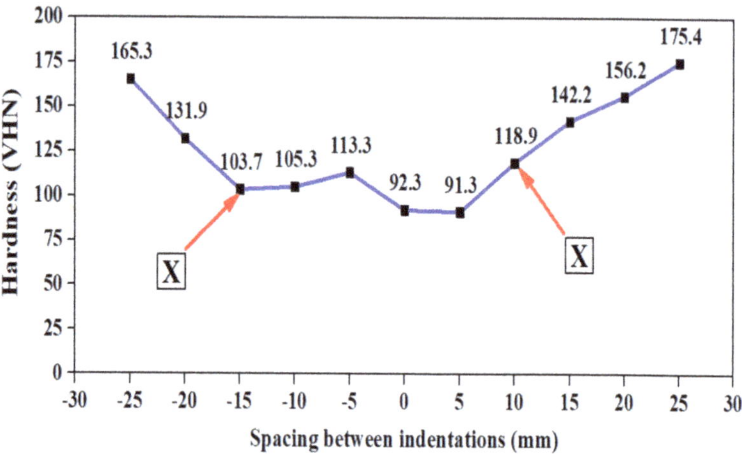

Figure 4. Hardness map for 900-110 joint. X–Hardness transition point.

Table 4. Hardness variation table in the vicinity of heat affected zone (HAZ).

Sl. No.	Description	Spacing mm	Side	Hardness Difference, VHN	Remarks	Reference
1.	Hardness profile for 900-90 joint	Between 10 mm and 15 mm	AS	38.1	The slope increases from 10 mm	Figure 3
2.	Hardness profile for 900-90 joint	Between 10 mm and 15 mm	RS	23.5	The slope increases from 10 mm	Figure 3
3.	Hardness profile for 900-110 joint	Between 15 mm and 20 mm	AS	28.2	The slope increases from 15 mm (Deviation observed)	Figure 4
4.	Hardness profile for 900-110 joint	Between 10 mm and 15 mm	RS	23.3	The slope increases from 10 mm.	Figure 4

AS—Advancing side RS—Retreating side.

It is observed from Figure 5a–c (EDS spectrum) and Table 5, that Aluminum (Al) is present in a major amount followed by Copper (Cu), Magnesium (Mg) and Iron (Fe) elements on both the base metal and in the weld joint. Fe formed the precipitate Al-Cu-Fe with Cu and Al, which created detrimental effect to the joint integrity in the grain boundary (eutectics) which is also clear from EDS spectrum. However, Cu and Mg formed Al-Cu-Mg precipitate escalated the joint integrity. Hence, the precipitate formation and grain boundary strengthening mechanisms assisted in the bond formation in the stir zone as shown in Figure 6a–d. Figure 6a,b shows the density of HAGB (High Angle Grain Boundary) and Low Angle Grain Boundary (LAGB) in the stir zone and the corresponding grain mis orientation. The HAGB is higher indicating the higher amount of stirring work in the weld zone and increased recrystallization. The grain boundary mis-orientation (>15°.) is observed in enormous amount which is the direct indication of recrystallized and rotated grains. The grain boundary strengthening will be carried out in the stir zone via accumulation of precipitates in the high angle and low angle grain boundary. Further, heat in the weld zone spreads to adjacent HAZ which increases the grain size and possibly precipitate size which is shown in Figure 6c,d. This increased grain boundary mis-orientation in the SZ dictates the fact that the orientation of the grains in the stir zone is not textured but randomized by the taper and threaded profile in the tool pin. The decreased grain mis-orientation in the HAZ dictates the fact that the zone experienced only heat without physical rotation. Particularly HAGB is very lesser in the HAZ. This phenomenon has decreased the hardness in the HAZ which is clearly shown in Figures 3 and 4.

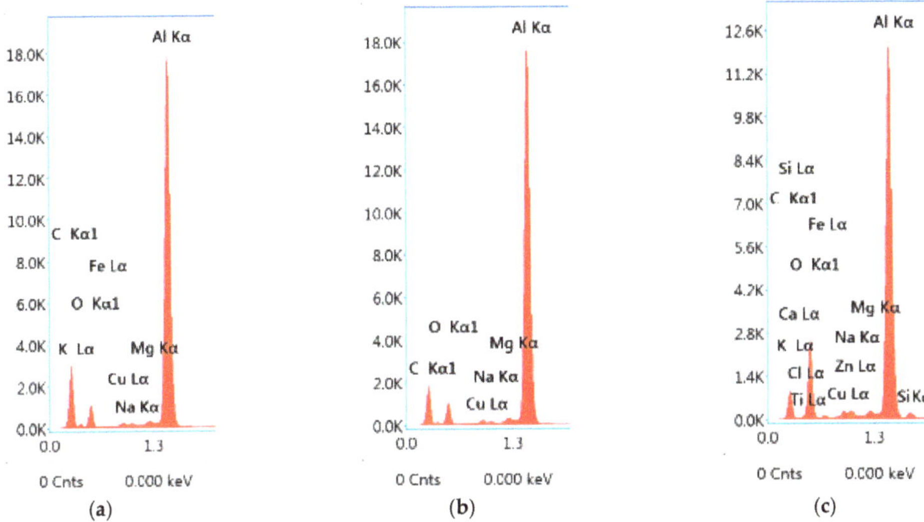

Figure 5. EDS spectrum analysis graph (**a**) Base metal (**b**) 900-90 joint (**c**) 900-110 joint.

Table 5. Chemical elements present in the base-metal and in the weld joint.

Sl. No.	Joint Description	Al (wt.%)	Cu (wt.%)	Mg (wt.%)	Fe (wt.%)
1.	Base metal	40.1	0.7	0.9	0.5
2.	900-90 Joint	47.1	0.9	1.1	-
3.	900-110 Joint	42.5	1.8	1.1	2.5

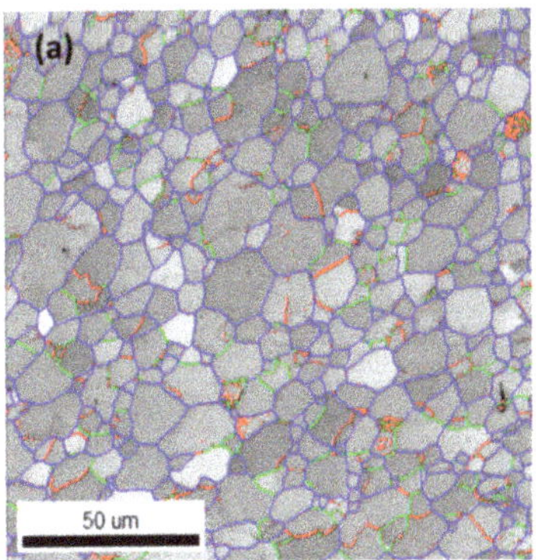

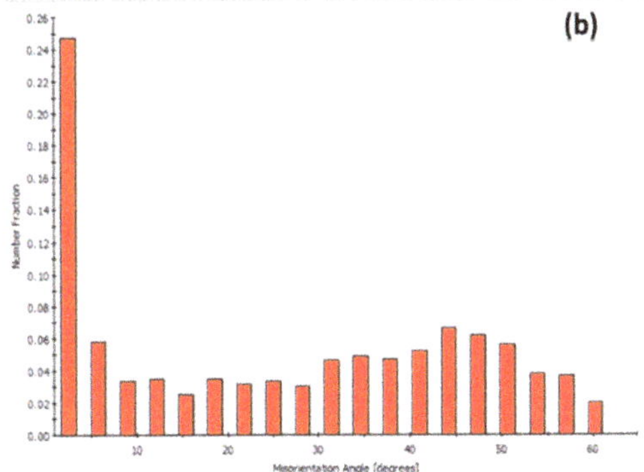

Figure 6. *Cont.*

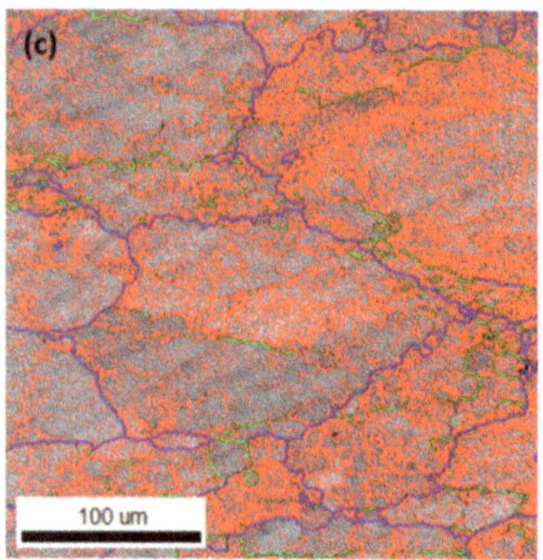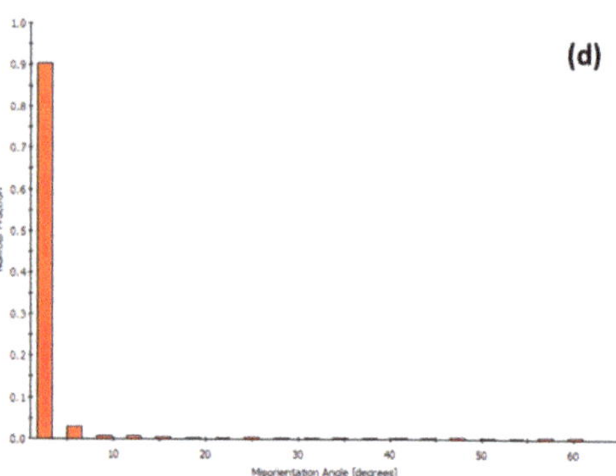

Figure 6. Grain Boundary maps for 900-90 joint (**a**) HAGB-LAGB map—S (**b**) Grain mis-orientation—SZ (**c**) HAGB-LAGB map—HAZ (**d**) Grain mis-orientation—HAZ.

3.2. Corrosion Analysis

This section deals with the corrosion analysis of 900-90 joint and 900-110 joint. Corrosion analysis is a necessary test for this alloy as the application of this alloy is in aerospace and in airplane structural construction which operates in corrosive marine atmosphere.

Immersion corrosion test has been conducted by immersing the work in 57 g/L NaCl (catalyst) + 10 mL/L H_2O_2 (strong oxidizer) solution for 6 h at 32 °C to evaluate the corrosion behavior of the alloy. In the above said solution used for corrosion, NaCl is used as a catalyst used to accelerate the corrosive action between H_2O_2 and the metal. The corroded surface is shown in Figure 7 and is highlighted in the blue-colored area. Figure 8a–j indicates the SEM and optical microscopy images before and after corrosion. Figure 8a,b clearly indicates the track marks produced by the FSW tool and the induced shear stress (induced in the pin of the tool due to threads in the pin).

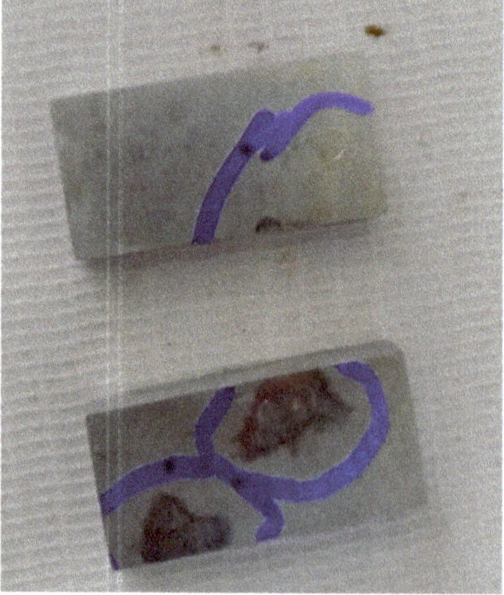

Figure 7. Corroded specimen after immersion test.

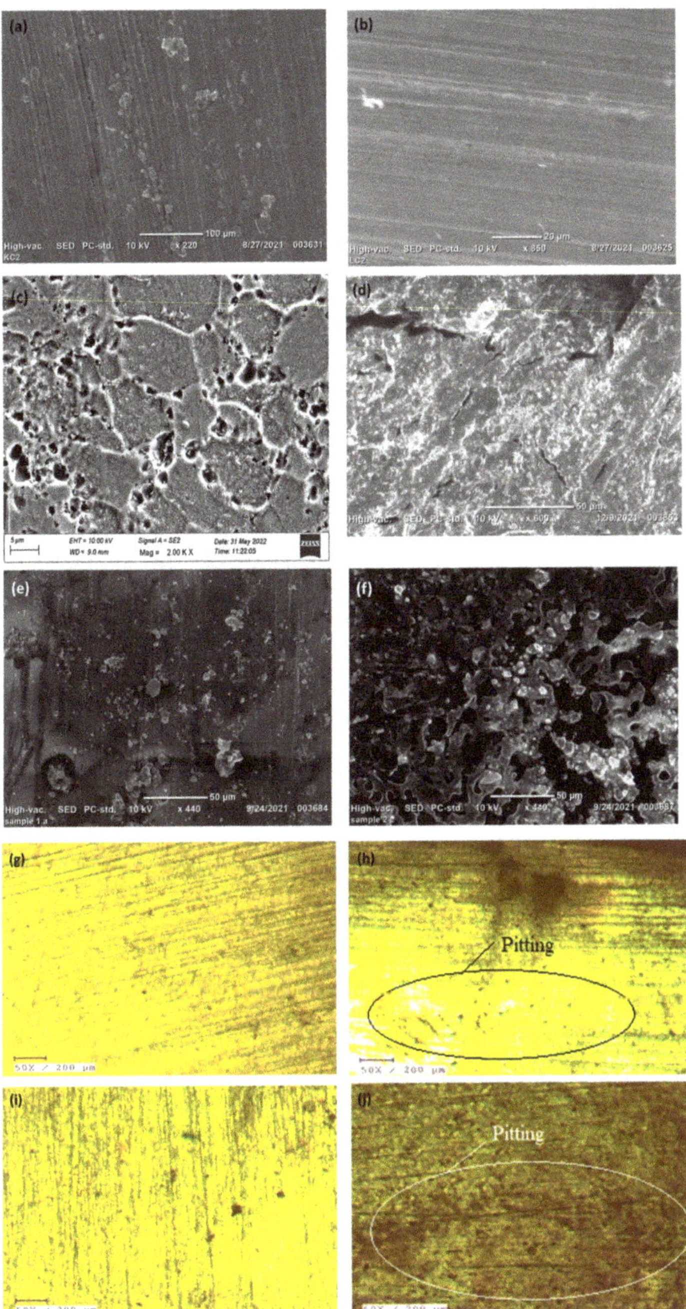

Figure 8. SEM micrographs (**a**) unetched 900-90 joint (**b**) unetched 900-110 joint (**c**) etched 900-90 joint (**d**) etched 900-110 joint (**e**) Corroded 900-90 joint (**f**) Corroded 900-110 joint (**g**) Optical microscopy image of 900-90 joint before corrosion (**h**) Optical microscopy image of 900-90 joint after corrosion (**i**) Optical microscopy image of 900-110 joint before corrosion (**j**) Optical microscopy image of 900-110 joint after corrosion.

Figure 8 shows the intensified physical action of the tool on the work, and it displays the various microstructural deformities that occur during the process. Figure 8c,d indicates the sub surface morphology after etching is carried out. Figure 8c displays the grain surrounded by precipitates (Al-Cu) which is confirmed from EDS spectrum analysis (Figure 5). The precipitate (Al_2Cu) formed on the grain boundaries are responsible for the hardness attained in the stir zone and in HAZ in 900-90 joint. Figure 8d shows the crack formed in the sub surface of the weld zone. Figure 8e,f presents pits formed in the weld zone after immersion corrosion test. These pits are the starting point of fracture on the surface. The fracture occurrence depends on the density of pits on the surface. The 900-90 joint shows lesser pits than 900-110 joint. This is due to the dissolution of precipitates in the stir zone in 900-90 joint due to heavy stirring. The higher number of precipitates formed in the 900-110 joint are responsible for the pits produced which is also confirmed from Figure 8g–j. Mass of the welded specimen before corrosion is 4.49 g. After corrosion, mass of the welded specimen is 4.479 g. for 900-90 joint. Hence, mass loss is 0.011 g. Mass of the welded specimen (900-110 Joint) before corrosion is 4.86 g. After corrosion, mass was reduced to 4.819 g. Mass loss is 0.041 g. This can be evidenced from the aforementioned SEM image. Hence, mass loss is more in the case of 900-110 joint. Cu wt.% in 900-110 joint is higher than in 900-90 joint which is evident from Table 6. This dictates the fact that Al_2Cu (the predominant precipitate) corroded the aluminum phases more in 900-110 joint.

Table 6. Chemical elements in the weld joint after corrosion test.

Sl. No.	Joint Description	Al (wt.%)	Cu (wt.%)	Mg (wt.%)
1.	900-90	84.9	1.1	2.0
2.	900-110	88.3	1.5	2.0

From Figure 9, Tables 6 and 7, it is understood that Cu and Mg co-exist in base metal and the same is reflected in the weld joint. Cu is the noble metal in corrosion perspective. Hence, Cu acts as cathode and Al matrix acts as anode. Hence, the aluminum matrix is corroded by the precipitate Al-Cu-Fe, Al-Cu (Table 5) and formed Cu and Mg rich phases (Table 7). Table 6 clearly shows wt.% of Cu present in the base metal. From Tables 5 and 7, it is inferred that Cu is higher in wt.% in 900-110 joint than in 900-90 joint which led to more corrosive impact in 900-110 joint. Hence, the rate of corrosion is directly proportional to the wt.% Cu present in the joint and to the corrosive medium. Another interesting observation is considerable increment in Mg (1.1% to 2%) after corrosion which induces considerable corrosion resistance. The precipitate Al-Cu-Mg has been exposed by corrosion activity. One common aspect can be noted from the chemical composition (from EDS spectrum) before and after corrosion is the presence of Cu and Mg which contributed not only for corrosion but also to the enhancement of hardness in the weld joint.

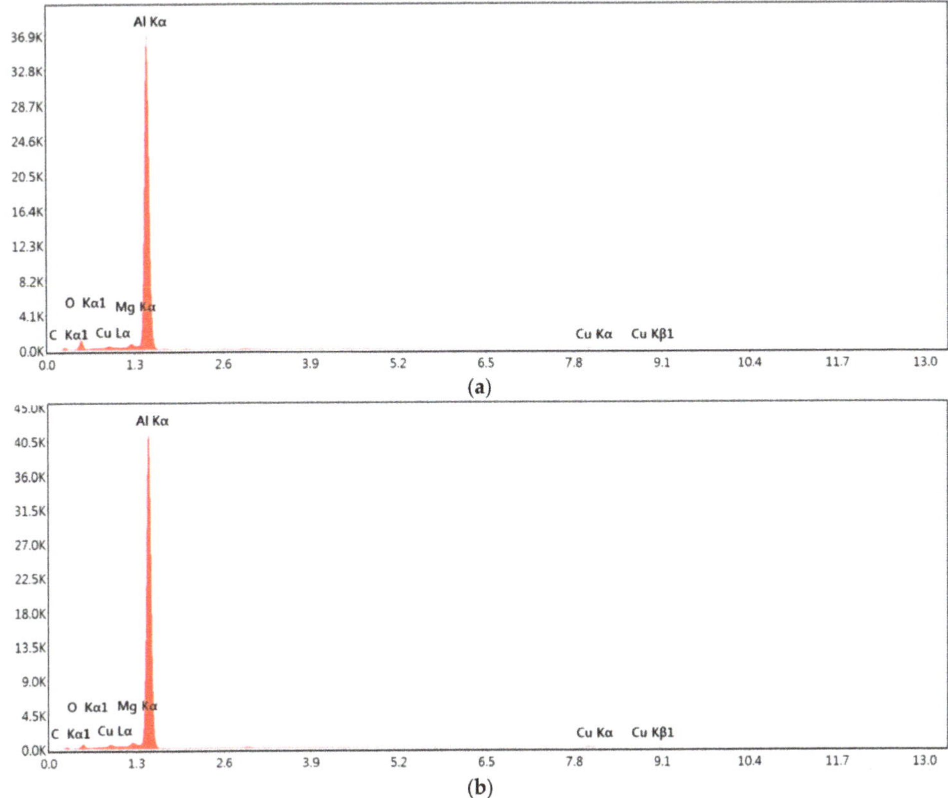

Figure 9. EDS spectrum analysis after immersion corrosion test (**a**) 900-90 Joint (**b**) 900-110 Joint.

Table 7. Influence of Cu on corrosion in the Base metal.

Alloy	Cu	Mg	Li	Cu/Mg
AA8090	1.4	0.8	2.3	1.75

4. X-ray Diffraction Analysis on Corroded Specimens

This section compares the X-ray diffraction analysis of 900-90 joint and 900-110 joint with that of base metal to appreciate the phase changes that happened after corrosion.

From Figure 10 of XRD analysis it is clear that all odd/even miller indices confirmed that the material analyzed has FCC structure. From EDS analysis and XRD analysis, the change in wt.% Al from 47.5% (uncorroded) to 84.9% (corroded) is reflected in the XRD peak of 900-90 joint and the change in wt.% Al from 42.5% (uncorroded) to 88.3% (corroded) is reflected in the XRD peak of 900-110 joint. The primary reason for the increase in the (111) peak is the evolution of $Al(OH)_3$ which is responsible for the corrosion pit formation. Primary source of $Al(OH)_3$ has evol3ved from H_2O_2 solution which has been used for the corrosion test. Secondary source of hydroxide is from dissolved oxygen in NaCl solution. Hence, these facts confirm that corrosion has occurred in the surface of the 900-90 and 900-110 joint which is also reflected in the SEM and optical microscopy analysis as previously mentioned.

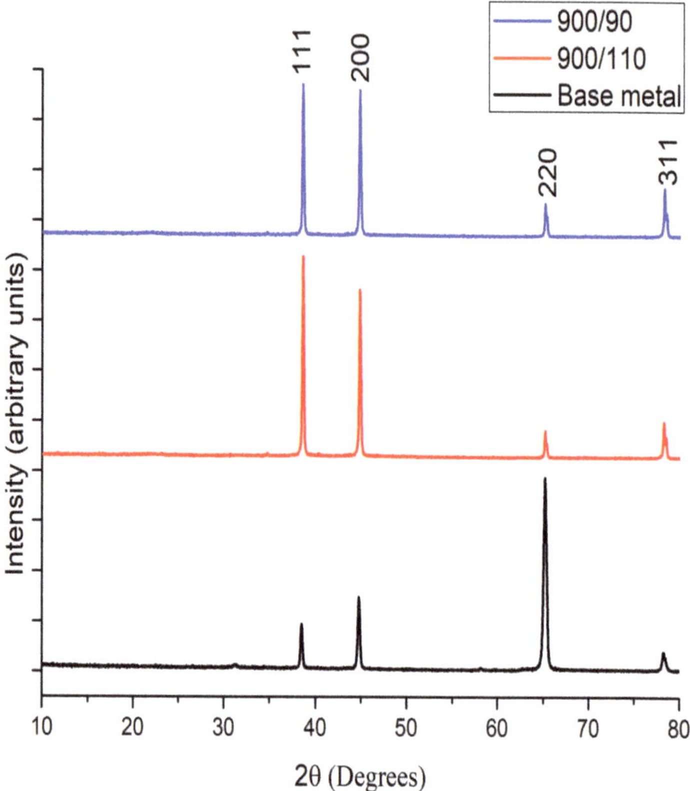

Figure 10. X-ray diffraction pattern of welded sample after corrosion analysis.

Effect of Corrosion on Surface Roughness and Hardness of the Joint

This section presents the influence of corrosion on change in surface roughness and hardness of the specimen. Surface roughness is one of the important properties to be analyzed in the perspective of the possibility of crack formation from the surface due to different forms of corrosion phenomena.

The surface roughness and hardness of the welded specimens before and after corrosion is analyzed and presented in Table 8. From the tabulated result, it is understood that the surface roughness values increased after corrosion. Hence, the presence of rougher surface is confirmed which is indicating the presence of small pits and also indicated the initiation of crack is possible from the surface. The hardness before the corrosion is 121.8 VHN for 900-90 joint and 98.3 VHN for 900-110 joint (Figures 3 and 4). The hardness values after the corrosion are 125.2 VHN in the stir zone for 900-90 joint and 117.5 VHN in the stir zone for 900-110 joint. This hardness increase after the corrosion could be due to increase in Cu and Mg wt.% as indicated in Table 7.

Table 8. Surface roughness and Hardness before and after corrosion.

Sl. No.	Joint Desc.	R_a before Corrosion (μm)	R_a after Corrosion (μm)	Hardness before Corrosion (VHN)	Hardness after Corrosion (VHN)
1.	900-90	4.067	5.064	121.8	125.2
2.	900-110	0.566	1.223	92.3	117.5

5. Conclusions

In this work, mechanical properties such as hardness of the alloy, surface roughness and microstructure of the joint are analyzed before and after corrosion using various testing methods. The following observations have been made from the tests conducted:

- The recrystallization behavior, grain boundary strengthening and precipitate formation in the grain boundaries were observed in the stir zone of the weld joint.
- The hardness variation in the weld stir zone was analyzed before and after corrosion and minor change in the hardness is noticed after corrosion. The hardness before corrosion for the 900-90 joint is 73.3% of the base metal and 54.5% of the base metal for 900-110 joint. The hardness increase has been observed after corrosion which was 2.79% for 900-90 joint and 27.3% for 900-110. This hardness variation is the indication of evolution of Al-Cu precipitates in the grain and in the grain boundaries.
- Corrosion analysis has been performed and found that higher density of pits was formed in the 900-110 joint than in the 900-90 joint which is due to the presence of increased wt.% of Cu in 900-110 joint and hydroxides formed during corrosion mechanism. The mass loss % per year is 0.2% for 900-90 joint and 0.8% for 900-110 joint. This confirms the fact that more no. of Cu containing precipitates were formed in 900-110 joint.
- From EDS spectrum before corrosion and after corrosion, it was observed that the alloying elements Al, Cu, Mg, Fe contributed to corrosion behavior.
- Surface roughness analysis has been carried out to analyze the irregularity in the surface after corrosion and found that the surface roughness values escalated to 24.5% and 116% for 900-90 and 900-110 joints after corrosion.
- The traverse speed variation had led to severe corrosion which further led to the deviation in the surface roughness and also hardness of the weld joint. The underlying phenomenon for all these variations was precipitate formation and grain boundary strengthening.

Author Contributions: Conceptualization, C.S., R.S., J.T.W.J., A.N., M.R., S.S.A., C.P., S.D. and N.I.V.; Data curation, C.S., R.S., J.T.W.J., A.N., M.R., S.S.A., C.P., S.D. and N.I.V.; Formal analysis, C.S., R.S., C.P., S.D. and N.I.V.; Funding acquisition, A.N., M.R., S.S.A., C.P., S.D. and N.I.V.; Investigation, C.S., R.S., J.T.W.J., A.N., M.R., S.S.A., C.P., S.D. and N.I.V.; Methodology, C.S., R.S., J.T.W.J., A.N., M.R., S.S.A., C.P., S.D. and N.I.V.; Project administration, C.P. and N.I.V.; Resources, A.N.; Software, A.N. and N.I.V.; Supervision, R.S., M.R., S.S.A., C.P. and N.I.V.; Validation, C.S., S.S.A., C.P. and S.D.; Visualization, J.T.W.J., S.D. and N.I.V.; Writing—original draft, C.S., R.S., J.T.W.J., A.N., M.R., S.S.A., C.P. and S.D.; Writing—review & editing, C.P., S.D. and N.I.V. All authors have read and agreed to the published version of the manuscript.

Funding: The research is partially funded by the Ministry of Science and Higher Education of the Russian Federation under the strategic academic leadership program 'Priority 2030' (Agreement 075-15-2021-1333 dated 09/30/2021).

Institutional Review Board Statement: Not applicable.

Informed Consent Statement: Not applicable.

Data Availability Statement: Not applicable.

Conflicts of Interest: The authors declare no conflict of interest.

References

1. Raheja, G.S.; Singh, S.; Prakash, C. Processing and characterization of Al5086-Gr-SiC hybrid surface composite using friction stir technique. *Mater. Today Proc.* **2020**, *28*, 1350–1354. [CrossRef]
2. Singh, S.; Singh, G.; Prakash, C.; Kumar, R. On the mechanical characteristics of friction stir welded dissimilar polymers: Statistical analysis of the processing parameters and morphological investigations of the weld joint. *J. Braz. Soc. Mech. Sci. Eng.* **2020**, *42*, 154. [CrossRef]
3. Singh, R.K.R.; Sharma, C.; Dwivedi, D.K.; Mehta, N.K.; Kumar, P. The microstructure and mechanical properties of friction stir welded Al-Zn-Mg alloy in as welded and heat treated conditions. *Mater. Des.* **2011**, *32*, 682–687. [CrossRef]

4. Devanathan, C.; Babu, A.S. Friction Stir Welding of Metal Matrix Composite Using Coated Tool. *Procedia Mater. Sci.* **2014**, *6*, 1470–1475. [CrossRef]

5. Attallah, M.M.; Salem, H.G. Friction stir welding parameters: A tool for controlling abnormal grain growth during subsequent heat treatment. *Mater. Sci. Eng. A* **2005**, *391*, 51–59. [CrossRef]

6. Mao, Y.; Ke, L.; Chen, Y.; Liu, F.; Xing, L. Inhomogeneity of microstructure and mechanical properties in the nugget of friction stir welded thick 7075 aluminum alloy joints. *J. Mater. Sci. Technol.* **2018**, *34*, 228–236. [CrossRef]

7. Moshtaghi, M.; Loder, B.; Safyari, M.; Willidal, T.; Hojo, T.; Mori, G. Hydrogen trapping and desorption affected by ferrite grain boundary types in shielded metal and flux-cored arc weldments with Ni addition. *Int. J. Hydrogen Energy* **2022**, *47*, 20676–20683. [CrossRef]

8. Liu, F.J.; Fu, L.; Chen, H.Y. Microstructure evolution and fracture behaviour of friction stir welded 6061-T6 thin plate joints under high rotational speed. *Sci. Technol. Weld. Join.* **2018**, *23*, 333–343. [CrossRef]

9. Vijay, S.J.; Murugan, N. Influence of tool pin profile on the metallurgical and mechanical properties of friction stir welded Al-10wt.% TiB2 metal matrix composite. *Mater. Des.* **2010**, *31*, 3585–3589. [CrossRef]

10. Moradi, M.M.; Jamshidi Aval, H.; Jamaati, R.; Amirkhanlou, S.; Ji, S. Microstructure and texture evolution of friction stir welded dissimilar aluminum alloys: AA2024 and AA6061. *J. Manuf. Process.* **2018**, *32*, 1–10. [CrossRef]

11. Gopkalo, O.; Liu, X.; Long, F.; Booth, M.; Gerlich, A.P.; Diak, B.J. Non-isothermal thermal cycle process model for predicting post-weld hardness in friction stir welding of dissimilar age-hardenable aluminum alloys. *Mater. Sci. Eng. A* **2019**, *754*, 205–215. [CrossRef]

12. Safarbali, B.; Shamanian, M.; Eslami, A. Effect of post-weld heat treatment on joint properties of dissimilar friction stir welded 2024-T4 and 7075-T6 aluminum alloys. *Trans. Nonferrous Met. Soc. China* **2018**, *28*, 1287–1297. [CrossRef]

13. Salih, O.S.; Neate, N.; Ou, H.; Sun, W. Influence of process parameters on the microstructural evolution and mechanical characterisations of friction stir welded Al-Mg-Si alloy. *J. Mater. Process. Technol.* **2020**, *275*, 116366. [CrossRef]

14. Kumar, P.V.; Reddy, G.M.; Rao, K.S. ScienceDirect Microstructure and pitting corrosion of armor grade AA7075 aluminum alloy friction stir weld nugget zone e Effect of post weld heat treatment and addition of boron carbide. *Def. Technol.* **2015**, *11*, 166–173. [CrossRef]

15. Naumov, A.A.; Isupov, F.Y.; Golubev, Y.A.; Morozova, Y.N. Effect of the Temperature of Friction Stir Welding on the Microstructure and Mechanical Properties of Welded Joints of an Al–Cu–Mg Alloy. *Met. Sci. Heat Treat.* **2019**, *60*, 695–700. [CrossRef]

16. El Mouhri, S.; Essoussi, H.; Ettaqi, S.; Benayoun, S. Relationship between Microstructure, Residual Stress and Thermal Aspect in Friction Stir Welding of Aluminum AA1050. *Procedia Manuf.* **2019**, *32*, 889–894. [CrossRef]

17. Fahimpour, V.; Sadrnezhaad, S.K.; Karimzadeh, F. Corrosion behavior of aluminum 6061 alloy joined by friction stir welding and gas tungsten arc welding methods. *Mater. Des.* **2012**, *39*, 329–333. [CrossRef]

18. Balaji Naik, D.; Venkata Rao, C.H.; Srinivasa Rao, K.; Madhusudan Reddy, G.; Rambabu, G. Optimization of friction stir welding parameters to improve corrosion resistance and hardness of AA2219 aluminum alloy welds. *Mater. Today Proc.* **2019**, *15*, 76–83. [CrossRef]

19. Sinhmar, S.; Dwivedi, D.K. A study on corrosion behavior of friction stir welded and tungsten inert gas welded AA2014 aluminium alloy. *Corros. Sci.* **2018**, *133*, 25–35. [CrossRef]

20. Meng, Q.; Liu, Y.; Kang, J.; Fu, R.D.; Guo, X.Y.; LI, Y.J. Effect of precipitate evolution on corrosion behavior of friction stir welded joints of AA2060-T8 alloy. *Trans. Nonferrous Met. Soc. China* **2019**, *29*, 701–709. [CrossRef]

21. Gharavi, F.; Amin, K.; Yunus, R. Corrosion behavior of Al6061 alloy weldment. *Integr. Med. Res.* **2015**, *4*, 314–322.

22. Baiyao, H.; Hua, Z.; Fuad, K. Mechanical behavior associated with metallurgical aspects of friction stir welded Al-Li alloy exposed to exfoliation corrosion test. *Mater. Res. Express* **2020**, *7*, 066502. [CrossRef]

23. Uddin, M.; Basak, A.; Pramanik, A.; Singh, S.; Krolczyk, G.M.; Prakash, C. Evaluating hole quality in drilling of Al 6061 alloys. *Materials* **2018**, *11*, 2443. [CrossRef]

24. Basak, A.K.; Pramanik, A.; Prakash, C. Deformation and strengthening of SiC reinforced Al-MMCs during in-situ micro-pillar compression. *Mater. Sci. Eng. A* **2019**, *763*, 138141. [CrossRef]

25. Prakash, C.; Singh, S.; Gupta, M.K.; Mia, M.; Królczyk, G.; Khanna, N. Synthesis, characterization, corrosion resistance and in-vitro bioactivity behavior of biodegradable Mg–Zn–Mn–(Si–HA) composite for orthopaedic applications. *Materials* **2018**, *11*, 1602. [CrossRef]

26. Prakash, C.; Singh, S.; Pabla, B.S.; Sidhu, S.S.; Uddin, M.S. Bio-inspired low elastic biodegradable Mg-Zn-Mn-Si-HA alloy fabricated by spark plasma sintering. *Mater. Manuf. Processes* **2019**, *34*, 357–368. [CrossRef]

27. Prakash, C.; Kansal, H.K.; Pabla, B.S.; Puri, S. Powder mixed electric discharge machining: An innovative surface modification technique to enhance fatigue performance and bioactivity of β-Ti implant for orthopedics application. *J. Comput. Inf. Sci. Eng.* **2016**, *16*, 041106. [CrossRef]

28. Shanmugavel, R.; Chinthakndi, N.; Selvam, M.; Madasamy, N.; Shanmugakani, S.K.; Nair, A.; Prakash, C.; Buddhi, D.; Dixit, S. Al-Mg-MoS2 Reinforced Metal Matrix Composites: Machinability Characteristics. *Materials* **2022**, *15*, 4548. [CrossRef] [PubMed]

29. Das, L.; Nayak, R.; Saxena, K.K.; Nanda, J.; Jena, S.P.; Behera, A.; Sehgal, S.; Prakash, C.; Dixit, S.; Abdul-Zahra, D.S. Determination of Optimum Machining Parameters for Face Milling Process of Ti6A14V Metal Matrix Composite. *Materials* **2022**, *15*, 4765. [CrossRef]

 *materials*

Article

A Numerical Study on the Effect of Tool Speeds on Temperatures and Material Flow Behaviour in Refill Friction Stir Spot Welding of Thin AA7075-T6 Sheets

Venkata Somi Reddy Janga, Mokhtar Awang * and Srinivasa Rao Pedapati

Department of Mechanical Engineering, Universiti Teknologi PETRONAS, Seri Iskandar 32610, Malaysia; venkata_19001587@utp.edu.my (V.S.R.J.); srinivasa.pedapati@utp.edu.my (S.R.P.)
* Correspondence: mokhtar_awang@utp.edu.my; Tel.: +60-53687204

Abstract: A three-dimensional (3D) numerical model was created to simulate and analyze the effect of tool rotational speeds (RS) and plunge rate (PR) on refill friction stir spot welding (refill FSSW) of AA7075-T6 sheets. The numerical model was validated by comparing the temperatures recorded at a subset of locations with those recorded at the exact locations in prior experimental studies from the literature. The peak temperature at the weld center obtained from the numerical model differed by an error of 2.2%. The results showed that with the rise in RS, there was an increase in weld temperatures, effective strains, and time-averaged material flow velocities. With the rise in PR, the temperatures and effective strains were reduced. Material movement in the stir zone (SZ) was improved with the increment of RS. With the rise in PR, the top sheet's material flow was improved, and the bottom sheet's material flow was reduced. A deep understanding of the effect of tool RS and PR on refill FSSW joint strength were achieved by correlating the thermal cycles and material flow velocity results obtained from the numerical models to the lap shear strength (LSS) from the literature.

Keywords: friction stir spot welding; numerical modeling; refill friction stir spot welding; thermal cycles; simulation of joining processes; material flow; thermomechanical characteristics

Citation: Janga, V.S.R.; Awang, M.; Pedapati, S.R. A Numerical Study on the Effect of Tool Speeds on Temperatures and Material Flow Behaviour in Refill Friction Stir Spot Welding of Thin AA7075-T6 Sheets. *Materials* 2023, *16*, 3108. https://doi.org/10.3390/ma16083108

Academic Editor: Raul D. S. G. Campilho

Received: 10 February 2023
Revised: 16 March 2023
Accepted: 21 March 2023
Published: 14 April 2023

1. Introduction

Solid-state joining techniques, friction stir welding (FSW) and friction stir spot welding (FSSW), are used to a great extent for the welding of aluminium and magnesium alloys and are capable of overcoming issues with fusion weldings such as porosity and liquid cracking [1]. In FSW, a rotating tool penetrates perpendicularly (plunging) and then moves transversely in the direction of the weld path. FSW and FSSW are based on the principle of frictional heating. However, FSSW can only plunge, dwell, and retract in an axial direction. The four main categories of FSSW approaches that are now in use are FSSW [2], swept FSSW [3], refill FSSW [4], and swing FSSW [5]. Helmholtz-Zentrum Hereon has developed and patented refill FSSW, a variation of FSSW that produces a spot weld without the exit hole that is unavoidable in FSSW [4].

Compared to FSSW, refill FSSW offers advantages since it improves weld volume and reduces corrosion cracking, fatigue, and stress concentration [6]. Refill FSSW's tool assembly comprises a clamping ring, a shoulder, and a probe that move independently. The moving tools, probe, and shoulder plasticize the material through frictional heating. The workpieces are tightly gripped by the clamping ring and a backing anvil, which prevents material from flashing near the tool's outer edge (shoulder) during the process. The process is classified as probe plunging and shoulder plunging versions subjected to the order of probe/shoulder vertical movement [7]. The shoulder plunging form is the most popular because it results in a more robust, larger-volume weld than the probe plunging type. Figure 1 illustrates the refill FSSW process (shoulder plunging variant).

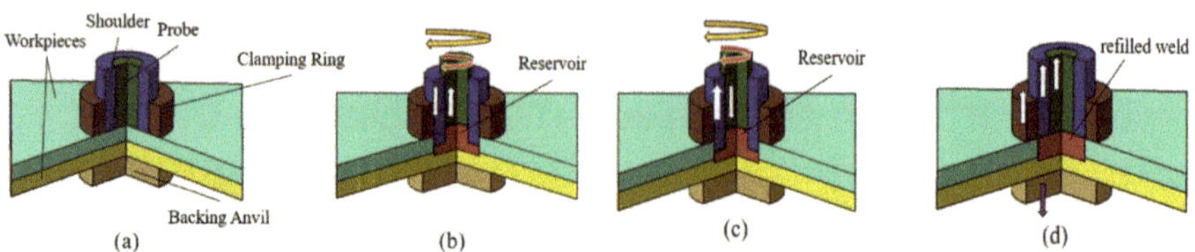

Figure 1. Schematic representation of refill FSSW procedure: (**a**) step 1—initiation; (**b**) step 2—shoulder plunging; (**c**) step 3—refilling; (**d**) step 4—finishing.

Step 1: In this step, the clamping force is applied to the workpieces, and the clamping ring and the backing anvil firmly hold the workpieces. The probe and shoulder do not initiate frictional contact with the workpiece and rotate above the workpieces.

Step 2: The rotating shoulder moves vertically downwards and penetrates the workpiece (plunging), and the probes advance vertically in the direction opposite the shoulder. This action initiates deformation and frictional heating. The softened material is drawn into the reservoir; see Figure 1b.

Step 3: The material (plasticized) is refilled/pushed into the workpiece by switching the vertical directions of the rotating tools in this refilling step.

Step 4: The shoulder and probe's vertical and rotational motions are stopped once Step 3 is finished, and the clamping force is released.

After plunge depth (PD), tool RS is also an essential factor influencing joint strength [8–17]. According to Shen et al. [8], the increase in weld volume and material mixing are influential in determining joint strength. Previous research on refilling FSSW of aluminium alloys shows that with an increase in RS, joint strength increases initially and then decreases [9,12,14,17–22]. Some studies present a decrease [23–26] and an increase [10,11,27,28] in joint strength with an increase in RS. According to research done by Zhou et al. [20], at lower RSs, voids are formed due to poor material flow, and with an increase in RS, vertical plastic flow is enhanced. Joint strength initially increases and gradually reduces with increased RS due to excessive heat induced by a further rise in RS. Li et al. [23] stated that poor metallurgical bonding exists at the interface of SZ and the thermomechanically affected zone (TMAZ) at lower RSs. Still, there is sufficient stirring at the lap interface. In addition to plasticization, material mixing significantly affects joint quality in friction-stir-based joining processes [29,30]. Increasing the RS increases material flow, enhancing material mixing and the metallurgical bonding at the SZ/TMAZ interface. Still, the dispersed material at the lap interface and coarsening of the microstructure are prone to cracks, reducing the joint strength. Kubit et al. [15] stated that there is insufficient plasticization at lower RSs, and the weakening of the joint is associated with less heat produced due to lower RSs. Overall, from the experimental studies and microstructural observations, it is observed that an increase in RS increases heat input, which induces more plasticity and enhances material flow in the vertical direction. The enhanced material flow improves joint strength and helps avoid defects like voids and incomplete refill in the weld. However, excessive heat input with a further increase in RS will weaken the joint because of the coarsening microstructure and the reduction of bonding ligament length (dispersed lap interface) because of excessive vertical material flow. PR/welding time (WT) is the parameter with the least influence on joint strength [18,22,24,31]. In most of the earlier work, it was observed that joint strength is enhanced with an increase in the WT/decrease in PR [8,10,27,32,33]. The increased WT allows a rise in temperature for proper phase change of the material and improves material flowability and mixing. Similar to the RS trend, in some cases, with an increase in WT, joint strength increases initially and then drops [15,28,31]. In some studies, joint strength decreases with a rise in WT [18,19,24]. The decrease in weld strength is due to the induced excessive temperatures with an increase in WT, similar to increasing RS.

Muci-Küchler et al. [34] used the Abaqus-Explicit code to simulate the probe plunging version of refill FSSW, but the simulation was restricted to the plunging step. The thermal and material flow results were reported, and the model was validated using experimental findings. Ji et al. [35,36] used Ansys Fluent to create a numerical model to examine the material flow velocities in refill FSSW. Since the simulation was conducted using various models at various PDs and presented mainly the material flow velocities, the actual material flow behavior and other quantitative results were not reported. According to their studies, increasing RSs and adding geometric features to the tools improve material flow. They also conclude that increasing the RS is an effective method to enhance material flow. Utilizing DEFORM-3D, Malik et al. [37] refilled the exit hole using multi-stage processes that differ from those of traditional refill FSSW. Quantitative results were not presented despite the exit hole being refilled. Kubit et al. [38] created a two-dimensional (2D) numerical model with Simufact Forming code. The 2D material flow from the simulation was validated with the joint's microstructures. Zhang et al. [39], using ABAQUS (coupled Eulerian–Lagrangian formulation), developed and validated a 3D thermomechanical model. The material flow characteristics and thermal cycles during refill FSSW of sheets made of magnesium alloy were presented. The numerical models of the processes involving high deformations and temperatures, like friction stir processing, FSW, and FSSW, were developed using DEFORM-3D (Lagrangian incremental formulation) [40–42]. The simulations can withstand significant deformations and predict material flow and thermal cycles. A thermomechanical model was created by Janga et al. [43,44] in DEFORM-3D, and the model was validated via temperatures obtained from experiments. The results of the material flow velocities, strain, and temperatures were shown and connected to the experimental findings. Xiaong et al. [45] developed an axisymmetric 2D model to simulate refill FSSW of AA7075-T6 in ForgeNxt3.2 software. The model was validated based on thermal results, and void formation was correlated to material flow. However, as the model was 2D, the outward material shearing under the influence of the rotating tool could not be visualized. Raza et al. [46] explored the evolution of intermetallic compounds propelled by chemical and mechanical forces, considering the effect of several driving forces in a multiphase-field framework numerical model developed in DEFORM-3D.

The RS and PR parameters' effect in refilling FSSW has been presented in prior experimental studies. However, measuring thermal cycles and understanding the local material flow behavior is challenging, even with micrographs, thermocouples, and start/stop experiments. Also, studying material flow behavior and strain locally experienced in joining thin sheets is quite difficult based on experimental results. In earlier research, a few computational studies demonstrated the material flow and thermal cycles during refill FSSW and correlated the numerical results to the joint characteristics. The thermal cycles and temperatures in the SZ and their influence on refill FSSW joint strength and material flow are still unclear when RS and PR are varied. Furthermore, no numerical investigations presented thermal cycles, strain, and material flow velocities when RS and PR parameters were varied. Therefore, a 3D numerical model was developed and validated in the current study to investigate the thermal cycles and local material flow behavior during the process by varying RSs and PRs.

2. Refill FSSW: Finite Element Modeling

The refill FSSW process was modeled in DEFORM-3D using an incremental Lagrangian formulation. The simulations in the current study replicated the prior experimental work done by Yamin et al. [31] for model validation. As a result, the material, tool geometry dimensions, and refill FSSW process parameters were all taken from the experimental investigation. Simulations were run with changing RSs and PRs while keeping all other parameters constant to gain insight into the impact of tool speeds on the process. In the first three models, M1, M2, and M3, RSs of 2000 rpm, 2500 rpm, and 3000 rpm, respectively, were varied, while the PD was 0.7 mm and PR was 0.5 mm/s, unchanged. In the following two simulations, in the models M4 and M5, the PRs were 0.25 mm/s and 0.75 mm/s, while the RS and PD were 3000 rpm and 0.7 mm, respectively.

2.1. Refill FSSW Geometry

A toolset (shoulder, probe, and clamping ring), workpiece, and backing anvil were the geometry utilized in the finite element model of refill FSSW. Figure 2 depicts the detailed dimensions of geometries. CATIA V5R20 was used to model the geometries. A fillet was added for the shoulder bottom edges to reduce penetration stresses during plunging and improve the contact area [34]. The geometries were then assembled in DEFORM-3D. In general, the sheets were firmly clamped. Thus, sheets in the simulation were treated as a combined sheet for simplification. The workpiece's length and the toolset's height were restricted to reduce computing time.

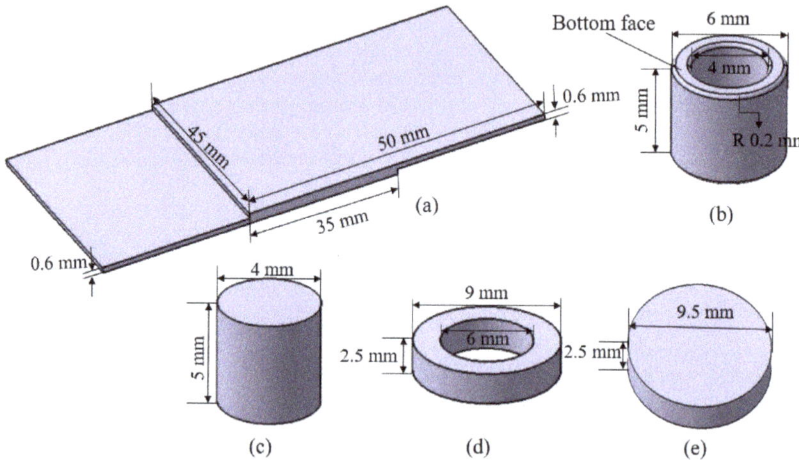

Figure 2. Geometries used for the numerical model with detailed dimensions (**a**) Workpiece; (**b**) Shoulder; (**c**) Probe; (**d**) Clamping Ring; (**e**) Banking Anvil.

2.2. Material Law, Materials, and Meshing

It is critical to choose an applicable strain-rate and temperature-dependent law to simulate a material's response in processes like refill FSSW. In the refill FSSW process and similar operations, the Johnson-Cook material law, which considers the impacts of strain rate, strain, and temperature, is frequently used [44,47]:

$$\sigma_y = \left[A + B \left[\varepsilon^{pl} \right]^n \right] \left[1 + C \, ln\left(\frac{\dot{\varepsilon}^{pl}}{\dot{\varepsilon}^0} \right) \right] \left[1 - \left[\frac{T - T_r}{T_m - T_r} \right]^m \right], \tag{1}$$

where σ_y denotes the material flow stress, the reference plastic strain rate is represented by $\dot{\varepsilon}^0$, ε^{pl} denotes the equivalent plastic strain, $\dot{\varepsilon}^{pl}$ denotes the plastic strain rate, and T_m and T_r—the melting and reference temperatures, respectively. The material constants n-coefficient of strain hardening, m-thermal softening coefficient, A-quasi-static yield strength, B-strain hardening constant, and the strain-rate dependency are described at the reference strain rate by the strengthening coefficient C.

The experiment used a commercial Al-Zn-Mg-Cu alloy AA 7075-T6 with a sheet thickness of 0.6 mm [31]. The material AA7075-T6 applied to the workpiece was loaded from DEFORM-3D's library and adhered to the Johnson–Cook material law. The Johnson–Cook material law coefficients and constants were taken from a study by Fang et al. [48]. Additional material parameters were Young's modulus E = 68.9 GPa, thermal conductivity = 180.175 W/m K, Poisson's ratio = 0.3, specific heat capacity c_p = 870 J/kg K, and $\alpha = 2.2 \times 10^{-5}$/K (thermal expansion coefficient).

Meshing details were as follows—70,000 elements meshed the shoulder and probe, 10,000 elements meshed the clamping ring, 15,000 meshed the backing anvil, and 180,000 meshed the workpiece. Due to the process' severe deformations, an extensive remeshing scheme was used. A condition connected to the interference depth (0.25 mm) was specified to trigger remeshing. However, remeshing also started when the elements significantly deformed and became unusable.

2.3. Boundary Conditions and Contact

For conduction between the workpiece and toolset, a conductive heat transfer coefficient of 11 $\frac{N}{mm\,s\,K}$ was used [43,44]. For convective heat transfer from tools/workpieces to the environment, the heat transfer coefficient h = 0.02 $\frac{N}{mm\,s\,K}$ was used [44,49]. The side faces of the workpiece were constrained in all degrees of freedom. The clamping ring and the backing anvil were made stationary. The rotation and translation movement for the shoulder and probe were defined according to the parameters. Incipient melting might cause tool slippage as temperatures increase. As a result, a friction coefficient that changes as a function of temperature is suggested [50]. Coulomb's friction law was used in the simulation, with a temperature-dependent friction coefficient [43,44], as shown in Table 1.

Table 1. Temperature-dependent Coulomb's friction coefficient used in the numerical model [44].

Temperature (°C)	20	160	200	400	500	580
Coefficient of Friction (μ)	0.35	0.3	0.26	0.08	0.03	0.01

The two significant steps in the simulation process were as follows:

1. Plunging stage: The shoulder plunged with the specified PR, RS, and PD. The probe went in the reverse direction vertically, with a speed of 1.25 times that of the shoulder's PR and the same RS as that of the shoulder. The softened/plasticized material was drawn inwards into the reservoir in this step.
2. Refilling stage: The probe and shoulder switched their vertical movements, maintaining the axial speed of the previous step (plunging step), allowing the material to be refilled.

3. Results and Discussion

The temperatures from the simulations for models M1, M2, and M3, which showed variations of temperatures with RSs, are presented in Figure 3a. These temperature values were recorded at the characteristic points via point tracking at T1, T2, and T3 at the weld center, 4 mm away and 7.5 mm away from the weld center, respectively. The model was validated by correlating model M3 temperature results to experimental temperature findings from previous work [31] at locations T1, T2, and T3. Table 2 compares the maximum temperatures obtained from the simulations with the experimental values. The maximum temperatures at T1, T2, and T3 locations differed by errors of 2.2%, 2.3%, and 6.4%, respectively. There was a strong correlation between numerical and experimental temperature results at all three locations. The temperatures from the simulations for models M4 and M5, which showed variations of temperatures with PRs, are presented in Figure 3b,c. All the models showed a sharp increase in temperature at the start of the plunging stage at measured location T1. The temperature was then steadily increased in the weld zone, after which it gradually decreased as the process ended. The behavior of the temperature curve at location T2 was similar to that at the weld center, although temperatures dropped further than at the center of a weld. A sharp increase seen at T1 was no longer there at T3, and the temperature rose gradually. In all the models, the peak temperatures were shown following the plunge phase and the beginning of the refilling phase, approximately at 0.6 t (t = process time). All the models showed that temperatures dropped further away from the tool's axis of rotation. Lower temperatures among the RS variation simulations were observed in model M1 because of the lowest RS among the three models. Model M3 had a higher RS than the other models, contributing to its higher temperatures than other models. Among the PR variation simulations from the models M3, M4, and M5, the temperatures at T1, T2, and T3 were higher in model M4 with a 0.25 mm/s PR due to an increased WT. With the increase in PR, the WT and the contact time were reduced; hence, the temperatures were lower in model M3 and model M5 than in model M4. The comparison of the temperatures measured at T1, T2, and T3 from the simulations is tabulated in Table 2.

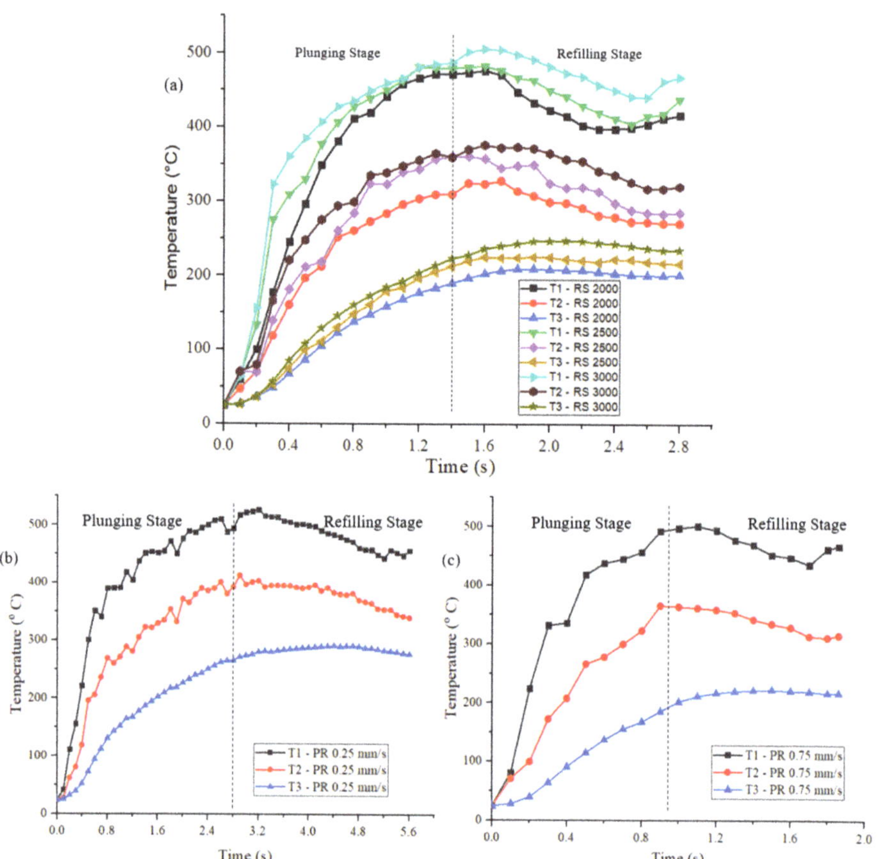

Figure 3. Temperature behavior in the numerical models at characteristic points T1, T2, and T3 of (**a**) models M1, M2, and M3; (**b**) model M4; (**c**) model M5.

Table 2. Comparison of maximum temperatures from the experiment and simulation.

	Maximum Temperature (°C) at T1	Maximum Temperature (°C) at T2	Maximum Temperature (°C) at T3	Maximum Temperature (°C) in SZ	% of Melting Point (635 °C)
Experiment	495	386	231	-	-
Model M1	476	328	210	490	77
Model M2	483	361	225	500	78
Model M3	506	377	246	520	81
Model M4	527	413	291	540	85
Model M5	501	363	222	510	80

The temperatures directly impacted the softening and stirring of the material during the process. Figure 4 displays the simulation temperature contours of the three models. The temperatures were distributed symmetrically around the weld center. All models' temperatures dropped as they moved outward from the weld center. The region where material stirring occurred, or the SZ, was where the highest temperatures were recorded. Models M1, M2, M3, M4, and M5 had maximum temperatures of 490, 500, 520, 540, and 510 °C, respectively, which were 77%, 78%, 81%, 85%, and 80%, respectively, of the material's melting point (T_m = 635 °C). The range of the temperatures in the SZ was from 430 to 490 °C (model M1), from 440 to 500 °C (model M2), from 460 to 520 °C (model M3), from 480 to 540 °C (model M4), and from 450 to 510 °C (model M5). It was evident from the contours that in the SZ, the temperatures were above the solidus temperature (475 °C) [51]. In the range of 475 to 540 °C, the base material's hardening precipitates ($MgZn_2$) dissolve rapidly [52]. As a result, the material in the SZ becomes softer and moves and shears more quickly as the temperatures increase.

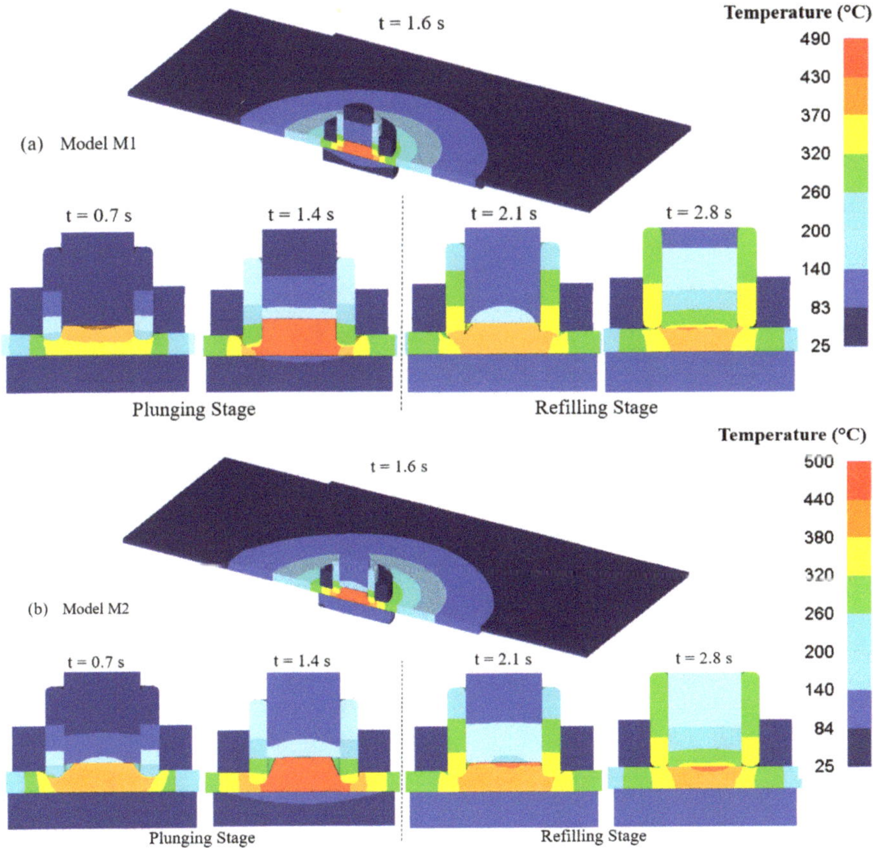

Figure 4. *Cont.*

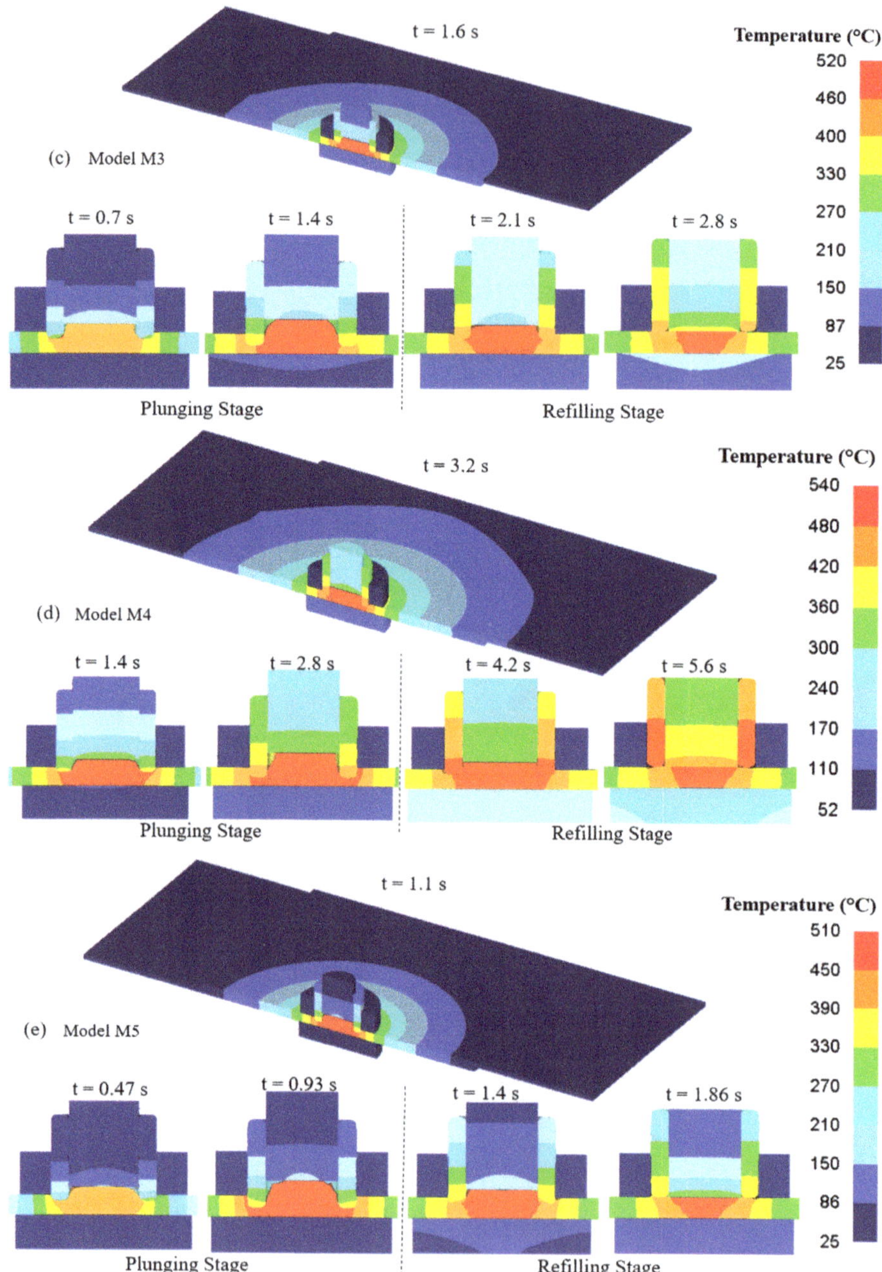

Figure 4. Temperature distribution in the process of refill FSSW from the simulations (**a**) Model M1; (**b**) Model M2; (**c**) Model M3; (**d**) Model M4; (**e**) Model M5.

The weld's microstructure and grain size depend on thermal cycles, strain rates, and plastic strains. Figure 5 shows the simulation contours of the effective strain during the process. Generally, microstructure results can be used to identify the distinct zones from the experiments. However, it is challenging to estimate the strain the material experienced locally. All the models had an effective strain contour symmetrically around the weld center. It was clear from the contours that as the RSs increased, effective strain also increased. Dynamic recrystallization, which occurs as a result of extreme plastic deformations and steep temperature gradients in the SZ, produces a finer microstructure [44,53] in the SZ. It was evident from the simulation results that the area of the high-strain zone grew with the rise in RS, indicating more plasticization of the material with an increase in RS [10,15,23]. At lower RSs, sharper and more concentrated effective strain was seen around the shoulder region compared to that at higher RSs. This was because with poor material flow at lower temperatures, the material could stick to the shoulder. However, as the plasticity increased at higher RSs, there was a slip phenomenon consistent with the literature [31]. Experimental observations state that incipient melting occurrs at higher RSs of 2500 and 3000 rpm, which is not seen at the RS of 2000 rpm [31]. The increased RS led to higher temperatures and higher strains, resulting in welding tool slippage in the SZ [31,51,54]. The experimental findings showed that when the temperature reached 495 °C, incipient melting started occurring in the grain boundaries. In contrast, at 480 °C, the sign of incipient melting was not visible. This can also be confirmed by the temperatures obtained from the simulations, see Figure 4. The maximum temperature of model M1 was less than 490 °C; hence, there was no incipient melting and tool slippage, and thus the concentrated, effective strains near the shoulder-affected SZ. A TMAZ adjacent to the SZ has moderate plastic strains and temperatures, due to which there is no recrystallization, and can be identified by elongated grains in the micrographs [44]. The long grains are formed due to the plasticization and tool movement at these moderate temperatures. The TMAZ was extended for a few microns adjacent to the SZ near the shoulder's outer edge. This range of TMAZ could be identified from the material movement, which will be discussed further. The heat-affected zone (HAZ) refers to the region only impacted by thermal cycles without the material moving mechanically. The weld's hardness characteristics depend on the thermal cycles and the plastic deformation that the base material undergoes. The condition of isomorphous precipitates that exists after welding becomes the major determinant for hardness because strong plasticization significantly impacts how much precipitates dissolve when stirred [55]. The thermal cycles in HAZ reduce hardness due to coarsening of precipitates. With the temperatures and the plasticization the SZ experiences, the hardening precipitates are dissolved, reducing the hardness and improving the SZ's capability of hardness by reprecipitation [56]. In the TMAZ, the material becomes the softest as it experiences higher temperatures than the HAZ does, and there is no reprecipitation. As discussed earlier, an increase in RS increases temperatures in the weldments, which affects the hardness of the base material differently in different zones. Therefore, although the increase in RS improves the plasticity/movement of the material, the excessive temperatures that the base material experiences can weaken the joint strength [11,15]. At lower PRs, i.e., with an increase in WT, the effective strains are higher and more distributed than the effective strains from the models with higher PRs; see Figure 5b. As discussed earlier, this was due to increased frictional heating which plasticized the material. Wider distribution of effective strains was observed as WT increased. In model M5, the quickest of the simulations, higher effective strain was seen in a small, concentrated region adjacent to the shoulder. It was observed that PR/WT significantly affected the strain around the weld region.

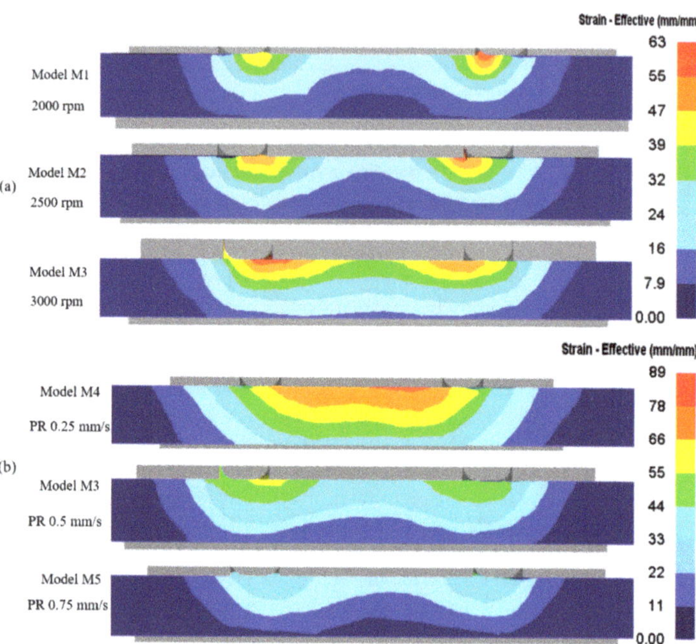

Figure 5. Comparison of effective strain contours at the process' completion (**a**) at different RSs; (**b**) at different PRs.

For a deeper understanding of local material flow behavior with changes in RS and PR, selected points were marked via point tracking, as shown in Figure 6. Considering the process' symmetrical nature, as observed from the above results, the initial characteristic points via point tracking were marked on the half section; see the zoomed view in Figure 6. The horizontal and vertical distances separating the characteristic points are 0.5 mm and 0.3 mm. Distinct points were P1–P9 (the surface of the workpiece) and P10–P18 (middle) in the top sheet. In the bottom sheet, the characteristic points were P19–P27 (the top of the bottom sheet/interface) and P28–P36 (the middle of the bottom sheet).

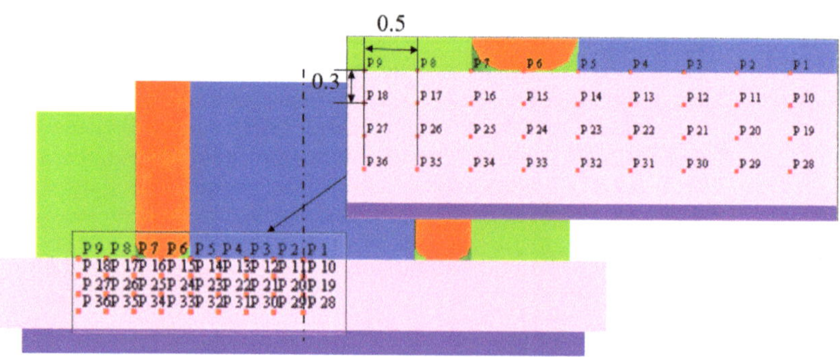

Figure 6. Initial positions of characteristic points using point tracking for material flow analysis.

The material flow patterns with similar parameters of model M3 are discussed in an earlier study by Janga et al. in detail [43]. Therefore, this study does not discuss material flow patterns but focuses on material flow velocities. The time-averaged velocities, derived from the characteristic points in Figure 6 of all the models, are presented in Figure 7. For model M1 with a 2000 rpm RS, the maximum average velocities were detected close to the inner periphery of the shoulder (2 mm from the tool's rotational axis). The maximum average velocity at this location could be due to the combination of strong outward shearing and inward squeezing according to the material flow patterns observed in a previous study [43]. For model M1, 5.6 mm/s was the maximum average velocity, followed by 5.4 mm/s; related locations were P14 and P23. This further confirmed sticking/no slippage of the shoulder due to lower temperatures, as discussed earlier. The graph shows that more material was squeezed inwards into the reservoir with the increase in RS, as it could be seen that the velocities of the material adjacent to the shoulder's outer edge were enhanced. The maximum average velocities from models M2 and M3 were seen at 2.5 mm away (material underneath the shoulder) and 3 mm away (material near the outer edge of the shoulder), indicating this phenomenon. The maximum average velocity for model M2 was 9.6 mm/s, and the next was 9 mm/s; related locations P24 and P15. The maximum time-averaged velocity value obtained from model M3 was 13.3 mm/s, and the next was 9.6 mm/s, corresponding to points P25 and P16. The averaged top sheet velocities within the SZ (points P1–P7 and P10–P16) increased by 2.3% and 34.6% from model M1 to model M2 and from model M1 to model M3, respectively. From model M1 to model M2 and model M1 to model M3, the time-averaged velocities from the characteristic points within the SZ in the bottom sheet (points P19–P25 and P28–P34) increased by 15.7% and 51.9%. This shows that the RS significantly affected material flow, and a higher RS improved the flow of materials in the SZ.

For model M4 with 0.25 mm/s PR, the maximum average velocities were seen near the shoulder's outer periphery. For model M4, 7.2 mm/s was the maximum average velocity, followed by 5.9 mm/s; related locations were P25 and P15. It was observed that material velocities rose as a result of a rise in PRs. The maximum average velocities from models M3 and M5 were seen at 2.5 mm away, i.e., at the material underneath the shoulder, and 3 mm away, i.e., at the material near the outer edge of the shoulder, respectively. A gradual rise of averaged velocities in the shoulder-affected region (between 2 and 3 mm from the central axis) was observed in model M4. The maximum average velocity for model M5 was 16 mm/s, and the next was 15.3 mm/s; related locations P7 and P24. The averaged top sheet velocities within the SZ (points P1–P7 and P10–P16) increased by 19.9% and 40.4% from model M4 to model M3 and from model M4 to model M5, respectively. This shows that PR significantly affected material flow velocities, especially at the material around the shoulder. From model M4 to model M3 and model M4 to model M5, respectively, the time-averaged velocities from the characteristic points within the SZ in the bottom sheet (points P19–P25 and P28–P34) reduced by 6.9% and 8.6%, respectively. This indicates that with the increase in PR, the material moment in the top sheet was enhanced, but the material velocities in the bottom sheet were best with a lower PR. This slight reduction in material velocities in the bottom sheets was due to decreased contact time of the rotating tool and the temperature drop. Additionally, it was evident from the material flow velocity results that there was no significant material movement after 3.5 mm from the rotational axis in all the models, which confirmed significant material movement only in the SZ. As discussed earlier, the TMAZ adjacent to the shoulder's outer edge was a narrow zone within 3.5 mm from the outer edge of the shoulder.

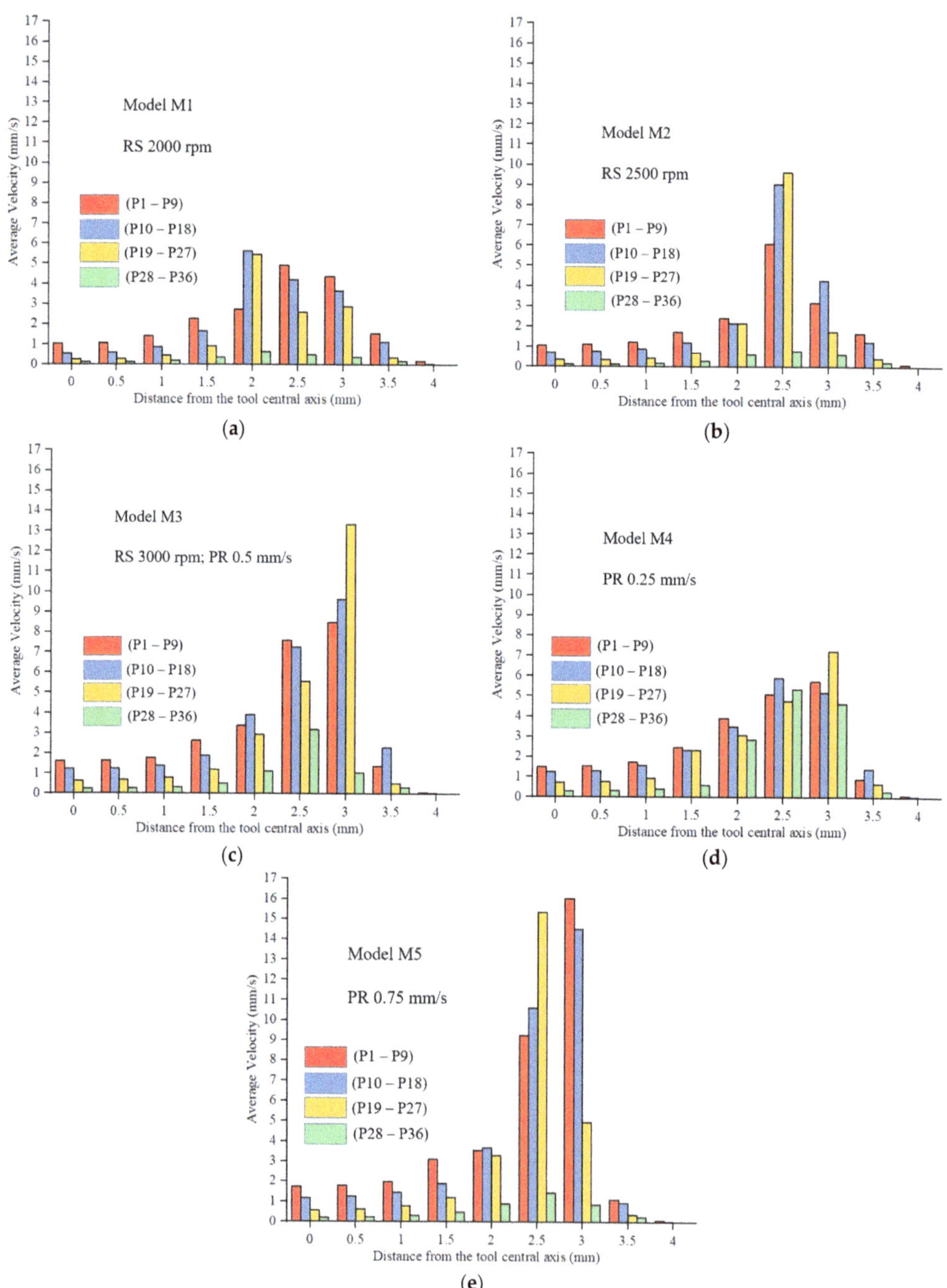

Figure 7. Time-averaged velocities from the characteristic points that are marked in Figure 6, (**a**) model M1; (**b**) model M2; (**c**) model M3; (**d**) model M4; (**e**) model M5.

The lap shear strengths (LSS) reported from the experiments with the same parameters of models M1, M2, M3, M4, and M5 were 2752 N, 2917 N, 3069 N, 2918 N, and 2957 N, respectively [31]. According to the results of the experimental LSS tests, RS had a considerable effect on joint strength after PD, which was the most significant factor. The LSS increased to 5.6% and 10.3% from model M1 to model M2 and from model M1 to model M3, respectively. It was evident from the material flow velocities that the material flow was improved with an increase in RS, hence the increase in LSS [8,20,20,21,23,35]. With the increase in RS, more material was squeezed inwards and involved in stirring and joint formation, particularly the material near the shoulder's outer periphery and underneath the shoulder. Accordingly, increased material movement was also seen in the bottom sheet with increased RS. Additionally, material flow was aided by the material's softening as a result of the increasing temperatures in the SZ with the increase in RS. Increased temperatures allow proper phase change and enhance metallurgical bonding between weldments [33,57]. According to the experimental results, the PR parameter had the least influence on joint strength [31]. The LSS slightly increased from model M4 to M3 by 4.9%, and from model M3 to model M5, it was reduced slightly by 3.6%. The lowest LSS in model M4 among the variant of PR's could be due to excessive temperatures and increased process time. Although the material was softened/plasticized more due to higher temperatures, the weakening of the joint, as discussed earlier, was due to a reduction of hardness in base materials in different zones, especially in the TMAZ, consistent with earlier literature [8,32,44], where minimal hardness was seen at the TMAZ/HAZ interface. In addition, the averaged velocities in the SZ, as discussed earlier, showed the least material movement in model M4. In model M5, material movement was better than in model M4, and the averaged velocities in the SZ were better than in model M3. Still, the contribution to the enhanced material movement in the SZ was due to increased material movement adjacent to the higher-PR shoulder. Overall, the peak temperatures in the SZ, which were close and around 80% of the melting temperature, provided better joint quality (reported in FSW commonly [58]), and 85% and above could be considered excessive temperatures weakening the material. Along with this, material flow was also a key factor for enhancing joint strength. It could be concluded that enhanced material movement and controlled temperatures produced high-quality and high-strength joints.

4. Conclusions

The effect of tool RS and PR during refill FSSW of thin AA7075-T6 sheets was studied using a validated 3D thermomechanical model. The thermal cycles were key factors determining the heat input and material softening, which affected material flow and joint quality. It could be concluded that temperatures around 80% of the T_m were ideal for joining, and excessive temperatures weakened the joint. In addition to the thermal cycles, material flow in the SZ was another major factor contributing to joint strength. With an increase in RS, the temperatures increased, and more material was squeezed inwards into the stirring, indicating enhanced movement in the SZ, especially adjacent to the shoulder's outer periphery. The temperatures were reduced with an increase in PR/decrease in WT. Among the PR variants, a more balanced and enhanced material flow was observed in the SZ in the model M3. The LSS of the joint was correlated with material movement and temperature data from the simulations. The numerical analysis provided a deeper insight into the influence of thermal cycles and material flow on joint strength. Overall, enhanced material movement under controlled temperatures leads to superior quality and strong joints.

Author Contributions: V.S.R.J. and M.A. conceived the idea for this study; and with M.A.'s and S.R.P.'s guidance, V.S.R.J. carried out the modeling and the simulation and prepared an outline for the paper. M.A. and S.R.P. reviewed and revised the paper. All authors have read and agreed to the published version of the manuscript.

Funding: The APC charges were covered by the Graduate Studies funding by Centre of Graduate Studies–Cost Centre 015BD1-001 and Institute of Transport Infrastructure–Cost Center 015NBO-001, Universiti Teknologi PETRONAS.

Institutional Review Board Statement: Not applicable.

Informed Consent Statement: Not applicable.

Data Availability Statement: Not applicable.

Acknowledgments: The authors sincerely appreciate the financial support provided by Universiti Teknologi PETRONAS through the Graduate Assistant Scheme.

Conflicts of Interest: The authors declare no conflict of interest.

References

1. Nandan, R.; DebRoy, T.; Bhadeshia, H.K.D.H. Recent Advances in Friction-Stir Welding—Process, Weldment Structure and Properties. *Prog. Mater. Sci.* **2008**, *53*, 980–1023. [CrossRef]
2. Iwashita, T. Method and Apparatus for Joining. U.S. Patent 6,601,751, 5 August 2003.
3. Witthar, K.; Brown, J.; Burford, D. Swept FSSW in Aluminum Alloys through Sealants and Surface Treatments. In *Friction Stir Welding and Processing VI*; John Wiley & Sons, Ltd.: Hoboken, NJ, USA, 2011; pp. 417–424. ISBN 978-1-118-06230-2.
4. Schilling, C.; dos Santos, J. Method and Device for Joining at Least Two Adjoining Work Pieces by Friction Welding. U.S. Patent US6,722,556, 20 April 2004.
5. Okamoto, K.; Hunt, F.; Hirano, S. *Development of Friction Stir Welding Technique and Machine for Aluminum Sheet Metal Assembly—Friction Stir Welding of Aluminum for Automotive Applications (2)*; No. 2005–01–1254; SAE: Warrendale, PA, USA, 2005.
6. Shen, Z.; Ding, Y.; Gerlich, A.P. Advances in Friction Stir Spot Welding. *Crit. Rev. Solid State Mater. Sci.* **2020**, *45*, 457–534. [CrossRef]
7. Mazzaferro, J.A.E.; Rosendo, T.d.S.; Mazzaferro, C.C.P.; Ramos, F.D.; Tier, M.A.D.; Strohaecker, T.R.; dos Santos, J.F. Preliminary Study on the Mechanical Behavior of Friction Spot Welds. *Soldag. Insp.* **2009**, *14*, 238–247. [CrossRef]
8. Shen, Z.; Chen, Y.; Hou, J.S.C.; Yang, X.; Gerlich, A.P. Influence of Processing Parameters on Microstructure and Mechanical Performance of Refill Friction Stir Spot Welded 7075-T6 Aluminium Alloy. *Sci. Technol. Weld. Join.* **2015**, *20*, 48–57. [CrossRef]
9. Luty, G.; Andres, J.; Wrońska, A.; Burek, R.; Gałaszczyński, T. Effect of process parameters on microstructure and mechanical properties of RFSSW lap joints of thin AL7075-T6 sheets. *Arch. Metall. Mater.* **2018**, *63*, 39–43.
10. Kluz, R.; Kubit, A.; Trzepiecinski, T.; Faes, K. Polyoptimisation of the Refill Friction Stir Spot Welding Parameters Applied in Joining 7075-T6 Alclad Aluminium Alloy Sheets Used in Aircraft Components. *Int. J. Adv. Manuf. Technol.* **2019**, *103*, 3443–3457. [CrossRef]
11. Kluz, R.; Kubit, A.; Trzepiecinski, T.; Faes, K.; Bochnowski, W. A Weighting Grade-Based Optimization Method for Determining Refill Friction Stir Spot Welding Process Parameters. *J. Mater. Eng. Perform.* **2019**, *28*, 6471–6482. [CrossRef]
12. Chai, P.; Wang, Y. Effect of Rotational Speed on Microstructure and Mechanical Properties of 2060 Aluminum Alloy RFSSW Joint. *Met. Mater. Int.* **2019**, *25*, 1574–1585. [CrossRef]
13. Silva, B.H.; Zepon, G.; Bolfarini, C.; dos Santos, J.F. Refill Friction Stir Spot Welding of AA6082-T6 Alloy: Hook Defect Formation and Its Influence on the Mechanical Properties and Fracture Behavior. *Mater. Sci. Eng. A* **2020**, *773*, 138724. [CrossRef]
14. Wang, Y.; Chai, P. Effects of Welding Parameters on Micro-Junction Structure and Fracture Behavior of Refill Friction Stir Spot Welded Joints for 2060 Aluminum Alloys. *Weld. World* **2020**, *64*, 2033–2051. [CrossRef]
15. Kubit, A.; Trzepieciński, T.; Gadalińska, E.; Slota, J.; Bochnowski, W. Investigation into the Effect of RFSSW Parameters on Tensile Shear Fracture Load of 7075-T6 Alclad Aluminium Alloy Joints. *Materials* **2021**, *14*, 3397. [CrossRef] [PubMed]
16. Effertz, P.S.; de Carvalho, W.S.; Guimarães, R.P.M.; Saria, G.; Amancio-Filho, S.T. Optimization of Refill Friction Stir Spot Welded AA2024-T3 Using Machine Learning. *Front. Mater.* **2022**, *9*, 864187. [CrossRef]
17. Zou, Y.; Li, W.; Yang, X.; Su, Y.; Chu, Q.; Shen, Z. Microstructure and Mechanical Properties of Refill Friction Stir Spot Welded Joints: Effects of Tool Size and Welding Parameters. *J. Mater. Res. Technol.* **2022**, *21*, 5066–5080. [CrossRef]
18. Zhou, L.; Luo, L.Y.; Wang, R.; Zhang, J.B.; Huang, Y.X.; Song, X.G. Process Parameter Optimization in Refill Friction Spot Welding of 6061 Aluminum Alloys Using Response Surface Methodology. *J. Mater. Eng. Perform.* **2018**, *27*, 4050–4058. [CrossRef]
19. Kubit, A.; Bucior, M.; Wydrzyński, D.; Trzepieciński, T.; Pytel, M. Failure Mechanisms of Refill Friction Stir Spot Welded 7075-T6 Aluminium Alloy Single-Lap Joints. *Int. J. Adv. Manuf. Technol.* **2018**, *94*, 4479–4491. [CrossRef]
20. Zhou, L.; Luo, L.Y.; Zhang, T.P.; He, W.X.; Huang, Y.X.; Feng, J.C. Effect of Rotation Speed on Microstructure and Mechanical Properties of Refill Friction Stir Spot Welded 6061-T6 Aluminum Alloy. *Int. J. Adv. Manuf. Technol.* **2017**, *92*, 3425–3433. [CrossRef]
21. Ji, S.; Wang, Y.; Zhang, J.; Li, Z. Influence of Rotating Speed on Microstructure and Peel Strength of Friction Spot Welded 2024-T4 Aluminum Alloy. *Int. J. Adv. Manuf. Technol.* **2017**, *90*, 717–723. [CrossRef]
22. de Castro, C.C.; Plaine, A.H.; de Alcântara, N.G.; dos Santos, J.F. Taguchi Approach for the Optimization of Refill Friction Stir Spot Welding Parameters for AA2198-T8 Aluminum Alloy. *Int. J. Adv. Manuf. Technol.* **2018**, *99*, 1927–1936. [CrossRef]
23. Li, Z.; Gao, S.; Ji, S.; Yue, Y.; Chai, P. Effect of Rotational Speed on Microstructure and Mechanical Properties of Refill Friction Stir Spot Welded 2024 Al Alloy. *J. Mater. Eng. Perform.* **2016**, *25*, 1673–1682. [CrossRef]
24. Tier, M.D.; Rosendo, T.S.; dos Santos, J.F.; Huber, N.; Mazzaferro, J.A.; Mazzaferro, C.P.; Strohaecker, T.R. The Influence of Refill FSSW Parameters on the Microstructure and Shear Strength of 5042 Aluminium Welds. *J. Mater. Process. Technol.* **2013**, *213*, 997–1005. [CrossRef]

25. de Castro, C.C.; Plaine, A.H.; Dias, G.P.; de Alcântara, N.G.; dos Santos, J.F. Investigation of Geometrical Features on Mechanical Properties of AA2198 Refill Friction Stir Spot Welds. *J. Manuf. Process.* **2018**, *36*, 330–339. [CrossRef]

26. Kwee, I.; de Waele, W.; Faes, K. Weldability of High-Strength Aluminium Alloy EN AW-7475-T761 Sheets for Aerospace Applications, Using Refill Friction Stir Spot Welding. *Weld. World* **2019**, *63*, 1001–1011. [CrossRef]

27. Yang, H.G.; Yang, H.J. Experimental Investigation on Refill Friction Stir Spot Welding Process of Aluminum Alloys. *Appl. Mech. Mater.* **2013**, *345*, 243–246. [CrossRef]

28. Kubit, A.; Kluz, R.; Trzepieciński, T.; Wydrzyński, D.; Bochnowski, W. Analysis of the Mechanical Properties and of Micrographs of Refill Friction Stir Spot Welded 7075-T6 Aluminium Sheets. *Arch. Civ. Mech. Eng.* **2018**, *18*, 235–244. [CrossRef]

29. Kar, A.; Yadav, D.; Suwas, S.; Kailas, S.V. Role of Plastic Deformation Mechanisms during the Microstructural Evolution and Intermetallics Formation in Dissimilar Friction Stir Weld. *Mater. Charact.* **2020**, *164*, 110371. [CrossRef]

30. Kar, A.; Kailas, S.V.; Suwas, S. Effect of Mechanical Mixing in Dissimilar Friction Stir Welding of Aluminum to Titanium with Zinc Interlayer. *Trans. Indian Inst. Met.* **2019**, *72*, 1533–1536. [CrossRef]

31. Yamin, M.F. Mechanical and Microstructure Properties Evaluation of Similar Thin AA7075-T6 Welds by Refill Friction Stir Spot Welding. Master's Thesis, Universiti Teknologi Petronas, Seri Iskandar, Malaysia, 2021.

32. Shen, Z.; Yang, X.; Zhang, Z.; Cui, L.; Li, T. Microstructure and Failure Mechanisms of Refill Friction Stir Spot Welded 7075-T6 Aluminum Alloy Joints. *Mater. Des.* **2013**, *44*, 476–486. [CrossRef]

33. Kluz, R.; Kubit, A.; Wydrzyński, D. Analysis of Structure and Shear/Peel Strength of Refill Friction Stir Spot Welded 7075-T6 Aluminium Alloy Joints. *Adv. Sci. Technol. Res. J.* **2017**, *11*, 297–303. [CrossRef]

34. Muci-Küchler, K.H.; Kalagara, S.; Arbegast, W.J. Simulation of a Refill Friction Stir Spot Welding Process Using a Fully Coupled Thermo-Mechanical FEM Model. *J. Manuf. Sci. Eng.* **2010**, *132*, 14503. [CrossRef]

35. Ji, S.; Li, Z.; Wang, Y.; Ma, L.; Zhang, L. Material Flow Behavior of Refill Friction Stir Spot Welded LY12 Aluminum Alloy. *High Temp. Mater. Process.* **2017**, *36*, 495–504. [CrossRef]

36. Ji, S.; Wang, Y.; Li, Z.; Yue, Y.; Chai, P. Effect of Tool Geometry on Material Flow Behavior of Refill Friction Stir Spot Welding. *Trans. Indian Inst. Met.* **2017**, *70*, 1417–1430. [CrossRef]

37. Malik, V.; Sanjeev, N.K.; Hebbar, H.S.; Kailas, S.V. Finite Element Simulation of Exit Hole Filling for Friction Stir Spot Welding—A Modified Technique to Apply Practically. *Procedia Eng.* **2014**, *97*, 1265–1273. [CrossRef]

38. Kubit, A.; Trzepiecinski, T. A Fully Coupled Thermo-Mechanical Numerical Modelling of the Refill Friction Stir Spot Welding Process in Alclad 7075-T6 Aluminium Alloy Sheets. *Arch. Civ. Mech. Eng.* **2020**, *20*, 117. [CrossRef]

39. Zhang, H.F.; Zhou, L.; Li, G.H.; Tang, Y.T.; Li, W.L.; Wang, R. Prediction and Validation of Temperature Distribution and Material Flow during Refill Friction Stir Spot Welding of AZ91D Magnesium Alloy. *Sci. Technol. Weld. Join.* **2021**, *26*, 153–160. [CrossRef]

40. Pashazadeh, H.; Masoumi, A.; Teimournezhad, J. A Study on Material Flow Pattern in Friction Stir Welding Using Finite Element Method. *Proc. Inst. Mech. Eng. Part B J. Eng. Manuf.* **2013**, *227*, 1453–1466. [CrossRef]

41. D'Urso, G.; Longo, M.; Giardini, C. Friction Stir Spot Welding (FSSW) of Aluminum Sheets: Experimental and Simulative Analysis. *Key Eng. Mater.* **2013**, *549*, 477–483. [CrossRef]

42. Asadi, P.; Mahdavinejad, R.A.; Tutunchilar, S. Simulation and Experimental Investigation of FSP of AZ91 Magnesium Alloy. *Mater. Sci. Eng. A* **2011**, *21*, 6469–6477. [CrossRef]

43. Janga, V.S.R.; Awang, M. Influence of Plunge Depth on Temperatures and Material Flow Behavior in Refill Friction Stir Spot Welding of Thin AA7075-T6 Sheets: A Numerical Study. *Metals* **2022**, *12*, 927. [CrossRef]

44. Janga, V.S.R.; Awang, M.; Yamin, M.F.; Suhuddin, U.F.H.; Klusemann, B.; dos Santos, J.F. Experimental and Numerical Analysis of Refill Friction Stir Spot Welding of Thin AA7075-T6 Sheets. *Materials* **2021**, *14*, 7485. [CrossRef]

45. Xiong, J.; Peng, X.; Shi, J.; Wang, Y.; Sun, J.; Liu, X.; Li, J. Numerical Simulation of Thermal Cycle and Void Closing during Friction Stir Spot Welding of AA-2524 at Different Rotational Speeds. *Mater. Charact.* **2021**, *174*, 110984. [CrossRef]

46. Raza, S.H.; Mittnacht, T.; Diyoke, G.; Schneider, D.; Nestler, B.; Klusemann, B. Modeling of Temperature- and Strain-Driven Intermetallic Compound Evolution in an Al–Mg System via a Multiphase-Field Approach with Application to Refill Friction Stir Spot Welding. *J. Mech. Phys. Solids* **2022**, *169*, 105059. [CrossRef]

47. Mandal, S.; Rice, J.; Elmustafa, A.A. Experimental and Numerical Investigation of the Plunge Stage in Friction Stir Welding. *J. Mater. Process. Technol.* **2008**, *203*, 411–419. [CrossRef]

48. Fang, N. A New Quantitative Sensitivity Analysis of the Flow Stress of 18 Engineering Materials in Machining. *J. Eng. Mater. Technol.* **2005**, *127*, 192–196. [CrossRef]

49. Akhunova, A.K.; Imayev, M.F.; Valeeva, A.K. Influence of the Pin Shape of the Tool during Friction Stir Welding on the Process Output Parameters. *Lett. Mater.* **2019**, *9*, 456–459. [CrossRef]

50. Awang, M. *Simulation of Friction Stir Spot Welding (FSSW) Process: Study of Friction Phenomena*; West Virginia University Libraries: Morgantown, WV, USA, 2007.

51. Gerlich, A.; Avramovic-Cingara, G.; North, T.H. Stir Zone Microstructure and Strain Rate during Al 7075-T6 Friction Stir Spot Welding. *Metall. Mater. Trans. A* **2006**, *37*, 2773–2786. [CrossRef]

52. Starink, M.J. Effect of Compositional Variations on Characteristics of Coarse Intermetallic Particles in Overaged 7000 Aluminium Alloys. *Mater. Sci. Technol.* **2001**, *17*, 1324–1328. [CrossRef]

53. Shen, Z.; Ding, Y.; Chen, J.; Fu, L.; Liu, X.C.; Chen, H.; Guo, W.; Gerlich, A.P. Microstructure, Static and Fatigue Properties of Refill Friction Stir Spot Welded 7075-T6 Aluminium Alloy Using a Modified Tool. *Sci. Technol. Weld. Join.* **2019**, *24*, 587–600. [CrossRef]

54. Gerlich, A.; Yamamoto, M.; North, T.H. Local Melting and Cracking in Al 7075-T6 and Al 2024-T3 Friction Stir Spot Welds. *Sci. Technol. Weld. Join.* **2007**, *12*, 472–480. [CrossRef]

55. Kalinenko, A.; Kim, K.; Vysotskiy, I.; Zuiko, I.; Malopheyev, S.; Mironov, S.; Kaibyshev, R. Microstructure-Strength Relationship in Friction-Stir Welded 6061-T6 Aluminum Alloy. *Mater. Sci. Eng. A* **2020**, *793*, 139858. [CrossRef]

56. Woo, W.; Choo, H.; Withers, P.J.; Feng, Z. Prediction of Hardness Minimum Locations during Natural Aging in an Aluminum Alloy 6061-T6 Friction Stir Weld. *J. Mater. Sci.* **2009**, *44*, 6302–6309. [CrossRef]

57. Deng, H.; Chen, Y.; Jia, Y.; Pang, Y.; Zhang, T.; Wang, S.; Yin, L. Microstructure and Mechanical Properties of Dissimilar NiTi/Ti6Al4V Joints via Back-Heating Assisted Friction Stir Welding. *J. Manuf. Process.* **2021**, *64*, 379–391. [CrossRef]

58. Tang, W.; Guo, X.; McClure, J.C.; Murr, L.E.; Nunes, A. Heat Input and Temperature Distribution in Friction Stir Welding. *J. Mater. Process. Manuf. Sci.* **1998**, *7*, 163–172. [CrossRef]

Article

Modeling of Probeless Friction Stir Spot Welding of AA2024/AISI304 Steel Lap Joint

Mariia Rashkovets , **Nicola Contuzzi** * and **Giuseppe Casalino**

Dipartimento di Meccanica, Matematica e Management, Polytechnic University of Bari, Via Orabona 4, 70125 Bari, Italy
* Correspondence: nicola.contuzzi@poliba.it

Abstract: In the present study, AA2024 aluminum alloy and AISI304 stainless steel were welded in a lap joint configuration by Probeless Friction Stir Spot Welding (P-FSSW) with a flat surface tool. A full factorial DOE plan was performed. The effect of the tool force (4900, 7350 N) and rotational speed (500, 1000, 1500, 2000 RPM) was analyzed regarding the microstructure and microhardness study. A two-dimensional arbitrary Eulerian–Lagrangian FEM model was used to clarify the temperature distribution and material flow within the welds. The experimental results for the weld microstructures were used to validate the temperature field of the numerical model. The results showed that the tool rotation speed had an extensive influence on the heat generation, whereas the load force mainly acted on the material flow.

Keywords: probeless friction stir spot welding; dissimilar weld; numerical analysis; material flow; microstructure

Citation: Rashkovets, M.; Contuzzi, N.; Casalino, G. Modeling of Probeless Friction Stir Spot Welding of AA2024/AISI304 Steel Lap Joint. *Materials* **2022**, *15*, 8205. https://doi.org/10.3390/ma15228205

Academic Editor: Raul D.S.G. Campilho

Received: 21 October 2022
Accepted: 11 November 2022
Published: 18 November 2022

Publisher's Note: MDPI stays neutral with regard to jurisdictional claims in published maps and institutional affiliations.

1. Introduction

The design of efficient and lightweight products encourages engineers to join dissimilar structures. One of the most promising material couples used in automotive industries is the aluminum–steel joint due to its specific strength and high corrosion resistance [1]. About 90% of the assembly in automotive or aerospace products is performed by spot welding, where Resistance Spot Welding (RSW) is actively used [2]. RSW typically involves compressed metal plates between two electrodes to produce an electrical impulse that heats and melts plates at the contact point, with subsequent crystallization. However, melting forms a layer of brittle intermetallic phases between aluminum and steel [3] or even unexpected phases between the same grade of materials [4,5]. Different modifications of RSW can reduce unwanted phases [6,7]; however, the welding process becomes more complicated and the mismatch in the aluminum–steel bonding remains [8,9].

Friction Stir Spot Welding (FSSW) is a solid-state process that allows for the joining of different materials by a combination of friction and plastic deformation, avoiding the problems that result from their melting and solidification. Spot welding is created by plunging a non-consumable rotating tool consisting of a cylindrical shoulder and a probe into the workpiece with further material stirring. Therefore, the total amount of heat input generated during the friction must be sufficient to soften, stir, and plastically deform materials. Although there are many reports on optimizing FSSW (e.g., [10,11]), the most pronounced keyhole and 'hooking' defects owing to the probe remain [12]. Moreover, the wear or failure of a probe made of expensive material (e.g., WC-Co or CBN) can cause defects such as a lack of penetration [13]. All probe-related defects are crucial for the welds' performance, which also affects the cost of the whole process. For these reasons, several different processes have been proposed.

Recently, the refill FSSW process was applied to eliminate keyholes due to the separate movement of the tool [14]. However, the process required a very complex machine, as well as high control and the regular cleaning of a tool-holder system. Along with these, defects such as voids, hooks, annular grooves, and insufficient refill are still observed [15]. D. Bakavos and P.B. Prangnell [16] produced results that went against common wisdom that the probe must penetrate at least 25% into the bottom plate to make an acceptable weld. Later, Y. Tozaki et.al [17] and other authors [18–21] demonstrated the possibility of producing quality welds using a probeless tool. The ability of the simple probeless FSSW (P-FSSW) process to create a full metallurgical bond, eliminate keyholes, and improve tool life was shown.

The movement of the material within the lap welds during friction processes is more important than the microstructure due to the interface between the plates [22]. Nowadays, complex numerical models estimate the material flow in FSSW and P-FSSW [23,24]. The results of various studies have shown variations in the failure modes attributed to the different material flow between the plates. Currently, authors suggest using a probeless tool with a grooved shoulder since it ensures a deeper material flow to the bottom plate and yields higher failure energies [25]. However, grooves filled with softened aluminum create the need for frequent tool cleaning and the excessive material flow can lead to microcracks and voids.

Most of the mentioned studies were based on the welding of light materials such as Al, Li, and Mg by a probeless tool with a grooved shoulder, whereas fewer studies focused on deep aluminum–steel examination. Based on previous results, the P-FSSW process for an Aluminum–steel couple using a probeless tool with a flat shoulder was the focus of this study. The main goal was to clarify the feasibility of a completely flat tool to weld a dissimilar AA2024-AISI304 lap configuration using different process parameters for P-FSSW, as well as the quality of welds. A two-dimensional arbitrary Eulerian–Lagrangian (ALE) FEM model was validated with experimental data to analyze the material flow and temperature distribution. The scope of the experimental and numerical study was the observation of four levels of rotation speed and two levels of tool vertical force.

2. Materials and Methods

2.1. Materials

Welds were made using commercial rolled aluminum plates of AA2024 (0.8 mm) and AISI304 stainless steel (6 mm) in a lap configuration. The chemical compositions of the used materials are presented in Table 1. The initial microstructure of the commercial rolled AA2024-T3 plate had equiaxed grains with an α-solid solution of Cu-Al (the bright contrast) and secondary phases (the dark contrast), whereas the rolled AISI304 plate was characterized by a typical microstructure with elongated grains in a rolling direction (Figure 1). The thermal properties of the used materials are listed in Table 2.

Table 1. Chemical compositions of used materials.

Materials	Elements [wt.%]										
	Fe	Ni	Al	Cu	Mg	Mn	Si	Cr	Zn	Ti	C
AA2024-T3	$\leq$0.5	-	Bal.	3.8–4.9	1.2–1.8	0.3–0.9	$\leq$0.5	$\leq$0.1	$\leq$0.25	$\leq$0.15	
AISI 304 *	Bal.	8–10.5	-	-	2	-	1	17.5–19.5	-	-	0.07

$\leq$0.05% P and $\leq$0.03% S *.

Table 2. Thermal properties of used materials.

Materials	Melting Point [°C]	Thermal Conductivity [W/m K]	Specific Heat Capacity [J/g °C]
AA2024-T3	502–638	121	0.875
AISI 304	1400–1455	16.2	0.5

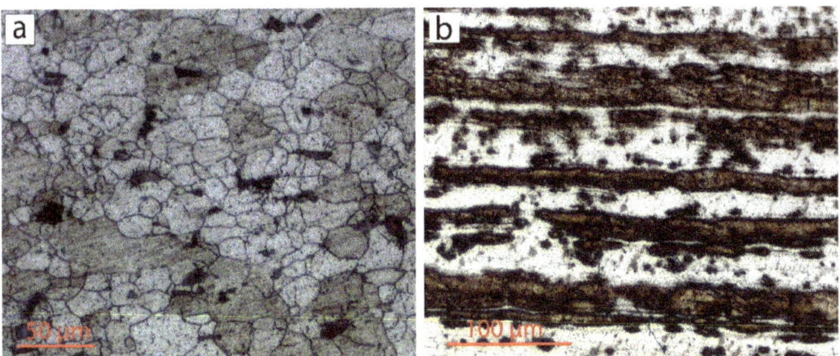

Figure 1. Initial microstructure of AA2024-T3 plate (**a**) and AISI304 plate (**b**).

2.2. Methods

Since acceptable friction welding cannot occur over a wide range of plunge depths, the P-FSSW process was performed in a force control mode: (1) as the rotating tool at the set rotation speed began to contact the top surface of the upper plate, the force control mode was activated and the control system monitored the loading force of the tool until the set force parameter was reached; (2) once the force load reached the set value, the tool started to experience micro-up-to-down movements to keep the set parameter constant throughout the P-FSSW process. The lap configuration is presented in Figure 2. The probeless tool was made of H13 steel with a 30 mm shoulder diameter and a tilt angle of 0 degrees. The parameters of the P-FSSW process were chosen based on the preliminary experiments (Table 3). The dwell time was set at 30 s for each specimen. The rotation speed of 500 RPM coupled with the 7350 N and 4900 N load forces did not generate a sufficient heat input to soften the workpiece material and welding was not achieved. Therefore, those samples are overlooked in the following discussion.

Samples for optical microscopy were cut perpendicular to the weld spot by an electrical discharge cutting machine. The aluminum microstructure was etched with Keller's reagent (1.5 mL HCl, 2.5 mL HNO_3, 1 mL HF, and 95 mL distilled water), whereas the mixture of 1 mL HNO_3 and 3 mL HCl was applied to etch the surface of the stainless steel.

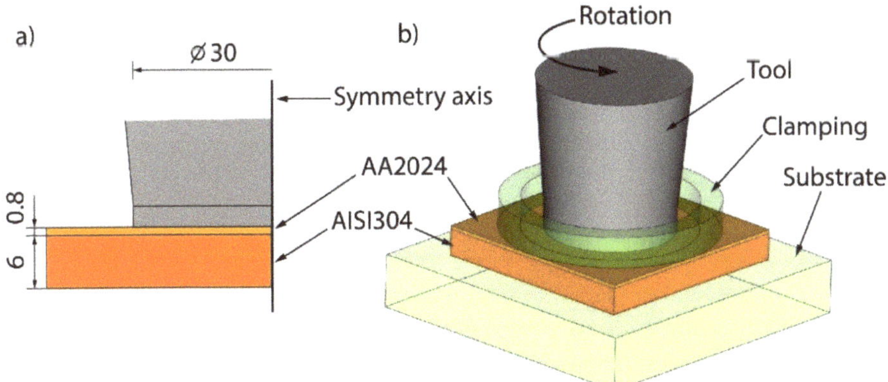

Figure 2. P-FSSW lap configuration: cross-sectional dimensions (**a**), CAD model (**b**).

Table 3. Process parameters.

Test	Rotation Speed [RPM]	Tool Force [N]
AS-01	2000	
AS-02	1500	
AS-03	1000	7350
AS-04	500	
AS-05	2000	
AS-06	1500	
AS-07	1000	4900
AS-08	500	

2.3. Numerical Model

For the prediction of the thermal field and material flow in the P-FSSW process for AA2024/AISI, 304 welds were performed using the finite element software Simufact Forming® 2021. Based on the previous studies [24,26], the metal flow was purely axisymmetric; therefore, quad 2D elements were used to mesh all the parts of the P-FSSW CAD model (Figure 2b). In the model, the rotating tool, clamping system, and substrate were defined as rigid bodies, whereas the plates to be welded were deformable bodies. The initial element sizes were chosen as 0.07 mm for the AA2024 plate (upper plate) and 0.2 mm for the AISI304 plate. The initial element sizes of the rigid bodies varied from 0.2 mm (tool, clamping system) to 0.7 mm (substrate). The friction coefficient applied between the workpiece and the tool governed by Coulomb's friction law with a simplified contact condition of sliding friction was 0.1.

The change in the contact pressure (P) and tool rotation speed (RS) over time consisted of the following stages:

1. Linear increase in the RS from zero (RS_0) to the rated value of RS_{set} for separate specimens ($P_0 = 0$).
2. Tool downward movement until it touched the plate top surface (RS_{set}; P_0).
3. Fast compression of the aluminum plate with an instant increase in the pressure up to the rated value of P_{set}. Frictional heat was generated, which was proportional to the specific frictional power.
4. Formation of welding seam for the DT (RS_{set}; P_{set}).
5. Tool upward movement, with a pressure decreasing to P_0 (RS_{set}).
6. Linear decrease in the RS from RS_{set} to RS_0 for separate specimens.

Stages 2–5 were implemented in the presented FEM. An arbitrary Eulerian–Lagrangian finite element method was used for the plates (the mesh moved independently from the material in a way that spanned the material at any point in time), whereas the tool, clamping system, and substrate were calculated as rigid Lagrangian bodies.

The material flow was simulated as a bi-dimensional fluid flow through an enclosed volume in accordance with the following equations of equilibrium:

$$\frac{D(\rho v_i)}{Dt} = \rho b_i + \frac{\partial \sigma_{ij}}{\partial x_j} \tag{1}$$

where $\frac{D}{Dt}$ is the material time derivative of a quantity, v_i is the velocity of the particle flowing through the mesh, σ_{ij} is the Truesdell rate of Cauchy stress, ρ is the material density, and b_i is the body force. For an incompressible fluid, Equation (1), along with the continuity equation (mass conservation), yields:

$$\rho \frac{\partial v_i}{\partial t} + \rho v_j \frac{\partial v_i}{\partial x_j} = \rho b_i + \frac{\partial \sigma_{ij}}{\partial x_j} \tag{2}$$

The left-hand side of the continuity equation, Equation (2), represents the local rate of change augmented by the convection effects.

Experimental data such as torque or efficient energy cannot be directly used for the validation of the results according to simplify assumptions of a built-in 2D model. However, a 2D model can predict many trends corresponding to the experimental data such as the material flow. Arrow plots were used to visualize the material flow.

3. Results and Discussion

3.1. Microstructure

The upper-surface view and corresponding cross-section for each sample are presented in Figure 3. Several conditions were observed among the upper surfaces and corresponding cross-sections: thinning with a small flash volume (AS-01, AS-05), partial distortion with a medium flash volume (AS-02 and AS-06), and a complete metallurgical bond with the highest (AS-03) and least (AS-07) volumes of flash.

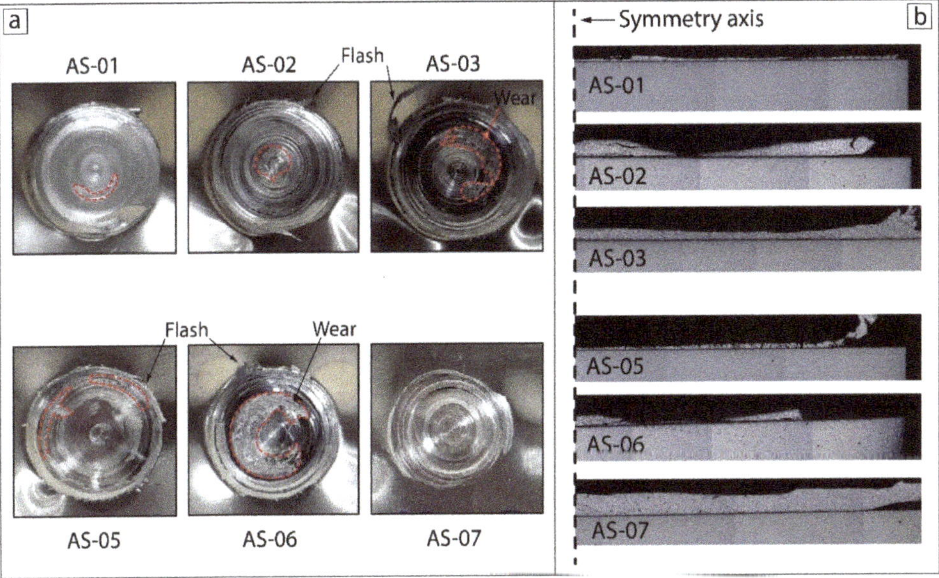

Figure 3. The upper-surface appearances (**a**) and cross-sections (**b**) of P-FSSW samples.

The degree of material softening was affected by the duration of the process at a certain temperature. The higher temperature generated at 1500 RPM coupled with both load forces of 7350 N and 4900 N was responsible for an excessive heat input with a great softening and distortion of the upper AA2024 plate, which also matched the equivalent stress result owing to fast material heating. A combination of the highest rotation speed (2000 RPM) and both cases of tool force produced the thinning of the upper plate in samples AS-01 and AS-05. Detailed images show that there were no refined grains (Figure 4). The samples made at the lowest rotation speed (1000 RPM) had a good metallurgical bonding due to the reduction in the heat input per unit length and lower dissipation of heat over a wider region of the workpieces, which made the heating slow and appropriate. The difference in the tool axial force was responsible for a large gap in the flash volume of samples AS-03 and AS-07. The higher amount of flash in sample AS-03 was due to stronger tool penetration into the workpiece, which was also demonstrated by the numerical results.

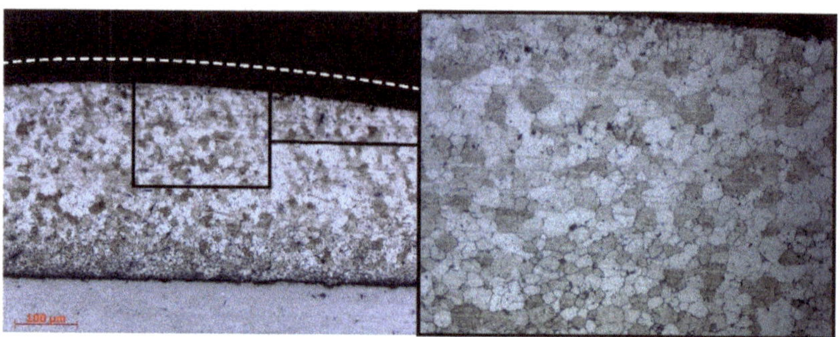

Figure 4. Microstructure of sample AS-01 under a 7350 N load force and at 2000 RPM.

Figure 5 show detailed cross-sections of P-FSSP welds at a rotation speed of 1000 RPM under load forces of 7350 N (AS-03) and 4900 N (AS-07), respectively. The effect of severe plastic deformation was seen in the upper part of the AA2024 plate, which experienced elevated temperatures and rapid recrystallization (Figure 5a).

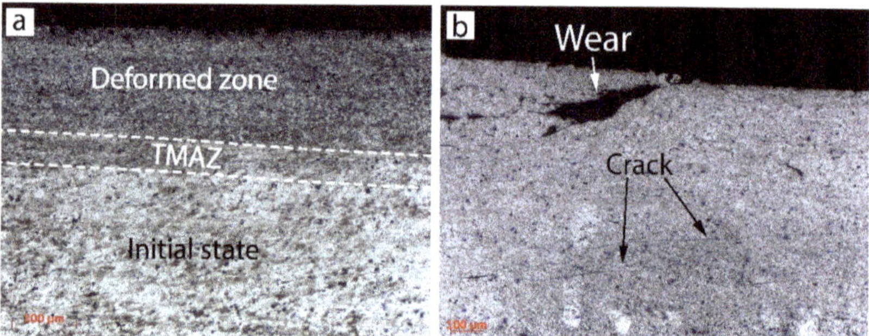

Figure 5. Microstructure in cross-section of sample AS-03: different areas of upper plate (**a**) and defects (**b**).

The middle part of the AA2024 plate had a very narrow zone with non-recrystallized and elongated grains (TMAZ) caused by the material flow and shear stress. The shear stress took place only in the upper and middle parts of the aluminum plate, whereas the material close to the interface between the AA2024 and AISI304 plates did not experience any plastic flow. This zone was influenced only by the heat. Furthermore, a slight thermal cycle did not permit the recrystallization of the aluminum. This was probably due to the lower conductivity of the steel and the presence of the Thermal Contact Conductance (TCC) at the interface, which represented the fraction of heat that was channeled from the aluminum and steel plates. The TCC did not allow for a complete heat exchange between the two plates and a fraction of the heat persisted in the AA2024 plate, decreasing the cooling rate [27].

The AISI304 side showed the rolled grain structure of the initial state (see Figure 1b) without microstructure changes.

The higher average temperature was accompanied by a longer heating time, which promoted a greater volume of material to be heated and a wider zone with dynamic recrystallization. The corresponding dynamic recrystallization zones were 125 µm (sample AS-03 with $T_{av.}$ = 400 °C) and 155 µm (sample AS-07 with $T_{av.}$ = 430 °C). The resistance of the material flow with a greater volume of refined grains decreased significantly [28], which explains the differences in the material flow patterns of the numerical and experimental results.

3.2. Material Flow

The material underneath the shoulder in both cases was pushed downward with further outward extrusions (Figure 6c,d), which correlates with [26]. Figure 6 represents the stages of the simulated material flow for samples AS-03 and AS-07, namely the start (5–10 s), middle (15 s), and final (30 s) stages and 10 sec after the FSSP process was completed. The main difference in the material flow between the two cases was observed at 10 sec (Figure 6b) when the higher force in sample AS-03 led to the obvious shift of hotter material outside the tool and created flash (Figure 7e,f), whereas the lower tool force in sample AS-07 contributed only to the material flow (Figure 7a–d). The subsequent increase in middle stage temperature improved the material flow in sample AS-03 (Figure 6c) and some similarityi in the material flow remained until the final stage of the P-FSSW process. After the tool was retracted, sample AS-07 had a more uniform return with a high concentration at the center compared to sample AS-03 (Figure 6e).

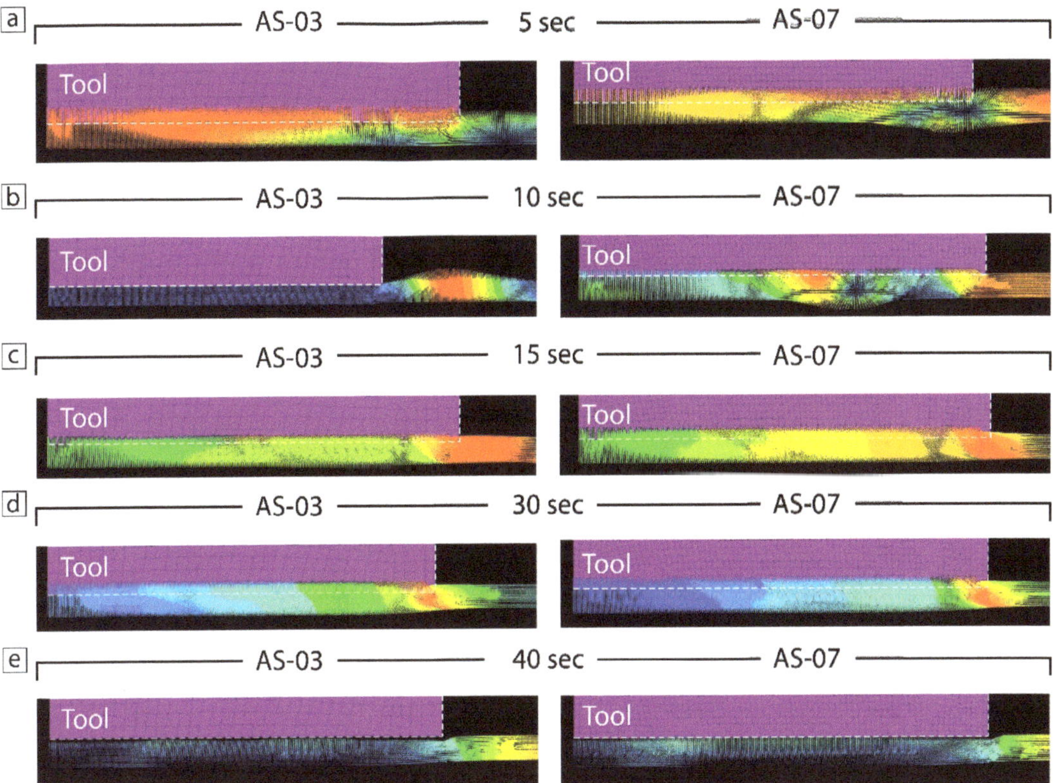

Figure 6. Material flow in samples AS-03 and AS-07 at the different time stages (**a–e**).

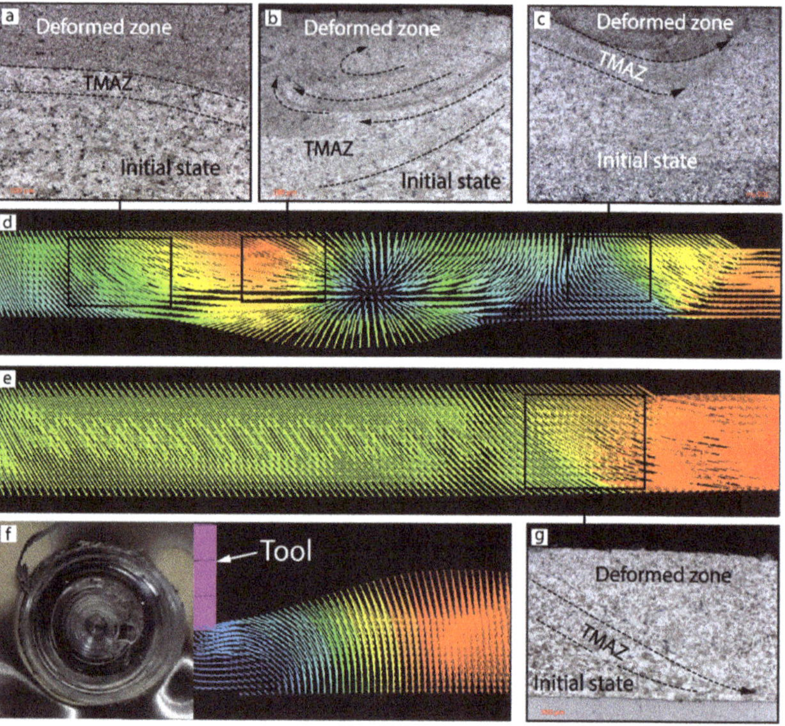

Figure 7. Microstructure (**a–c**) and material flow (**d**) of AA2024 plate in sample AS-07 at 1000 RPM and a 4900 N load force; material flow (**e,f**) and microstructure (**g**) of AA2024 plate in sample AS-03 at 1000 RPM and a 7350 N load force. The blue color represent low material flow, while the red one high material flow.

3.3. FEM Analysis: Temperature and Stress

The graphs of the average temperature and equivalent stress along the tool shoulder–workpiece interface during the P-FSSW process for different process parameters are shown in Figure 8. The average temperature for all the samples in the tool shoulder–workpiece interface was about 400 °C, whereas an evident difference in temperature was seen in the samples made using 7350 N of tool vertical force. The equivalent stress decreased at a low rotation speed due to the dissipation of heat over a wider area of the workpiece under slower material heating.

Figure 9 shows the numerical results on the influence of the welding parameters regarding the temperature field in the cross-sections of weld spots. The temperature had a uniform distribution with the maximum value found along the tool shoulder–workpiece interface. The maximum temperature of a friction solid process can range from 80% to 90% of the melting point as measured by Tang W. et al. [29] and the average temperature at the shoulder–workpiece interface can reach 90–95% of the melting temperature [30]. The maximum {peak} temperature of 630–640 °C at the AA2024-AISI304 interface was generated at 2000 RPM (samples AS-01 and AS-05), whereas the minimum temperature of about 610 °C was observed at 1000 RPM (samples AS-03 and AS-07). Both peak temperatures were greater than the melting point of the AA2024 alloy (602 °C) due to the great differences in the materials' thermal properties (see Table 2) and the thermal contact resistance between the two plates (Figure 9b).

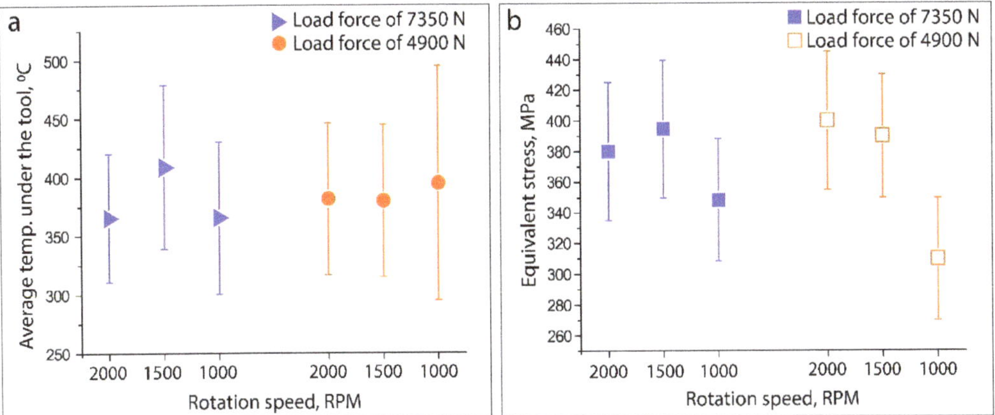

Figure 8. Numerical results of average temperature (**a**) and equivalent stress (**b**) along the tool shoulder–workpiece interface as a function of load force and rotation speed combination.

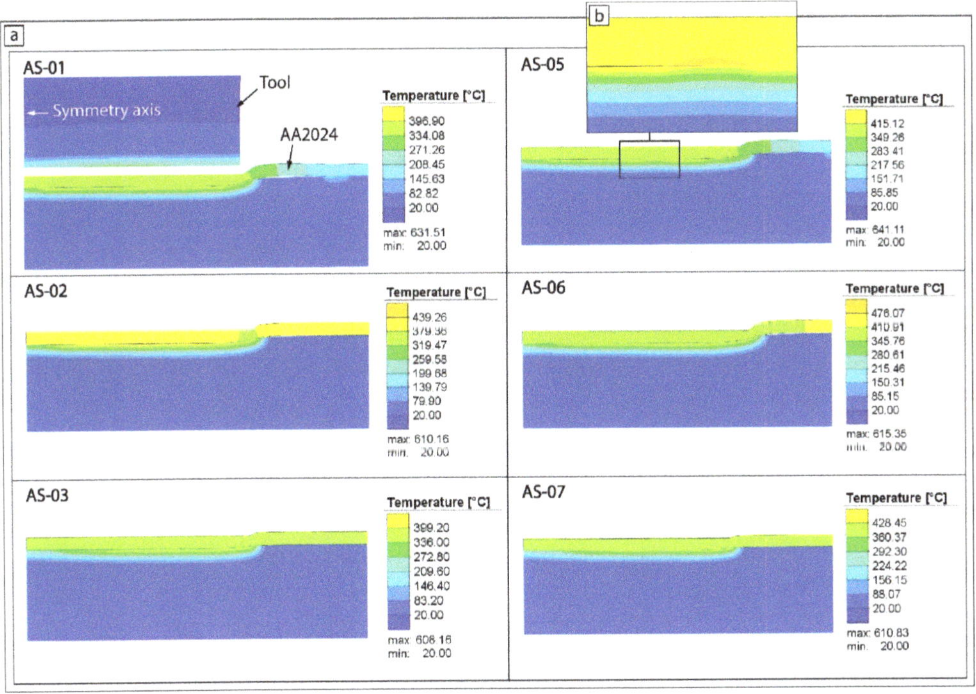

Figure 9. The average temperature distribution profiles (**a**) and peak temperatures on the interface (**b**).

3.4. Microhardness

All the samples showed similar results for the average microhardness profiles (Figure 10). Figure 11 presents the microhardness trends in the cross-sections of each P-FSSW weld. The upper area of the aluminum plate was characterized by a higher microhardness with respect to the base metal due to the grain refinement and a reprecipitation process that might have taken place during the cooling of the heat treatable AA2024 alloy, whereas the TMAZ and HAZ were low due to the coarsening or dissolution of the hardening precipitates. The main differences in the microhardness distributions of samples AS-03 and AS-07 were consistent with the material high plastic deformation flow, and the most uniform microhardness distribution was seen in sample AS-07.

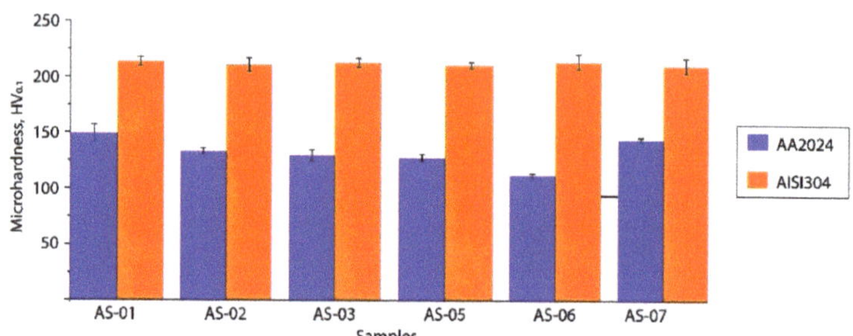

Figure 10. Microhardness distribution.

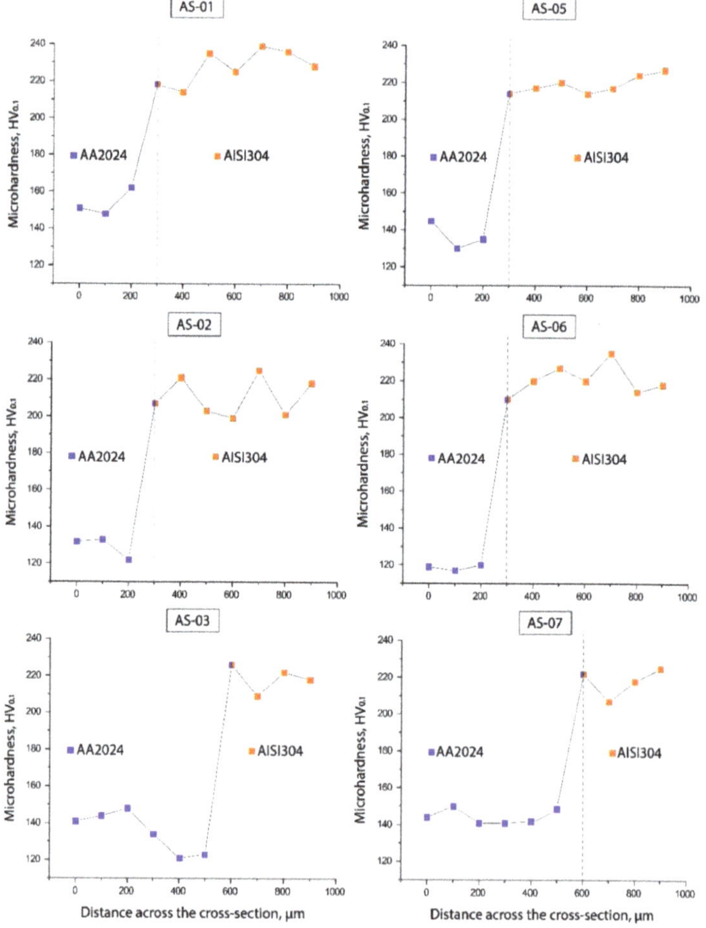

Figure 11. Microhardness profiles in cross-sections of the P-FSSW welds.

4. Conclusions

In the current study, commercial rolled plates of AA2024 aluminum alloy (0.8 mm) and AISI304 stainless steel (6 mm) were lap-welded with a Probeless Friction Stir Spot Welding (P-FSSW) process at two levels of axial force and four levels of rotation speed. There were two main goals for this study. The first goal was to perform and investigate the effect of the rotation speed and tool force on the microstructure and microhardness of spot-dissimilar AA2024-AISI304 welds in a lap configuration made with a flat tool. The second goal was to validate the numerical results and clarify the temperature distribution and material flow within the workpiece during P-FSSW with a flat tool.

This study demonstrated the feasibility of the use of a probeless tool with a flat shoulder for solid-state aluminum–steel welding. The micrographic analysis did not reveal intermetallic compounds typical in fusion welding. The results suggest that the rotating tool transports the material only at the upper part of the AA2024 plate with high plastic deformation. The increase in microhardness is accompanied by dynamic recrystallization and spreading of HAZ at depths throughout the entire cross-section of the workpiece. The tool rotation speed has a significant influence on heat generation, which is supported by the numerical results on the temperature distribution, whereas the load force provides the main contribution to the material flow. The numerical results show that a completely different and strong material flow occurs in the first stage of the P-FSSW process. Overall, variations in the average temperature and stress during the P-FSSW process were seen to be relatively low for all welds compared to the variation in the stress of the samples obtained at 1000 RPM and with both load forces. Consequently, it appears that the material flow during the P-FSSW process is primarily driven by the load force. The optimal rotation speed of 1000 RPM, coupled with 4900 N and 7350 N load force, provides sufficient forging pressure and successful welding with a noticeable difference in the material flow. To consider the balance of the tool force at 1000 RPM and to estimate the material flow at the lap joints, further studies are needed to optimize the welding parameters

Author Contributions: Conceptualization, M.R., N.C. and G.C.; Data curation, M.R. and N.C.; Formal analysis, M.R. and N.C.; Methodology, N.C. and G.C.; Software, M.R.; Supervision, N.C.; Validation, G.C.; Writing—original draft, M.R.; Writing—review & editing, N.C., M.R. and G.C. All authors have read and agreed to the published version of the manuscript.

Funding: DRIMeG PhD program at DMMM—Polytechnic University of Bari.

Informed Consent Statement: Informed consent was obtained from all subjects involved in the study.

Conflicts of Interest: The authors declare no conflict of interest.

References

1. Gullino, A.; Matteis, P.; D'Aiuto, F. Review of Aluminum-To-Steel Welding Technologies. *Metals* **2019**, *9*, 315. [CrossRef]
2. Manladan, S.M.; Yusof, F.; Ramesh, S.; Fadzil, M.; Luo, Z.; Ao, S. A review on resistance spot welding of aluminum alloys. *Int. J. Adv. Manuf. Technol.* **2017**, *90*, 605–634. [CrossRef]
3. Kobayashi, S.; Yakou, T. Control of intermetallic compound layers at interface between steel and aluminum by diffu-sion-treatment. *Mater. Sci. Eng. A* **2002**, *338*, 44–53. [CrossRef]
4. Casalino, G.; Angelastro, A.; Perulli, P.; Casavola, C.; Moramarco, V. Study on the fiber laser/TIG weldability of AISI 304 and AISI 410 dissimilar weld. *J. Manuf. Processes* **2018**, *35*, 216–225. [CrossRef]
5. Adin, M.Ş.; Okumuş, M. Investigation of Microstructural and Mechanical Properties of Dissimilar Metal Weld Between AISI 420 and AISI 1018 Steels. *Arab. J. Sci. Eng.* **2022**, *47*, 8341–8350. [CrossRef]
6. Li, M.; Yang, S.; Wang, Y.; Tao, W. Joining aluminum to steel dissimilar metals using novel resistance spot welding process. *Mater. Lett.* **2011**, *318*, 132215. [CrossRef]
7. Li, M.; Tao, W.; Zhang, J.; Wang, Y.; Yang, S. Hybrid resistance-laser spot welding of aluminum to steel dissimilar materials: Microstructure and mechanical properties. *Mater. Des.* **2022**, *221*, 111022. [CrossRef]
8. Atabaki, M.M.; Nikodinovski, M.; Chenier, P.; Ma, J.; Harooni, M.; Kovacevic, R. Welding of Aluminum Alloys to Steels: An Overview. *J. Manuf. Sci. Prod.* **2014**, *14*, 59–78. [CrossRef]
9. Casalino, G.; Leo, P.; Mortello, M.; Perulli, P.; Varone, A. Effects of laser offset and hybrid welding on microstructure and IMC in Fe-Al dissimilar welding. *Metals* **2017**, *7*, 282. [CrossRef]

10. Suryanarayanan, R.; Sridhar, V. Studies on the influence of process parameters in friction stir spot welded joints—A review. *Mater. Today Proc.* **2021**, *37*, 2695–2702. [CrossRef]

11. Adin, M.Ş.; Işcan, B. Optimization of Welding Parameters of AISI 431 and AISI 1020 Joints Joined by Friction Welding Using Taguchi Method. *Adv. Mater. Process. Technol.* **2022**, *9*, 453–470. [CrossRef]

12. Sharma, C.; Tripathi, A.; Upadhyay, V.; Verma, V.; Sharma, S.K. Friction Stir Spot Welding-Process and Weld Properties: A Review. *J. Inst. Eng. Ser. D Vol.* **2021**, *102*, 549–565. [CrossRef]

13. Siddiquee, A.N.; Pandey, S. Experimental investigation on deformation and wear of WC tool during friction stir weld-ing (FSW) of stainless steel. *Int. J. Adv. Manuf. Technol.* **2014**, *73*, 479–486. [CrossRef]

14. Shen, J.; Lage, S.B.M.; Suhuddin, U.F.H.; Bolfarini, C.; dos Santos, J.F. Exture development and material flow behavior during refill friction stir spot welding of AlMgSc. *Metall. Mater. Trans. A* **2018**, *49*, 241–254. [CrossRef]

15. Silva, B.H.; Zepon, G.; Bolfarini, C.; dos Santos, J.F. Refill friction stir spot welding of AA6082-T6 alloy: Hook defect formation and its influence on the mechanical properties and fracture behavior. *Mater. Sci. Eng. A* **2009**, *773*, 743–756. [CrossRef]

16. Bakavos, D.; Prangnell, P. Effect of reduced or zero pin length and anvil insulation on friction stir spot welding thin gauge 6111 automotive sheet. *Sci. Technol. Weld. Join.* **2009**, *14*, 743–756. [CrossRef]

17. Tozaki, Y.; Uematsu, Y.; Tokaji, K. A newly developed tool without probe for friction stir spot welding and its performance. *J. Mater. Process. Technol.* **2010**, *210*, 844–851. [CrossRef]

18. Mortello, M.; Pedemonte, M.; Contuzzi, N.; Casalino, G. Experimental investigation of material properties in FSW dissimilar aluminum-steel lap joints. *Metals* **2021**, *11*, 1474. [CrossRef]

19. Yang, X.-W.; Feng, W.-Y.; Li, W.-Y.; Chu, Q.; Xu, Y.-X.; Ma, T.-J.; Wang, W.-B. Microstructure and mechanical properties of dissimilar pinless friction. *J. Cent. South Univ.* **2018**, *25*, 3075–3084. [CrossRef]

20. Chu, Q.; Hao, S.; Li, W.; Yang, X.; Zou, Y.; Wu, D. Impact of shoulder morphology on macrostructural forming and the texture development during probeless friction stir spot welding. *J. Mater. Res. Technol.* **2021**, *12*, 2042–2054. [CrossRef]

21. Chu, Q.; Hao, S.; Li, W.; Yang, X.; Zou, Y.; Wu, D.; Vairis, A. On the association between microhardness, corrosion resistance and microstructure of probeless friction stir spot welded Al-Li joint. *J. Manuf. Processes* **2021**, *14*, 2394–2405. [CrossRef]

22. Cederqvist, L.; Reynolds, A.P. Factors affecting the properties of friction stir welded aluminum lap joints. *Weld. J. N. Y.* **2021**, *80*, 281.

23. Yang, X.; Feng, W.; Li, W.; Xu, Y.; Chu, Q.; Ma, T.; Wang, W. Numerical modelling and experimental investigation of thermal and material flow in probeless friction stir spot welding process of Al 2198-T8. *Sci. Technol. Weld. Join.* **2018**, *23*, 704–714. [CrossRef]

24. Chu, Q.; Yang, X.; Li, W.; Vairis, A.; Wang, W. Numerical analysis of material flow in the probeless friction stir spot welding based on Coupled Eulerian-Lagrangian approach. *J. Manuf. Process.* **2018**, *36*, 181–187. [CrossRef]

25. Bakavos, D.; Chen, Y.; Babout, L.; Prangnell, P. Material Interactions in a Novel Pinless Tool Approach to Friction Stir Spot Welding Thin Alu-minum Sheet. *Metall. Mater. Trans. A Vol.* **2011**, *42*, 1266–1282. [CrossRef]

26. Reilly, A.; Shercliff, H.; Chen, Y.; Prangnell, P. Modelling and visualisation of material flow in friction stir spot. *J. Mater. Process. Technol.* **2015**, *225*, 473–484. [CrossRef]

27. Contuzzi, N.; Campanelli, S.; Casalino, G.; Ludovico, A. On the role of the Thermal Contact Conductance during the Friction Stir Welding of an AA5754-H111 butt joint. *Appl. Therm. Eng.* **2016**, *104*, 263–273. [CrossRef]

28. Wang, K.; Liu, F.C.; Xue, P.; Wang, D.; Xiao, B.L.; Ma, Z.Y. Superplastic constitutive equation including percentage of high-angle grain boundaries as a microstructural parameter. *Metall. Mater. Trans. A* **2015**, *47*, 546–559. [CrossRef]

29. Tang, W.; Guo, X.; McClure, J.C.; Murr, L.; Nunes, A. Heat Input and Temperature Distribution in Friction Stir Welding. *J. Mater. Process. Manuf. Sci.* **1998**, *7*, 163–172. [CrossRef]

30. Nandan, R.; Roy, G.G.; Debroy, T. Numerical simulation of three-dimensional heat transfer and plastic flow during friction stir welding. *Metall. Mater. Trans. A* **2006**, *37*, 1247–1259. [CrossRef]

materials

A Review of Numerical Simulation of Laser–Arc Hybrid Welding

Zhaoyang Wang [1], Mengcheng Gong [1], Longzao Zhou [2,*] and Ming Gao [1,*]

[1] Wuhan National Laboratory for Optoelectronics (WNLO), Huazhong University of Science and Technology, Wuhan 430074, China
[2] School of Materials Science and Engineering, Huazhong University of Science and Technology, Wuhan 430074, China
*   Correspondence: lzzhou@hust.edu.cn (L.Z.); mgao@mail.hust.edu.cn (M.G.); Tel.: +86-027-87541423 (M.G.)

Abstract: Laser–arc hybrid welding (LAHW) is known to achieve more stable processes, better mechanical properties, and greater adaptability through the synergy of a laser and an arc. Numerical simulations play a crucial role in deepening our understanding of this interaction mechanism. In this paper, we review the current work on numerical simulations of LAHW, including heat source selection laws, temperature field, flow field, and stress field results. We also discuss the influence of laser–arc interaction on weld defects and mechanical properties and provide suggestions for the development of numerical simulations of LAHW.

Keywords: numerical simulation; laser–arc; interaction; welding

Citation: Wang, Z.; Gong, M.; Zhou, L.; Gao, M. A Review of Numerical Simulation of Laser–Arc Hybrid Welding. *Materials* **2023**, *16*, 3561. https://doi.org/10.3390/ma16093561

Academic Editors: Frank Czerwinski and Raul D. S. G. Campilho

Received: 12 March 2023
Revised: 11 April 2023
Accepted: 18 April 2023
Published: 6 May 2023

1. Introduction

International Welding Commission defines welding standards as a process involving the application of heat or pressure (often localized) to join parts together in a non-detachable manner, either through the addition or absence of filler material or by overlaying the substrate surface [1]. Theodore Maiman developed the world's first ruby laser in 1960, which attracted a lot of research interest due to its high brightness, directivity, monochromaticity, and coherence. However, it was not until the advent of high-power lasers and special optical fibers that lasers became widely used in the field of welding [2]. The laser method is based on Einstein's theory of stimulated radiation. The process involves exciting electrons to high energy levels via the pump, the pump substance, and the resonant cavity. These electrons transition back to lower energy levels and release photons, which are then amplified to create a laser [3]. Compared to arc welding, laser welding has irreplaceable advantages in thick plate welding due to its high efficiency and penetration. However, it also has limitations, such as high assembly requirements, easy generation of porosity and cracks, and low clearance tolerance, which hinder its practical application [4].

Around 1980, Steen [5] introduced the concept of hybrid welding for the first time, and since then hybrid welding has been widely studied around the world. Laser–arc hybrid welding, using the combined heat source of laser and arc, enables high-speed welding with high filling efficiency, better shape, fewer weld defects, and excellent mechanical properties. It is widely used in fields such as transportation and aerospace [6–10]. Although hybrid welding has many advantages, it is difficult to achieve high-quality welds in practical applications due to the complex interaction mechanism of laser and arc in space–time. The simplest and most universal method to achieve optimal welding process and results are by adjusting welding parameters to observe and macroeconomically regulate and control the optimal process interval. This approach results in fewer defects and high-performance welds [11–14]. However, macroeconomic regulation and control based on parameter optimization cannot solve the problem in essence and cannot cope with the intense interaction between laser and material in welding. With the development of physics and technology, X-ray detection, high-speed cameras, and so on appear one after another. People can observe and verify by any means (such as high-speed photography, spectral analysis, and X-ray observation, etc.), and the physical mechanism of laser–arc interaction is investigated [15–18]. However, the shortcomings of high preconditions, high resource consumption, and cumbersome steps have greatly slowed down the pace of scientific research. The emergence of numerical analysis provides a low-cost, high-efficiency means of selecting parameters, and the visual description of multi-physical fields in the welding process also provides a great help in completing the physical mechanism of laser–arc interaction.

While numerical simulation has been effective at predicting and verifying experiments, the advent of finite element analysis and computers brought about rapid development in this field [19]. The step-by-step method of numerical simulation allows for the gradual visualization of high-speed and violent reactions in laser–arc hybrid welding (LAHW), which meets the requirements of this process [20]. In the numerical simulation of welding, to explain the interaction mechanism between complex temperature fields and material, laser and arc heat sources are optimized to adapt to different welding environments [21–24], it has become a common phenomenon to adopt different heat source models for different emphasis. To save on computational costs and improve computational efficiency, the existing numerical simulation mainly focuses on the influence of the temperature field on the flow field and stress field [25–27]. The accuracy of the model has been evaluated by the temperature field profile and the law of flow field and stress field [28–30].

This paper discusses the effects of LAHW on the simulation results and the microstructure and properties of weld defects in three parts. The first part mainly describes the law of heat source selection, the second part mainly describes the change law of the multi-physical field in LAHW, and the third part mainly describes the law of its influence on the microstructure and properties of weld defects.

2. Selection of Laser and Arc Heat Source

The heat source equation, as a basic thermal physical condition in the numerical simulation of LAHW, has received much attention from researchers. The more mature heat source models are the double ellipsoid heat source, the Gaussian heat source, and the variant heat source based on these two types of heat sources [31–33], in which arc heat source is usually represented by a double ellipsoid heat source or surface heat source, and laser heat sources are often represented by Gaussian heat sources. The interaction between the heat source and the material in the numerical simulation of laser–arc welding is presented by the coupling between them, and its mechanism is explained. As shown in Figure 1a, the double ellipsoid heat source is mainly composed of two identical hemi-ellipsoids with different short axes on the long axis, and the heat source formula is shown in Formulas (1) and (2) [34]. As shown in Figure 1b, since the energy distribution of the laser spot conforms to the Gaussian distribution, the Gaussian heat source has been used as the main representation of the laser heat source in recent years. The main distribution pattern is shown in Formula (3) [35], and in the stress simulation, the main distribution pattern is shown in Formula (4).

Compared to arc heat sources, more work is required to improve the accuracy of numerical models for the laser heat sources. In laser deep penetration welding, the incident laser is repeatedly reflected and absorbed by the keyhole wall. In order to account for the effect of wall reflection and absorption, this effect has been introduced by changing the absorption efficiency of the material to the laser [36]. The formation of the keyhole is closely related to the recoil pressure of metal vapor, and the heat loss due to metal vapor evaporation must be taken into account [37]. For example, Li et al. [38] considered the external pressure, internal evaporation, and recoil pressure of metal vapor by introducing the Wilson equation and the gas–liquid equilibrium equation, the numerical simulation of molten pool flow in laser welding of the aluminum alloy under different pressure was realized. Cho et al. [39] used Flow-3D v.11.2 (2017) numerical simulation software to simulate the process of welding aluminum alloy with sinusoidal oscillating laser wire filler.

By combining the Gauss heat source with the sinusoidal oscillating trajectory, the oscillating laser Gaussian heat source was realized. Under the conditions of recoil pressure Formula (5), fender reflection absorption and evaporation, and steam pressure, the numerical simulation of aluminum alloy wire-filled welding with sinusoidal oscillation track of laser heat source was completed. For example, Yu et al. [40] used a laser Gaussian heat source to realize the stress simulation of welding 5A90Al-Li alloy, and the experimental results have a high degree of coherence. Tsirkas et al. [41] realized the numerical simulation of CO_2 laser welding of aircraft aluminum alloy components using numerical analysis software under the condition of introducing the "Life and Death Unit".

As shown in Figure 1c, based on the Gaussian heat source, the shape and energy distribution of the laser heat source can be adjusted by changing the radius and energy distribution of the heat source at the top, middle, and bottom. The heat source formula is shown in Formulas (6) and (7). For example, Farrokhi et al. [42] introduced r_t, r_b, r_i, z_b, z_t, z_i and other physical quantities into the numerical simulation of temperature–stress by changing the shape and energy distribution of the Gaussian heat source; thus, the error between the simulated weld and the actual shape was smaller and more accurate. Geng et al. [43] introduced the deformed Gaussian heat source into the numerical simulation of the temperature-flow field and obtained the hourglass-shaped keyhole and weld. It was concluded that Marangoni convection was mainly distributed near the keyhole and decreased with the increase in the distance from the keyhole.

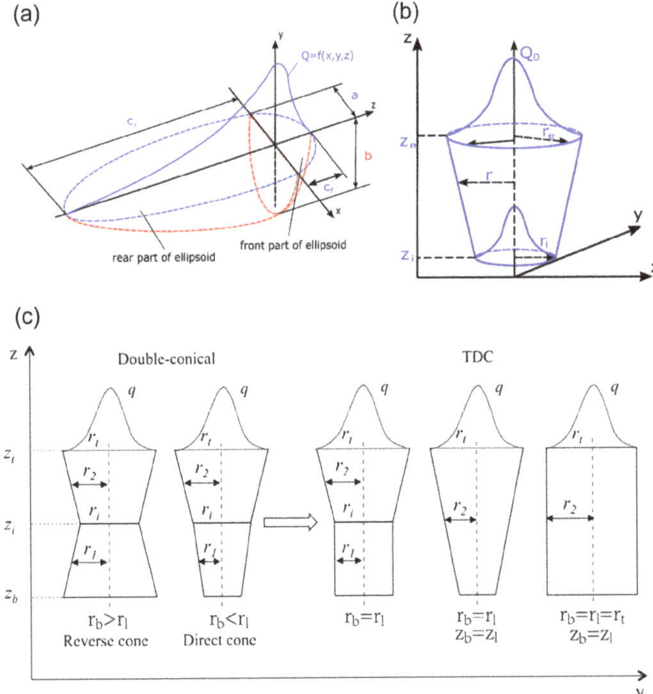

Figure 1. (**a**) double ellipsoid heat source; (**b**) Gauss heat source [44]; (**c**) variant Gaussian heat source [42].

$$Q_f = \frac{6\sqrt{3}Ak_f\delta_1 P}{c_f ba\pi\sqrt{\pi}}\exp\left\{-3\left[\frac{(x-x_0)^2}{a^2} + \frac{(y-y_0)^2}{b^2} + \frac{(z-z_0)^2}{c_f^2}\right]\right\} \qquad (1)$$
$$-b \le y \le 0, z \ge z_0$$

$$Q_r = \frac{6\sqrt{3}Ak_r\delta_2 P}{c_r ba\pi\sqrt{\pi}}\exp\left\{-3\left[\frac{(x-x_0)^2}{a^2}+\frac{(y-y_0)^2}{b^2}+\frac{(z-z_0)^2}{c_r^2}\right]\right\}$$
$$-b \leq y \leq 0, z \leq z_0 \tag{2}$$

$$Q = \frac{9Ak_s P}{R_0^2\pi(1-e^{-3})}\exp\left\{-9\frac{\left[(x-x_0)^2+(y-y_0)^2\right]}{R_0^2\log[H/(z-z_0)]}\right\} \tag{3}$$

$$Q = \frac{9Ak_s P}{R_0^2\pi H(1-e^{-3})}\exp\left\{-9\frac{\left[(x-x_0)^2+(y-y_0)^2\right]}{R_0^2\log[H/(z-z_0)]}+\frac{(z-z_0)^2}{H^2}\right\} \tag{4}$$

$$P_r \cong 0.54P_0\exp\left(H_{lv}\frac{T-T_g}{RTT_g}\right) \tag{5}$$

$$Q_{l1} = \frac{9P_{l1}e^3}{\pi(e^3-1)(z_i-z_b)(r_i^2+r_i r_b+r_b^2)}\exp\left\{-3\frac{\left[(x-x_0)^2+(y-y_0)^2\right]}{r_1^2}\right\} \tag{6}$$

$$Q_{l2} = \frac{9P_{l2}e^3}{\pi(e^3-1)(z_t-z_i)(r_t^2+r_i r_t+r_i^2)}\exp\left\{-3\frac{\left[(x-x_0)^2+(y-y_0)^2\right]}{r_2^2}\right\} \tag{7}$$

$$\begin{cases} r_1(z) = r_i - (r_i-r_b)\frac{z_i-z}{z_i-z_b} \\ r_2(z) = r_t - (r_t-r_i)\frac{z_t-z_1}{z_t-z_i} \end{cases} \tag{8}$$

$$\begin{cases} P_{l1} = \frac{z_i-z_b}{z_t-z_b}P \\ P_{l2} = \frac{z_t-z_i}{z_t-z_b}P \end{cases} \tag{9}$$

where a is the laser absorptivity, k_S is the ratio of laser energy to Gauss body heat source ($k_S = 1$ when there is only a Gauss body heat source), P is the laser power, R_0 is the radius of the laser spot, H is the height of the heat source, (x_0, y_0, z_0) is the center coordinate of the heat source, k_f, k_r is the ratio of laser energy to the cover heat source, δ_1 is the ratio of the first half of the double ellipsoid heat source, δ_2 is the ratio of the second half of the double ellipsoid heat source, and a, b, and c_f, c_r are the parameters used to define the size and shape of the double ellipsoid. The parameters of r_t, r_b, r_i, z_b, z_t and z_i are shown in Figure 1c.

In the numerical simulation of laser–arc hybrid welding, researchers usually make assumptions about the model to reduce the difficulty of modeling and reduce the calculation time of the model. In fluid modeling, common assumptions are: (i) the molten metal is assumed to be incompressible Newtonian fluid; (ii) the deposition of laser heat is in conduction mode due to the high welding speed as well as high oscillating frequency of the laser beam; (iii) the small gap between the two sheets in the lap joint to help the aluminum vapor to escape is ignored in the simulation, since the gap has insignificant effects on the weld geometry and fluid flow in the weld pool; (iv) average current and voltage are used to calculate the arc power, and the pulse effect is neglected [36–38,45]. The common assumptions used in stress simulation are: (i) phenomena such as vaporization, ejection of material, circulation of the molten metal, formation of ions, etc., were left out of this work. (ii) Both the initial specimen temperature and the ambient temperature were 20 °C. (iii) The welded steel was considered a flat plate. (iv) The phase change and material flow in the welding zone is not considered. Heat source, heat transfer, and boundary condition of radiation and convection were considered to analyze the thermal data [40,41,46–48].

3. Influence of Laser–Arc Interaction on Multi-Physical Fields

When two kinds of heat sources (laser and arc heat source) act on the weld, they will have a great effect on the temperature field, increase the pool volume and surface area, promote the formation of a complex pool in the pool, and then affect the weld microstructure and mechanical properties. Additionally, the different coupling modes between the laser and the arc will have a great impact on the influence of multi-physics; for example, different coupling and spatial distribution, such as paraxial coupling, coaxial coupling, laser defocus, and the relative position of the laser and arc will have a great influence on multi-physics in LAHW.

In coaxial coupling, the energy density of the laser and arc will be superimposed on the same central point, and the interaction between heat and matter will undergo a series of different changes under the condition of high energy density, the most intuitive performance in the temperature field temperature gradient changes. The coaxial coupling of arc and laser not only increases the laser power at the center point but also increases the heating area around the center point. As shown in Figure 2, in the coaxial coupling [49], by comparing the temperature field and the melt flow field of arc welding, it can be found that the coaxial coupling is caused by the addition of a central laser heat source. In the paraxial coupling, the laser and arc are located at different center points and move to the welding direction at the same speed. The different heat source center points lead to an increase in the heating area and the sequence of the laser and arc heat source acting at the same point. Therefore, the relative position of the laser and arc has a great influence on the multi-physical fields and deformation of welding in paraxial recombination. Bakir et al. [32] found that the absolute value of laser defocusing was positively correlated with the number of weld cracks and concentrated at the root of the weld by adjusting the ratio of laser to arc heat source heat input. As shown in Figure 2b, Kim et al. [50] found that when arc heat input is dominant, the angular deformation is v-shaped, and when laser heat input is dominant, the angular deformation is inverted v-shaped; as shown in Figure 2c, it was found by Cai et al. [51] that the relative position of laser and arc can affect the velocity and direction of melt flow, and then affect the grain size and mechanical properties of the weld, the minimum grain size and the highest mechanical properties can be obtained via laser at the front of the arc region.

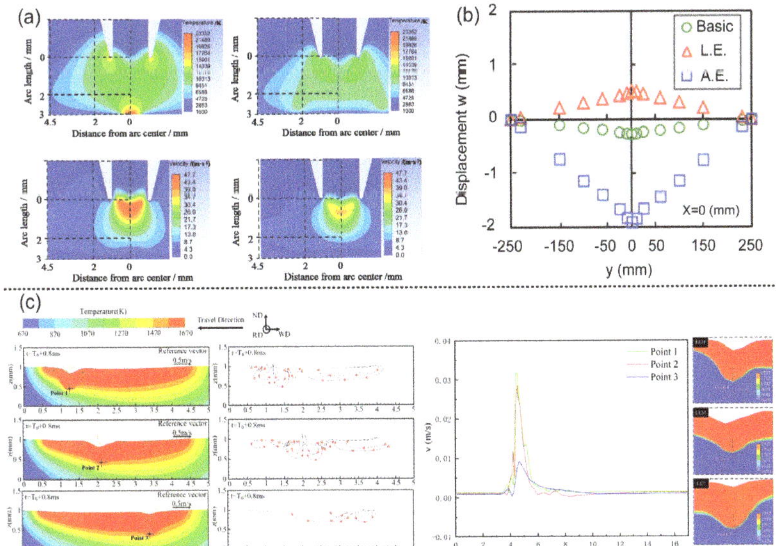

Figure 2. (**a**) Comparison of temperature field and velocity field in coaxial coupling and arc welding [49]; (**b**) influence of laser defocusing on deformation angle after welding [50]; (**c**) influence of the relative position of laser and arc on temperature field, flow field, and flow velocity [51].

The different energy characteristics of laser and arc lead to the different shape characteristics of weld. The increase in arc current often leads to a faster increase in weld width than penetration, and the increase in laser power will have more advantages in the depth of penetration. The different ratio of laser to arc has great influence on the weld shape during laser–arc hybrid welding. As shown in Figure 3a, a low-power laser and high-current arc are common in the coaxial coupling, the weld shape is v-shaped, and the penetration depth increases with the increase in laser power. As shown in Figure 3b, in the paraxial coupling, the weld is wide and shallow like a bowl in the upper arc region, narrow and deep in the lower laser region, and the weld width decreases sharply as a thin neck because of the higher penetration of the laser relative to the arc. As shown in Figure 3c, when the laser power is further increased to penetration, the weld appearance is mainly an hourglass shape. At this time, a wider molten pool will appear at the bottom under the action of the surface tension of the molten pool and the backing force of the protective gas at the bottom and keep it steady.

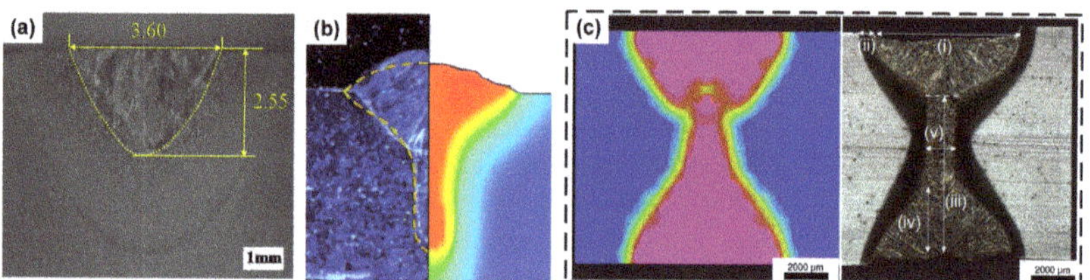

Figure 3. (**a**) coaxial coupling weld profile [52]; (**b**) paraxial coupling weld profile, light blue and red are experimental and simulated weld shapes [53]; (**c**) penetration weld profile [54]. (i) Weld width on the upper surface; (ii) HAZ; (iii) Weld width on the middle zone; (iv) laser zone penetration; (v) the lower surface width.

4. Explaining the Mechanism of Laser–Arc Interaction

4.1. Defect Suppression

In LAHW, welding parameters (such as laser power, arc current, and wire spacing), heat source parameters, and process parameters (such as shielding gas flow rate and welding speed) have great influence on laser–arc interaction, and help to achieve the control of weld formation, welding process stability, and welding defects and other purposes [55–57]. Gas hole defect is the key research object of welding-related subjects because it can reduce the effective working section of the weld seam, lead to stress concentration, and then reduce the mechanical properties of the weld seam. The keyhole is easy to collapse in laser welding, so it is easy to form process-type pores. However, the temperature gradient of laser–arc hybrid welding decreases due to the addition of an arc heat source, which leads to the increase in the molten pool area and a longer solidification time. Gao et al. [58] found that AZ31B magnesium alloy can be undercut and blowhole defects can be suppressed by LAHW; by comparing the microstructures and mechanical properties of MIG and MIG-Laser-hybrid welding pure copper, Zhang et al. [12] found that hybrid welding can achieve narrower HAZ, finer grain structure, and higher conductivity. The fundamental reason for this that there is less porosity and impurity in the composite welding.

In laser–arc hybrid welding, the arc droplet will strike the surface of the molten pool and cause oscillation of the molten pool and keyhole, which easily forms humps and dent on the weld surface. At this point, the way the laser and the ARC are guided will have a more important effect on the impact force. Tang et al. [59] considered that the balance between gravity and surface tension of a molten pool is an important factor to influence the appearance of the hump; however, the arc-guided LAHW has a stronger restraining effect on the root hump defects because of its smaller droplet diameter and longer transition period. Zhang et al. [60] found that compared with arc welding, the spatter of LAHW is reduced, employing spectral analysis and high-speed photography of plasma plume in LAHW; moreover, the welding process is more stable, and the laser–arc interaction can promote the photon energy level transition to enhance the heat input, which results in a lower droplet transfer force. To clarify and supplement the mechanism of laser–arc interaction, researchers have conducted a lot of research using numerical analysis.

As an inert gas protective layer used to prevent metal oxidation in the molten pool, the related parameters affect the melting effect and even the depth of molten metal. For example, during the welding of carbon steel, carbon dioxide gas is mixed into the inert gas to achieve the effect of increasing penetration [61]. In laser welding, the plasma above the weld produces the absorption effect on the laser. The plasma is heated and excited to expand, which increases the absorption efficiency, and then forms positive feedback [62]. At this time, the protective gas not only plays the role of protecting the molten pool metal but also plays the role of blowing away the plasma above the weld, which plays a greater role in increasing the penetration of laser welding [63]. As shown in Figure 4a, Yang et al. [64] found that the effect of high shielding gas flow rate on LAHW was less than that of single-arc welding, and a high gas flow rate had a positive effect on the spread of molten metal. It is shown that the high-speed and high-pressure gas field was caused by the high shielding gas flow rate above the weld, and the plasma was easily affected by this field in arc welding. The explanation for this is that the effect of this field and the binding of the plasma were effectively restrained by the high-pressure steam ejected from the keyhole during the hybrid welding.

In laser–arc hybrid welding, the welding stability is the direct-viewing factor that affects the welding defects. In the welding dynamic stability stage, when the laser or the arc appears to be an out-of-cycle fluctuation, it will probably cause the welding process to destabilize. The stability of the laser keyhole and arc droplet transfer process affects the weld quality. A more stable droplet transfer process, such as jet transfer, reduces spatter and minimizes the impact on the molten pool. This process is crucial to maintaining the stability of the molten pool. As shown in Figure 4b, Huo et al. [65] performed numerical simulation of pulsed LAHW of magnesium alloy and found that the maximum energy efficiency can be achieved by keeping the pulse frequency at 20 Hz with the optimum range of laser excitation current being 150–175 A. If the pulse frequency is less than 20 Hz, the longitudinal profile of the weld will be discontinuous, and if the laser excitation current is too high, serious spatter will occur. As shown in Figure 4c, Xue et al. [66] discovered that compound welding effectively inhibits hump defects compared to laser welding. They explained that the TIG heat source introduces arc force and increases the area of the molten pool, which can be efficiently suppressed by high-speed melt flow in laser welding, leading to the formation of the hump.

In laser–arc hybrid welding, laser heating directly heats the arc as it passes through it [61]. Metal vapor ejected from the keyhole can alter the composition of the arc plasma and influence fluid flow and heat transfer [62]. The combined plasma above the weld absorbs the laser energy, which reduces the laser's thermal efficiency [63]. The interaction between laser-induced metal vapor and arc plasma is affected by laser power, arc current, and the relative positions of laser and arc. Mu et al. [67] discovered through experiments and numerical simulations that high-speed melt flow in laser welding can be effectively inhibited. In LAHW, the arc tail oscillates at a frequency of 1–3 kHz, and the filament spacing affects the arc plasma area significantly. The high-speed metal vapor ejected from the laser keyhole exerts physical shielding, which compresses the arc plasma. The high-pressure metal vapor in the keyhole maintains the stability of the Keyhole, and the keyhole wall handles the equilibrium state of steam recoil pressure, surface tension, and gravity, etc. Each bulge is a potential cause of keyhole collapse, where the bulge is more susceptible to force imbalance, cutting off the keyhole cavity and creating bubbles of gas below. Finally, it is captured by the solidified molten pool and becomes a gas hole defect. The probability of keyhole collapse increases with the number of protrusions in the molten metal. As shown in Figure 4d, Chen et al. [68] observed that the oscillating frequency of the keyhole in hybrid welding is several kilohertz, and the keyhole change can be divided into three stages: initial establishment, rapid deployment, and oscillatory dynamic stability. The instability of the welding process primarily occurs during the stage of keyhole closing and re-opening. The formula for oscillation frequency is given $f_k = (\gamma/r^3\rho)^{1/2}$, and the surface tension primarily causes keyhole instability [69].

Laser–arc hybrid welding can increase the likelihood of bubble overflow by increasing the molten pool's area and solidification time, but it does not effectively suppress porosity in aluminum alloy welding. Keyhole collapse and arc droplet impact in high-power lasers still significantly influence welding spatter and hump. Unstable welding processes and welding defects affect the solidification and mechanical properties of welds, leading to coarse grain, variable element distribution, and decreased mechanical properties.

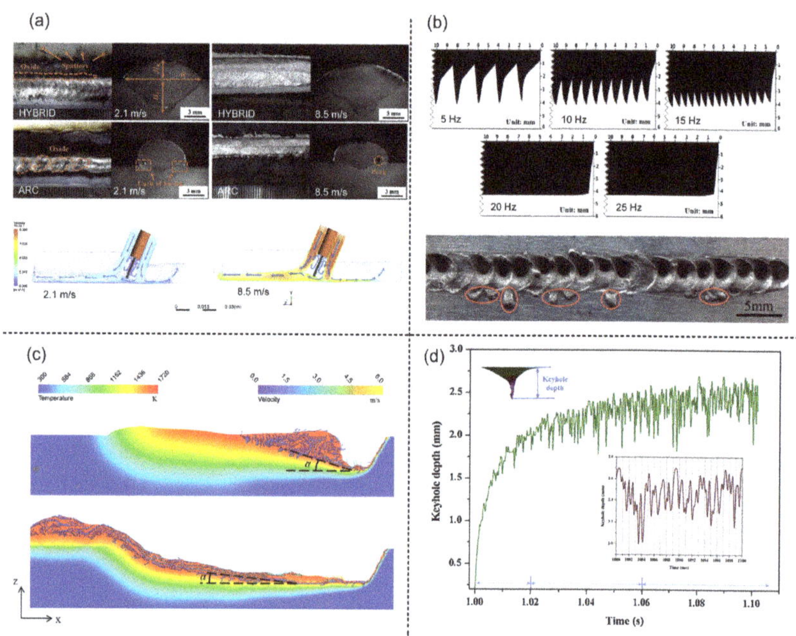

Figure 4. (**a**) the effect of gas flow rate on weld profile (top) and gas flow field (bottom) [64]; (**b**) the effect of pulse frequency on weld profile (top) and weld surface formation (bottom) under a high excitation current [65]; (**c**) the influence of filament spacing on arc plasma area [66]; (**d**) the variation rule of the keyhole in LAHW [68].

4.2. Stress and Mechanical Properties

Mechanical properties are key factors for assessing weld quality and are influenced by factors including weld microstructure, defects, and stresses [70–72]. Strengthening elements in welding, on one hand, play an important role in enhancing the properties of welds. For example, Si in Al alloy enhances the fluidity of the molten metal while also forming new compounds with the base metal, ultimately reinforcing the properties of the weld. Hao et al. [73] observed that Si and Sn in LAHW of pure copper improved the tensile strength and corrosion resistance but lowered electrical conductivity. They explained the promotion of columnar crystal growth in the fusion zone by Si and Sn, thereby enhancing tensile properties using the solid solution strengthening mechanism. The use of strengthening elements is an effective method of improving the mechanical properties of welds, but in laser–arc welding alone, the element distribution is non-uniform, leading to differences in mechanical properties at various positions. Although laser–arc hybrid welding enhances element diffusion due to longer solidification time and higher-speed pool flow, it still cannot ensure uniform distribution of elements in the molten pool, leading to differences in mechanical properties at different positions.

Yan et al. [74] investigated the effect of gravity on weld properties and residual stress by modifying the angle of the base metal (within the range of 0–90°). As per their study results, the microstructure and residual stress were symmetrically distributed only when the weld was vertically inverted. As the angle increased from 0° to 90°, the micro-grain size of the upper side of the weld increased from 66.9 μm to 113.9 μm, which resulted in a decrease in tensile strength from 352.8 MPa to 319.8 MPa. The weld metal flowed to the lower side, resulting in uneven temperature distribution on both sides of the weld. Additionally, non-uniform solidification on both sides of the weld led to asymmetric distribution of longitudinal and transverse residual stress. Numerical simulation can obtain and verify the impact of the laser–arc interaction on the weld seam mechanical properties, making it a cost-effective and simplified alternative to experimental testing.

Zhan et al. [12] found that the welding time, material consumption, and deformation peak value of MIG were 8, 16, and 3 times those of MIG and laser-MIG hybrid welding, respectively. It was also shown that hybrid welding can achieve smaller heat input with higher filling efficiency and welding speed, and finally achieve smaller welding deformation. Liu et al. [47] concluded that the elasticity of the martensite formed during welding had little impact on residual stress. Furthermore, they found that residual stress increased with higher annealing temperatures. Jiang et al. [55] found that because of the smaller heat input and weld cross-sectional area compared with traditional MIG welding, laser-MIG hybrid welding has great advantages in grain refinement and porosity suppression. The smaller thermal input means the lower temperature gradient and peak temperature of hybrid welding, and there is a close relationship between the solidification rate, phase transformation, and temperature field; therefore, the crystallization and phase transformation in laser–arc hybrid welding deserve our attention. As shown in Figure 5a, Qi et al. [75] studied the effect of secondary peak temperature on the microstructure and the mechanical properties of X100 pipeline steel using a numerical simulation. The impact energy of HAZ increased gradually (the maximum was 37.2 j) and the dimple size at tensile fracture decreased gradually (minimum is 6.1 μm). The increase in the amount of reversion austenite leads to the decrease in the carbon concentration and the formation of different microstructures. As shown in Table 1, the different peak temperatures (740, 790 and 840 °C) corresponded to the microstructures of necklace-type M-A, necklace-type martensite, and granular bainite + acicular ferrite, respectively.

Table 1. Secondary thermal cycle products at austenite grain boundaries at different temperatures [75].

Temperature	740 °C	790 °C	840 °C
Products of the secondary thermal cycle	Necklace-type M-A constituent	Necklace-type lath martensite	Granular bainite + acicular ferrite

As shown in Figure 5b, by simulating the temperature and stress fields of NV E690 Steel, Sun et al. [46] found that the residual stress near the fusion zone and the heat-affected zone was high and showed tensile stress. As explained, the large grain size in HAZ is the main factor causing the large residual stress; as shown in Figure 5c, Churiaque et al. [76] optimized the parameters of fillet welds of 8 mm-thick EH36 steel plate using a numerical simulation, and showed that the large grain size in HAZ is the main factor causing the larger residual stress. Finally, the full penetration welds with displacement disturbance of 1.29 mm for the bottom plate, 1.92 mm for Rib Plate and 380 MPA for maximum residual stress are obtained, the welding speed (2.2 m/min) is 1.76 times higher than that of Valdaytseva et al. [77] (1.25 m/min). Because of the difference in action point between the laser and arc center in the paraxial coupling, both of them will produce a molten pool and converge, and then the molten pool will be affected by both. The molten pool with shallow and wide arc [78] and the molten pool with narrow and deep laser flow influence each other. Cho et al. [79] found that the liquid in the arc bath can rapidly expand to the laser heating point, which increases the transverse diffusion of the bath, produces a larger front beam angle, and then reduces the stress concentration; under the influence of non-uniform temperature field, the difference between the deformation caused by thermal expansion at different positions and the constraint caused by the thermal expansion will eventually affect the residual stress and mechanical properties in this area. Qian et al. [80] found that in LAHW, the sequence of residual stress from large to small is the fusion line and heat-affected zone, laser zone, arc zone, mixed zone, bottom of weld, and top of weld. Zhang et al. [81] found that the fluid flow has a great influence on the temperature field in the hybrid welding; this was determined based on a numerical simulation of the melt flow near the keyhole of TCS stainless steel.

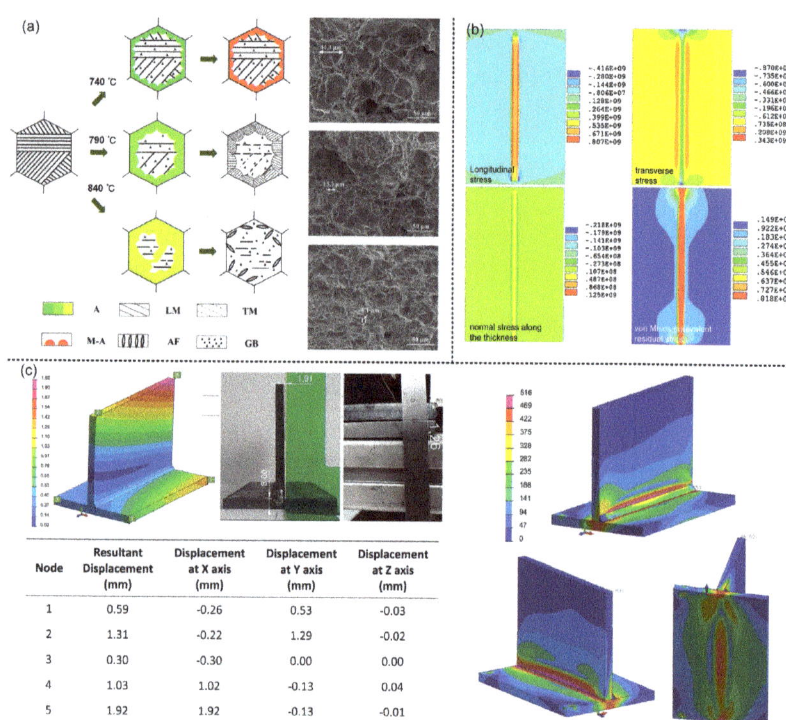

Node	Resultant Displacement (mm)	Displacement at X axis (mm)	Displacement at Y axis (mm)	Displacement at Z axis (mm)
1	0.59	-0.26	0.53	-0.03
2	1.31	-0.22	1.29	-0.02
3	0.30	-0.30	0.00	0.00
4	1.03	1.02	-0.13	0.04
5	1.92	1.92	-0.13	-0.01

Figure 5. (**a**) Left: Schematic diagram of microstructure transformation at different secondary peak temperatures (A: austenite, LM: granular bainite, TM: tempered martensite, M-A: necklace-type M-A component, AF: acicular ferrite, GB: parent material), right: tensile fracture micrograph [75]; (**b**) stress and residual stress in different directions in hybrid welding [46]; (**c**) left: multi-position deformation in simulation and experiment, right: residual stress distribution in multi-view [76].

In laser–arc hybrid welding, because of the wider pool and more complex pool, the solidification time of the pool is much higher than that of single-laser or arc welding. The longer solidification time increases the interaction between laser and arc. The results show that this method can reduce the impact force of droplet on the molten pool, increase the time and velocity of bubble overflow due to keyhole collapse, and reduce the plastic strain and residual stress due to uneven thermal expansion. The reduction in the occurrence probability and the degree of influence on the molten pool can refine the grain structure and strengthen the mechanical properties.

4.3. Laser–Arc Hybrid Additive Manufacturing

Compared to both laser and arc additive manufacturing, laser–arc hybrid additive manufacturing offers nearly all the advantages of hybrid welding. Its high deposition rate makes it advantageous in the production of large-scale aviation and transportation components. The wider pool area and longer solidification time are beneficial for uniform element distribution and bubble overflow. Complex flow interactions can also effectively enhance pool stability. Liu et al. [82] compared the microstructure and mechanical properties in different regions of arc additive manufacturing and laser–arc hybrid additive manufacturing and found that the periodic changes in the microstructure are the same. The coarse columnar, fine columnar and fine equiaxed grain regions are found at the bottom, middle, and top, respectively, but the grain refinement regions appear in the manufacturing of hybrid materials due to the laser thermal effect. In addition, there are fewer Al-Si-Sr phases and more homogeneous Sr elements in the hybrids. These factors increased the microhardness by 4.2 $HV_{0.05}$ and the tensile strength by 20.8 MPa. In additive manufacturing, preheating has an important effect on the microstructure phase transition, as shown in Figure 6a. Cui et al. [83] performed numerical simulation, and presented the effect of preheating temperature on the microstructure phase transition of materials. When the preheating temperature increases to 400 °C, the welding area decreases by 600 μm compared with that without preheating. The preheating can promote the formation of secondary α phase, which can effectively increase the mechanical properties of the members as a strengthening phase. By optimizing the heat source model, Sun et al. [46] presented the distribution characteristics of residual stress in 7A52 aluminum alloy members welded by laser–arc hybrid welding, and found that the variation of beam height only affected the longitudinal residual stress in the substrate and the beam; it had little effect on the transverse residual stress of the substrate, and the corresponding substrate constraints had little effect on the transverse residual stress of the beam.

Laser–arc hybrid additive manufacturing inherits the benefits of laser–arc hybrid welding, including enhanced deposition efficiency, improved formation, stabilized molten pool, and grain refinement. However, it also inherits and potentially intensifies the negative aspects of hybrid welding, such as the high level of difficulty, post-processing requirements, and internal defects after forming. The introduction of oscillating laser technology offers a new approach to address these limitations. As shown in Figure 6b, Gong et al. [84] introduced laser oscillation into the manufacturing of hybrid additives and studied the influence of an oscillating laser on the manufacturing of hybrid additive. They found that the addition of an oscillating laser has great advantages in stabilizing droplet transition, reducing surface roughness and grain refinement, etc. The oscillating laser–arc hybrid welding also has its great advantages in deposition effect and shaping. As shown in Table 2, the section of arc increasing material is crescent-shaped, while the cross section of the laser–arc hybrid is in the shape of a wine glass. The section shape of the oscillating laser–arc hybrid is between the arc and laser–arc hybrid. Its remelting depth is the smallest, and its remelting rate reaches 70%, which is 1.4 times that of arc-increasing material.

Table 2. The single-layer penetration depth (Δh_1) and layer height (Δh_2) in three increasing materials [84].

	Arc	Laser–Arc Hybrid	Oscillating Laser–Arc Hybrid
Δh_1 (mm)	4.1	3.5	2.7
Δh_2 (mm)	2	0.8	0.8

The potential benefits of oscillating laser–arc hybrid additive manufacturing, including improved forming, reduced porosity defects, refined grains, and strengthened properties, are not yet fully understood due to limited research on the internal mechanisms. There is a particular lack of studies on numerical simulation of molten pool flow and element distribution, stress, strain, and residual stress. Further research is needed to fully comprehend this technology.

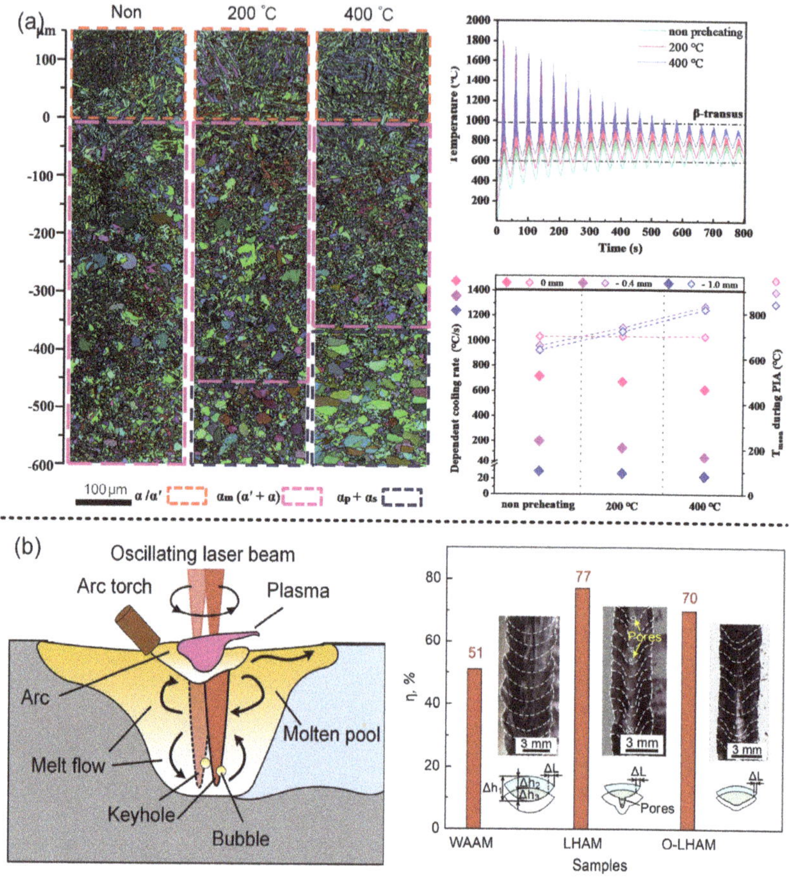

Figure 6. (**a**) microstructure (left) and temperature distribution (right) in the welding region [83]; (**b**) keyhole bubble trapping schematic diagram (left) and cross-sectional morphology under different welding methods (right) [84].

4.4. Oscillating Laser–Arc Hybrid Welding

In oscillating laser welding, it has been proved that the oscillatory behavior of a laser heat source has the advantages of improving forming, suppressing defects, refining grains and strengthening properties because of its special locus of motion [53,85]. Oscillating laser–arc hybrid welding (O-LAHW) has shown excellent metallurgical ability and wide adaptability in many types of material welding due to the advantages of both O-LAHW and arc integration [84,86]. For example, Meng et al. [33] found that weld formation and non-fusion defects in steel/aluminum dissimilar joints can be effectively suppressed at an oscillating frequency of 150 Hz. The physical model and coupling mechanism of the interaction between the oscillating scanning laser and the electric arc can be clearly described by numerical simulation, and its model and mechanism can be optimized and perfected [87].

In Gao et al.'s [45] numerical simulation of oscillating scanning laser–arc hybrid welding using double ellipsoidal and cylindrical heat sources on lap plates, it was observed that the width and depth of the weld line increased with the arc droplet's entry into the weld pool, as shown in Figure 7a. At the moment of droplet contact with the molten pool, the liquid metal flow direction can be divided into up and down, and the sinusoidal oscillation of the laser is uniformly distributed throughout the molten pool. Due to the varying heat transfer between the upper and lower molten pools in oscillating scanning laser–arc hybrid welding, the temperature gradient and heat transfer efficiency of the upper plate is higher than that of the lower plate. As depicted in Figure 7b, at x = 33 mm, the shallow molten pool under droplet impact causes the entire molten flow to divide into two directions. Whilst the sinusoidal oscillating scanning laser produces a larger and more uniform energy and temperature field, the impact angle of the droplet results in greater heat input to the plate. As a result of the combined action of impact force, gravity, and Marangoni force, resultant forces on the upper and lower parts appear in different directions.

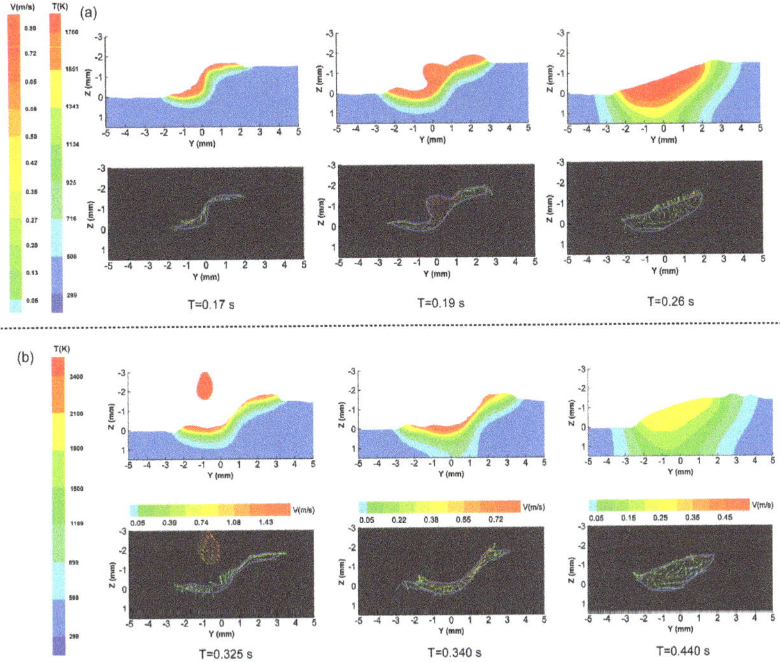

Figure 7. (**a**) The change rule of temperature field and flow field at x = 20 mm; (**b**) the change rule of temperature field and flow field at X = 33 mm [45].

Shi et al. [53] carried out a numerical analysis of circular oscillating laser–arc hybrid welding of 304SUS stainless steel. As shown in Figure 8a, with the increase in the oscillation frequency, the keyhole depth in the weld pool becomes shallow, the keyhole opening increases, the weld pool becomes shallow, and the weld porosity decreases before disappearing entirely. As shown in Figure 8b, when the oscillation frequency is low, the eddy current direction in the molten pool changes with the change in the laser beam direction. When the laser beam oscillates forward, it forms an anticlockwise vortex, and when the laser beam oscillates backward, it forms a clockwise vortex. Because the oscillating-induced eddy current can significantly improve the temperature and flow fields, circular oscillating laser–arc hybrid welding has a more stable welding process, fewer defects, and a lower weld depth-to-width ratio. When the oscillating frequency is higher than 150 Hz, the eddy current velocity can reach 0.7 ms to control the flow field, stabilize the molten pool, and realize the aim of restraining spatter and porosity.

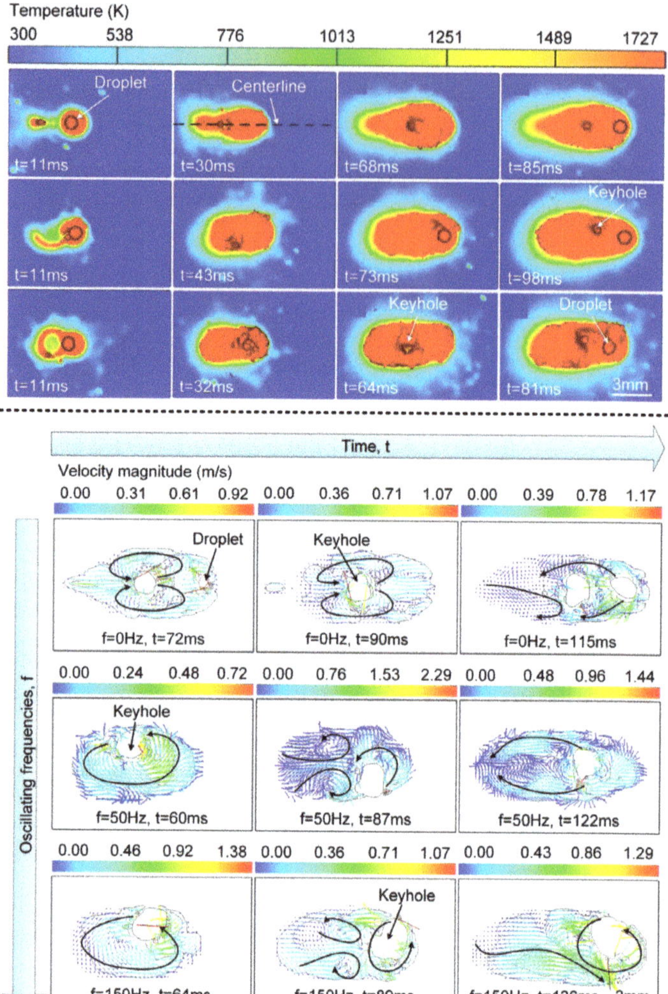

Figure 8. Top: The variation in temperature field with time at oscillating frequencies of 0, 50, and 150 Hz; **bottom:** the variation in the flow field with time at oscillating frequencies of 0, 50, and 150 Hz [53].

5. Conclusions

In this paper, the effects of laser–arc interaction on weld formation, multi-physical fields (temperature field, flow field, and stress field), stability of welding process, defects, and microstructure and properties are reviewed. The strengthening mechanism and element distribution of different elements in welding wire are widely concerned, but the research on microstructure evolution and solute element distribution in LAHW is lacking. Although there is a lot of literature on the research of oscillating laser welding numerical simulation, there are few types of research on the numerical simulation of O-LAHW, the physical mechanism of the interaction between the oscillating laser and the arc is not perfect, and its effect on the evolution of multi-physical fields, welding defects, microstructure, and mechanical properties is not clear. To enhance the understanding of microstructure strengthening and mass transfer crystallization, future numerical simulations of oscillating laser–arc hybrid welding (O-LAHW) should be conducted. Additionally, the benefits of O-LAHW, including the influence of the oscillating laser on the molten pool, keyhole stability, and performance enhancement, should be explored further. Understanding these key areas of O-LAHW will be critical to the future advancement of this technology.

Author Contributions: Writing: Z.W.; Review of literature: Z.W. and M.G. (Mengcheng Gong); investigation: L.Z.; Analysis: M.G. (Ming Gao) All authors have read and agreed to the published version of the manuscript.

Funding: Supported by 'the Fundamental Research Funds for the Central Universities', HUST: 2022JYCXJJ027 and the National Natural Science Foundation of China (52275335 and 52205360).

Institutional Review Board Statement: Not applicable.

Informed Consent Statement: Not applicable.

Data Availability Statement: Not applicable.

Acknowledgments: Thanks to the financial support of Huazhong University of Science and Technology and the careful guidance of Ming Gao.

Conflicts of Interest: The authors declare no conflict of interest.

References

1. Riley, J.J.; Smith, C.E. Influence of magnetic materials on the welding characteristics of resistance welding machines. *Electr. Eng.* **1946**, *65*, 852–860. [CrossRef]
2. Vassie, L.H.; Tyrer, J.R.; Soufi, B.; Clarke, A.A. Lasers and laser applications in the 1990s: A survey of laser safety schemes. *Opt. Lasers Eng.* **1993**, *18*, 339–347. [CrossRef]
3. Wise, J.B. Errors in Laser Spot Size in Laser Trabeculolasty. *Ophthalmology* **1984**, *91*, 186–190. [CrossRef]
4. Zhang, S.; Sun, J.; Zhu, M.; Zhang, L.; Nie, P.; Li, Z. Fiber laser welding of HSLA steel by autogenous laser welding and autogenous laser welding with cold wire methods. *J. Mater. Process. Technol.* **2020**, *275*, 116353. [CrossRef]
5. Steen, W.M. Arc augmented laser processing of materials. *J. Appl. Phys.* **1980**, *51*, 5636–5641. [CrossRef]
6. Wang, L.; Gao, M.; Hao, Z. A pathway to mitigate macrosegregation of laser-arc hybrid Al-Si welds through beam oscillation. *Int. J. Heat Mass Transf.* **2020**, *151*, 119467. [CrossRef]
7. Wu, J.; Zhang, C.; Lian, K.; Cao, H.; Li, C. Carbon emission modeling and mechanical properties of laser, arc and laser–arc hybrid welded aluminum alloy joints. *J. Clean. Prod.* **2022**, *378*, 134437. [CrossRef]
8. Hao, K.; Gao, Y.; Xu, L.; Han, Y.; Zhao, L.; Ren, W. Plasticity improvement coupled by carbon nanotubes and beam oscillation in laser-arc hybrid welding of magnesium alloy. *Mater. Sci. Eng. A* **2022**, *857*, 144093. [CrossRef]
9. Bunaziv, I.; Akselsen, O.M.; Frostevarg, J.; Kaplan, A.F.H. Laser arc hybrid welding of thick HSLA steel. *J. Mater. Process. Technol* **2018**, *259*, 75–87. [CrossRef]
10. Zhan, X.; Li, Y.; Ou, W.; Yu, F.; Chen, J.; Wei, Y. Comparison between hybrid laser-MIG welding and MIG welding for the invar36 alloy. *Opt. Laser Technol.* **2016**, *85*, 75–84. [CrossRef]
11. Huang, H.; Zhang, P.; Yan, H.; Liu, Z.; Yu, Z.; Wu, D.; Shi, H.; Tian, Y. Research on weld formation mechanism of laser-MIG arc hybrid welding with butt gap. *Opt. Laser Technol.* **2021**, *133*, 106530. [CrossRef]
12. Zhang, L.J.; Bai, Q.L.; Ning, J.; Wang, A.; Yang, J.N.; Yin, X.Q.; Zhang, J.X. A comparative study on the microstructure and properties of copper joint between MIG welding and laser-MIG hybrid welding. *Mater. Des.* **2016**, *110*, 35–50. [CrossRef]
13. Vorontsov, A.; Zykova, A.; Chumaevskii, A.; Osipovich, K.; Rubtsov, V.; Astafurova, E.; Kolubaev, E. Advanced high-strength AA5083 welds by high-speed hybrid laser-arc welding. *Mater. Lett.* **2021**, *291*, 129594. [CrossRef]
14. Liu, S.; Li, Y.; Liu, F.; Zhang, H.; Ding, H. Effects of relative positioning of energy sources on weld integrity for hybrid laser arc welding. *Opt. Lasers Eng.* **2016**, *81*, 87–96. [CrossRef]
15. Gao, X.; Wang, Y.; Chen, Z.; Ma, B.; Zhang, Y. Analysis of welding process stability and weld quality by droplet transfer and explosion in MAG-laser hybrid welding process. *J. Manuf. Process.* **2018**, *32*, 522–529. [CrossRef]
16. Li, Y.; Geng, S.; Zhu, Z.; Wang, Y.; Mi, G.; Jiang, P. Stability evaluation of laser-MAG hybrid welding process. *Opt. Laser Technol.* **2019**, *116*, 284–292.
17. Li, Y.; Geng, S.; Zhu, Z.; Wang, Y.; Mi, G.; Jiang, P. Effects of heat source configuration on the welding process and joint formation in ultra-high power laser-MAG hybrid welding. *J. Manuf. Process.* **2022**, *77*, 40–53. [CrossRef]
18. Yang, T.; Chen, L.; Zhuang, Y.; Liu, J.; Chen, W. Arcs interaction mechanism in Plasma-MIG hybrid welding of 2219 aluminium alloy. *J. Manuf. Process.* **2020**, *56*, 635–642. [CrossRef]
19. Kruse, R.; Schwecke, E.; Heinsohn, J. *Uncertainty and Vagueness in knowledge Based Systems: Numerical Methods*; Springer Science & Business Media: Berlin/Heidelberg, Germany, 2012.
20. Adomaitis, R.A.; Çinar, A. Numerical singularity analysis. *Chem. Eng. Sci.* **1991**, *46*, 1055–1062. [CrossRef]
21. Ai, Y.; Jiang, P.; Shao, X.; Li, P.; Wang, C. A three-dimensional numerical simulation model for weld characteristics analysis in fiber laser keyhole welding. *Int. J. Heat Mass Transf.* **2017**, *108*, 614–626. [CrossRef]

22. Ai, Y.; Jiang, P.; Shao, X.; Li, P.; Wang, C.; Mi, G.; Geng, S.; Liu, Y.; Liu, W. The prediction of the whole weld in fiber laser keyhole welding based on numerical simulation. *Appl. Therm. Eng.* **2017**, *113*, 980–993. [CrossRef]

23. Wang, H.; Jing, H.Y.; Zhao, L.; Han, Y.D.; Xu, L.Y. Study on residual stress in socket weld by numerical simulation and experiment. *Sci. Technol. Weld. Join.* **2016**, *21*, 504–514. [CrossRef]

24. Wang, X.; Shao, M.; Gao, S.; Gau, J.-T.; Tang, H.; Jin, H.; Liu, H. Numerical simulation of laser impact spot welding. *J. Manuf. Process.* **2018**, *35*, 396–406. [CrossRef]

25. Wang, Q.; Liu, X.; Wang, P.; Xiong, X.; Fang, H. Numerical simulation of residual stress in 10Ni5CrMoV steel weldments. *J. Mater. Process. Technol.* **2016**, *240*, 77–86. [CrossRef]

26. Bai, S.; Fang, G.; Zhou, J. Integrated physical and numerical simulations of weld seam formation during extrusion of magnesium alloy. *J. Mater. Process. Technol.* **2019**, *266*, 82–95. [CrossRef]

27. Ma, M.; Lai, R.; Qin, J.; Wang, B.; Liu, H.; Yi, D. Effect of weld reinforcement on tensile and fatigue properties of 5083 aluminum metal inert gas (MIG) welded joint: Experiments and numerical simulations. *Int. J. Fatigue* **2021**, *144*, 106046. [CrossRef]

28. Chen, L.; Guo, Z.; Zhang, C.; Li, Y.; Jia, Y.; Liu, G. Experiments and numerical simulations on joint formation and material flow during resistance upset welding of WC-10Co and B318 steel. *J. Mater. Process. Technol.* **2021**, *296*, 117164. [CrossRef]

29. Zhang, Q.-H.; Ma, Y.; Cui, C.; Chai, X.-Y.; Han, S.-H. Experimental investigation and numerical simulation on welding residual stress of innovative double-side welded rib-to-deck joints of orthotropic steel decks. *J. Constr. Steel Res.* **2021**, *179*, 106544. [CrossRef]

30. Ai, Y.; Zheng, K.; Shin, Y.C.; Wu, B. Analysis of weld geometry and liquid flow in laser transmission welding between polyethylene terephthalate (PET) and Ti6Al4V based on numerical simulation. *Opt. Laser Technol.* **2018**, *103*, 99–108. [CrossRef]

31. Wang, M.; Guo, K.; Wei, Y.; Cao, C.; Tong, Z. Welding process optimization for the inner tank of the electric water heater by numerical simulation and experimental study. *J. Manuf. Process.* **2023**, *85*, 52–68. [CrossRef]

32. Bakir, N.; Üstündağ, O.; Gumenyuk, A.; Rethmeier, M. Experimental and numerical study on the influence of the laser hybrid parameters in partial penetration welding on the solidification cracking in the weld root. *Weld. World* **2020**, *64*, 501–511. [CrossRef]

33. Meng, Y.; Jiang, L.; Cen, L.; Gao, M. Improved mechanical properties of laser-arc hybrid welded Al/steel dissimilar butt-joint through beam oscillation. *Sci. Technol. Weld. Join.* **2021**, *26*, 487–492. [CrossRef]

34. Mirakhorli, F.; Nadeau, F.; Guillemette, G.C. Single pass laser cold-wire welding of thick section AA6061-T6 aluminum alloy. *J. Laser Appl.* **2018**, *30*, 032421. [CrossRef]

35. Shi, L.; Li, X.; Jiang, L.; Gao, M. Numerical study of keyhole-induced porosity suppression mechanism in laser welding with beam oscillation. *Sci. Technol. Weld. Join.* **2021**, *26*, 349–355. [CrossRef]

36. Cho, J.H.; Na, S.J. Implementation of real-time multiple reflection and Fresnel absorption of laser beam in keyhole. *J. Phys. D Appl. Phys.* **2006**, *39*, 5372. [CrossRef]

37. Lin, R.; Wang, H.-P.; Lu, F.; Solomon, J.; Carlson, B.E. Numerical study of keyhole dynamics and keyhole-induced porosity formation in remote laser welding of Al alloys. *Int. J. Heat Mass Transf.* **2017**, *108*, 244–256. [CrossRef]

38. Li, L.; Peng, G.; Wang, J.; Gong, J.; Meng, S. Numerical and experimental study on keyhole and melt flow dynamics during laser welding of aluminium alloys under subatmospheric pressures. *Int. J. Heat Mass Transf.* **2019**, *133*, 812–826. [CrossRef]

39. Cho, W.I.; Woizeschke, P. Analysis of molten pool dynamics in laser welding with beam oscillation and filler wire feeding. *Int. J. Heat Mass Transf.* **2021**, *164*, 120623. [CrossRef]

40. Yu, H.; Zhan, X.; Kang, Y.; Xia, P.; Feng, X. Numerical simulation optimization for laser welding parameter of 5A90 Al-Li alloy and its experiment verification. *J. Adhes. Sci. Technol.* **2019**, *33*, 137–155. [CrossRef]

41. Tsirkas, S.A. Numerical simulation of the laser welding process for the prediction of temperature distribution on welded aluminium aircraft components. *Opt. Laser Technol.* **2018**, *100*, 45–56. [CrossRef]

42. Farrokhi, F.; Endelt, B.; Kristiansen, M. A numerical model for full and partial penetration hybrid laser welding of thick-section steels. *Opt. Laser Technol.* **2019**, *111*, 671–686. [CrossRef]

43. Geng, S.; Jiang, P.; Shao, X.; Guo, L.; Gao, X. Heat transfer and fluid flow and their effects on the solidification microstructure in full-penetration laser welding of aluminum sheet. *J. Mater. Sci. Technol.* **2020**, *46*, 50–63. [CrossRef]

44. Kik, T.; Górka, J. Numerical Simulations of Laser and Hybrid S700MC T-Joint Welding. *Materials* **2019**, *12*, 516. [CrossRef]

45. Gao, X.S.; Wu, C.S.; Goecke, S.F.; Kügler, H. Numerical simulation of temperature field, fluid flow and weld bead formation in oscillating single mode laser-GMA hybrid welding. *J. Mater. Process. Technol.* **2017**, *242*, 147–159. [CrossRef]

46. Sun, G.F.; Wang, Z.D.; Lu, Y.; Zhou, R.; Ni, Z.H.; Gu, X.; Wang, Z.G. Numerical and experimental investigation of thermal field and residual stress in laser-MIG hybrid welded NV E690 steel plates. *J. Manuf. Process.* **2018**, *34*, 106–120. [CrossRef]

47. Liu, S.; Kouadri-Henni, A.; Gavrus, A. DP600 dual phase steel thermo-elasto-plastic constitutive model considering strain rate and temperature influence on FEM residual stress analysis of laser welding. *J. Manuf. Process.* **2018**, *35*, 407–419. [CrossRef]

48. Liu, S.; Kouadri-Henni, A.; Gavrus, A. Numerical simulation and experimental investigation on the residual stresses in a laser beam welded dual phase DP600 steel plate: Thermo-mechanical material plasticity model. *Int. J. Mech. Sci.* **2017**, *122*, 235–243. [CrossRef]

49. Lei, Z.; Zhu, Z.; Chen, H.; Li, Y. Fusion enhancement of hollow tungsten arc coaxially assisted by fiber laser. *Opt. Laser Technol.* **2022**, *150*, 107905. [CrossRef]

50. Kim, Y.C.; Hirohata, M.; Murakami, M.; Inose, K. Effects of heat input ratio of laser–arc hybrid welding on welding distortion and residual stress. *Weld. Int.* **2015**, *29*, 245–253. [CrossRef]

51. Cai, Y.; Luo, Y.; Tang, F.; Wang, X.; Zhang, F.; Peng, Y.; Yang, S. Effect of laser energy excitation position on the microstructure in laser-arc hybrid heat source processing. *Weld. World* **2022**, *66*, 879–894. [CrossRef]
52. Zhu, Z.; Lei, Z.; Li, F.; Zhang, X.; Li, Y.; Chen, H. Formation mechanism of double-critical effect in hollow tungsten arc coaxially assisted by fiber laser welding. *Opt. Laser Technol.* **2023**, *159*, 109031. [CrossRef]
53. Shi, L.; Jiang, L.; Gao, M. Numerical research on melt pool dynamics of oscillating laser-arc hybrid welding. *Int. J. Heat Mass Transf.* **2022**, *185*, 122421. [CrossRef]
54. Du, Z.; Sun, X.; Ng, F.L.; Chew, Y.; Tan, C.; Bi, G. Thermo-metallurgical simulation and performance evaluation of hybrid laser arc welding of chromium-molybdenum steel. *Mater. Des.* **2021**, *210*, 110029. [CrossRef]
55. Jiang, Z.; Hua, X.; Huang, L.; Wu, D.; Li, F.; Zhang, Y. Double-sided hybrid laser-MIG welding plus MIG welding of 30-mm-thick aluminium alloy. *Int. J. Adv. Manuf. Technol.* **2018**, *97*, 903–913. [CrossRef]
56. Xin, Z.; Yang, Z.; Zhao, H.; Chen, Y. Comparative Study on Welding Characteristics of Laser-CMT and Plasma-CMT Hybrid Welded AA6082-T6 Aluminum Alloy Butt Joints. *Materials* **2019**, *12*, 3300. [CrossRef] [PubMed]
57. Han, J.; Han, Y.; Sun, Z.; Hong, H. Effect of plasma welding current on heat source penetration ability of plasma-GMAW hybrid welding. *Int. J. Adv. Manuf. Technol.* **2022**, *123*, 1835–1844. [CrossRef]
58. Gao, Y.; Hao, K.; Xu, L.; Han, Y.; Zhao, L.; Ren, W.; Jing, H. Microstructure homogeneity and mechanical properties of laser-arc hybrid welded AZ31B magnesium alloy. *J. Magnes. Alloy.* **2022**, *in press*. [CrossRef]
59. Tang, G.; Zhao, X.; Li, R.; Liang, Y.; Jiang, Y.; Chen, H. The effect of arc position on laser-arc hybrid welding of 12-mm-thick high strength bainitic steel. *Opt. Laser Technol.* **2019**, *121*, 105780. [CrossRef]
60. Zhang, C.; Gao, M.; Zeng, X. Influences of synergy effect between laser and arc on laser-arc hybrid welding of aluminum alloys. *Opt. Laser Technol.* **2019**, *120*, 105766. [CrossRef]
61. Cho, Y.T.; Na, S.J. Numerical analysis of plasma in CO 2 laser and arc hybrid welding. *Int. J. Precis. Eng. Manuf.* **2015**, *16*, 787–795. [CrossRef]
62. Wu, D.; Zhang, P.; Yu, Z.; Gao, Y.; Zhang, H.; Chen, H.; Chen, S.; Tian, Y. Progress and perspectives of in-situ optical monitoring in laser beam welding: Sensing, characterization and modeling. *J. Manuf. Process.* **2022**, *75*, 767–791. [CrossRef]
63. Gao, Z.; Jiang, P.; Shao, X.; Cao, L.; Mi, G.; Wang, Y. Numerical analysis of hybrid plasma in fiber laser-arc welding. *J. Phys. D Appl. Phys.* **2018**, *52*, 025206. [CrossRef]
64. Yang, X.; Chen, H.; Zhu, Z.; Cai, C.; Zhang, C. Effect of shielding gas flow on welding process of laser-arc hybrid welding and MIG welding. *J. Manuf. Process.* **2019**, *38*, 530–542. [CrossRef]
65. Hou, Z.-L.; Liu, L.-M.; Lv, X.-Z.; Qiao, J.; Wang, H.-Y. Numerical simulation for pulsed laser–gas tungsten arc hybrid welding of magnesium alloy. *J. Iron Steel Res. Int.* **2018**, *25*, 995–1002. [CrossRef]
66. Xue, B.; Chang, B.; Wang, S.; Hou, R.; Wen, P.; Du, D. Humping Formation and Suppression in High-Speed Laser Welding. *Materials* **2022**, *15*, 2420. [CrossRef]
67. Mu, Z.; Chen, X.; Hu, R.; Lin, S.; Pang, S. Laser induced arc dynamics destabilization in laser-arc hybrid welding. *J. Phys. D Appl. Phys.* **2019**, *53*, 075202. [CrossRef]
68. Chen, X.; Mu, Z.; Hu, R.; Liang, L.; Murphy, A.B.; Pang, S. A unified model for coupling mesoscopic dynamics of keyhole, metal vapor, arc plasma, and weld pool in laser-arc hybrid welding. *J. Manuf. Process.* **2019**, *41*, 119–134. [CrossRef]
69. Wu, C.S.; Zhang, H.T.; Chen, J. Numerical simulation of keyhole behaviors and fluid dynamics in laser–gas metal arc hybrid welding of ferrite stainless steel plates. *J. Manuf. Process.* **2017**, *25*, 235–245. [CrossRef]
70. Gui, X.; Gao, X.; Zhang, Y.; Wu, J. Investigation of welding parameters effects on temperature field and structure field during laser-arc hybrid welding. *Mod. Phys. Lett. B* **2022**, *36*, 2150467. [CrossRef]
71. Chen, C.; Gao, M.; Mu, H.; Zeng, X. Microstructure and mechanical properties in three-dimensional laser-arc hybrid welding of AA2219 aluminum alloy. *J. Laser Appl.* **2019**, *31*, 032005. [CrossRef]
72. Zhang, C.; Zhang, H.; Wang, L.; Gao, M.; Zeng, X. Microcracking and mechanical properties in laser-arc hybrid welding of wrought Al-6Cu aluminum alloy. *Met. Mater. Trans. A* **2018**, *49*, 4441–4445. [CrossRef]
73. Hao, K.; Gong, M.; Xie, Y.; Gao, M.; Zeng, X. Effects of alloying element on weld characterization of laser-arc hybrid welding of pure copper. *Opt. Laser Technol.* **2018**, *102*, 124–129. [CrossRef]
74. Yan, Z.; Chen, S.; Jiang, F.; Tian, O.; Huang, N.; Zhang, S. Weld properties and residual stresses of VPPA Al welds at varying welding positions. *J. Mater. Res. Technol.* **2020**, *9*, 2892–2902. [CrossRef]
75. Qi, X.; Di, H.; Wang, X.; Liu, Z.; Misra, R.; Huan, P.; Gao, Y. Effect of secondary peak temperature on microstructure and toughness in ICCGHAZ of laser-arc hybrid welded X100 pipeline steel joints. *J. Mater. Res. Technol.* **2020**, *9*, 7838–7849. [CrossRef]
76. Churiaque, C.; Sánchez-Amaya, J.M.; Üstündağ, Ö.; Porrua-Lara, M.; Gumenyuk, A.; Rethmeier, M. Improvements of hybrid laser arc welding for shipbuilding T-joints with 2F position of 8 mm thick steel. *Opt. Laser Technol.* **2021**, *143*, 107284. [CrossRef]
77. Valdaytseva, E.A.; Udin, I.N. Determination of the heat source parameters for the case of simultaneous two-sided laser-arc welding of extended T-joints. In *Journal of Physics: Conference Series*; IOP Publishing: Bristol, UK, 2018; Volume 1109, p. 012009.
78. Sun, Z.; Han, Y.; Du, M.; Hong, H.; Tong, J. Numerical simulation of VPPA-GMAW hybrid welding of thick aluminum alloy plates considering variable heat input and droplet kinetic energy. *J. Manuf. Process.* **2018**, *34*, 688–696. [CrossRef]
79. Cho, M.H.; Farson, D.; Lim, Y.C.; Choi, H.W. Hybrid laser/arc welding process for controlling bead profile. *Sci. Technol. Weld. Join.* **2007**, *12*, 677–688. [CrossRef]

80. Qian, X.; Ye, X.; Hou, X.; Jin, H.; Zhang, P.; Yu, Z.; Wu, D.; Fu, K. Research on Residual Stress Distribution in Different Areas of Laser-MAG Arc Hybrid Welding by Numerical Simulation. In *Journal of Physics: Conference Series*; IOP Publishing: Bristol, UK, 2022; Volume 2160, p. 012026.
81. Zhang, Z.; Wu, C.S. Effect of fluid flow in the weld pool on the numerical simulation accuracy of the thermal field in hybrid welding. *J. Manuf. Process.* **2015**, *20*, 215–223. [CrossRef]
82. Liu, M.; Ma, G.; Liu, D.; Yu, J.; Niu, F.; Wu, D. Microstructure and mechanical properties of aluminum alloy prepared by laser-arc hybrid additive manufacturing. *J. Laser Appl.* **2020**, *32*, 022052. [CrossRef]
83. Cui, D.; Zhang, Y.; He, F.; Ma, J.; Zhang, K.; Yang, Z.; Li, J.; Wang, Z.; Kai, J.-J.; Wang, J.; et al. Heterogeneous microstructure of the bonding zone and its dependence on preheating in hybrid manufactured Ti-6Al-4V. *Mater. Res. Lett.* **2021**, *9*, 422–428. [CrossRef]
84. Gong, M.; Meng, Y.; Zhang, S.; Zhang, Y.; Zeng, X.; Gao, M. Laser-arc hybrid additive manufacturing of stainless steel with beam oscillation. *Addit. Manuf.* **2020**, *33*, 101180. [CrossRef]
85. Wang, L.; Gao, M.; Zeng, X. Experiment and prediction of weld morphology for laser oscillating welding of AA6061 aluminium alloy. *Sci. Technol. Weld. Join.* **2019**, *24*, 334–341. [CrossRef]
86. Hao, K.; Li, G.; Gao, M.; Zeng, X. Weld formation mechanism of fiber laser oscillating welding of austenitic stainless steel. *J. Mater. Process. Technol.* **2015**, *225*, 77–83. [CrossRef]
87. Wu, M.; Luo, Z.; Li, Y.; Liu, L.; Ao, S. Effect of heat source parameters on weld formation and defects of oscillating laser-TIG hybrid welding in horizontal position. *J. Manuf. Process.* **2022**, *83*, 512–521. [CrossRef]

Article

Experimental Study of Steel–Aluminum Joints Made by RSW with Insert Element and Adhesive Bonding

Anna Guzanová [1,*], Janette Brezinová [1], Ján Varga [1], Miroslav Džupon [2], Marek Vojtko [2], Erik Janoško [1], Ján Viňáš [1], Dagmar Draganovská [1] and Ján Hašuľ [1]

1 Department of Technology, Materials and Computer Supported Production, Faculty of Mechanical Engineering, Technical University of Košice, Mäsiarska 74, 040 01 Košice, Slovakia
2 Institute of Materials Research, Slovak Academy of Sciences, Watsonova 1935/47, 040 01 Košice, Slovakia
* Correspondence: anna.guzanova@tuke.sk

Abstract: This work focuses on joining steel to aluminum alloy using a novel method of joining by resistance spot welding with an insert element based on anticorrosive steel in combination with adhesive bonding. The method aims to reduce the formation of brittle intermetallic compounds by using short welding times and a different chemical composition of the insert element. In the experiment, deep-drawing low-carbon steel, HSLA zinc-coated steel and precipitation-hardened aluminum alloy 6082 T6 were used. Two types of adhesives—one based on rubber and the other based on epoxy resin—were used for adhesive bonding, while the surfaces of the materials joined were treated with a unique adhesion-improving agent based on organosilanes. The surface treatment improved the chemical bonding between the substrate and adhesive. It was proved, that the use of an insert element in combination with adhesive bonding is only relevant for those adhesives that have a load capacity just below the yield strength of the substrates. For bonded joints with higher load capacities, plastic deformation of the substrates occurs, which is unacceptable, and thus, the overall contribution of the insert element to the load capacity of the joint becomes negligible. The results also show that the combination of the resistance spot welding of the insert element and adhesive bonding facilitates the joining process of galvanized and nongalvanized steels with aluminum alloys and suppresses the effect of brittle intermetallic phases by minimizing the joining area and welding time. It is possible to use the synergistic effect of insert element welding and adhesive bonding to achieve increased energy absorption of the joint under stress.

Keywords: hybrid joining of dissimilar materials; spot resistance welding; adhesive bonding; load-carrying capacity of joint

Citation: Guzanová, A.; Brezinová, J.; Varga, J.; Džupon, M.; Vojtko, M.; Janoško, E.; Viňáš, J.; Draganovská, D.; Hašuľ, J. Experimental Study of Steel–Aluminum Joints Made by RSW with Insert Element and Adhesive Bonding. *Materials* **2023**, *16*, 864. https://doi.org/10.3390/ma16020864

Academic Editor: Raul D.S.G. Campilho

Received: 13 December 2022
Revised: 9 January 2023
Accepted: 12 January 2023
Published: 16 January 2023

1. Introduction

The issue of joining dissimilar materials is an area of interest, particularly in the automotive industry, sustainably focused on a reduction in weight and emissions and performance enhancement. The reason for this is the high degree of automation of the process [1,2]. Joining different types of materials involves two significant factors, namely, welding and mechanical joining, and each of these methods has its own advantages and disadvantages.

Welding involves the different melting temperatures of the materials, resulting in the formation of brittle intermetallic compounds at the steel–Al alloy interface [3–11]. Many authors [8–14] have taken the route of reducing the heat input during welding [3–7,15,16], thereby reducing the thickness of brittle IMCs, or have taken advantage of the better solubilities of Al and Zn compared to the very low solubilities of Fe and Al and have used the zinc interlayer and the associated eutectic reaction for joining [8–13]. Experiments were carried out focusing on a reduction in the IMC layer thickness using an ultrasonic resistance welding process or a novel projection welding technology, consisting of joining an insert element (a short piece of welding filler of suitable composition) to an aluminum plate and then joining the aluminum plate to the steel plate using RSW technology [17,18]. It is also appropriate to support the strength of the joints with, for example, adhesive bonding. Adhesive bonding technology provides many advantages; however, in the bonding process, we encounter the need for surface preparation, which can cause a delay in the production line. For this reason, adhesives requiring no surface preparation or adhesion promoters that are easily applied at room temperature are being developed [19,20]. In the bonding process, it should be kept in mind that the application of adhesives requires curing time, which may cause deformations between Al and steel parts due to different coefficients of thermal expansion. To avoid this undesirable effect, combinations of adhesive and thermal or mechanical bonding are beginning to be applied. Gullino et al. [14] in their experiment showed the advantages of combining spot welding and adhesive bonding using fusion welding and solid-state welding, with a focus on steel–aluminum alloy joining. In addition to the formation of IMCs, another problem arises when joining dissimilar materials with unequal thicknesses, namely, the problem of the subsequent formability of these tailored blanks. The solid-state welding of dissimilar materials (e.g., FSW), compared to fusion welding methods, is more protective of the joint from reducing the formability of the individual materials [21]. Similarly, Aminzadeh in [22] investigated the differences in the formability of joints of unequal-thickness materials formed by laser welding and the TIG method. In practice, however, the automotive industry prefers existing established technologies, such as resistance spot welding (RSW), mainly because of its short welding time, process stability and cost-effectiveness [23].

To achieve a high-quality resistance spot weld in Al steel, some obstructions, such as oxides formed on the surface of the aluminum alloy, must also be dealt with. The oxides formed can result in an increase in the contact resistance between the Al sheet and the welding electrode, thereby limiting the effectiveness of interface heating in the weld area [24,25]. Another problem to be considered in the spot resistance welding of Al on steel is the possibility of electrode wear due to the adhesion of Al to the copper electrode [26]. The author Madhusudana [27] investigated the joining of two materials, 2 mm thick 5054 Al alloy and 1 mm thick galvanized steel, using resistance spot welding. The welds were made with IMC thicknesses of less than 5.5 μm, where the needle-like shape of $FeAl_3$ near the Al sheet and the toot-like shape of Fe_2Al_5 near the galvanized steel were observed.

When joining hybrid structures, such as aluminum alloy with steel, it is necessary to solve technical problems. Since the properties of the materials to be joined are different, especially the density or melting temperature, during welding, solidification segregation occurs in the weld bath, when inferior joints between aluminum alloy and steel are formed. Residual stresses in the joint occur due to large differences in the coefficient of expansion between the materials being joined, resulting in cracks or the distortion of the structure. The formation of intermetallic compounds and thus probable cracks is also assumed based on the Fe-Al binary phase diagram [28,29]. In some studies, we are informed that the formed Al-Fe IMCs at the interface are an important element to achieve a high-strength joint formed by friction stir welding. In a study of the tensile shear strength of a spot weld of galvanized steel, the value was lower than that of uncoated steel [30,31].

For steel–aluminum joining, mechanical joining methods are widely used due to the low thermal change in the microstructure, as reported in [16,32], and are thus suitable for the bonding process. Joining materials by self-piercing riveting (SPR) is among the most advanced cold-forming methods used for joining aluminum alloys to steel. The advantage is that in mechanical joining, the different melting temperatures of the materials being joined are not a problem. Similarly, brazing and friction stir welding can successfully be used for joining dissimilar materials. Mechanical joining techniques can be very easily combined with adhesive bonding. Ezzine [33,34] successfully focused on hybrid joining using adhesive bonding and riveting. When the requirement is to join dissimilar materials over a larger continuous area, cold-rolling joining comes into consideration. Rahmatabadi [35] tested this in the fabrication of three-layer Al/Mg/Al composites.

The authors were inspired by Zvorykina et al. [18] and designed an experiment focused on joining zinc-coated and deep-drawn uncoated steel to aluminum alloy by spot resistance welding with an insert element and also combined this process with adhesive bonding using rubber- and epoxy-based adhesives. The results obtained could provide more information about joining these materials and the application of the process in the automotive industry.

2. Experimental Section

2.1. Materials

This experimental research was carried out with the following materials:

-   DC04: Deep-drawn, uncoated, cold-rolled low-carbon steel for bodywork, manufactured in compliance with EN 10,130:2006 and EN 10,131:2006. The thickness of the DC04 substrate was 0.8 mm. Hereafter: DC.
-   TL 1550-220+Z: Zinc-galvanized fine-grained high-strength low-alloy steel with increased cold formability, manufactured in compliance with VW TL 1550:2008-12 and EN 10143:2006-12. The zinc layer applied was 104 g/m^2. The thickness of the TL 1550-220+Z substrate was 0.8 mm. Hereafter: TL.
-   EN AW-6082 T6: Precipitation-hardened aluminum alloy AlSi1MgMn, manufactured in compliance with EN 573-3, EN-485-1+A1, EN-485-2+A1 and EN-485-4. The substrate thickness of the test specimen was 1 mm. Hereafter: Al.

The mechanical characteristics of the materials used are given in Table 1, and the chemical properties are described in Table 2. The mechanical properties are from a tensile test performed by the manufacturer, and the chemical composition is from the metallurgical certificate provided by the material manufacturer, valid for both the lot and cast number used in experiments.

Table 1. Mechanical properties and some specific conditions of materials.

	YS [MPa]	UTS [MPa]	Elongation [%]	Thickness [mm]	Conditions
DC	197	327	39	0.8	Electrostatically oiled
TL	292	373	34	0.8	Zn-coated
Al	290	340	14	1.0	Solution-treated, artificially aged

Table 2. Chemical composition of materials, wt. %.

DC				
C	Mn	P	S	Fe
0.040	0.250	0.009	0.008	bal.

TL									
C	Mn	Si	P	S	Al	Nb	Ti	Cu	Fe
0.100	1.000	0.500	0.080	0.030	0.015	0.100	0.150	0.200	bal.

Al								
Si	Fe	Cu	Mn	Mg	Cr	Zn	Ti	Al
1.00	0.40	0.06	0.44	0.70	0.02	0.08	0.03	bal.

2.2. Shape and Dimensions of Test Samples

Materials were cut to 40×110 mm test samples. Next, the test samples were degreased, coated with an adhesion promoter and joined together by welding, with welding and adhesive bonding having an overlap of 30 mm in the configurations DC-Al and TL-Al

(Figure 1). Five test joints were made for each material combination, joining technology and adhesive used (Table 3).

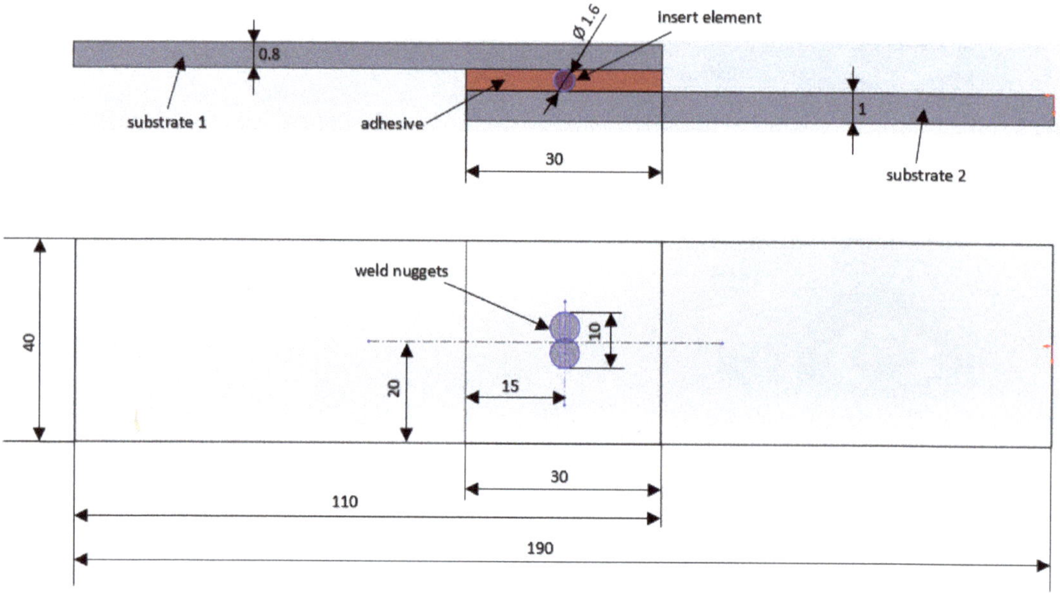

Figure 1. Dimensions of the test joint.

Table 3. Number of joints made by resistance spot welding (RSW) and hybrid RSW and adhesive bonding (AB) technology.

	RSW	RSW + Adhesive 1	RSW + Adhesive 2
DC-Al	5	5	5
TL-Al	5	5	5

2.3. Surface Preparation

The samples were cleaned and degreased in laboratory conditions before joining. Surface preparation consisted of the following immediately sequential steps, applied to both sides of the substrates:

- Degreasing using 5% solution of alkaline degreasing agent (Pragolod 57 N, Pragochema spol. s r. o., Prague, Czech Republic), degreasing time: 10 min; temperature of solution: 60 °C; method: immersion in stirred solution.
- Rinsing in flowing service water, room temperature, 20 s.
- Rinsing in flowing demi water, room temperature, 20 s.
- Application of adhesion promoter based on organosilanes by immersion in solution, immersion time: 10 min.
- Hot-air drying.

The individual surface preparation steps were carried out in five-liter beakers. The individual substrates had a hole made near the edge and were suspended in the solutions on insulated steel wire, 10 pieces per batch. The solutions were sized so that the active ingredients in the solutions were not depleted during preparation.

After this procedure, materials immediately proceeded to adhesive bonding and spot resistance welding.

2.4. Microgeometry of the Contact Surface

The contact surfaces of the individual materials were evaluated with a Surftest SJ-301 stylus profilometer in accordance with STN EN ISO 21920-2:2022—Geometric Specification of Products (GPS). The arithmetic mean deviation of the profile Ra, the maximum height of the profile Rz, the mean width of the profile elements RSm and the number of peaks per centimeter RPc were used as monitored parameters. The evaluation length was 4.0 mm, and the λc–profile filter was set to 0.8 mm. Roughness measurements were carried out 10 times on each type of surface, and the arithmetic means of these measurements are presented. The surfaces of the materials were also observed by SEM. EDX planar element analysis was performed in different areas of the joints on a scanning electron microscope EVO MA15 EDX/WDX (Oxford Instruments, Abingdon, UK).

2.5. Joining of Materials by Resistance Welding with Insert Element

A novel method of joining by resistance spot welding using an insert element was implemented. In the first stage, the insert element, 308 LSI PR welding wire with a diameter of 1.6 mm and a length of 10 mm, was welded to the Al sample; in the second stage, the welded insert element was covered by a steel plate, and resistance welding was performed again. The joining procedure is shown in Figure 2. Spot resistance welding was carried out on a Nimak Magnetic Drive machine (NIMAK GmbH, Wissen, Germany) with the welding parameters listed in Table 4.

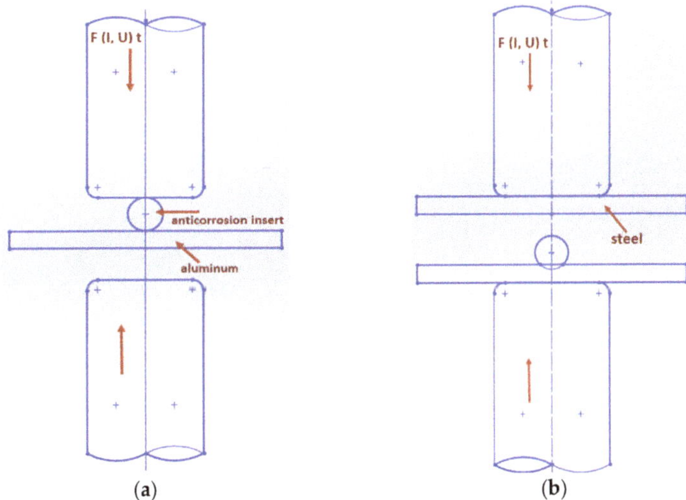

Figure 2. Schematic representation of joint formation by resistance spot welding: (**a**) welding of insert element to aluminum and (**b**) welding of steel plate to insert–aluminum joint.

Table 4. Spot resistance welding parameters.

	Stage 1	Stage 2
Welding force, F [kN]	2	4
Welding time, t [ms]	10	16
Welding power, I [kA]	8	12

Table 5 shows the chemical properties of the insert element, which is 308 LSI PR welding wire.

Table 5. Chemical composition of insert element, wt. %.

	C	Mn	Si	Cr	Ni	Mo	Cu	Fe
ER 308LSi	0.02	1.8	0.85	20	10	0.2	0.2	bal.

2.6. Joining of Materials by Resistance Welding with Insert Element and Adhesive Bonding

This type of joint was created by welding the insert element to the aluminum plate. Next, a layer of the adhesive was applied on the aluminum substrate manually using a plastic blade. The thickness of the adhesive was at the level of the insert element height. The joint was then covered by the steel plate, and resistance welding was performed again. As the last step, when double RSW was completed, the curing of the adhesive took place at 175 °C for 25 min.

The principle of making joints with an adhesive is shown in Figure 3.

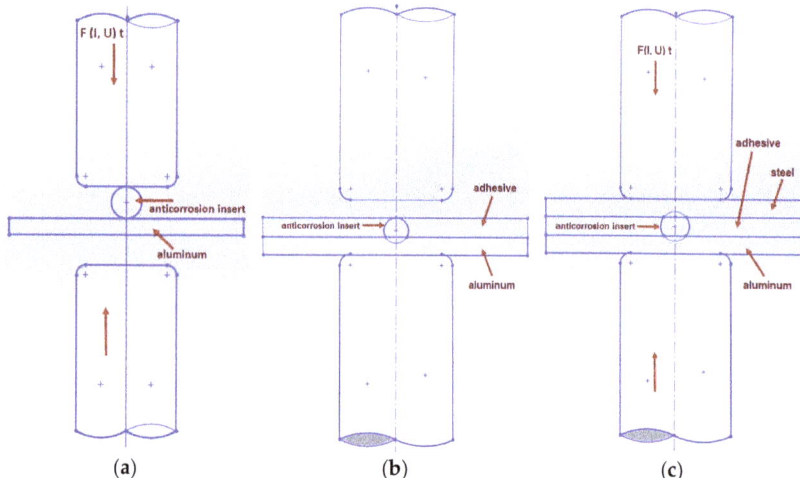

Figure 3. Schematic representation of joint formation using adhesive: (**a**) welding of insert element to aluminum, (**b**) application of adhesive on aluminum substrate, (**c**) welding of steel plate to insert–aluminum joint.

The following types of adhesives were chosen for the experiments, and their properties are described in more detail in Table 6.

Table 6. Selected properties of adhesives, given by adhesive producer (Henkel AG & Co., KGaA, Düsseldorf, Germany).

	TEROSON RB 5197	**TEROSON EP 5090**
E-module	880 MPa	2 GPa
Tensile strength	12 MPa	35 MPa
Elongation at break	10%	10%
Poisson's ratio	0.4	0.4
Shear strength (DIN EN 1465)	at 20 °C >15 MPa	>30 MPa
Layer thickness	0.2 mm	0.2 mm

- Adhesive 1: TEROSON RB 5197, which is a heat-curing, one-component, rubber-based adhesive with no added solvents.
- Adhesive 2: TEROSON EP 5090, which is a solvent-free, one-component, heat-setting adhesive based on epoxy resins.

Both selected adhesives are used in automotive construction, and both can be combined with other joining technologies, such as spot welding.

2.7. Testing of Load-Carrying Capacity of Joints

The load-carrying capacity of the overlapped welded/bonded joints was then tested under tensile stress on a TIRA test 2300 universal testing machine (TIRA GmbH, Schalkau, Germany) at a testing machine ram speed of 10 mm/min, which corresponds to a quasistatic strain rate of 0.0033 s^{-1}. After the test, the fracture surfaces of the joints and the type of fracture were evaluated. The quality of the joints was also evaluated by analyzing the contact areas (fracture surfaces) of the joints using the scanning electron microscope specified above.

3. Results

3.1. Evaluation of Substrate Microgeometry

Based on the methodology described in Section 2.4, the microgeometry parameters of the substrate surfaces were evaluated. The selected parameters of microgeometry are presented in Table 7. Figure 4 shows the surface profiles and the load-bearing profile curves (Abbott curves), and Figure 5 shows the visual appearance of the materials after the application of the adhesion promoter.

Table 7. Selected parameters of surface microgeometry.

	Ra [μm]	Rz [μm]	RSm [μm]	RPc [-/cm]
DC initial surface	0.87	5.12	300.30	34.20
DC + adhesion promoter	0.94	4.66	256.80	39.86
TL initial surface	1.00	5.11	137.8	73.76
TL + adhesion promoter	0.62	4.03	83.0	120.94
Al initial surface	0.15	1.04	142	74.00
Al + adhesion promoter	0.24	1.61	202	51.00

From the measured data, it can be concluded that DC and TL steels have very similar vertical roughness parameters (Ra and Rz), while the Al alloy sheet has a significantly lower initial roughness. From the measured values of the horizontal roughness parameters and from the profilographs, it is evident that after the application of the adhesion promoter, the greatest changes in terms of the number of peaks per centimeter occurred on the TL substrate, where the number of peaks increased by 64%; on the DC substrate, it increased by 16%, and on the Al substrate, a decrease in the number of peaks per centimeter of 31% was recorded. The number, height and distribution of peaks and valleys on substrates have a positive effect on adhesion during adhesive bond formation, as they increase the contact area for the formation of chemical bonds to the substrate and, finally, increase the likelihood of the mechanical anchoring of the adhesive as well. For very smooth substrates, it is necessary to either roughen the surface prior to adhesive bonding or to provide adhesion using various adhesion promoters. Organosilane-based adhesion promoters applied on substrates contain molecules able to bond to the metallic substrate on one end and bond to paints or adhesives on the other end. This 'click chemistry' can enhance chemical bonding between the substrate and adhesive and also provide resistance against corrosion.

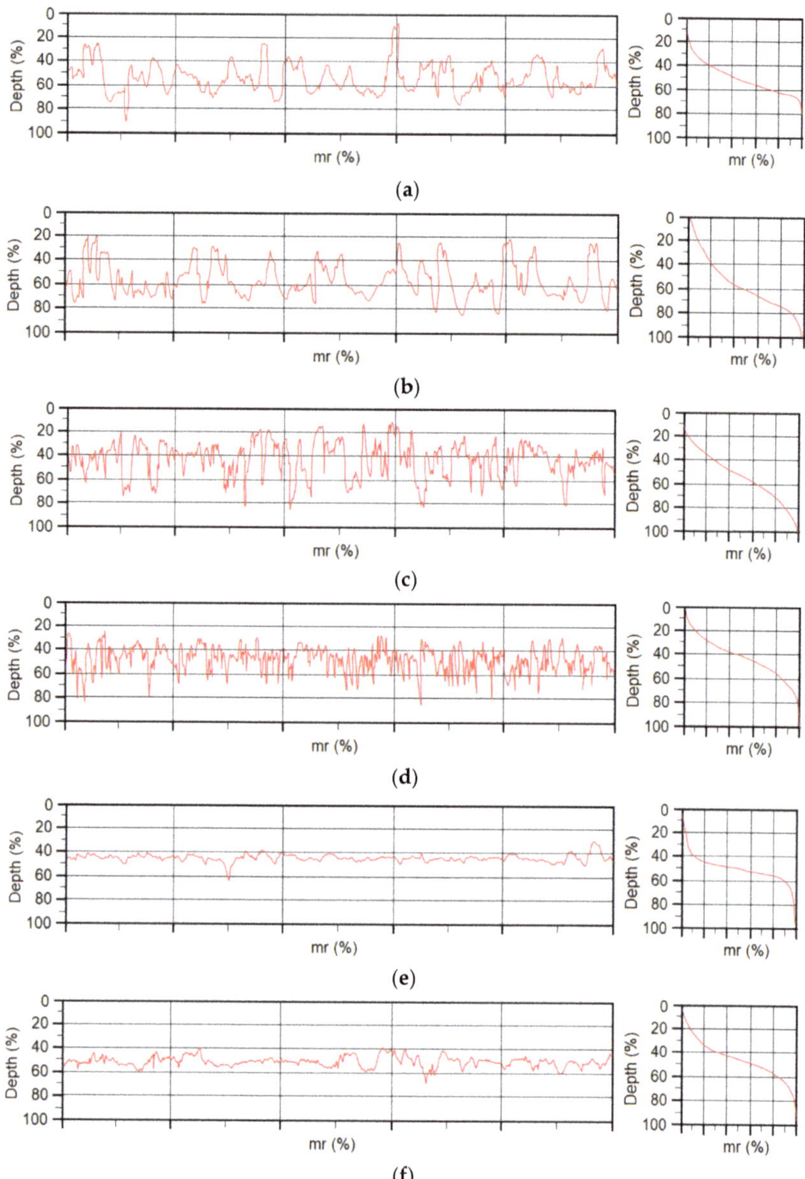

Figure 4. Profilograph of materials in initial state as well as treated with adhesion promoter: (**a**) DC initial surface, (**b**) DC + adhesion promoter, (**c**) TL initial surface, (**d**) TL + adhesion promoter, (**e**) Al initial surface and (**f**) Al + adhesion promoter.

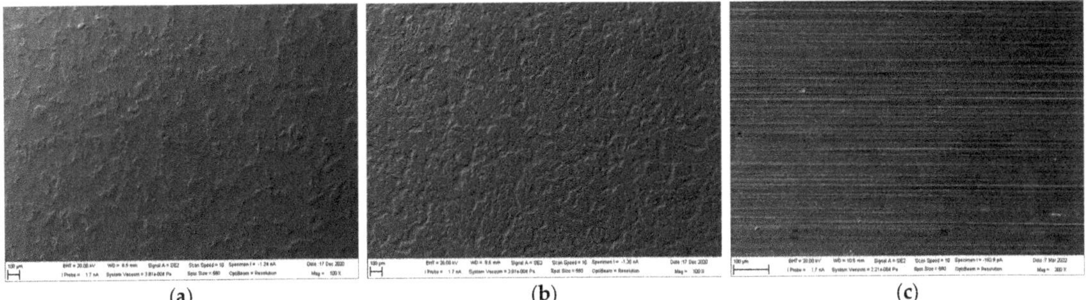

Figure 5. Appearance of materials treated with adhesion promoter: (**a**) DC, (**b**) TL and (**c**) Al.

3.2. Load-Carrying Capacity of the Joints

3.2.1. Joints Formed by Resistance Spot Welding

Figure 6 shows the appearance of the test joints made by resistance spot welding (Al-DC and Al-TL).

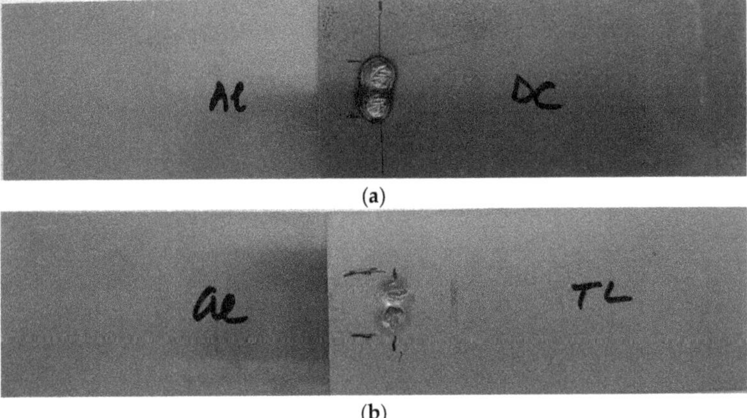

Figure 6. Appearance of test joints made by resistance spot welding: (**a**) Al-DC and (**b**) Al-TL.

Figure 7 presents representatives of fracture surfaces of Al-DC test joints after the tensile shear test, since the process of joint destruction was the same for all five samples. It can be seen that the molten insert element remained on the aluminum substrate after fracture in four out of five samples (an exception is shown in Figure 7b). The red arrows indicate the locations where the molten insert element remained after destruction. Around the spot weld on the aluminum substrate, the extrusion of the aluminum material from the weld location is visible. Plastic deformation of the samples did not occur in any of the tested joints.

Similar to the Al-DC test samples, only two representative samples are shown after the tensile shear test of Al-TL joints (Figure 8). The same pattern of joint destruction was observed for the four samples, with the greater part of the bonding element remaining welded to the Al substrate, as is shown in Figure 8a. An exception is sample 4 (Figure 8b), where most of the bonding element remained on the TL substrate. On the TL substrate, in the area around the spot weld, extrusion of the Zn layer occurred. As with Al-DC, we do not observe plastic deformation in any sample.

The load–displacement curves obtained in the tensile shear test of Al-DC and Al-TL test joints are shown in Figure 9. The load curves are nearly linear up to the point of failure and then cascade downward, corresponding to the gradual detachment of the substrates from the insert element. A slightly higher load at failure was recorded for the Al-TL joint compared to the Al-DC joint.

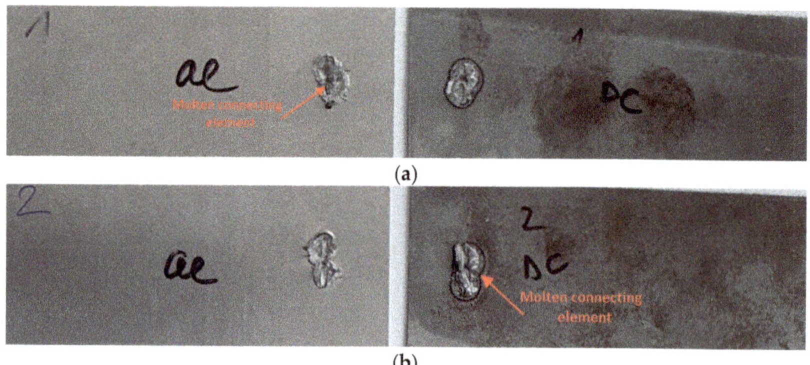

Figure 7. Fracture surfaces of some Al-DC test joints after tensile shear test: (**a**) sample 1 and (**b**) sample 2.

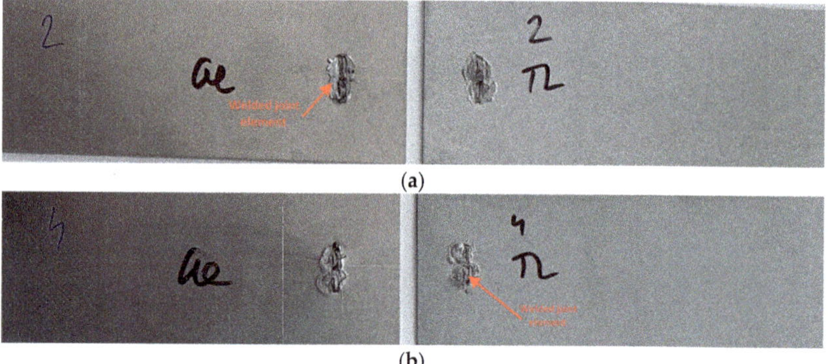

Figure 8. Fracture surfaces of some Al-TL test joints after tensile shear test: (**a**) sample 2 and (**b**) sample 4.

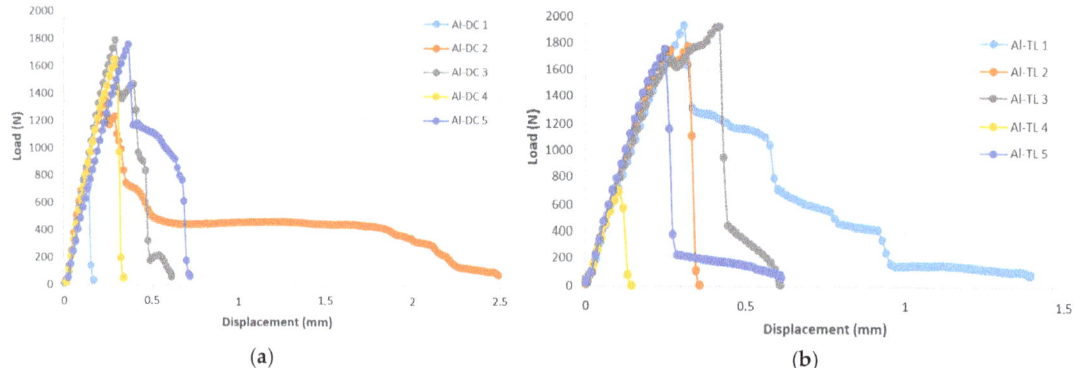

Figure 9. Load–displacement curves for (**a**) Al-DC and (**b**) Al-TL joints.

From the maximum load F_{max} at failure and the cross-sectional area A of individual substrates, the stress σ in each substrate was calculated using Formula (1). The aim of this calculation was to prove whether the yield strength (YS) had been overcome in any of the substrates and thus whether plastic deformation in any of the substrates had occurred. The results are listed in Table 8.

$$\sigma = \frac{F_{max}}{A} \quad [\text{MPa}] \tag{1}$$

Table 8. Maximum load Fmax [N] and stress σ [MPa] in substrates.

Sample No.	Al-DC			Al-TL		
	Fmax (Al-DC)	σ (DC)	σ (Al)	Fmax (Al-TL)	σ (TL)	σ (Al)
1	848	26.52	21.22	1933	60.42	48.33
2	1351	42.22	33.77	1778	55.56	44.44
3	1785	55.78	44.62	1923	60.10	48.08
4	1645	51.39	41.11	700	21.86	17.48
5	1755	54.84	43.87	1758	54.94	43.95

If we consider the YS of DC steel to be equal to 197 MPa and the width of the substrate is 40 mm with a thickness of 0.8 mm, the load at the yield point equals 6304 N. For the YS of TL steel (292 MPa) and the same dimensions of the substrate, the load at the yield point equals 9344 N, and, finally, for the YS of Al alloy (290 MPa) and sample dimensions of 40 × 1 mm, the load at the yield point equals 11 600 N.

Since the stresses in the substrates (Table 8) are smaller than the yield strength values of Al, DC and TL, no plastic deformation has occurred in any of the substrates.

3.2.2. Joints Formed by Resistance Spot Welding and Rubber-Based Adhesive Bonding

Figure 10 shows a test joint made by resistance spot welding with the insert element and adhesive bonding with a rubber-based adhesive (black in color).

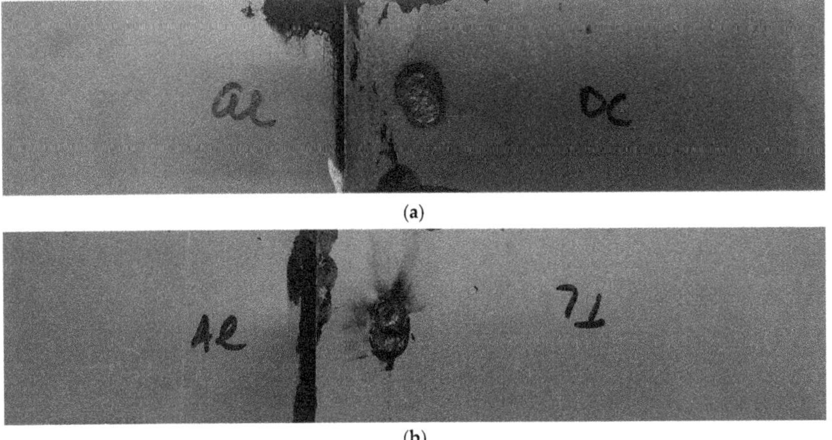

Figure 10. Appearance of test joints made by resistance spot welding and AB: (**a**) Al-DC and (**b**) Al-TL, rubber based adhesive.

Figure 11 presents the test joints after the tensile shear test. The failure process was the same for all samples, so only two samples are documented. The red arrows point to the locations of the insert element, which always remained on the DC substrate. From a macroscopic point of view, no plastic deformation of the substrates is visible. The Al substrate on each of the samples remained perforated after the test, as the joining element, along with a piece of the Al material, was torn away from the Al substrate and remained attached to the DC substrate. This proved the good weld connection between the iron-based insert element and the iron-based DC substrate. However, the bond between the iron-based insert element and Al alloy is also very strong, as the element has been torn out together with part of the Al substrate. The adhesive failure for all samples was 100% cohesive.

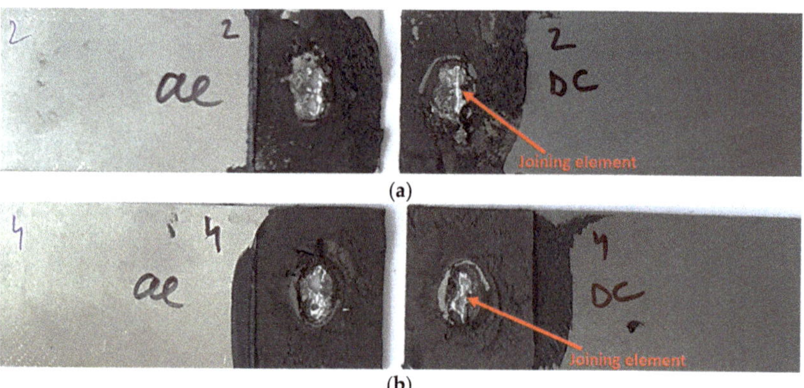

Figure 11. Fracture surfaces of Al-DC test joints made by resistance spot welding and AB after tensile shear test: (**a**) sample 2 and (**b**) sample 4, rubber based adhesive.

The fracture surfaces of Al-TL test joints after the tensile shear test are shown in Figure 12. The red arrows point to the locations where the insert element remained. In this case, due to some, even if limited, solubility of Al and Zn, the insert element remained welded alternately to both the Al (samples 3 and 4) and TL substrates (samples 1, 2 and 5). This means that the joining of the insert element to both surfaces is approximately at the same level. No visible plastic deformation was observed macroscopically. The mode of the adhesive failure is 100% cohesive for all samples.

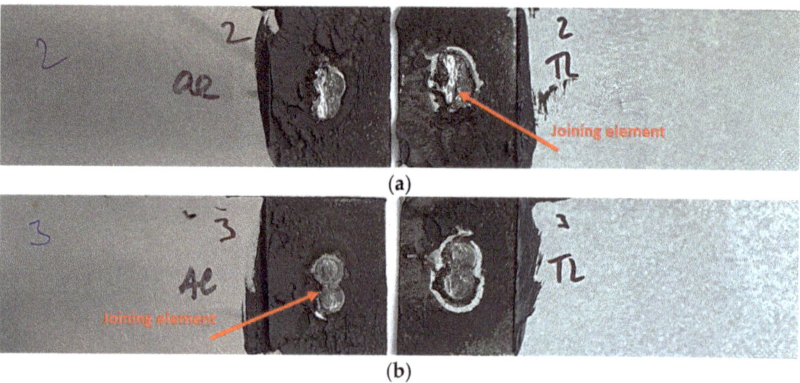

Figure 12. Fracture surfaces of Al-TL test joints made by resistance spot welding and AB after tensile shear test: (**a**) sample 2 and (**b**) sample 3, rubber based adhesive.

The load–displacement curves obtained in the tensile shear test of Al-DC and Al-TL test joints are shown in Figure 13. It can be concluded that, compared to joints made by resistance spot welding only, a higher load was necessary to break the joints due to the adhesive bonding contribution. Until failure, the character of the process is nearly linear. For specimen 5 of Al-DC joints, the load exceeded the yield strength of one of the substrates and caused it to strengthen. The load–displacement relationship resembles a stress–strain diagram of the substrate. At some point in the strengthening, the load exceeded the cohesion of the adhesive, and the failure of the joint occurred. The load at the failure of Al-TL joints is similar to that of Al-DC joints with the rubber adhesive, and it varies from 5 to 6.5 kN.

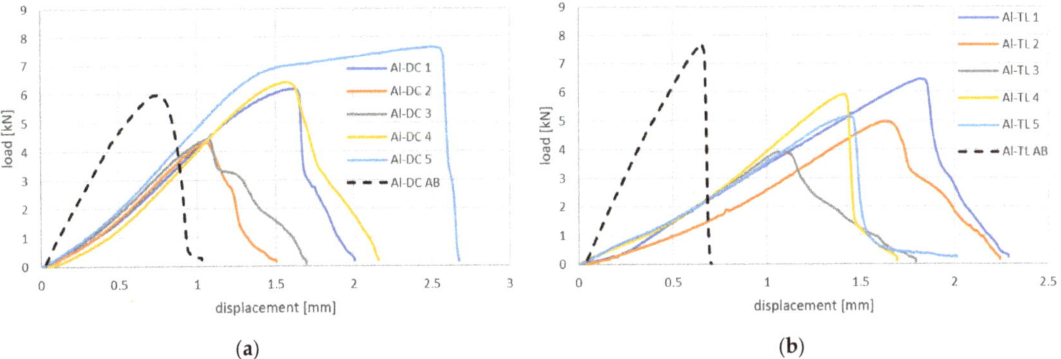

Figure 13. Load–displacement curves for (**a**) Al-DC and (**b**) Al-TL joints made by resistance spot welding + AB, compared with adhesive bonding only (dashed black line).

The dashed black lines in Figure 13 indicate the load–displacement curves of adhesive joints only (without welding or the insert element). It can be seen that using hybrid joining technology–resistance spot welding with the insert element in combination with adhesive bonding is manifested by a larger area under the load–displacement curve, i.e., an increase in joint energy absorption is proven.

A typical load–displacement curve of a purely bonded joint has a typical triangular shape, where the load increases linearly with displacement up to the failure of the joint. After a joint failure, the force immediately decreases almost perpendicular to the x-axis (Figure 14a). When using an insert element, a gradual decrease in force from Fmax to 0 N can be seen in the load–displacement curves, achieved under quasistatic loading. This gradual disappearance of force is caused by the insert element. The failure process can best be seen from the comparison shown in Figure 14b.

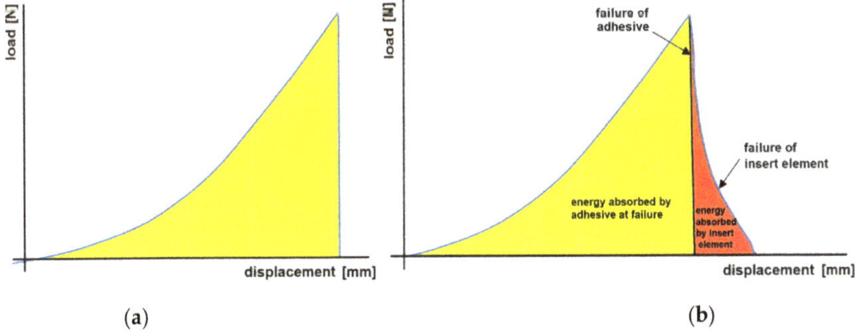

Figure 14. Load–displacement curve for (**a**) pure adhesive-bonded joint and (**b**) hybrid welded joint combined with adhesive bonding.

In Figure 14b, on the descending portion of the curve, the adhesive failure region and the insert element failure region are clearly distinguishable. If we calculate the area under the curve, we can quantify the contribution of the adhesive and insert element to the energy absorbed by the joint. Hence, the importance of using an insert element is in increasing the area under the load–displacement curve and hence not increasing the load capacity of the connection but the energy absorbed by the connection, which is of importance in a crash event.

Table 9 shows the level of stresses in the substrates, from which it can be determined whether the yield strength was overcome and therefore plastic deformation in any of the substrates occurred.

Table 9. Maximum load Fmax [N] and stress σ [MPa] in substrates used in welded-bonded joints (rubber-based adhesive).

Sample No.	Al-DC + AB			Al-TL + AB		
	Fmax (Al-DC)	σ (DC)	σ (Al)	Fmax (Al-TL)	σ (TL)	σ (Al)
1	6180	193.15	154.52	6462	201.95	161.56
2	4356	136.11	108.89	4979	155.58	124.46
3	4431	138.47	110.77	3923	122.60	98.08
4	6419	**200.59**	160.47	5927	185.21	148.17
5	7677	**239.90**	191.92	5149	160.91	128.73

From the above stresses, as well as from the previous load–displacement curves, it is clear that plastic deformation occurred in the DC substrate and in the Al-DC joint (joints 4 and 5) formed by resistance spot welding and adhesion bonding with the rubber-based adhesive.

3.2.3. Joints Formed by Resistance Spot Welding and Epoxy-Based Adhesive Bonding

Figure 15 shows test joints made by resistance spot welding with the insert element and adhesive bonding with an epoxy-based adhesive (purple in color).

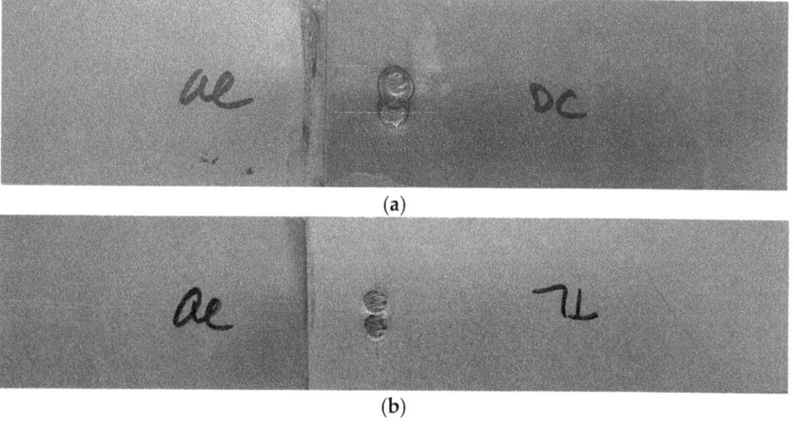

Figure 15. Appearance of test joints made by resistance spot welding and AB: (**a**) Al-DC and (**b**) Al-TL, epoxy based adhesive.

Figure 16 presents Al-DC joints after the tensile shear test. The failure process was the same for all samples, so only two samples are documented. The red arrows point to the locations of the insert element, which always remained on the DC substrate. From a macroscopic point of view, the plastic deformation of DC substrates is visible. Blue arrows indicate the contraction of the DC substrate. The Al substrate on each of the samples remained perforated after the test, as the joining element, along with a piece of the Al material, was torn away from the Al substrate and remained attached to the DC substrate. This proved the good weld connection between the iron-based insert element and iron-based DC substrate. However, the bond between the iron-based insert element and Al alloy is also very strong, as the element has been torn out together with part of the Al substrate. An adhesive–cohesive failure of the bonded area for all samples was identified, but the adhesive failure appears to be dominant over approximately 80% of the bonded area. The adhesive was separated from the DC substrate, always remaining on the Al substrate.

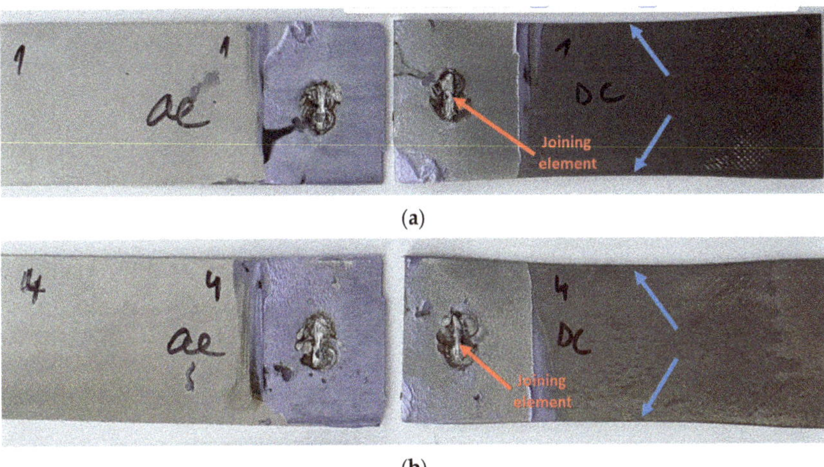

(a)

(b)

Figure 16. Fracture surfaces of Al-DC test joints made by resistance spot welding and AB after tensile shear test: (**a**) sample 1 and (**b**) sample 4, epoxy based adhesive.

Selected fracture surfaces of Al-TL test joints after the tensile shear test are shown in Figure 17. In this case, the insert element always remained welded to the TL substrate. In some Al-TL joints, the insert element detached from the Al substrate without disrupting it, while for other Al-TL joints, a part of the Al substrate along with the insert was torn off during the loading of the joint, causing Al substrate perforation and leaving the insert element attached to the TL substrate. The exception is joint 3, where the joint remained intact, but the TL substrate broke outside the joint, which means that the joint strength exceeds the tensile strength of the TL substrate. Macroscopic plastic deformation was observed in the TL substrate in all cases. The adhesive failure was mixed adhesive–cohesive for all samples, with the adhesive mode prevailing in approximately 90% of the bonded area.

The load–displacement curves obtained in the tensile shear test of Al-DC test joints are shown in Figure 18a. It can be concluded that, compared to Al-DC joints with a rubber-based adhesive, a higher load was necessary to break the joints due to the adhesive bonding contribution. The loading curves indicate that the load-carrying capacity of the joint is high, and the joint resisted in the region of the elastic deformation of the substrates, as well as in the region of the significant plastic deformation of the substrates. The failure of the connection occurred just before the maximum force was reached in the stress–strain diagram. Thus, it can be concluded that the bearing capacity of the connections is almost at the level of the maximum force that the DC substrate can withstand. At some point in substrate strengthening, the load exceeded the adhesion of the adhesive to the DC substrate, and the failure of the joint occurred at different displacement values.

A graphical representation of the load–displacement curves of Al-TL joints is shown in Figure 18b. In this case, the highest loads required to fracture the joints were applied, compared to Al-DC joints with the epoxy-based adhesive, as well as Al-TL joints with the rubber-based adhesive. The load–displacement curves are identical to the stress–strain diagram of TL steel, with a clearly recognizable yield phenomenon in TL steel. The failure of the joint occurred near the maximum force corresponding to TL steel, but at different displacement values.

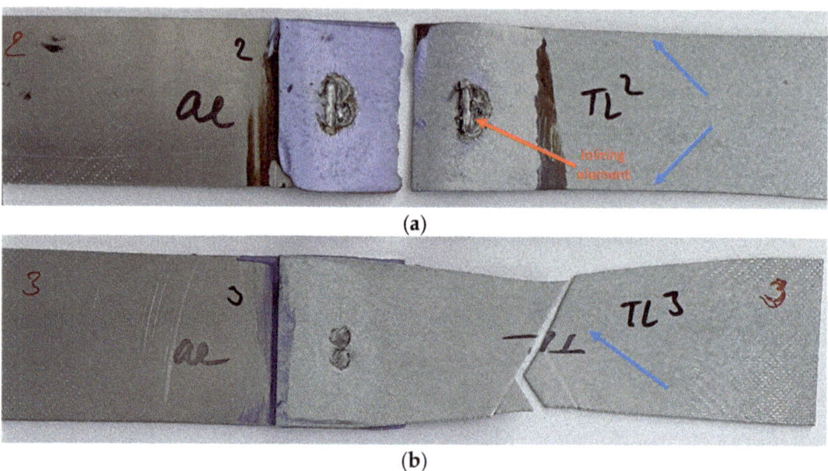

(a)

(b)

Figure 17. Fracture surfaces of Al-TL test joints made by resistance spot welding and AB after tensile shear test: (**a**) sample 2 and (**b**) sample 3, epoxy based adhesive.

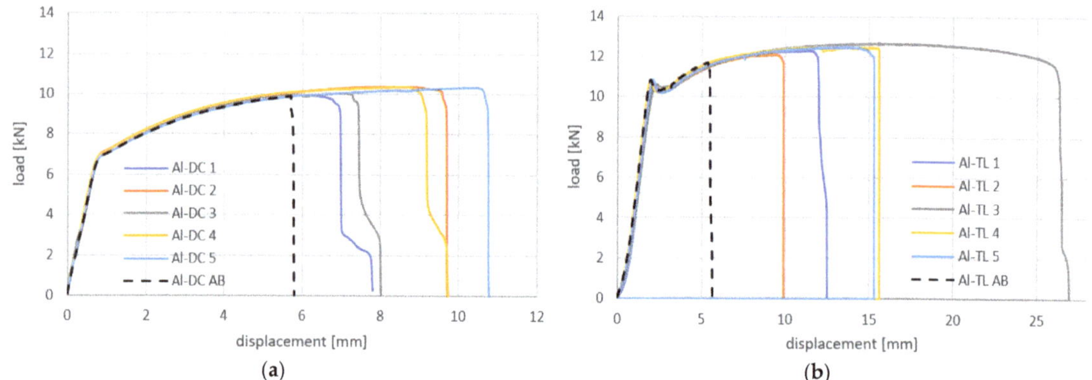

Figure 18. Load–displacement curves for (**a**) Al-DC and (**b**) Al-TL joints made by resistance spot welding + AB, compared with adhesive bonding only (black dashed line).

Figure 18 also shows that the contribution of the insert element to the total energy absorption during joint loading (deflection on the downward portions of the curves) is negligible compared to the large total area under the loading curve.

Again, dashed black lines in Figure 18 indicate the load–displacement curves of the adhesive joint only (without welding and the insert element). Compared to the pure adhesive joint, a significant increase in the energy absorption of the joint made by hybrid joining was again shown.

Table 10 shows the values of stresses in individual substrates, calculated from the load at failure. In Al-DC joints with an epoxy adhesive, the plastic deformation of the DC substrate only was proved in all tested joints, while in Al-TL joints with the epoxy adhesive, plastic deformation was proved for both the Tl and Al substrates. Macroscopic deformation and failure in the TL-Al pair occurred primarily in the TL substrate due to the higher stress values indicated in Table 10. The stresses in the Al substrate in the Al-TL joint are just above the yield strength ($YS_{Al} = 290$ MPa), and macroscopic deformation cannot be recognized yet.

Table 10. Maximum load Fmax [N] and stress σ [MPa] in substrates used in welded-bonded joints (epoxy-based adhesive).

Sample No.	Al-DC + AB			Al-TL + AB		
	Fmax (Al-DC)	σ (DC)	σ (Al)	Fmax (Al-TL)	σ (TL)	σ (Al)
1	9951	**310.99**	248.79	12,082	377.57	302.05
2	10,368	**324.01**	259.20	12,351	385.97	308.77
3	10,039	**313.71**	250.96	12,117	378.65	302.92
4	10,389	**324.65**	259.72	12,716	397.38	317.90
5	10,336	**323.01**	258.40	12,506	390.80	312.64

3.2.4. Overall Evaluation of the Load-Carrying Capacity of the Joints

Based on the comparison of the individual loads for the different joint types, it can be concluded that the joints formed with the rubber-based adhesive displayed higher values than the joints where no adhesive was used, but even higher values were observed for the joints bonded with the epoxy-resin-based adhesive.

Figure 19 shows the average loads at failure for Al-DC and AL-TL joints without an adhesive and with rubber-based and epoxy-based adhesives.

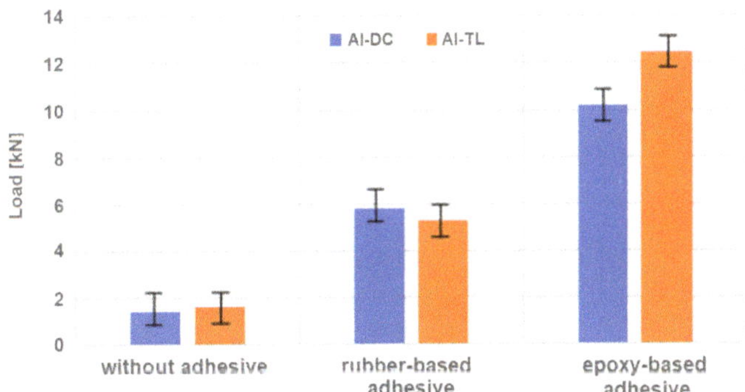

Figure 19. Tensile shear test results of Al-DC and AL-TL samples without adhesive and with two types of adhesives.

The rubber-based adhesive showed better adhesion to all three types of substrates, where 100% cohesive failure was always observed, and it seems to be suitable to combine this adhesive with insert element welding, because the contribution of the insert element to the overall load-bearing capacity of the test joint is more significant compared to the epoxy-based adhesive. The epoxy-based adhesive gives too much strength, and it is pointless for the strength of the bonded joint to exceed the yield strength of the substrate.

4. SEM analysis of Joints

Due to the large difference in the melting points of the insert element and the Al alloy, lower welding parameters were chosen in the first stage of joining. Actually, one cannot even speak of true welding. During welding in the first stage, the Al alloy melted, while the insert element remained unmolten. The liquid Al alloy wets and spreads on the solid steel surface, which creates a special brazed joint (Table 11, left). The insert element has been pressed into the molten Al sheet, and the molten Al alloy has splashed out into the surrounding area (see Table 11, pictures in the center and right). In addition to the metallurgical joint, the insert element is mechanically wedged into melted and solidified Al alloy.

Table 11. SEM analysis of cross-sections and fracture surfaces of welded joints.

Joint	Cross-Section of the Joint	Al Sheet Plate after Destruction, Embedded Insert Element	Detail View of Insert Element Embedded in Al Alloy Melt
Al-DC	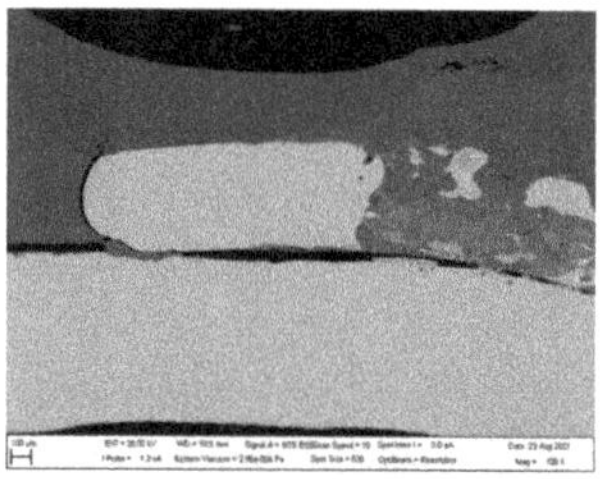	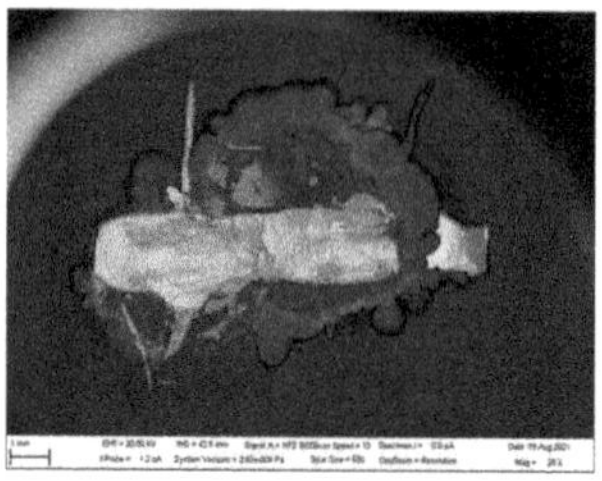	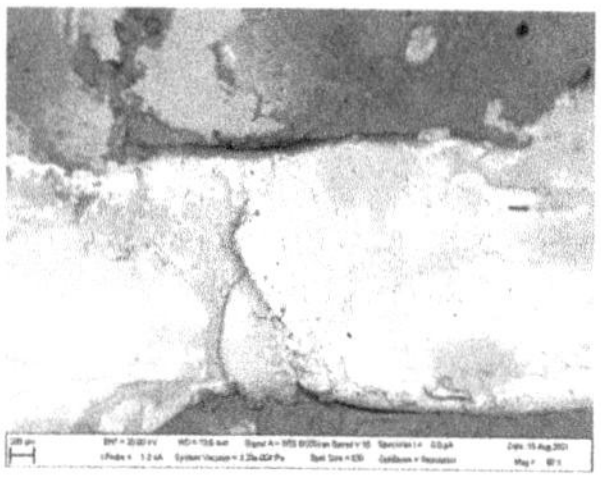
Al-TL	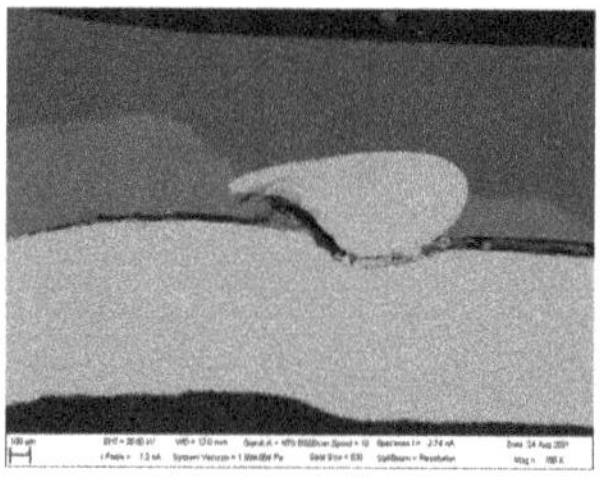	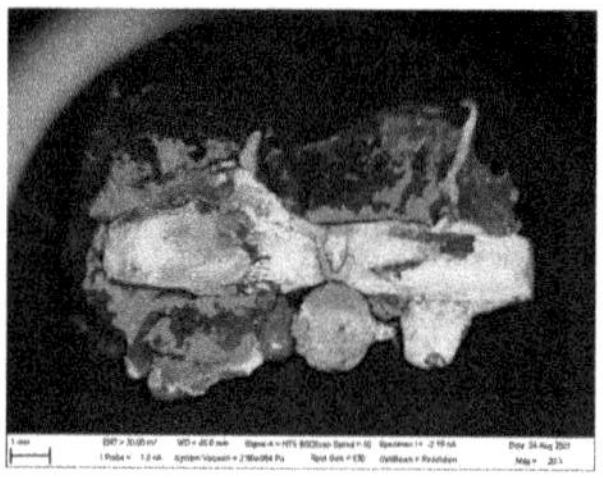	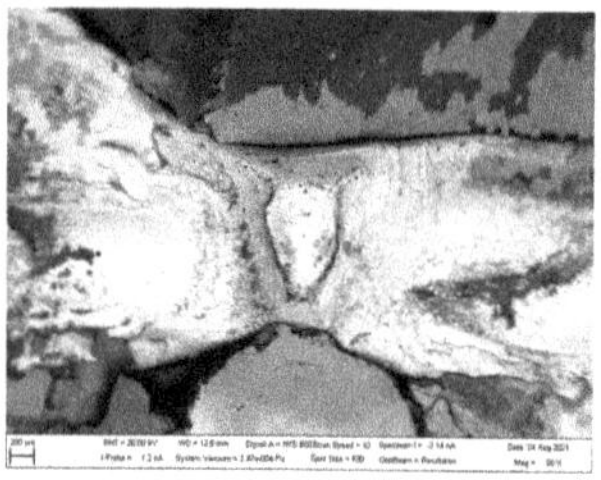

4.1. SEM Analysis of Al-DC Connection

Figure 20 shows the distribution maps of the individual elements on the cross-section of the joint, as well as the results of EDX analysis at selected locations.

The metallographic section of the intact joint showed the insert element (spectra 10, 16 and 17) sealed in the Al sheet. Spectrum 9 confirms the basic chemical composition of the Al sheet, and spectrum 11 confirms the basic chemistry of deep-drawn low-carbon DC steel. The other spectra span different phases at the Fe-Al interface, ranging from the more aluminum-rich phases located farther away from the steel (spectra 13 and 14) to the more iron-rich phases lying close to the steel (spectra 15 and 18). The SEM analysis of the joint detail at the DC-Al–insert-element interface (Figure 21) more clearly shows this gradient change in Al and Fe contents at the joint interface (see Al and Fe elemental distribution map).

At the interface between DC and the Al melt, regions with a needle-like structure are present (spectrum 28), which, according to [36,37], could correspond to Fe_2Al_5 or other phases, e.g., $FeAl_3$, $FeAl_2$, $FeAl$ or Fe_4Al_{13}. From the point of view of bond strength, it is desirable that there are no IMCs at the Fe-Al interface, or that they are as thin and discontinuous as possible.

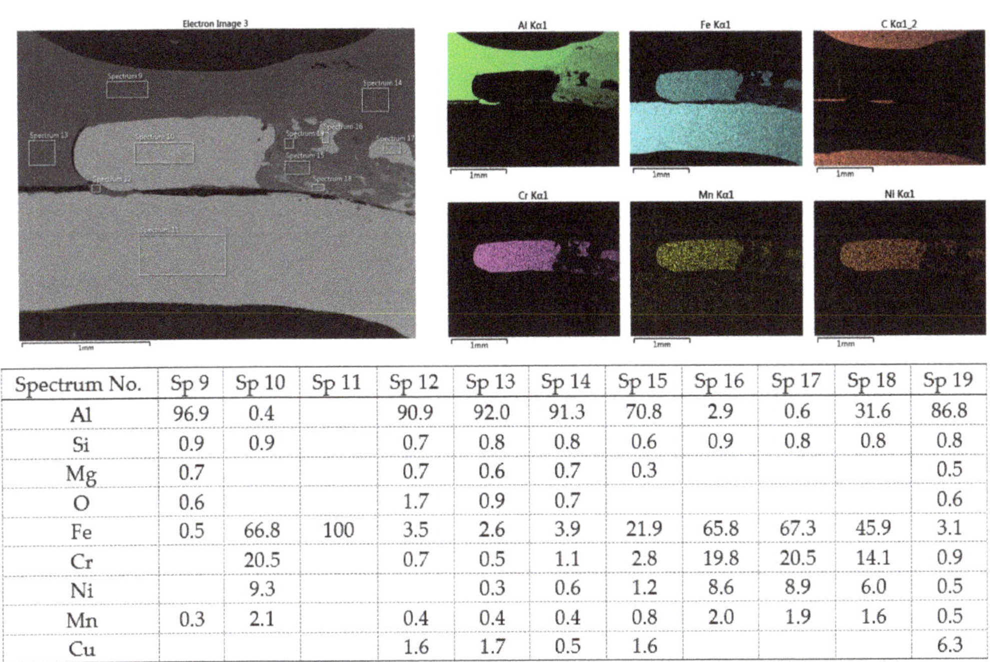

Spectrum No.	Sp 9	Sp 10	Sp 11	Sp 12	Sp 13	Sp 14	Sp 15	Sp 16	Sp 17	Sp 18	Sp 19
Al	96.9	0.4		90.9	92.0	91.3	70.8	2.9	0.6	31.6	86.8
Si	0.9	0.9		0.7	0.8	0.8	0.6	0.9	0.8	0.8	0.8
Mg	0.7			0.7	0.6	0.7	0.3				0.5
O	0.6			1.7	0.9	0.7					0.6
Fe	0.5	66.8	100	3.5	2.6	3.9	21.9	65.8	67.3	45.9	3.1
Cr		20.5		0.7	0.5	1.1	2.8	19.8	20.5	14.1	0.9
Ni		9.3			0.3	0.6	1.2	8.6	8.9	6.0	0.5
Mn	0.3	2.1		0.4	0.4	0.4	0.8	2.0	1.9	1.6	0.5
Cu				1.6	1.7	0.5	1.6				6.3

Figure 20. SEM analysis of Al-DC connection, distribution element maps and EDX spectra.

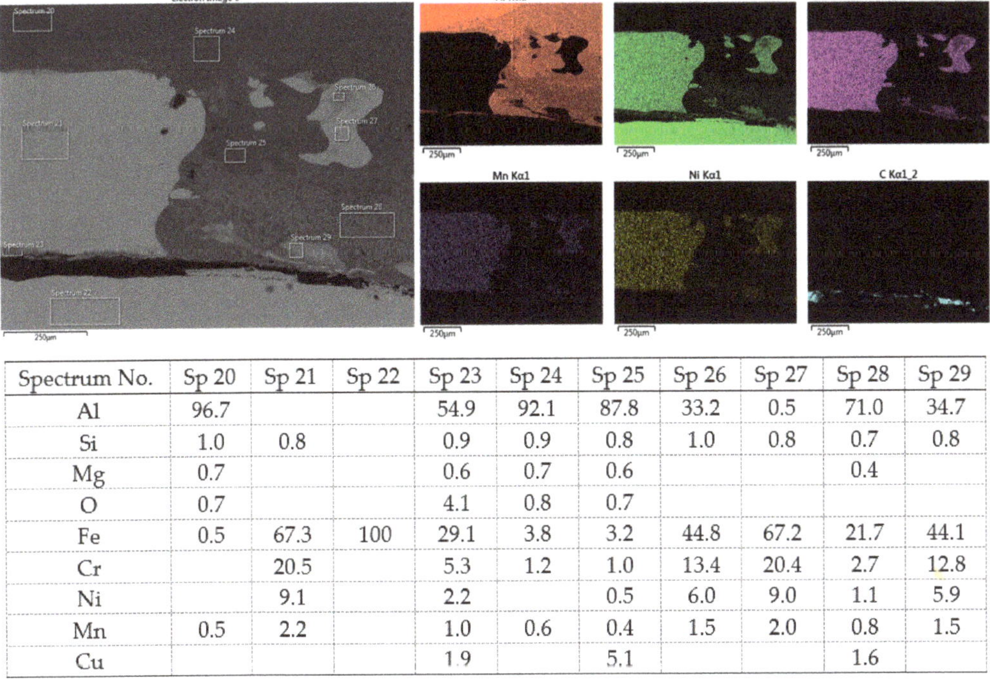

Spectrum No.	Sp 20	Sp 21	Sp 22	Sp 23	Sp 24	Sp 25	Sp 26	Sp 27	Sp 28	Sp 29
Al	96.7			54.9	92.1	87.8	33.2	0.5	71.0	34.7
Si	1.0	0.8		0.9	0.9	0.8	1.0	0.8	0.7	0.8
Mg	0.7			0.6	0.7	0.6			0.4	
O	0.7			4.1	0.8	0.7				
Fe	0.5	67.3	100	29.1	3.8	3.2	44.8	67.2	21.7	44.1
Cr		20.5		5.3	1.2	1.0	13.4	20.4	2.7	12.8
Ni		9.1		2.2		0.5	6.0	9.0	1.1	5.9
Mn	0.5	2.2		1.0	0.6	0.4	1.5	2.0	0.8	1.5
Cu				1.9		5.1			1.6	

Figure 21. SEM analysis of Al-DC connection, distribution element maps and EDX spectra-detail.

4.2. SEM Analysis of Al-TL Connection

A simpler interface structure without needle-like formations can be observed when joining Al to galvanized HSLA steel (Figure 22).

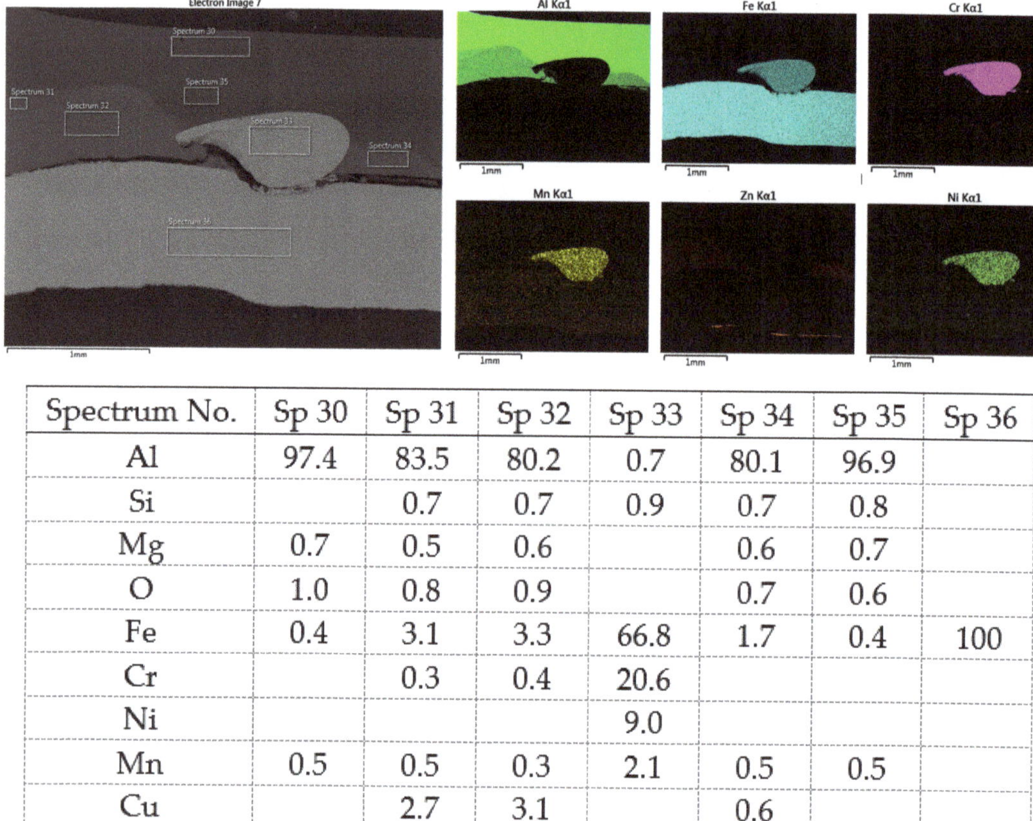

Spectrum No.	Sp 30	Sp 31	Sp 32	Sp 33	Sp 34	Sp 35	Sp 36
Al	97.4	83.5	80.2	0.7	80.1	96.9	
Si		0.7	0.7	0.9	0.7	0.8	
Mg	0.7	0.5	0.6		0.6	0.7	
O	1.0	0.8	0.9		0.7	0.6	
Fe	0.4	3.1	3.3	66.8	1.7	0.4	100
Cr		0.3	0.4	20.6			
Ni				9.0			
Mn	0.5	0.5	0.3	2.1	0.5	0.5	
Cu		2.7	3.1		0.6		
Zn		8.0	10.6		15.0		

Figure 22. SEM analysis of Al-TL connection, distribution element maps and EDX spectra.

Again, the insert element is pressed into the molten Al alloy during welding. Changes in the structure of the steel are not visible, as no melting has occurred, or if it has, the weld nugget is usually in the core of the steel. However, a relatively thick layer of Al-Zn alloy (spectra 31, 32 and 34)—the Al-rich α phase—is evident in the Al substrate toward the steel–Al alloy interface. It can be seen in more detail in Figure 23, where the distribution map of Zn and Al is particularly interesting, documenting its dissolution in Al.

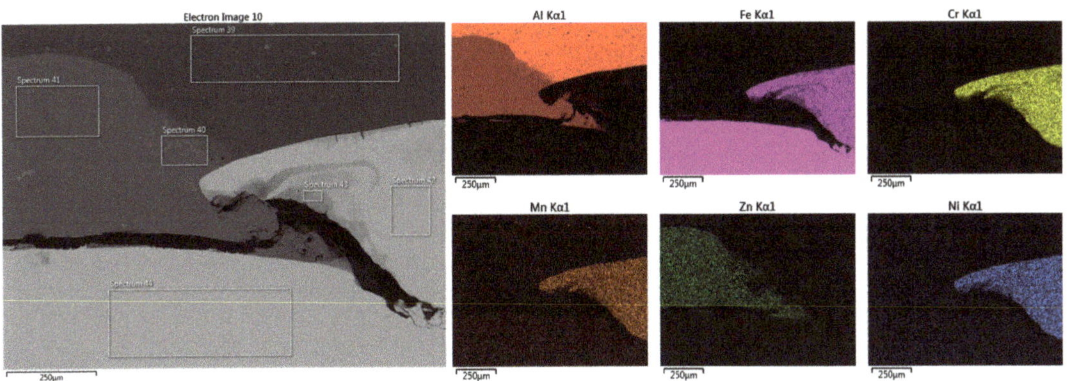

Spectrum No.	Sp 39	Sp 40	Sp 41	Sp 42	Sp 43	Sp 44
Al	96.6	78.8	80.6		28.0	
Si	1.1	0.8	0.8	0.9	0.9	
Mg	0.7	0.6	0.6			
O	0.7	1.1	0.6			
Fe	0.6	6.5	2.8	67.2	48.3	100
Cr		0.8	0.4	20.6	14.6	
Ni			0.3	9.3	6.5	
Mn	0.3	0.7	0.3	2.1	1.7	
Cu		2.7	2.8			
Zn		8.1	10.9			

Figure 23. SEM analysis of Al-TL connection, distribution element maps and EDX spectra-detail.

5. Conclusions

The aim of this research was to verify a new method for joining dissimilar materials—resistance spot welding using a joining element and adhesive bonding.

From the results, the following conclusions can be drawn:

- The surface preparation of steels and aluminum alloys with an organosilane-based adhesion promoter led to the excellent adhesion of both adhesives to all substrates, reflected in the high load-bearing capacity of connections.
- Due to the small contact area, connections formed by resistance welding with the insert element alone achieved a load-bearing capacity of up to 1800 N, irrespective of whether they were Al-DC or Al-TL joints. The joint formed is actually a brazed joint, where the molten Al alloy just wets the surface of the insert element and steel plate.
- Joints formed by spot resistance welding and a rubber-based adhesive had a load capacity of approximately 6 kN, while joints with an epoxy-based adhesive had a load capacity at the level of the maximum load capacity of the substrate (Al-DC: 10 kN; Al-TL: 12 kN)
- The importance of the insert element lies in the fact that it increases the energy absorption of the joint during breakage. This is manifested by a change in the slope of the downward part of the load–displacement curve.

- Although the load-carrying capacity of bonded joints and joints formed by hybrid joining technology (RSW+AB) is approximately the same, the use of an insert element causes a significant increase in the energy absorbed by the joint under stress.
- SEM analysis confirmed a strong bond between the insert element and both substrates during load testing. The element remained largely welded to the steel substrate, pulling out a portion of the volume from the Al substrate, which remained firmly attached to the element.
- The spot welding process is fast, which blocks the formation process of IMCs.
- However, the use of an insert element in combination with adhesive bonding is only relevant for those adhesives that have a load capacity just below the yield strength of the substrates. For bonded joints with higher load capacities, the plastic deformation of the substrates occurs, which is unacceptable, and thus, the overall contribution of the insert element to the load capacity of the joint becomes negligible.
- The results of the presented research show that the combination of the resistance spot welding of an insert element and adhesive bonding facilitates the joining process of galvanized and nongalvanized steels with aluminum alloys and suppresses the effect of brittle intermetallic phases by minimizing the joining area and welding time. It is possible to use the synergistic effect of insert element welding and adhesive bonding to achieve increased energy absorption of the joint under stress.

Author Contributions: Conceptualization, A.G., J.B. and J.V. (Ján Viňáš); methodology, A.G. and J.B.; validation, E.J., M.D. and M.V.; formal analysis, E.J. and J.H.; investigation, A.G., E.J. and J.H.; data curation, D.D.; writing—original draft preparation, J.V. (Ján Varga) and A.G.; writing—review and editing, A.G.; visualization, J.V. (Ján Varga); supervision, J.B.; project administration, A.G. All authors have read and agreed to the published version of the manuscript.

Funding: This research was funded by the Scientific Grant Agency of the Ministry of Education, Science, Research and Sports of the Slovak Republic under project VEGA 1/0154/19: Research of the combined technologies of joining dissimilar materials for automotive industry; KEGA 046TUKE-4/2022: Innovation of the educational process by implementing adaptive hypermedia systems in the teaching of subjects in the field of coating technology and welding of materials; and KEGA 036TUKE-4/2021: Distance form of education for foreign students in the field of manufacturing technologies using modern IT tools.

Institutional Review Board Statement: Not applicable.

Informed Consent Statement: Not applicable.

Data Availability Statement: Not applicable.

Conflicts of Interest: The authors declare no conflict of interest.

References

1. Kawajiri, K.; Kobayashi, M.; Sakamoto, K. Lightweight materials equal lightweight greenhouse gas emissions: A historical analysis of greenhouse gases of vehicle material substitution. *J. Clean. Prod.* **2019**, *253*, 119805. [CrossRef]
2. Brezinová, J.; Sailer, H. Technologies for joining dissimilar materials in the automotive industry. *Machines. Technologies. Mater.* **2021**, *15*, 117–120.
3. Guzanová, A.; Janoško, E.; Draganovská, D.; Brezinová, J.; Viňáš, J.; Tomáš, M.; Maláková, S.; Džupon, M.; Vojtko, M. Metallographic Study of Overlapped Laser Welds of Dissimilar Materials. *Met. Open Access Metall. J.* **2020**, *12*, 1682. [CrossRef]
4. Kashani, H.T.; Kah, P.; Martikainen, J. Laser overlap welding of zinc-coated steel on aluminum alloy. *Phys. Proc.* **2015**, *78*, 265–271. [CrossRef]
5. Malikov, A.; Vitoshkin, I.; Orishich, A.; Filippov, A.; Karpov, E. Effect of the aluminum alloy composition (Al-Cu-Li or Al-Mg-Li) on structure and mechanical properties of dissimilar laser welds with the Ti-Al-V alloy. *Opt. Laser Technol.* **2020**, *126*, 106135. [CrossRef]
6. Chen, L.; Wang, C.; Xiong, L.; Zhang, X.; Mi, G. Microstructural, porosity and mechanical properties of lap joint laser welding for 5182 and 6061 dissimilar aluminum alloys under different place configurations. *Mater. Des.* **2020**, *191*, 108625. [CrossRef]
7. Yang, J.; Li, Y.L.; Zhang, H. Microstructure and mechanical properties of pulsed laser welded Al/steel dissimilar joint. *Trans. Nonferrous Met. Soc. China* **2016**, *26*, 994–1002. [CrossRef]

8. Keisuke, U.; Tomo, O.; Shumpei, N.; Kenji, M.; Toshikazu, N.; Akio, H. Effects of Zn-Based Alloys Coating on Mechanical Properties and Interfacial Microstructures of Steel/Aluminum Alloy Dissimilar Metals Joints Using Resistance Spot Welding. *Mater. Trans.* **2011**, *52*, 967–973. [CrossRef]
9. Kenji, M.; Shigeyuki, N.; Chika, S.; Hiroshi, S.; Akio, H. Dissimilar Joining of Aluminum Alloy and Steel by Resistance Spot Welding. *Int. J. Mater. Manuf.* **2009**, *2*, 58–67. [CrossRef]
10. Jank, N.; Staufer, h.; Bruckner, J. Schweißverbindungen von Stahl mit Aluminium–eine Perspektive für die Zukunft. *BHM Berg-Hüttenmännische Mon.* **2008**, *153*, 189–192. [CrossRef]
11. Pouranvari, M. Critical assessment 27: Dissimilar resistance spot welding of aluminium/steel: Challenges and opportunities. *Mater. Sci. Technol.* **2017**, *33*, 1705–1712. [CrossRef]
12. Hendrawan, M.A.; Purboputro, P.I. Influence of zinc on mechanical behavior of resistance spot welding of aluminum and stainless-steel. *Conf. Proc.* **2018**, *1977*, 1–7. [CrossRef]
13. Oikawa, H.; Ohmiya, S.; Yoshimura, T.; Saitoh, T. Resistance spot welding of steel and aluminium sheet using insert metal sheet. *Mater. Sci.* **1999**, *4*, 80–88. [CrossRef]
14. Gullino, A.; Matteis, P.; D'aiuto, F. Review of Aluminum-to-Steel Welding Technologies for Car-Body Applications. *Metals* **2019**, *9*, 315. [CrossRef]
15. Suder, W.; Ganguly, S.; Yudodibroto, B. Penetration and mixing of filler wire in hybrid laser welding. In: Penetration and mixing of filler wire in hybrid laser welding. *J. Mater. Process. Technol.* **2021**, *291*, 117040. [CrossRef]
16. Evdokimov, A.; Doynov, N.; Ossenbrink, R.; Obrosov, A.; Weiss, S.; Michailov, V. Thermomechanical laser welding simulation of dissimilar Steel-Aluminum overlap joints. *Int. J. Mech. Sci.* **2020**, *190*, 106019. [CrossRef]
17. Cao, X.; Li, Z.; Zhou, X.; Luo, Z.; Duan, J. Modeling and optimization of resistance spot welded aluminum to Al-Si coated boron steel using response surface methodology and genetic algorithm. *Measurement* **2021**, *171*, 108766. [CrossRef]
18. Zvorykina, A.; Sherepenko, O.; Neubauer, M.; Jüttner, S. Dissimilar metal joining of aluminum to steel by hybrid process of adhesive bonding and projection welding using a novel insert element. *J. Mater. Process. Technol.* **2020**, *282*, 116680. [CrossRef]
19. Banea, M.D.; Rosioara, M.; Carbas, R.J.C.; da Silva, L.F.M. Multi-material adhesive joints for automotive industry. *Compos. B Eng.* **2018**, *151*, 71–77. [CrossRef]
20. Guzanová, A.; Draganovská, D.; Brezinová, J.; Viňáš, J.; Janoško, E.; Moro, R.; Szelag, P.; Vojtko, M.; Tomáš, M. Application of organosilanes in the preparation of metal surfaces for adhesive bonding. *J. Adhes. Sci. Technol.* **2022**, *36*, 1153–1175. [CrossRef]
21. Karami, V.; Mollaei Dariani, B.; Hashemi, R. Investigation of forming limit curves and mechanical properties of 316 stainless steel/St37 steel tailor-welded blanks produced by tungsten inert gas and friction stir welding method. *CIRP J. Manuf. Sci. Technol.* **2021**, *32*, 437–446. [CrossRef]
22. Aminzadeh, A.; Parvizi, A.; Safdarian, R.; Rahmatabadi, D. Comparison between laser beam and gas tungsten arc tailored welded blanks via deep drawing. *Proc. Inst. Mech. Eng. Part B J. Eng. Manuf.* **2021**, *235*, 583–760. [CrossRef]
23. Niu, S.; Ma, Y.; Lou, M.; Zhang, C.; Li, Y. Joint Formation Mechanism and Performance of Resistance Rivet Welding (RRW) for Aluminum Alloy and Press Hardened Steel. *J. Mater. Process. Technol.* **2020**, *286*, 116830. [CrossRef]
24. Hamelin, C.; Muránsky, O.; Smith, M.C.; Holden, T.M. Validation of a numerical model used to predict phase distribution and residual stress in ferritic steel weldments. *Acta Mater.* **2014**, *75*, 1–19. [CrossRef]
25. Ezazi, M.A.; Yusof, F.; Sarhan, A.A.D.; Shukor, M.H.A. Employment of fiber laser technology to weld austenitic stainless steel 304 l with aluminum alloy 5083 using pre-placed activating flux. *Mater. Des.* **2015**, *87*, 105–123. [CrossRef]
26. Yuce, C.; Karpat, F.; Yavuz, N. Investigations on the microstructure and mechanical properties of laser welded dissimilar galvanized steel–aluminum joints. *Int. J. Adv. Manuf. Technol.* **2019**, *104*, 5–8. [CrossRef]
27. Madhusudana, C.V. Thermal Contact Conductance. In *Part of the Book Series: Mechanical Engineering Series*; Springer: New York, NY, USA, 2014; pp. 9–23.
28. Li, P.; Chen, S.; Dong, H. Interfacial microstructure and mechanical properties of dissimilar aluminum/steel joint fabricated via refilled friction stir spot welding. *J. Manuf. Process.* **2020**, *49*, 385–396. [CrossRef]
29. Singh, P.; Deepak, D.; Brar, G.S. Optical micrograph and micro-hardness behavior of dissimilar welded joints of aluminum (Al 6061-T6) and stainless steel (SS 304) with friction crush welding. *Mater. Today Proc.* **2021**, *44*, 1000–1004. [CrossRef]
30. Matsuda, T.; Adachi, H.; Yoshida, R.; Sano, T.; Hori, H.; Hirose, A. Formation of interfacial reaction layer for stainless steel/aluminum alloy dissimilar joint in linear friction welding. *Mater. Sci.* **2020**, *26*, 101700. [CrossRef]
31. Jordon, B.; Amaro, R.; Allison, P.; Rao, H. Fatigue Damage in Engineering Structures. In *Fatigue in Friction Stir Welding*, 1st ed.; Butterworth-Heinemann Elsevier Ltd.: Oxford, UK, 2019; pp. 87–117.
32. Ma, Y.; Lou, M.; Li, Y.; Lin, Z. Effect of rivet and die on self-piercing rivetability of AA6061-T6 and mild steel CR4 of different gauges. *J. Mater. Process. Technol.* **2018**, *251*, 282–294. [CrossRef]
33. Ezzine, M.C.; Madani, K.; Tarfaoui, M.; Touzain, S.; Mallarino, S. Comparative study of the resistance of bonded, riveted and hybrid assemblies. Experimental and numerical analyses. *Struct. Eng. Mech.* **2019**, *70*, 467–477. [CrossRef]
34. Ezzine, M.C.; Amiri, A.; Tarfaoui, M.; Madani, K. Damage of Bonded, Riveted and Hybrid (Bonded/Riveted) Joints, Experimental and Numerical Study Using CZM and XFEM Methods. *Adv. Aircr. Spacecr. Sci.* **2018**, *5*, 595–613. [CrossRef]
35. Rahmatabadi, D.; Pahlavani, M.; Marzbanrad, J.; Hashemi, R.; Bayati, A. Manufacturing of three-layered sandwich composite of AA1050/LZ91/AA1050 using cold roll bonding process. *Proc. Inst. Mech. Eng. Part B J. Eng. Manuf.* **2021**, *235*, 1363–1372. [CrossRef]

36. Wang, J.; Wang, H.P.; Lu, F.; Carlson, B.E.; Sigler, D.R. Analysis of Al-steel resistance spot welding process by developing a fully coupled multi-physics simulation model. *Int. J. Heat Mass Transf.* **2015**, *89*, 1061–1072. [CrossRef]
37. Novák, P.; Michalcová, A.; Marek, I.; Mudrová, M.; Saksl, K.; Bednarčík, J.; Zikmund, P.; Vojtěch, D. On the formation of intermetallics in Fe-Al system-An in situ XRD study. *Intermetallics* **2013**, *32*, 127–136. [CrossRef]

Article

Simulation and Experimental of Infiltration and Solidification Process for Al₂O₃₍₃D₎/5083Al Interpenetrating Phase Composite for High Speed Train Prepared by Low-Pressure Infiltration

Yanli Jiang [1], Pianpian Xu [1], Chen Zhang [1], Fengjun Jin [1], Yichao Li [1], Xiuling Cao [2,3,*] and Liang Yu [1,4,5,*]

[1] Key Laboratory of New Processing Technology for Nonferrous Metals & Materials,
Guilin University of Technology, Guilin 541004, China; 2010043@glut.edu.cn (Y.J.); 17375063927@163.com (P.X.);
zc405716298@163.com (C.Z.); jfj_6868@163.com (F.J.); y13833243036@163.com (Y.L.)

[2] Hebei Technology Innovation Center for Intelligent Development and Control of Underground Built Environment,
Shijiazhuang 050031, China

[3] School of Exploration Technology and Engineering, Hebei GEO University, Shijiazhuang 050031, China

[4] Collaborative Innovation Center for Exploration of Nonferrous Metal Deposits and Efficient Utilization
of Resources, Guilin University of Technology, Guilin 541004, China

[5] Guangxi Modern Industry College of Innovative Development in Nonferrous Metal Material,
Guilin 541004, China

* Correspondence: caoxlhbdz@163.com (X.C.); 2010054@glut.edu.cn (L.Y.);
Tel.: +86-135-1331-0032 (X.C.); +86-191-7734-7551(L.Y.)

Abstract: Understanding the infiltration and solidification processes of liquid 5083Al alloy into Al₂O₃ three-dimensional reticulated porous ceramic (Al₂O₃₍₃D₎ RPC) is essential for optimizing the microstructure and properties of Al₂O₃₍₃D₎/5083Al interpenetrating phase composites (IPCs) prepared by low-pressure infiltration process (LPIP). This study employs ProCAST software to simulate the infiltration and solidification processes of liquid 5083Al with pouring velocities (PV) of 0.4 m/s infiltrating into Al₂O₃₍₃D₎ RPC preforms with varying porosities at different pouring temperatures (PT) to prepare Al₂O₃₍₃D₎/5083Al IPCs using LPIP. The results demonstrate that pore diameter of Al₂O₃₍₃D₎ RPC preforms and PT of liquid 5083Al significantly influence the of the infiltration. Solidification process analysis reveals that the Al₂O₃₍₃D₎ RPC preform with smaller pore diameters allows the lower pouring velocity of 5083Al to solidify faster compared to the preform with larger pore diameters. Al₂O₃₍₃D₎/5083Al IPCs were prepared successfully from Al₂O₃₍₃D₎ RPC porosity of 15 PPI with liquid 5083Al at PV 0.4 m/s and PT 800 °C using LPIP, resulting in nearly fully dense composites, where both Al₂O₃₍₃D₎ RPCs and 5083Al interpenetrate throughout the microstructure. The infiltration and solidification defects were reduced under air pressure of 0.3 MPa (corresponding to PV of 0.4 m/s) during LPIP. Finite volume method simulations are in good agreement with experimental data, validating the suitability of the simplified model for Al₂O₃₍₃D₎ RPCs in the infiltration simulation.

Keywords: Al₂O₃3D/5083Al; numerical simulation; Infiltration; solidification; ProCAST

Citation: Jiang, Y.; Xu, P.; Zhang, C.; Jin, F.; Li, Y.; Cao, X.; Yu, L. Simulation and Experimental of Infiltration and Solidification Process for Al₂O₃₍₃D₎/5083Al Interpenetrating Phase Composite for High Speed Train Prepared by Low-Pressure Infiltration. *Materials* 2023, 16, 6634. https://doi.org/10.3390/ma16206634

Academic Editor: Raul D. S. G. Campilho

Received: 2 September 2023
Revised: 29 September 2023
Accepted: 30 September 2023
Published: 11 October 2023

1. Introduction

Interpenetrating Phase Composites (IPCs), often referred to as co-continuous composites, are a class of materials where both metal and ceramic phases are topologically co-continuous and three-dimensionally interconnected, forming an intricate network structure [1]. In these composites, the continuous metallic network efficiently bridges cracks, while the ceramic phase redistributes stress, facilitates load transfer, and maintains dimensional stability at elevated temperatures [2]. Metal/ceramic IPCs are known for their exceptional strength, toughness, low thermal expansion coefficient, resistance to fatigue, wear, and corrosion [3]. The fabrication of metal/ceramic IPCs typically involves creating open-porous ceramic preforms and infiltrating them with molten metal [4].

Various techniques, including replica templates, direct foaming, freeze-casting, and sacrificial pore-forming agents, have been employed to prepare ceramic preforms with specific pore geometries [5].

One of the common methods for creating metal/ceramic IPCs is the Low-Pressure Infiltration Process (LPIP), where liquid metal is injected and solidified within open-porous ceramic preforms [6,7]. However, the LPIP process is complex, involving heat transfer, fluid mechanics with phase change, and the occurrence of defects such as shrinkage and porosity [5]. Quality control in metal/ceramic IPCs prepared via LPIP depends on various factors, including the geometry of the ceramic preforms, applied pressure, pouring velocities, pouring temperatures, and the behavior of the liquid metal [8]. Understanding and predicting infiltration and solidification defects are crucial for ensuring the quality of IPCs. To gain insights into the LPIP process and improve quality control, numerical simulations have been employed to study the infiltration of open-porous ceramic preforms with metallic alloys and predict solidification defects in IPCs. Previous research has used various numerical methods, such as the volume of fluid method, porous medium model, and finite element method, to describe the flow and solidification phenomena [9–17]. Despite these efforts, current numerical models often simplify the preform as a single-scale porous medium and describe LPIP in simple configurations, lacking free surface tracking and comprehensive solidification modeling. Thus, the development of advanced 3D models that consider these aspects is imperative.

The 5XXX series Al-Mg alloys, known for their excellent mechanical properties, lightweight nature, corrosion resistance, and weldability, are commonly used in shipbuilding for top structures and hulls [18]. The 5083Al alloy, in particular, contains supersaturated Mg (>3.5 wt%) to enhance solid solution strengthening [19]. In previous work, Al_2O_3 three-dimensional reticulated porous ceramic ($Al_2O_{3(3D)}$ RPC) preforms were prepared and used in $Al_2O_{3(3D)}$/5083Al IPCs, demonstrating exceptional corrosion resistance attributed to the interface between $Al_2O_{3(3D)}$ RPC preforms and the 5083Al matrix [20]. Key parameters affecting the interface, including the rheology of the Al_2O_3 ceramic slurry, adhesion with the organic sponge replica, and cell size of the replica, were identified as critical. To optimize the interface and reduce defects in $Al_2O_{3(3D)}$/5083 IPCs, it is essential to thoroughly study the infiltration and solidification processes during their manufacture. Although previous research has extensively modeled metal infiltration and solidification processes, limited work has focused on the evolution of $Al_2O_{3(3D)}$/5083Al IPCs.

In this study, $Al_2O_{3(3D)}$ RPC preforms were simplified into periodic geometric shape arrays using Kelvin cells, and the infiltration and solidification processes of liquid 5083Al into $Al_2O_{3(3D)}$ RPC via LPIP were simulated using ProCAST software. The results were combined with experimental data to investigate factors influencing the infiltration and solidification of $Al_2O_{3(3D)}$/5083 IPCs, ultimately leading to process optimization.

2. Digital Analogue

2.1. Geometrical Model

The $Al_2O_{3(3D)}$/5083Al IPCs three-dimensional models in Figure 1 have been generated with the software SolidWorks 2018. The Kelvin cell model was utilized to represent $Al_2O_{3(3D)}$ RPC preforms, as illustrated in Figure 1a. The cell length, pore size and struct diameter of the $Al_2O_{3(3D)}$ RPC preforms is 3 mm, 2.3 mm (approximately equivalent to 15 PPI) and 3 mm, respectively. The infiltration process involves the flow of liquid metal (5083Al) through a porous $Al_2O_{3(3D)}$ RPC preform. The smaller the pore size of the preform, the greater the resistance to the flow of liquid metal. As a result, there is a loss of energy of the liquid metal as it moves through the preform. This loss of energy can lead to variations in porosity within the preform, with smaller pores being filled earlier than larger ones. Higher porosities result in uneven distribution or significant variations in pore size, which can negatively affect the uniformity of penetration, so that parts of the pore area will not be well filled, resulting in reduced performance of the test sample. An increase in porosity may lead to changes in the curing behaviour within the preform, with large porosities tending to accelerate the curing behaviour, resulting in the creation of internal porosity. The infiltration domain of $Al_2O_{3(3D)}$ RPC model was made up with 32 Kelvin cells obtained by array processing of the infiltration cell along the x, y, z-direction, respectively [21]. $Al_2O_{3(3D)}$ RPC Kelvin cells represented as a network of open cells with typical 12–14 pentagonal or hexagonal faces. The infiltration unit with blue was combined with 5083Al with pink to form a single infiltration unit as depicted in Figure 1c. The $Al_2O_{3(3D)}$/5083 IPCs model was obtained by array processing of the infiltration unit as depicted in Figure 1d. The chemical composition of 5083Al is presented in Table 1.

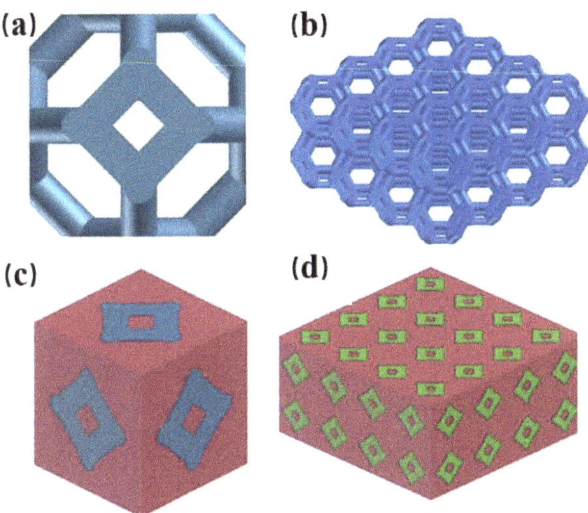

Figure 1. $Al_2O_{3(3D)}$/5083Al IPCs model. (**a**) Kelvin cell model; (**b**) $Al_2O_{3(3D)}$ RPC model; (**c**) infiltration unit; (**d**) $Al_2O_{3(3D)}$/5083Al IPCs model.

Table 1. Chemical composition of 5083Aluminum alloy.

Elements	Si	Cu	Mg	Zn	Mn	Ti	Cr	Fe	Al
Wt.%	0.4	0.03	4.5	0.27	0.50	0.15	0.07	0.15	Balance

A three-dimensional model generated with the software SolidWorks 2018 was used to simulate the infiltration and solidification process of $Al_2O_{3(3D)}$/5083Al IPCs during LPIP as depicted in Figure 2. Figure 2a shows the $Al_2O_{3(3D)}$ RPC of Kelvin model with dimensions of $12 \times 12 \times 6$ mm^3. Figure 2b displays the $Al_2O_{3(3D)}$ RPC of Kelvin model was placed in graphite upper mold with dimensions of $16 \times 16 \times 5$ mm^3. Figure 2c presents a schematic diagram of the infiltration process. The blue purple part represents 5083Al. After merging the upper and lower molds with dimensions of $16 \times 16 \times 10$ mm^3 was represented in Figure 2d. The clamping model in Figure 2d includes an impregnation mouth at the bottom with a diameter of 10 mm and two vents at the top with a diameter of 2 mm.

2.2. Governing Equations

The transient temperature distribution and solidification velocities were calculated by finite volume method using the momentum conservation equation, mass conservation equation, and energy conservation equation expressed in the literatures [9,15,22]. In order to achieve a complete description of the infiltrating process, the flow velocities of liquid 5083Al at various positions were provided by solving the Navier–Stokes equations given in ref. [9]. The Navier–Stokes equations in matrix zone are given by

$$\rho \frac{Dv}{Dt} = -\nabla\tau - \nabla P + \rho G \tag{1}$$

where ρ is density of molten metal (In this work assumed the liquid metal to be incompressible, which means ρ keeps constant); t is flow time; P is pressure at certain position; u, v, w is the velocity in x, y, z direction, g_x, g_y, g_z is accelerated velocity in x, y, z direction, respectively [9].

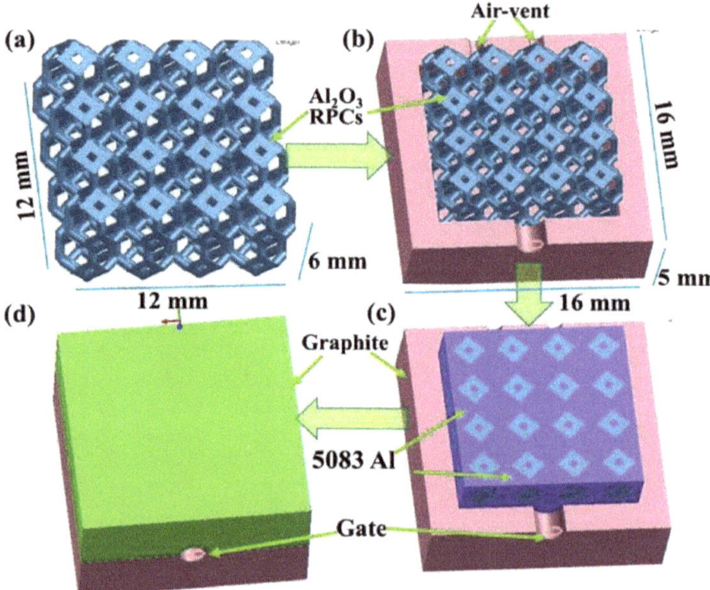

Figure 2. $Al_2O_{3(3D)}$/5083Al IPCs model for infiltration and solidification simulation during LPIP. (**a**) $Al_2O_{3(3D)}$ RPC of Kelvin model; (**b**) $Al_2O_{3(3D)}$ RPC of Kelvin model was placed in graphite upper mold; (**c**) schematic diagram of infiltration process; (**d**) $Al_2O_{3(3D)}$/5083Al IPCs clamping model.

The heat exchange between the graphite mould, $Al_2O_{3(3D)}$ RPC and liquid 5083Al resulted in decreasing temperature during LPIP, which changed the liquid 5083Al thermophysical parameters, such as specific heat and viscosity. The thermophysical parameters material data of 5083Al, graphite mould and $Al_2O_{3(3D)}$ RPC were obtained directly from the database of PROCAST software [23,24]. The governing equations given of the heat and mass transfer in REF [21] were solved using ProCAST software in this paper. Density and specific heat of Al_2O_3/water nanofluid are evaluated by means of the correlations proposed by Khanafer et al.

$$\rho_{nf} = \varphi\rho_{np} + (1-\varphi)\rho_{bf} \tag{2}$$

And

$$c_{p,nf} = \frac{\varphi(\rho c_p)_{np} + (1-\varphi)(\rho c_p)_{bf}}{\rho_{nf}} \tag{3}$$

where the subscripts np and bf specify the nanoparticles and the base fluid, respectively, and φ indicates the nanoparticle volumetric concentration [9].

2.3. Numerical Procedure

Figure 3 shows the boundary and mesh of $Al_2O_{3(3D)}$/5083 IPCs models during LPIP. The integrity surface mesh was composed of triangles, which was divided into

$160 \times 160 \times 100$ cells, resulting in 25,600 more surface cells and 16,000 less surface cells shown in Figure 3a,b. The model was divided into approximately 180,000 volume mesh cells, which provides sufficient calculation accuracy shown in Figure 3c,d. The inlet was defined with a uniform velocity boundary condition, while all other solid surfaces were set as nonslip and nonpenetrating boundaries. As showed in Figure 3e,f, the pouring velocities (PV) of liquid 5083Al was set to 0.4 m/s, corresponding to an infiltration pressure (inlet pressure) of about 0.3 MPa, which was the pressure commonly used in low-pressure casting machines [12]. The outlet pressure is 0 Pa (absolute pressure minus atmospheric pressure is 0 Pa). The initial temperature and heat transfer coefficients (HTC) applied to each volume and boundary are listed in Table 2. Pouring temperature (PT) of liquid 5083Al was set to 740–800 °C. Initial temperature of graphite inlet, graphite gate, and graphite mold were set to 250 °C. Initial temperature of $Al_2O_{3(3D)}$ RPC was set to 540 °C. Liquid 5083Al was considered an ideal fluid for density calculations, and the effect of gravity was included in the momentum equation. Due to the low PT of liquid 5083Al, the radiation of liquid 5083Al into infiltrating $Al_2O_{3(3D)}$ RPC preform was not considered. The tip resistance can be ignored for liquid 5083Al flow in the pores is in the form of steady-state flow. Considering the liquid 5083Al as an incompressible homogeneous fluid and assuming the 5083Al flow in $Al_2O_{3(3D)}$ RPC preforms was a laminar flow. There was no residual air in the $Al_2O_{3(3D)}$ RPC preform, so the gas anti pressure was not considered. Fluid properties were defined as variables, and the momentum equation was coupled to the energy equation. The simulation employed a double precision coupling algorithm to couple the velocities. The second-order upwind advection model was used for the momentum equation, turbulent kinetic energy equation, and turbulent energy dissipation equation. The convergence criterion was set to 10^{-5}.

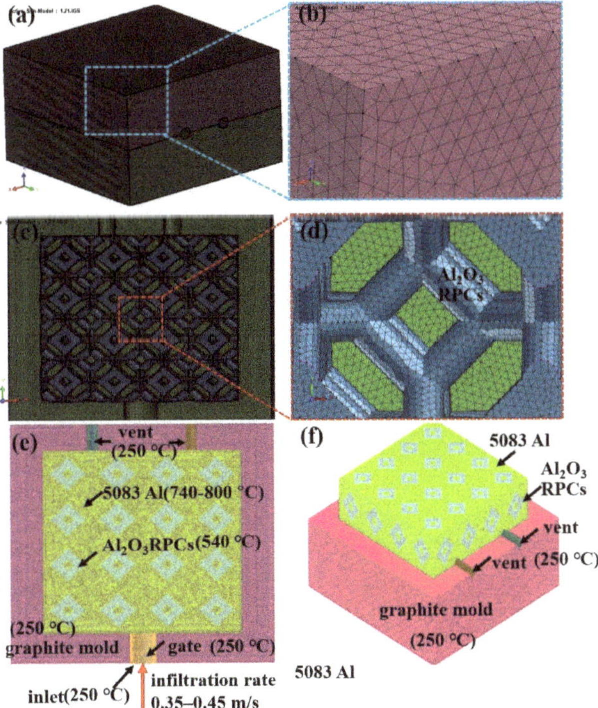

Figure 3. Boundary and mesh of $Al_2O_{3(3D)}$/5083 IPCs simulated model during LPIP. (**a**) surface mesh; (**b**) zoom of mesh; (**c**) volume mesh; (**d**) zoom of volume mesh; (**e**) front view of boundary; (**f**) side view of boundary.

Table 2. Boundary conditions of $Al_2O_{3(3D)}/5083$ IPCs simulated model during LPIP.

Volumes	Initial Temperature/°C	Boundary	HTC/ $(W \cdot m^{-2} \cdot °C^{-1})$
graphite inlet	250		EQUIV
graphite gate	250	graphite inlet & graphite gate	1416
graphite mold	250	liquid 5083Al & graphite mould	
liquid 5083Al	740, 760, 800	liquid 5083Al & graphite inlet and gate	1000
$Al_2O_{3(3D)}$		liquid 5083Al & $Al_2O_{3(3D)}$ RPC	480
RPC	540		

3. Material Preparation and Material Characterization

3.1. Material Preparation

$Al_2O_{3(3D)}$ RPC was prepared using replica methods in this paper. Replica methods often referred to as the lost mold process or Schwartzwalder method, have been frequently utilized to produce reticulated porous ceramics with large interconnected pores [25]. The detailed steps are as follows: (1) A three-dimensional mesh polyurethane sponge from Shenzhen Lvchuang Environmental Protection Filter Materials Co., Ltd. (Shenzhen, China) was immersed in a NaOH solution for 18 h to remove the interlayer film and increase surface roughness. The purpose is to improve the adhesion between the polyurethane sponge surface and the Al_2O_3 slurry. (2) The sponges used as templates were cut into a circle with a diameter of 100 mm and a thickness of 8 mm. (3) The sponge was impregnated into Al_2O_3 slurry. The impregnated sponge body was then passed through rollers to drain the surplus slurry and maintain the ceramic content in the infiltrated body. (4) The ceramic-coated template was subsequently dried in a microwave oven for 15 min to obtain a green alumina mesh porous body with a well-defined structure. (5) The pyrolyzed through careful heating to 400 °C for 2 h decomposed or burned out the polyurethane sponge templates. (6) In a graphite resistance furnace from Jinzhou Santai Electric Furnace Factory, China, with argon gas as the sintering atmosphere at 1600 °C for 3 h, the ceramic layers were sintered to obtain $Al_2O_{3(3D)}$ RPC with the same morphology as that of the original cellular polyurethane sponge template, which was approximately 15 PPI (pores per inch).

Figure 4 shows the schematic diagram of liquid 5083Al infiltrating into $Al_2O_{3(3D)}$ RPC using LPIP. The 5083Al was in the form of nuggets and placed in the graphite crucibles and heated from 25 °C to 800 °C for 2 h in the crucible furnace from Zhengzhou Xinhan Instrument Equipment Co., Ltd. (Zhengzhou, China). The liquid 5083Al in the graphite crucible was regularly stirred to ensure a uniform composition. $Al_2O_{3(3D)}$ RPC were heated to 540 °C. The $Al_2O_{3(3D)}$ RPC was placed on the liquid 5083Al, and pressurized gas was applied for about 20 min, as shown in Figure 4a. The liquid 5083Al completely infiltrated the $Al_2O_{3(3D)}$ RPC and cooled to obtain $Al_2O_{3(3D)})/5083$ IPCs in Figure 4b. The simulation results obtained from ProCAST were compared and verified.

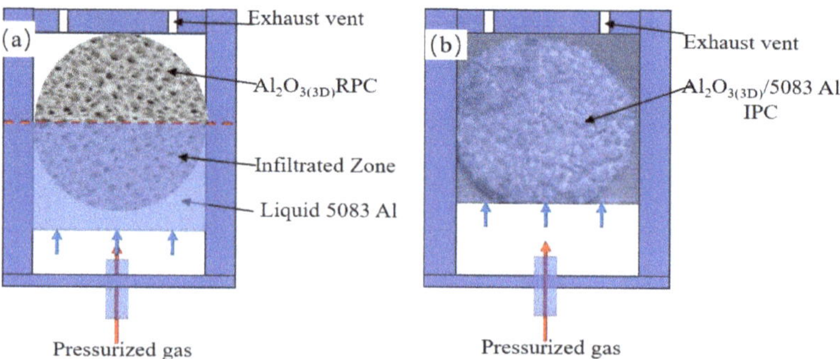

Figure 4. Schematic diagram of liquid 5083Al infiltrating into $Al_2O_{3(3D)}$ RPC using LPIP. (**a**) low-pressure infiltration process, (**b**) solidification process.

3.2. Material Characterization

The obtained samples of $Al_2O_{3(3D)}/5083$ IPCs were subjected to X-ray diffraction (XRD) analysis using Cu Kα radiation at 40 kV and 100 mA, employing a computer-controlled diffractometer (PANALYTICL B.V/PW3040/60, Netherlands). The XRD data were recorded in continuous scanning mode with a scanning angle (2θ) ranging from 10° to 90° and a scanning rate of 0.02°/s. The microstructure of the samples of $Al_2O_{3(3D)}/5083$ was characterized using scanning electron microscopy (SEM) at 15 kV and 10 mA. The composition of the material was analyzed using energy-dispersive spectroscopy (EDS).

4. Results and Discussion

4.1. Effect of Pouring Temperature on Infiltration Depth

Figure 5 shows the simulated results of infiltration depth of liquid 5083Al with pouring velocities (PV) of 0.4 m/s infiltrating and pouring temperature (PT) of 740 °C into $Al_2O_{3(3D)}$ RPC with different times in infiltrating stage using LPIP. During the initial infiltrating stage, the liquid 5083Al flowed freely upward along the vertical inlet under the influence of pressurized gas. At 0.767 s, the mold was filled to about 25%, the temperature of liquid 5083Al was 696 °C (Figure 5a). At 1.505 s, the mold was filled to about 50%, temperature of liquid 5083Al was higher than 644 °C (Figure 5b). The black arrow indicated the position where the 644 °C isotherm was located showed in Figure 5b–d. Temperature of liquid 5083Al was still above its solidus temperature, and the infiltrating process could continue. However, at 1.922 s, the mold was filled to about 70%, temperature of the liquid 5083Al was below 644 °C (Figure 5c). Temperature of liquid 5083Al was lower than the solidus temperature, and liquid 5083Al began to solidify. The mold infiltrating could not continue. The final infiltration depth was defined as the maximum length of $Al_2O_{3(3D)}$ preform which liquid 5083Al can percolate before the channel was completely blocked by liquid 5083Al solidification. It was evident that overall fill time was approximately 1.984 s, full impregnation was not achieved at 740 °C as well as the final infiltration depth was about 70% (Figure 5d).

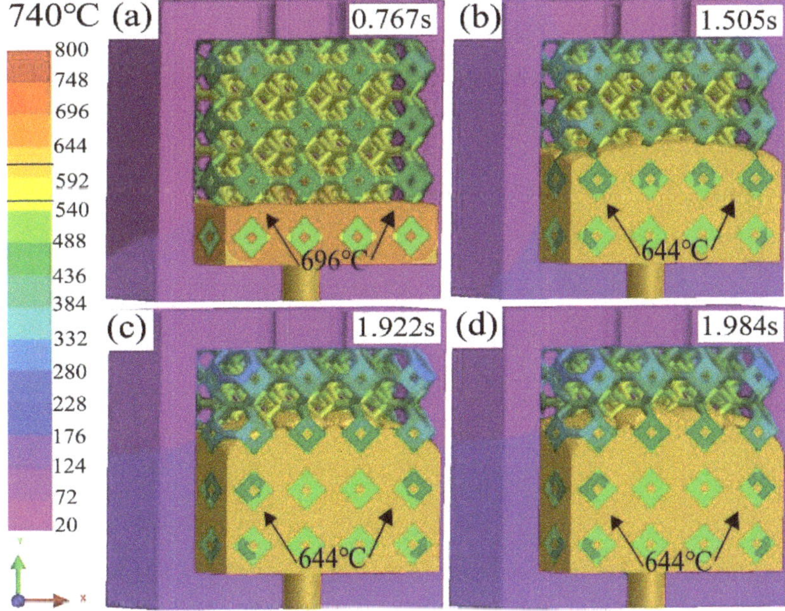

Figure 5. Infiltration depth of liquid 5083Al with PV of 0.4 m/s and PT of 740 °C infiltrating into $Al_2O_{3(3D)}$ with different infiltration times using LPIP. (**a**) 0.767 s; (**b**) 1.505 s; (**c**) 1.922s; (**d**) 1.984 s.

Figure 6 shows the simulated results of infiltration depth of liquid 5083Al with PV of 0.4 m/s and PT of 760 °C infiltrating into $Al_2O_{3(3D)}$ with different times in infiltrating stage using LPIP. The black arrow indicated the position where the 644 °C isotherm was located. Compared with infiltration depth of 644 °C isotherm of liquid 5083Al with filling 20%, 50%, 70% indicated by black arrow in Figure 5, the infiltration depth of 644 °C isotherm of liquid 5083Al in Figure 6 in infiltration direction was increased by about 10%, 20%, and 30% with filling 20%, 50%, 70%, respectively. At 0.715 s, the mold was filled to about 25% (Figure 6a). At 1.559 s, the mold was filled to about 50% (Figure 6b). However, at 2.852 s, the mold was filled to about 70% (Figure 6c). The overall fill time was about 3.018 s, the final infiltration depth was about 100%, complete impregnation was achieved (Figure 6d). The microporosities of infiltration gaps at the interface between 5083Al and $Al_2O_{3(3D)}$ RPC or the segregation of the 5083Al matrix were observed during infiltration.

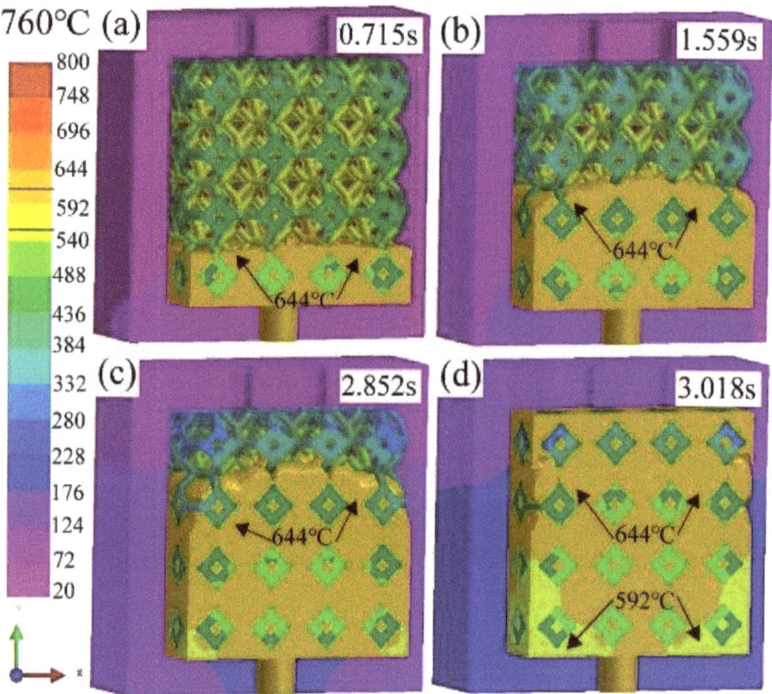

Figure 6. Infiltration depth of liquid 5083Al with PV of 0.4 m/s and PT of 760 °C infiltrating into $Al_2O_{3(3D)}$ with different infiltration times using LPIP. (**a**) 0.715 s; (**b**) 1.559 s; (**c**) 2.852 ;(**d**) 3.018 s.

Figure 7 shows the simulated results of infiltration depth of liquid 5083Al with PV of 0.4 m/s and PT of 800 °C infiltrating into $Al_2O_{3(3D)}$ with different times in infiltrating stage using LPIP. Compared with the depth of the 644 °C and 592 °C isotherms of liquid 5083Al with filling 20%, 50%, 70% indicated by black arrow in Figure 6, the depth of the 644 °C and 592 °C isotherms of liquid 5083Al in Figure 7 in the infiltration direction was increased by about 5%, 10%, and 12% with filling 20%, 50%, 70%, respectively. At 0.834 s, the mold was filled to about 25% (Figure 7a). At 1.488 s, the mold was filled to about 50% (Figure 7b). However, at 2.279 s, the mold was filled to about 70% (Figure 7c). The overall fill time was about 2.913 s and the final infiltration depth was about 100% (Figure 6d). Comparing the infiltration effects at these temperatures 740 °C and 760 °C, no obvious defects, and full impregnation was obtained at 800 °C. It can be observed that, the lower the PV, the more significant solidification and the lower the final infiltration depth. Increasing PT to 800 °C, predicting results showed that the interfaces of $Al_2O_{3(3D)}$ RPC–liquid 5083Al, and liquid 5083Al–mold experience higher temperature gradients. The viscosity of liquid 5083 decreased, result in higher infiltration velocities and shorter fill completion time.

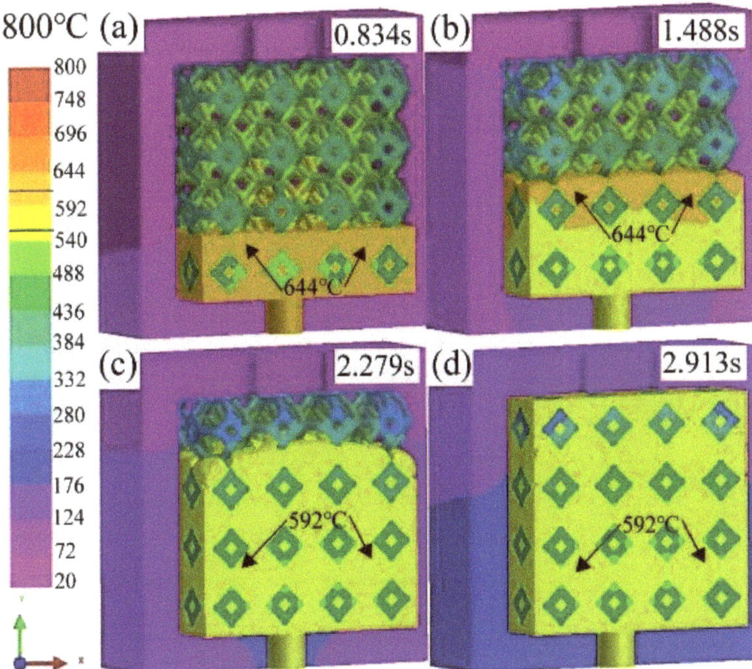

Figure 7. Infiltration depth of liquid 5083Al with PV of 0.4 m/s and PT of 800 °C infiltrating into $Al_2O_{3(3D)}$ with different infiltration times using LPIP. (**a**) 0.834 s; (**b**) 1.488 s; (**c**) 2.279 s; (**d**) 2.913 s.

4.2. Flow Field and Temperature Field of Liquid 5083Al at PT 800 °C

Figure 8 illustrates the infiltration velocities along the flow direction of liquid 5083Al with PV of 0.4 m/s and PT of 800 °C infiltrating into $Al_2O_{3(3D)}$ using LPIP. Liquid 5083Al was Infiltrated continuously through the bottom face of the channel at constant PV of 0.4 m/s and at constant PT of 800 °C. Because the placement of $Al_2O_{3(3D)}$ preform was not close to the wall of mold, the infiltration process was actually a three-dimensional multi-directional infiltration. Due to the viscous loss caused by the porous medium, the flow front became very flat. The infiltration process was relatively stable with small fluctuation, and $Al_2O_{3(3D)}$ preform was infiltrated completely in a very short time. During the infiltration process, the smaller the pore size of $Al_2O_{3(3D)}$ preform would cause the more work of resistance, the more loss of the energy of liquid 5083Al and the smaller PV. This correlation favors filling of larger pore prior to the smaller pores when the $Al_2O_{3(3D)}$ and liquid 5083Al system was poorly wetting. The PV decreased to 0.27 m/s at the place with the smallest pore size of the $Al_2O_{3(3D)}$ preform. Before liquid 5083Al reached the $Al_2O_{3(3D)}$ preforms, the flow front had tiny fluctuations and was not flat. This kind of flow can easily cause gas entrapment and casting defects. The reason for this is that the layered transition in temperature indicates that different regions of the material are solidifying at different rates. During solidification, as the material transitions from a liquid to a solid state, temperature gradients can develop within the material. Regions that solidify earlier will have lower temperatures, while those that solidify later will remain at higher temperatures. These temperature differences can lead to variations in the rate of solidification. The layered transition in infiltration time suggests that certain regions within the material are experiencing slower solidification rates. In some cases, this can result from a slower advancement of the solidification front in specific areas. Slower solidification rates can lead to incomplete filling of voids, creating porosity and gaps in the material. Non-uniform solidification can also contribute to the formation of shrinkage defects. As different regions solidify at different times and rates, they will undergo volume changes associated with the phase transition from liquid to solid. This non-simultaneous volume change can create internal stresses and voids, leading to shrinkage defects.

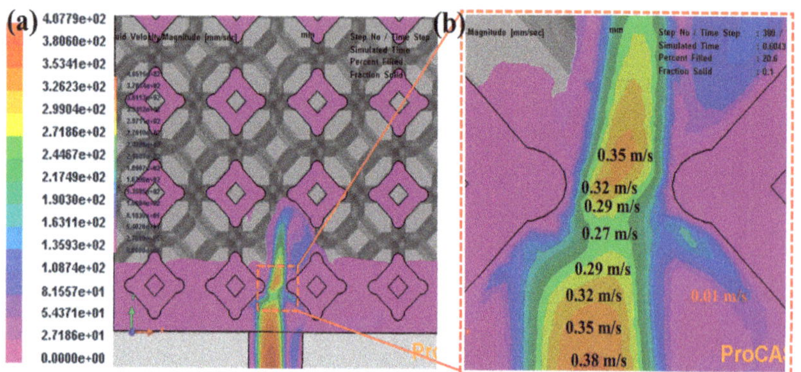

Figure 8. infiltration velocities along the flow direction of liquid 5083Al with PV of 0.4 m/s and PT of 800 °C infiltrating into $Al_2O_{3(3D)}$ using LPIP. (**a**) Overhead view; (**b**) zoom.

Figure 9 presents the temperature along the flow direction. The results indicate PV and PT played a crucial role in determining the velocity of liquid 5083Al through the clearance and the degree of pore shrinkage at the end of infiltration. Temperature of the liquid 5083Al decreased along the flow direction in Figure 9a. The section view in Figure 9b shows the temperature in the middle was higher, while the temperature around the $Al_2O_{3(3D)}$ dropped. This temperature distribution may affect the different solidification rates between the middle and the surrounding parts of the casting, resulting in defects in the middle of the casting. The viscosity and flow velocity of the liquid 5083Al undergo significant changes when there is a large temperature gradient in the region.

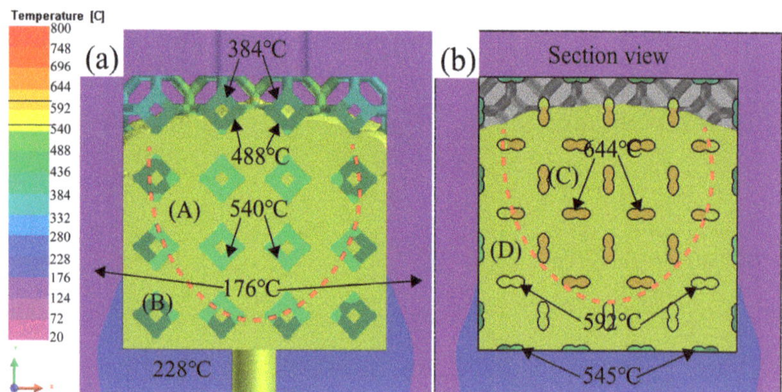

Figure 9. The temperature flow direction of liquid 5083Al with PV of 0.4 m/s and PT of 800 °C infiltrating into $Al_2O_{3(3D)}$ using LPIP. (**a**) overhead view; (**b**) section view.

4.3. Effect of Porosity of Al$_2$O$_{3(3D)}$ on Liquid 5083Al with PV 0.4 m/s and PT 800 °C in LPIP

The mesh numbers for 1, 2, 3, and 4 times Al$_2$O$_{3(3D)}$ RPC impregnating body were divided into 3,836,942, 3,514,000, 3,407,296, and 3,442,287, respectively. The impregnation time for 1, 2, 3, and 4 times was 3.00 s, 2.84 s, 2.94 s, and 3.490 s show in Figure 10. In most tests, the penetration rate of the whole cavity can be completed at around 0.4 m/s. The infiltration rate is calculated using Equation (1)

$$V = F/T \tag{4}$$

where V is the infiltration rate, F is the infiltration percentage, and T is the infiltration time. When the porosity is greater than 80%, the volume rate changes at 86.4%, 91.3%, and 95.1%, and the infiltration time is 3.13 s, 3.26 s, and 3.21 s, respectively, with infiltration rates of 26.8 %/s, 27.9 %/s, and 29.5 %/s. The optimal porosity exists in the range of 65% to 80%, with the volume rate changes at 65.1%, 73.3%, and 80.4%, the infiltration time 2.28 s, 2.51 s, and 2.83 s, and the infiltration rate 28.5 %/s, 29.1 %/s, and 28.3 %/s at an infiltration rate of 0.4 m/s. Due to the large pores of Al$_2$O$_{3(3D)}$ RPC with 5 PPI, infiltration becomes easier and the flowable area of the pores increases. As the porosity increased, the flowable area of the pores becomes wider, and the pore space is quickly occupied. Thus, the fluid volume increased within the porous diameter increased. It gradually loses its guiding effect on liquid 5083Al, leading to turbulent phenomena in Figure 10. Al$_2$O$_{3(3D)}$ RPC with positively influences infiltration, improving the infiltration effect [15]. Using porosity of 65~80% improved the infiltration effect and better prepare Al$_2$O$_{3(3D)}$/5083Al.

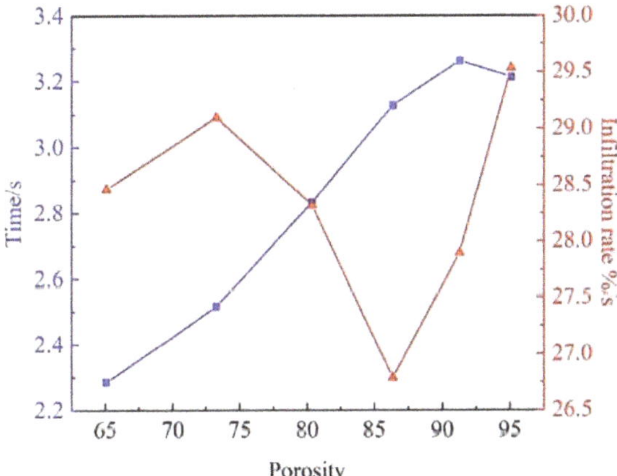

Figure 10. Effect of Al$_2$O$_{3(3D)}$ porosity on infiltration time and infiltration rate of liquid 5083Al with PV of 0.4 m/s and PT of 800 °C during LPIP.

Figure 11 of Al$_2$O$_{3(3D)}$ with 15 PPI and 5 PPI and infiltration. Al$_2$O$_{3(3D)}$ RPC porosity closely related to the infiltration rate. at PT 800 °C, laminar flow and turbulent flow were observed in Figure 11a,b, respectively. Al$_2$O$_{3(3D)}$ with 15 PPI in in Figure 11a could guide the infiltration, which is conducive to Al$_2$O$_{3(3D)}$/5083 infiltration forming and reducing infiltration defects [26]. The infiltration decreases first and then increases using Al$_2$O$_{3(3D)}$ RPC with 5 PPI in Figure 11b.

Figure 12 shows the temperature changes of graphite model with liquid 5083Al at PV of 0.4 m/s and PT of 800 °C during LPIP. The temperature change of Al$_2$O$_{3(3D)}$ RPC was slower than that of the graphite mold. The thermal conductivity of the graphite mold was better than that of the Al$_2$O$_{3(3D)}$ RPC, resulting in the Al$_2$O$_{3(3D)}$ RPC having a thermal insulation effect on liquid 5083Al compared to the graphite mold. The liquid 5083Al was divided into zones A, B, C, and D. The liquid 5083Al temperature in zone A was 644 °C, in zone B was 592 °C as shown in Figure 12a, and in zone C was 592 °C compared to zone D as shown in Figure 12b. The liquid 5083Al in zone A had high temperature, low viscosity, and fast infiltration rate, while the liquid 5083Al in zones B, C, and D had lower temperature, higher viscosity, and lower infiltration rate. The velocity field exhibited large fluctuations, leading to turbulence and low porosity. It is expected the results with PV of 0.4 m/s and PT of 800 °C during LPIP would help to improve the quality of combination of interfaces of $Al_2O_{3(3D)}$ and the 5083Al matrix.

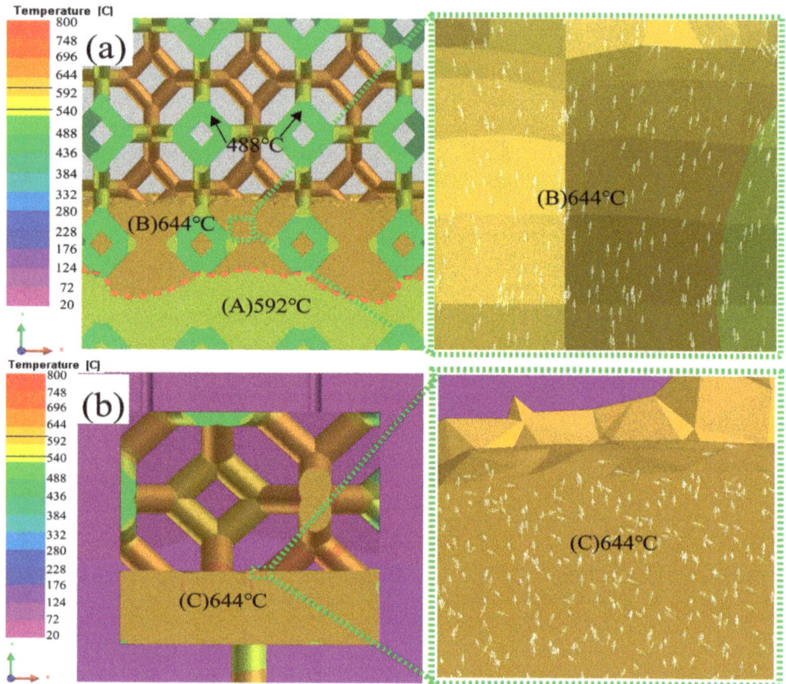

Figure 11. Effect of $Al_2O_{3(3D)}$ porosity on liquid 5083Al with PV of 0.4 m/s and PT of 800 °C during LPIP; (**a**) 15 PPI; (**b**) 5 PPI.

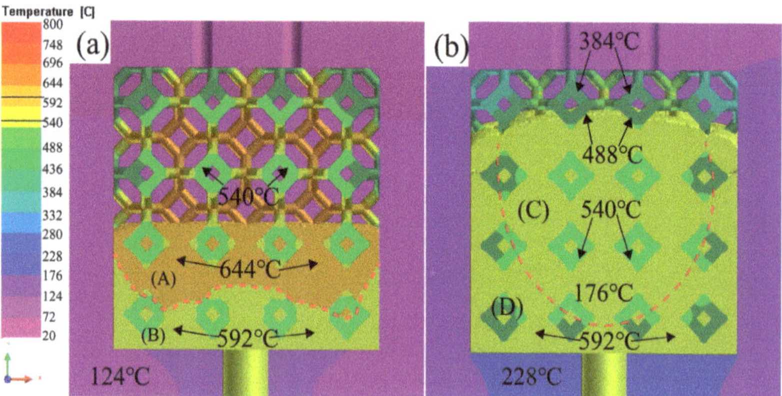

Figure 12. Temperature changes of graphite model with liquid 5083Al at PV of 0.4 m/s and PT of 800 °C during LPIP; (**a**) zones A, B; (**b**) zones C, D.

4.4. Solidification Process

Figure 13 shows the simulation result of mold temperature fields during solidification process with liquid 5083Al at PV of 0.4 m/s and PT of 800 °C. When infiltration was completed, the temperature of the whole mold dropped. The casting was divided into zones A and B according to the temperature zone. At the completion of infiltration, the temperature of $Al_2O_{3(3D)}$ RPC in zone A was 384 °C, and in zone B was 332 °C. The maximum temperature of the mold surface was 228 °C. The inner temperature of the casting was higher than that of the casting.

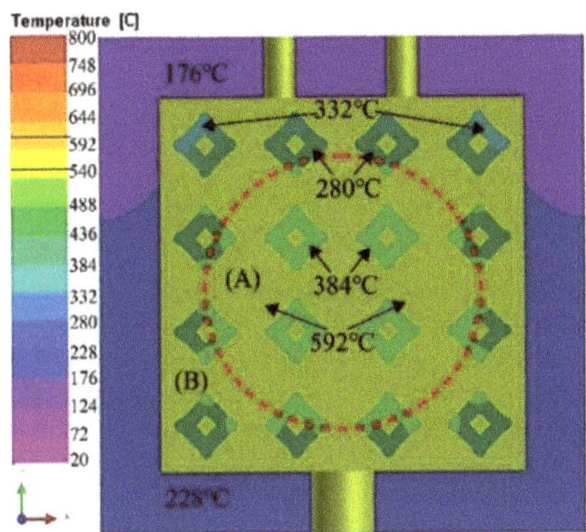

Figure 13. mold temperature fields during solidification process with liquid 5083Al at PV of 0.4 m/s and PT of 800 °C.

Figure 14 shows the simulation changes of infiltration time and temperature after infiltration completion with liquid 5083Al at PV of 0.4 m/s and PT of 800 °C. Both the infiltration time and the infiltration temperature presented a layered transition as shown in Figure 14a,b. The infiltration time could be divided into 15 layers in Figure 14a. The first five layers of infiltration time were short, corresponding to the 0 s~0.98 s stage of stable infiltration. The middle five layers from 0.98 s to 1.97 s showed a certain upward bulge in the two layers near the top, indicating that the liquid 5083Al flow velocity slowed down in this region. At 1.79~2.96 s, in regions with significant temperature gradients, the viscosity may increase, leading to slower infiltration rates and deformations in the flow front. the overall infiltration time bar has a large deformation and bulges upward, and the infiltration time bar thickens, indicating slower infiltration at this time. The infiltration rate of liquid 5083Al decreased under the influence of gravity, making shrinkage and loosening phenomena more likely to occur. Infiltration temperature divided into three layers as shown in Figure 14b. The temperature at the bottom where 5083Al was impregnated dropped rapidly and was close to the preset temperature of $Al_2O_{3(3D)}$ RPC. The middle layer maintained a stable temperature between 614 °C and 598 °C, indicating stable 5083Al infiltration. The top layer had a temperature ranging from about 566 °C to 582 °C, closed to the solidification temperature of liquid 5083Al. At this stage, liquid 5083Al became sticky, and the infiltration rate decreases rapidly. The velocities of liquid 5083Al at the bottom could not meet the stable infiltration at the top, resulting in faster infiltration time in the middle than on both sides.

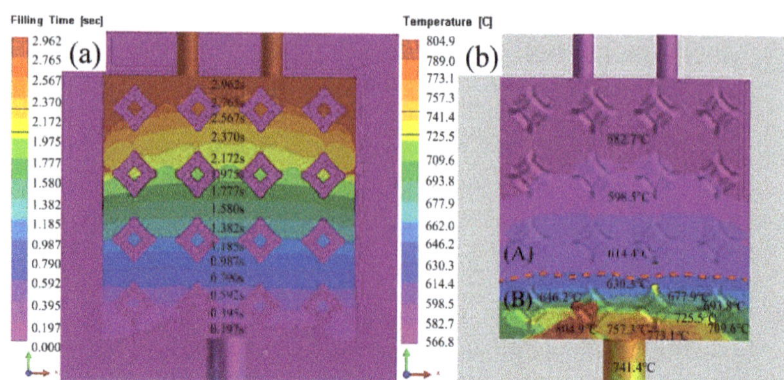

Figure 14. Simulation changes of time and temperature after infiltration completion with liquid 5083Al at PV of 0.4 m/s and PT of 800 °C; (**a**) time; (**b**) temperature.

Figure 15 shows the solidification velocities of different parts and the solidification curve with liquid 5083Al at PV of 0.4 m/s and PT of 800 °C. The iteration step size was 1100, and the solidification state was centered towards the periphery. During infiltration, liquid Al flows from the bottom center to the periphery [27]. As liquid 5083Al infiltrated upward, the flow rate of liquid 5083Al slowed down. Solidification rate of the Al liquid near the inner wall of the model with heat conduction of the graphite model was faster than that of the Al liquid under $Al_2O_{3(3D)}$ RPC insulation. This resulted in funnel-shaped solidification of liquid 5083Al. Figure 15c,d show the solidification temperature curves of marked points (c) and (d) in Figure 15b, respectively [28]. The solidification temperature curve in Figure 15c shows two changes in velocities, and the driving force of solidification was temperature change. As the solidification developed from the inner wall of the model to the center of the casting, the solidification in Figure 15c was controlled by the heat transfer of $Al_2O_{3(3D)}$ RPC, resulting in a faster solidification rate [29]. When $Al_2O_{3(3D)}$ RPC temperature was consistent with the temperature of the aluminized liquid, the solidification changed to be controlled by the air cooling of the outer mold [2]. Figure 15d shows in the first stage, when the liquid Al contacted the $Al_2O_{3(3D)}$ RPC, it was controlled by the heat transfer of the $Al_2O_{3(3D)}$ RPC, resulting in a faster solidification rate. In the second stage, because the temperature of $Al_2O_{3(3D)}$ RPC was not consistent with that of liquid 5083Al, solidification was controlled by the air cooling of $Al_2O_{3(3D)}$ RPC and outer mold. In the third stage, the temperature of $Al_2O_{3(3D)}$ RPC was the same as that of liquid 5083Al, and the solidification changed to be controlled by the air cooling of the graphite mold [30].

Figure 16a,b present a comparison of the solidification time and solid-phase transition completed time, revealing that the solidification time in the center was longer than that around it. The overall solidification process was influenced by $Al_2O_{3(3D)}$ RPC, resulting in a concentration of solidification time and solid-liquid phase in the center, forming a spherical diffusion pattern. The solid-liquid phase could be divided into three distinct parts [31]. The central part of the solidification processed and the time taken for the liquid phase to solidify were relatively long, indicating that the velocities of liquid 5083Al in this region was insufficient, and there was a probability of incomplete solidification leading to porosity. In contrast, the solidification time was more uniform in the peripheral regions due to the influence of the input of liquid Al and $Al_2O_{3(3D)}$ RPC. As a result, the time range for solid-liquid phase transition was larger than the solidification time range. Specifically, the second layer experienced a solidification time ranging from 3.7 s to 4.0 s, and the transition time from solid-liquid phase to solid was from 4.4 s to 4.7 s for the entire solidification process, which aligned with the characteristics of this part. The third layer was mainly affected by the inner wall of the model, and the infiltration rate had little impact. At about 5.0 s, solid phase transition completed Additionally, $Al_2O_{3(3D)}$ RPC resulted in a shorter solidification time, and the transition from solid-liquid phase to solid occurs earlier in this region [16].

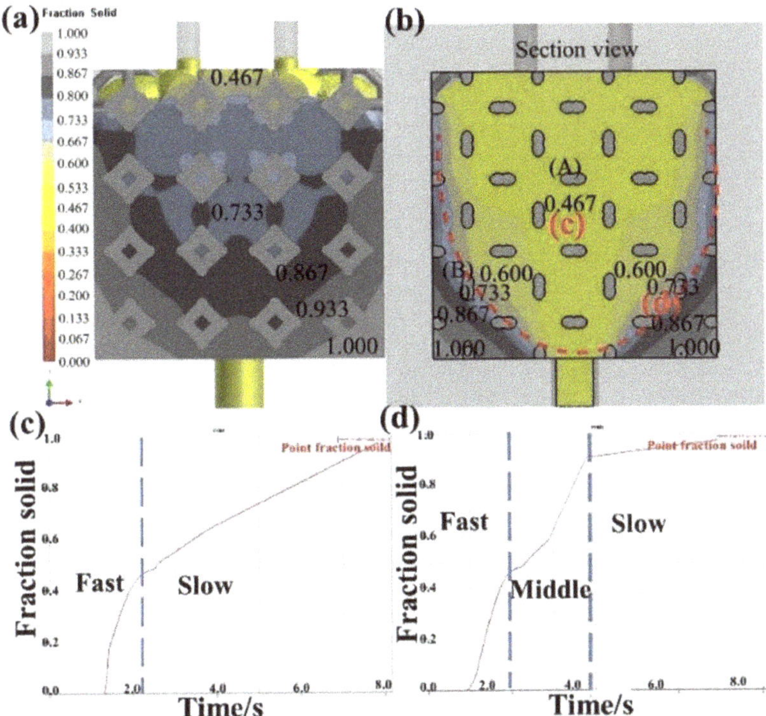

Figure 15. The solidification velocities of different parts and the solidification curve with liquid 5083Al at PV of 0.4 m/s and PT of 800 °C; (**a**) overhead view of fraction solid; (**b**) section view of fraction solid; (**c**) Solidification temperature curves of marked points c; (**d**) Solidification temperature curves of marked points d.

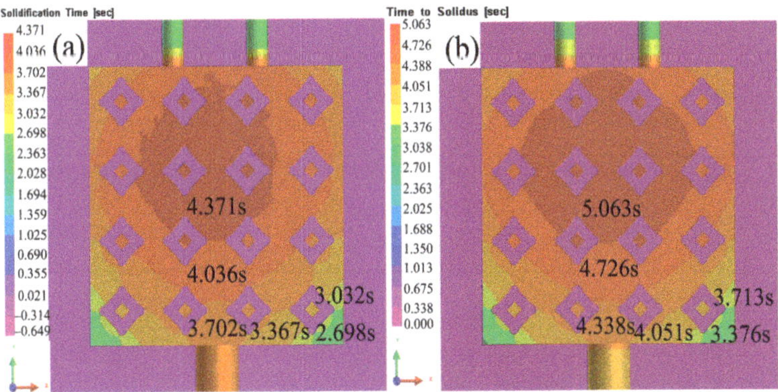

Figure 16. Solidification completion time and solid-phase transition completed time with liquid 5083Al at PV of 0.4 m/s and PT of 800 °C; (**a**) solidification completion time; (**b**) solid-phase transition completed time.

Figure 17 shows simulated prediction of porosity and shrinkage with liquid 5083Al at PV of 0.4 m/s and PT of 800 °C. Shrinkage pore distribution was more uniform, and the probability of shrinkage pore occurrence was small. The shrinkage porosity distribution is low and concentrated in $Al_2O_{3(3D)}$RPC center [17]. There was a probability of shrinkage at interface of $Al_2O_{3(3D)}$ RPC and 5083Al matrix, as well as certain probability of shrinkage in 5083Al matrix. The largest probability of shrinkage was the inlet part of liquid 5083Al. Due to the influence of many factors, such as residual stress concentration, the solidification temperature of liquid 5083Al in $Al_2O_{3(3D)}$ RPC, the increase of viscosity of liquid $Al_2O_{3(3D)}$ RPC, the shrinkage percentage was 13.33%, and the probability of shrinkage is small [1].

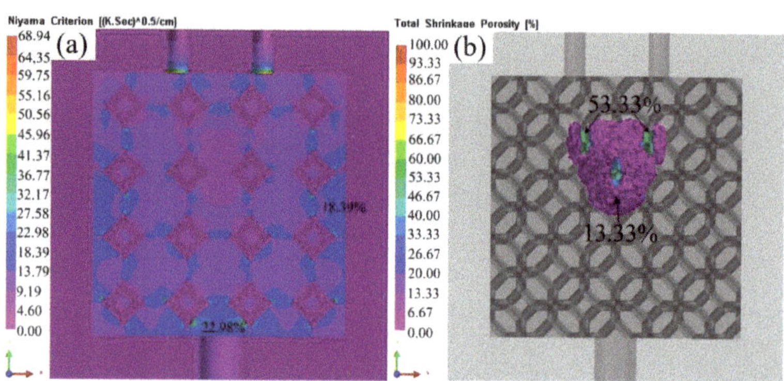

Figure 17. Porosity prediction and shrinkage prediction with liquid 5083Al at PV of 0.4 m/s and PT of 800 °C; (**a**) porosity prediction; (**b**) shrinkage prediction.

4.5. IPC Casting Process

Liquid 5083Al was infiltrated into the as-prepared $Al_2O_{3(3D)}$ RPC preforms with high uniform open porosity (58–74%), pore size (3.5 mm) to fabricate $Al_2O_{3(3D)}$/5083Al IPCs by LPIP. For successful melt infiltration to prepare $Al_2O_{3(3D)}$/5083Al IPCs, the $Al_2O_{3(3D)}$ RPC preforms must be predominantly open porous and sufficiently strong struts without cracks or other defects. It was observed that the infiltration of the liquid 5083Al at PV of 0.4 m/s and PT of 740 °C was incomplete, and a significant amount of liquid 5083Al remained trapped inside the $Al_2O_{3(3D)}$ RPC. This was likely due to the excellent heat dissipation ability of $Al_2O_{3(3D)}$ RPC, as depicted in Figure 18a, which resulted in rapid cooling of the liquid inside $Al_2O_{3(3D)}$ RPC. Consequently, the infiltration inlet was obstructed by the cooled 5083Al, preventing further infiltration, as shown in Figure 18b. To address this issue, $Al_2O_{3(3D)}$/5083 IPCs were prepared by LPIP with liquid 5083Al at PV of 0.4 m/s and PT of 800 °C, and the infiltration process was repeated. Test sample was successfully obtained in Figure 18c. The test sample exhibited certain characteristics, such as a considerable weight, a reflective silver luster, and a solid sound without any hollow sensation upon gentle tapping. After the successful infiltration, the obtained sample, $Al_2O_{3(3D)}$/5083Al, was further polished, as shown in Figure 18d. The surface of the polished sample exhibited distinct features: the gray parts corresponded to $Al_2O_{3(3D)}$ RPC, while the metal luster indicated 5083Al.

Figure 19 X-ray Diffraction (XRD) testing was conducted on $Al_2O_{3(3D)}$/5083Al IPCs. The results were compared with standard reference cards. The XRD analysis confirmed that the $Al_2O_{3(3D)}$/5083Al IPCs was composed of Al_2O_3 and Al alloy.

Figure 20 shows the SEM of surface morphology of $Al_2O_{3(3D)}$/5083Al. The dark color was $Al_2O_{3(3D)}$ RPC, which contains fine pores. The $Al_2O_{3(3D)}$/5083 appeared to be well bonded with no large pore defects, and the interface between the two phases was closely bonded [13]. The infiltration and solidification defects were reduced under air pressure of 0.3 MPa (corresponding to an inlet pressure of about 0.3 MPa or PV of 0.4 m/s) during LPIP. In addition, the $Al_2O_{3(3D)}$ RPC exhibited excellent affinity and good wettability with the liquid 5083Al under pressure, fine air bubbles were effectively minimized at the interface between the two materials until solidification crystallization completed. As the result, the interface between $Al_2O_{3(3D)}$ RPC and 5083Al demonstrated a strong bonding. This reduction in air bubbles helped to eliminate voids, leading to a more homogenous and structurally sound $Al_2O_{3(3D)}$/5083Al composite [13]. This property made it suitable for low-pressure casting applications.

Figure 18. $Al_2O_{3(3D)}$/5083Al IPCs prepared by LPIP; (**a,b**) 740 °C; (**c,d**) 800 °C.

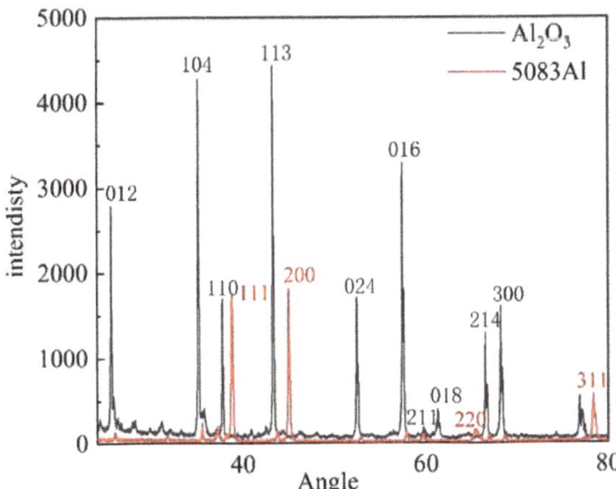

Figure 19. XRD patterns of $Al_2O_{3(3D)}$/5083Al.

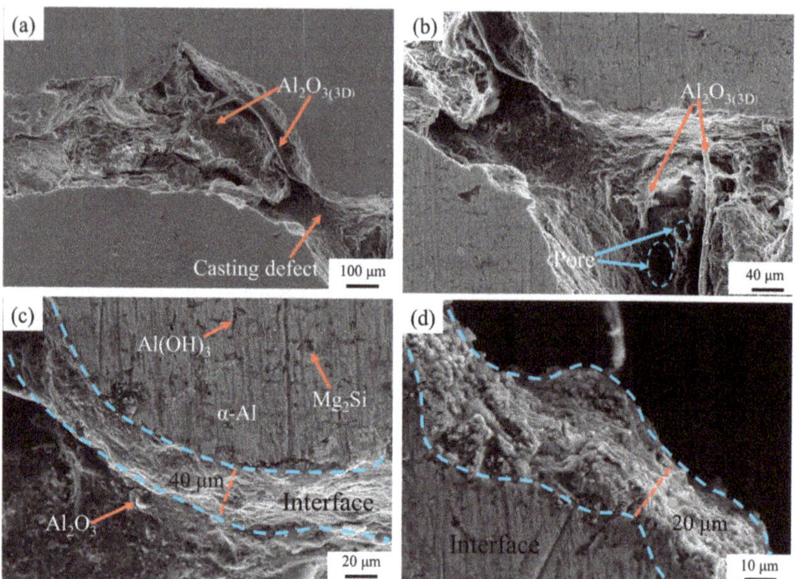

Figure 20. SEM images of Al$_2$O$_{3(3D)}$/5083Al from Al$_2$O$_{3(3D)}$ porosity of 15 PPI with liquid 5083Al at PV 0.4 m/s and PT 800 °C; (**a**) SEM of Al$_2$O$_{3(3D)}$/5083Al; (**b**) zoom of Al$_2$O$_{3(3D)}$/5083Al; (**c**) interface; (**d**) zoom of interface.

Figure 21 shows EDS map scanning shows the element distribution in Al$_2$O$_{3(3D)}$/5083Al. Al element was clearly divided at the phase interface [31]. Mg element was enriched in Al$_2$O$_{3(3D)}$ RPC compared to 5083Al, indicating Mg diffusion towards Al$_2$O$_{3(3D)}$ RPC. Si element was precipitated on 5083, and O element formed a full and uniform oxide film on the Al$_2$O$_{3(3D)}$/5083 surface [32].

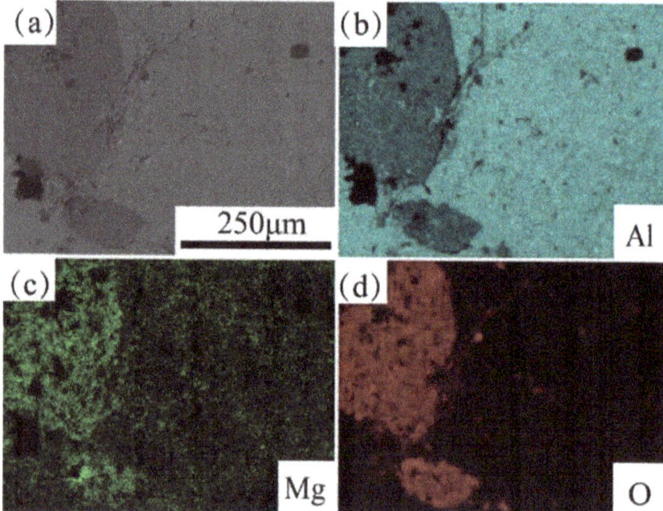

Figure 21. EDS map scanning of Al$_2$O$_{3(3D)}$/5083Al from Al$_2$O$_{3(3D)}$ porosity of 15 PPI with liquid 5083Al with PV 0.4 m/s at PT 800 °C. (**a**) SEM of Al$_2$O$_{3(3D)}$/5083Al; (**b**) Al; (**c**) Mg; (**d**) O.

Figure 22 shows SEM images and EDS results of Al$_2$O$_{3(3D)}$/5083 from Al$_2$O$_{3(3D)}$ porosity of 15 PPI with liquid 5083Al at PV 0.4 m/s and PT 800 °C. Point 1 contains 98.49% Al and 1.51% Mg, indicating that the material tested is an Al-Mg. Point 2, 50.18% Al, 23.19% C, 20.78% Fe, 4.1% Si, 0.82% Cu, and 0.94% Mn, indicating the presence of precipitates mainly containing Fe [33]. Infiltration kinetics was better in the case of the RMP route with liquid 5083Al with PV 0.4 m/s at PT 800 °C. The reactive infiltration was carried out at PV 0.4 m/s to prepare IPCs by reactive infiltration of liquid 5083Al into Al$_2$O$_{3(3D)}$ at 800 °C. The free surface tracking and the solidification phenomena for the infiltration of open-porous preforms was studied using both numerical simulation and experimental methods. The results provided insights into the optimal parameters for successful infiltration [34]. in this study may provide essential implication for the simulation and optimization of processing parameters in various infiltration casting systems.

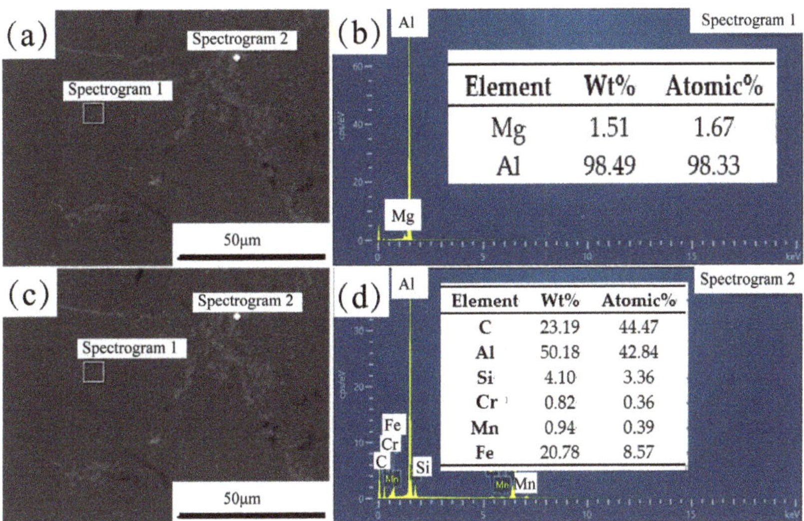

Element	Wt%	Atomic%
Mg	1.51	1.67
Al	98.49	98.33

Element	Wt%	Atomic%
C	23.19	44.47
Al	50.18	42.84
Si	4.10	3.36
Cr	0.82	0.36
Mn	0.94	0.39
Fe	20.78	8.57

Figure 22. SEM image and EDS results of Al$_2$O$_{3(3D)}$/5083 from Al$_2$O$_{3(3D)}$ porosity of 15 PPI with liquid 5083Al at PV 0.4 m/s and PT 800 °C; (**a**) SEM image; (**b**) EDS of spectrogram 1; (**c**) SEM image; (**d**) EDS of spectrogram 2.

5. Conclusions

In this investigation, in order to gain a deeper understanding of the infiltration and solidification processes of liquid 5083Al alloy into Al$_2$O$_{3(3D)}$ RPC during low-pressure infiltration process (LPIP), Al$_2$O$_{3(3D)}$ RPC preforms were simplified as Kelvin cells, and the infiltration and solidification processes of liquid 5083Al with pouring velocities (PV) of 0.4 m/s infiltrating into Al$_2$O$_{3(3D)}$ RPC preforms with varying porosities at different pouring temperatures (PT) were simulated using ProCAST software. The conclusions are the following:

1. During the infiltration process, the smaller the pore size of Al$_2$O$_{3(3D)}$ preform would cause the more work of resistance, the more loss of the energy of liquid 5083Al and the smaller PV.

2. The porosity of 65%~80% obtained better infiltration effect, indicating infiltration rate could be improved with reasonable porosity, leading to better preparation of Al$_2$O$_{3(3D)}$/5083 IPCs.

3. The PV and the PT of the liquid 5083Al alloy was too low, and the liquid metal cannot fully penetrate into the Al$_2$O$_{3(3D)}$ RPC. Shrinkage pore distribution was more uniform, and the probability of shrinkage pore occurrence was small. The shrinkage porosity distribution is low and concentrated in Al$_2$O$_{3(3D)}$ RPC center.

4. Process parameters were optimized to obtain the proper PT of 800 °C and PV of 0.4 m/s, a composited casting has been infiltrated completely without any defects, such as not full filling, porosity or shrinkage. The predicted results show good agreement with the experimental data. It can be a useful method to the preparation of other metal matrix composites reinforced by RPC using infiltration casting.

Author Contributions: Conceptualization, C.Z., Y.J., X.C. and L.Y.; writing original draft preparation, P.X., C.Z. and Y.J.; writing review and editing, C.Z., L.Y., X.C., P.X., F.J., Y.L. and Y.J.; supervision, Y.J., P.X. and Y.L.; project administration, L.Y., X.C. and Y.J.; funding acquisition, L.Y., X.C. and Y.J. All authors have read and agreed to the published version of the manuscript.

Funding: The work was supported by Guangxi Innovation Driven Development Project (Grant No. AA17204021), the foundation of Guangxi Key Laboratory of Optical and Electronic Materials and Devices (No. 20KF-4), and Foundation of introduction of senior talents in Hebei Province (H192003015).

Institutional Review Board Statement: Not applicable.

Informed Consent Statement: Not applicable.

Data Availability Statement: Data sharing is not applicable for this article.

Conflicts of Interest: The authors declare no conflict of interest.

References

1. Kota, N.; Charan, M.S.; Laha, T.; Roy, S. Review on development of metal/ceramic interpenetrating phase composites and critical analysis of their properties. *Ceram. Int.* **2022**, *48*, 1451–1483.
2. Jiang, L.; Jiang, Y.; Yu, L.; Yang, H.; Li, Z.; Ding, Y. Thermo-Mechanical Coupling Analyses for Al Alloy Brake Discs with Al_2O_3-$SiC_{(3D)}$/Al Alloy Composite Wear-Resisting Surface Layer for High-Speed Trains. *Materials* **2019**, *12*, 3155. [PubMed]
3. Yu, L.; Hao, S.; Nong, X.; Cao, X.; Zhang, C.; Liu, Y.; Yan, Y.; Jiang, Y. Comparative Study on the Corrosion Resistance of 6061Al and SiC(3D)/6061Al Composite in a Chloride Environment. *Materials* **2021**, *14*, 7730.
4. Schukraft, J.; Horny, D.; Schulz, K.; Weidenmann, K.A. 3D modeling and experimental investigation on the damage behavior of an interpenetrating metal ceramic composite (IMCC) under compression. *Mater Sci. Eng. A.* **2022**, *844*, 143147.
5. Zhao, D.; Haijun, S.; Liu, Y.; Shen, Z.; Liu, H.; Guo, Y.; Li, X.; Dong, D.; Jiang, H.; Liu, C.; et al. Ultrahigh-Strength Porous Ceramic Composites via a Simple Directional Solidification Process. *Nano Lett.* **2022**, *22*, 2405–2411.
6. Etemadi, R.; Wang, B.; Pillai, K.M.; Niroumand, B.; Omrani, E.; Rohatgi, P. Pressure infiltration processes to synthesize metal matrix composites—A review of metal matrix composites, the technology and process simulation. *Mater. Manuf. Process.* **2018**, *33*, 1261–1290.
7. da Silva, C.C.; Volpato, G.M.; Fredel, M.C.; Tetzlaff, U. Low-pressure processing and microstructural evaluation of unidirectional carbon fiber-reinforced aluminum-nickel matrix composites. *J. Mater. Process.Tech.* **2019**, *269*, 10–15.
8. Akbarnejad, S.; Tilliander, A.; Sheng, D.-Y.; Jönsson, P.G. Effect of Batch Dissimilarity on Permeability of Stacked Ceramic Foam Filters and Incompressible Fluid Flow: Experimental and Numerical Investigation. *Metals* **2022**, *12*, 1001.
9. Du, J.; Chong, X.; Jiang, Y.; Feng, J. Numerical simulation of mold filling process for high chromium cast iron matrix composite reinforced by ZTA ceramic particles. *Int. J. Heat Mass Trans.* **2015**, *89*, 872–883.
10. Chang, C.-Y. Numerical simulation of the pressure infiltration of fibrous preforms during MMC processing. *Adv. Compos. Mater.* **2006**, *15*, 287–300.
11. Chang, C.-Y. Simulation of molten metal through a unidirectional fibrous preform during MMC processing. *J. Mater. Process. Tech.* **2009**, *209*, 4337–4342.
12. Guan, J.T.; Qi, L.H.; Jian, L.I.U.; Zhou, J.M.; Wei, X.L. Threshold pressure and infiltration behavior of liquid metal into fibrous preform. *Trans. Nonferrous Met. Soc. China* **2013**, *23*, 3173–3179.
13. Regulski, W.; Szumbarski, J.; Łaniewski-Wołłk, Ł.; Gumowski, K.; Skibiński, J.; Wichrowski, M.; Wejrzanowski, T. Pressure drop in flow across ceramic foams—A numerical and experimental study. *Chem. Eng. Sci.* **2015**, *137*, 320–337.
14. Zabaras, N.; Samanta, D. A stabilized volume-averaging for flow in porous media and binary alloy solidification processes. *Int. J. Numer. Meth. Eng.* **2004**, *60*, 1103–1138. [CrossRef]
15. Wehinger, G.D.; Heitmann, H.; Kraume, M. An artificial structure modeler for 3D CFD simulations of catalytic foams. *Chem. Eng. J.* **2016**, *284*, 543–556.
16. Nie, Z.; Lin, Y.; Tong, Q. Numerical investigation of pressure drop and heat transfer through open cell foams with 3D Laguerre-Voronoi model. *Int. J. Heat Mass. Trans.* **2017**, *113*, 819–839.
17. Buonomo, B.; di Pasqua, A.; Manca, O.; Nappo, S.; Nardini, S. Entropy generation analysis of laminar forced convection with nanofluids at pore length scale in porous structures with Kelvin cells. *Int. Commun. Heat Mass. Trans.* **2022**, *132*, 105883.

18. Li, Y.; Yang, B.; Zhang, M.; Wang, H.; Gong, W.; Lai, R.; Li, Y.; Teng, J. The corrosion behavior and mechanical properties of 5083Al-Mg alloy manufactured by additive friction stir deposition. *Corros. Sci.* **2023**, *213*, 110972.

19. Nkoua, C.; Josse, C.; Proietti, A.; Basseguy, R.; Blanc, C. Corrosion behaviour of the microbially modified surface of 5083Aluminium alloy. *Corros.Sci.* **2023**, *210*, 110812.

20. Yu, L.; Zhang, C.; Liu, Y.; Yan, Y.; Xu, P.; Jiang, Y.; Cao, X. Comparing the Corrosion Resistance of 5083Al and $Al_2O_3$3D/5083Al Composite in a Chloride Environment. *Materials* **2022**, *16*, 86. [CrossRef] [PubMed]

21. Buonomo, B.; Pasqua, A.D.; Manca, O.; Sekrani, G.; Poncet, S. Numerical Analysis on Pressure Drop and Heat Transfer in Nanofluids at Pore Length Scale in Open Metal Porous Structures with Kelvin Cells. *Heat Trans. Eng.* **2020**, *42*, 1614–1624.

22. Dong, C. Numerical Simulation of Metal Melt Flows in Mold Cavity with Ceramic Porous Media. *Ceram-Silik.* **2016**, *60*, 129–135. [CrossRef]

23. Lu, S.-L.; Xiao, F.-R.; Zhang, S.-J.; Mao, Y.-W.; Liao, B. Simulation study on the centrifugal casting wet-type cylinder liner based on ProCAST. *Appl. Therm. Eng.* **2014**, *73*, 512–521.

24. Liu, L.-B.; Hu, C.-H.; Zhang, Y.-H.; Song, C.-J.; Zhai, Q.-J. Melt flow, solidification structures, and defects in 316 L steel strips produced by vertical centrifugal casting. *Adv.Manuf.* **2023**.

25. Hammel, E.C.; Ighodaro, O.L.R.; Okoli, O.I. Processing and properties of advanced porous ceramics: An application based review. *Ceram. Int.* **2014**, *40*, 15351–15370.

26. Zhang, S.; Zhu, M.; Zhao, X.; Xiong, D.; Wan, H.; Bai, S.; Wang, X. A pore-scale, two-phase numerical model for describing the infiltration behaviour of SiC p /Al composites. *Compos. Part A Appl. Sci. Manuf.* **2016**, *90*, 71–81.

27. Prakash, S.A.; Hariharan, C.; Arivazhagan, R.; Sheeja, R.; Raj, V.A.A.; Velraj, R. Review on numerical algorithms for melting and solidification studies and their implementation in general purpose computational fluid dynamic software. *J. Energy Storage* **2021**, *36*, 102341.

28. Kaur, I.; Singh, P. Numerical investigation on conjugate heat transfer in octet-shape-based single unit cell thick metal foam. *Int. Commun. Heat Mass Transf.* **2021**, *121*, 105090.

29. Guo, X.; Liu, R.; Wang, J.; Shuai, S.; Xiong, D.; Bai, S.; Zhang, N.; Gong, X.; Wang, X. 3D actual microstructure-based modeling of non-isothermal infiltration behavior and void formation in liquid composite molding. *Appl. Math. Model.* **2021**, *94*, 388–402.

30. Lacoste, E.; Arvieu, C.; Mantaux, O. Numerical Modeling of Fiber-Reinforced Metal Matrix Composite Processing by the Liquid Route: Literature Contribution. *Metall. Mater. Trans. B.* **2018**, *49*, 831–838.

31. Nong, X.D.; Jiang, Y.L.; Fang, M.; Yu, L.; Liu, C.Y. Numerical analysis of novel SiC_{3D}/Al alloy co-continuous composites ventilated brake disc. *Int. J. Heat Mass.Trans.* **2017**, *108*, 1374–1382.

32. Xue, L.; Wang, F.; Ma, Z.; Wang, Y. Effects of surface-oxidation modification and heat treatment on silicon carbide 3D/AlCu 5 MgTi composites during vacuum-pressure infiltration. *Appl. Surf. Sci.* **2015**, *356*, 795–803.

33. Ma, X.; Zhao, Y.F.; Tian, W.J.; Qian, Z.; Chen, H.W.; Wu, Y.Y.; Liu, X.F. A novel Al matrix composite reinforced by nano-AlN(p) network. *Sci. Rep.* **2016**, *6*, 34919. [CrossRef] [PubMed]

34. Potoczek, M.; Śliwa, R. Microstructure and Physical Properties of AlMg/Al_2O_3 Interpenetrating Composites Fabricated by Metal Infiltration into Ceramic Foams. *Arch. Metall. Mater.* **2011**, *56*. [CrossRef]

Review

The Characteristic Microstructures and Properties of Steel-Based Alloy via Additive Manufacturing

Chunlei Shang [1], Honghui Wu [1,*], Guangfei Pan [1], Jiaqi Zhu [2], Shuize Wang [1,*], Guilin Wu [1], Junheng Gao [1,*], Zhiyuan Liu [3], Ruidi Li [4] and Xinping Mao [1]

1   Beijing Advanced Innovation Center for Materials Genome Engineering, Innovation Research Institute for Carbon Neutrality, University of Science and Technology Beijing, Beijing 100083, China
2   Department of Environmental and Municipal Engineering, North China University of Water Resources and Electric Power, Zhengzhou 450046, China
3   Additive Manufacturing Institute, College of Mechatronics and Control Engineering, Shenzhen University, Shenzhen 518060, China
4   State Key Laboratory of Powder Metallurgy, Central South University, Changsha 410083, China
*   Correspondence: wuhonghui@ustb.edu.cn (H.W.); wangshuize@ustb.edu.cn (S.W.); junhenggao@ustb.edu.cn (J.G.)

Abstract: Differing from metal alloys produced by conventional techniques, metallic products prepared by additive manufacturing experience distinct solidification thermal histories and solid−state phase transformation processes, resulting in unique microstructures and superior performance. This review starts with commonly used additive manufacturing techniques in steel−based alloy and then some typical microstructures produced by metal additive manufacturing technologies with different components and processes are summarized, including porosity, dislocation cells, dendrite structures, residual stress, element segregation, etc. The characteristic microstructures may exert a significant influence on the properties of additively manufactured products, and thus it is important to tune the components and additive manufacturing process parameters to achieve the desired microstructures. Finally, the future development and prospects of additive manufacturing technology in steel are discussed.

Keywords: additive manufacturing; characteristic microstructure; steel−based materials; phase transformation; heat treatment

Citation: Shang, C.; Wu, H.; Pan, G.; Zhu, J.; Wang, S.; Wu, G.; Gao, J.; Liu, Z.; Li, R.; Mao, X. The Characteristic Microstructures and Properties of Steel-Based Alloy via Additive Manufacturing. *Materials* **2023**, *16*, 2696. https://doi.org/10.3390/ma16072696

Academic Editor: Raul D. S. G. Campilho

Received: 9 February 2023
Revised: 4 March 2023
Accepted: 13 March 2023
Published: 28 March 2023

1. Introduction

In recent decades, metal additive manufacturing techniques have received more and more attention [1,2]. The material products fabricated by metal additive manufacturing have been extended, but are not limited, to steel [3,4], aluminum alloy [5,6], magnesium alloy [7,8], titanium alloy [9], high−entropy alloy [10,11], etc. As one of the most fundamental metal materials, extensive research on additive manufacturing of iron−based alloys has been reported in the literature, mainly including 18Ni300 mold steel, 316 stainless steel, 304 stainless steel, etc. Compared with traditional manufacturing techniques (TMT), additive manufacturing methods can effectively save processing time and improve material utilization [12]. Due to the distinguished advantages of additive manufacturing (AM), the technique has been widely applied in industries [13–15].

The basic principle of AM is similar to that of multi−pass welding technology in that the material powder melts once the heat source passes through and solidifies before the next heat source [16]. In additive manufacturing, the printing parameters exert a significant effect on the characteristic microstructure. For instance, laser−generated melt pools and thermal gradients during solidification may produce columnar grains and textures in the products. In addition, the large temperature gradients cause a non−equilibrium microstructure, which is distinct from the microstructure produced by TMT. After cooling, the secondary heat cycle caused by the newly melted layer covering the solidified layer may further facilitate component diffusion and microstructure evolution [17,18]. The deposition path is another important influencing factor on the performance of AM samples [19,20]. To study the effect of deposition paths on sample performance and productivity, Veiga et al. [21] conducted several strategies. It was found that waving and cross−waving were the most beneficial strategies in terms of productivity, which achieved a 50% torch utilization rate compared to the total time.

The process parameters of additive manufacturing, such as scanning rate, powder particle size, pre−heat temperature, etc., significantly affect the performance of the additively manufactured products [22,23]. To investigate the effect of process parameters on the microstructure evolution and tensile performance of 304L stainless steel by additive manufacturing, Wang et al. [24] prepared two samples with different heat inputs by additive manufacturing. It was found that the samples with low linear heat input displayed higher strength and elongation than the samples with high linear heat input. Helmer et al. [25] also found that the transition of columnar grains to equiaxed grains could be achieved by rapidly switching the orientation of the additive manufacturing heat source and keeping the trajectories of the melted regions overlapped. To investigate the corrosion performance of 316L stainless steel during selective laser melting (SLM) additive manufacturing, Zhao et al. [26] employed SLM equipment to manufacture 316L stainless steel with distinct scanning tactics and then soaked the samples in NaCl aqueous solution (3.5 wt%) for electrochemical tests. It was found that although the passivation film on the side of the 316L stainless steel sample was thicker than the head of the surface, there was more pitting and faster corrosion rates occurred on the sides with more molten pool boundaries. To explore a way to reduce the cost of fine powder in additive manufacturing, Yang et al. [27] mixed the 316L fine and coarse powders with different mass ratios via ball milling, finding that the mechanical properties of the fabricated SLM samples with a mass ratio of 80:20 were comparable to those of the SLM samples fabricated with pure fine powder, which was closely related to the complex coupling of temperature gradients and surface tension gradients during additive manufacturing.

Based on the information retrieved with the keywords "additive manufacturing" and "steel" in the Web of Science database, as shown in Figure 1a, more than 9500 relevant articles on the topic of AM with steel materials have been published in the past 22 years, and the citation frequency of relevant additive manufacturing literature is increasing year by year. In the present work, we introduce the rapid development and advantages of AM and the utilization of additive manufacturing in steels, and then some commonly used compositions and processes of additive manufacturing are summarized. Furthermore, as shown in Figure 1b, steel−based alloys printed by additive manufacturing display some common characteristic microstructures, such as porosity, dendrites, residual stress, composition segregation, etc. These microstructures can be tuned by composition and process parameters, which in turn affect the service performance of the products prepared by additive manufacturing. Finally, the future development prospects and directions of AM in steel materials are discussed.

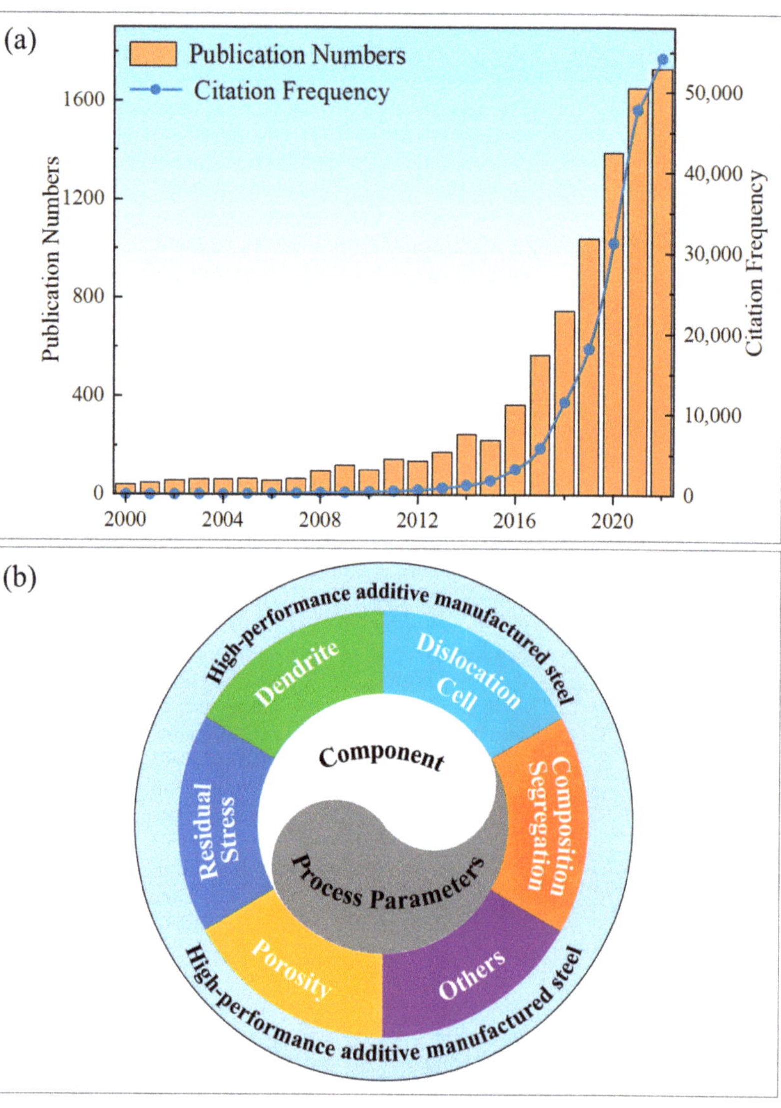

Figure 1. Utilization of additive manufacturing in steel−based alloys. (**a**) Histogram of the number and citation frequency of relevant articles retrieved with the keywords "additive manufacturing" and "steel" in the Web of Science database. (**b**) Some typical microstructures produced by additive manufacturing in steels, and the relationship among the component, process, microstructure and desired performance.

2. Advanced Additive Manufacturing Techniques

AM is a bottom−up, layer−by−layer manufacturing technique based on 3D model data, and the fabricated samples can be constructed by melting powder, wire, etc. using a laser or electron beam [24]. Additive manufacturing techniques with optimized process parameters can produce many superior properties, such as increased strength, improved ductility, and effectively enhanced service performance [28]. These technical methods can produce geometrically complex samples that are difficult to produce by TMT, which is advantageous in integrating digital design and product production. However, it is worth mentioning that high cost is one of the factors limiting the development of AM. The vision of additive manufacturing is to use these techniques to produce complex metal products for critical functions such as turbine blades and jet engines in aerospace [29]. In recent years, the development of AM techniques has been very rapid [30], including DED, powder bed fusion (PBF), binder jetting [31], metal extrusion [32], atomic diffusion additive manufacturing (ADAM) [33], sheet lamination [34,35], material jetting [36], arc additive manufacturing, [37] etc. For steel materials, PBF and DED additive manufacturing techniques are the most extensively used.

The printing parameter settings of AM influence the melting and solidification process of the powder and then determine the microstructures of the AM sample [38]. For instance, excessively high melt pool temperature and the presence of thermal gradients is the main reason for the formation of columnar grains and textures in the samples. Traditionally, casting and forging produce near−equilibrium microstructures, but excessive cooling rates during solidification in additive manufacturing produce non−equilibrium microstructures. Thermal cycling induced by the duplicate deposition of new molten layers on the solidified layer also leads to iterative microstructural evolution [39]. Some key parameters affect specimen quality, including building layer thickness, laser power, hatch spacing, scanning speed, etc. For the additive manufacturing of steel specimens, some commonly used process parameters are listed in Table 1.

Table 1. List of several typical deposition techniques and the relevant parameters in AM steel.

Process	Material Shape	Travel Speed (mm s^{-1})	Spot Size (mm)	Layer Height (mm)	Heat Input (W)	Material Feed Speed (mm s^{-1})	Ref.
Laser DED	powder	2.5–20	1.2–2	0.25–0.5	360–2600	2–20.4	[40–45]
PTA		1.3–1.7	/	/	/	25–35	[46]
GMAW	wire	2.5–30	/	0.5–2	3500–8400	28–166	[47–54]
GTAW		2.92–7	/	/	1920	16.67–58	[55]
PTA		0.6–2	/	/	350–3510	9–28	[55]

2.1. Powder Bed Fusion

In additive manufacturing, powder bed fusion is one of the most prevalent techniques that selectively melts/sinters areas of powder beds using thermal energy. Its heat sources mainly include lasers and electron beams. According to these two heat sources, PBF can be separated into two main techniques: SLM using high−intensity lasers and electron beam melting (EBM) using electron beams. In both processes, the powder needs to be held on a build platform [56].

2.1.1. Selective Laser Melting

SLM is a powder bed AM that applies a high−energy laser beam to selectively melt successive layers of powder in order to fabricate a sample [57–59]. During irradiation with laser beam energy, the irradiated powder melts and forms a tiny molten pool [60,61]. The thermal history experienced by samples manufactured by SLM technology is different from that of samples that have undergone traditional technologies [62,63]. In contrast to TMT, this technique combines fast melting and solidification, circular heating, and reciprocal cooling of the deposited layer to produce a characteristic microstructure that differs from that obtained by TMT [24,64,65]. The SLM manufacturing process displays the characteristics of a high utilization rate of raw material powder, which has great advantages in the production of complex samples [66]. SLM production sample quality is influenced by many factors, for instance, laser scanning speed, powder and shape, energy input, and so on. The SLM production process involves complex physical processes, such as fast melting and solidification [67,68], absorption and transmission of laser energy [69], as well as material flow in the molten pool [70].

2.1.2. Electron Beam Melting

Selective EBM additive manufacturing builds a sample layer−by−layer in a powder bed by selectively melting the powder using a beam of high−energy electrons [71,72]. Unlike SLM, electron beam application requires electrical conductivity and is therefore only suitable for metallic materials. When printing samples, the speed of the electron beam is as high as 10^5 m/s, that is, the electron beam can jump from one point to another in an instant. Therefore, by taking advantage of this feature, electron beam additive manufacturing can realize innovative heating and melting strategies.

2.2. Direct Energy Deposition

Another commonly used AM is DED. Differing from the SLM process, the DED production process employs a metal wire or metal powder flow instead of a powder bed as the raw material injection, and then it melts and deposits the material on a substrate using an electron beam or laser. Laser−engineered net shape (LENS) is a representative technique of DED technology [73]. LENS is an AM that uses a laser beam to feed metal powders of different compositions and properties into a molten pool for melting. The difference between LENS and SLM is the metal powder addition process. LENS technology adopts a synchronous powder feeding process in the molding process, wherein the metal powder is sprayed and heated by the laser beam at the same time. The LENS process, with the advantages of high molding efficiency and high sample density, can also be sprayed on the surface of the sample [74]. Another development of DED technology is the combination with topology optimization technology to design samples with excellent performance, wherein topology optimization is a mathematical method to allocate materials within a given design range according to specific physical problems and optimization targets [75].

3. Characteristic Microstructures of Steel Prepared via AM

Due to the complex phase structure in steel materials, its structure and performance are usually related to the solidification process and thermal history. Consequently, compared with other metal materials, the subsequent heat treatment strongly affects the microstructure and performance of the 3D−printed steel material, and thus further process optimization is very important [76]. Table 2 lists some commonly printed steels in additive manufacturing, including stainless steel, tool steel, die steel, etc. The characteristics of additively manufactured samples are summarized from the aspects of composition, process, and performance.

Table 2. List of composition, process, and properties of metal materials printed by AM. (SS: stainless steel, AP: as−produced, SA: solution annealed, AH: aging heat−treated, BP: base plate temperature during the preparation process.).

Steel Type	Elements (wt%)									3D Printing Techniques	Heat Treatment Process	Mechanical Properties				Ref.
	C	Cr	Ni	Mo	Mn	Si	Ti	Al	Others			YS (MPa)	UTS (MPa)	Elongation (%)	Hardness (HV or HRC)	
316L	<0.03	16–18	10–14	2–3	<2	<0.75	/	/	N < 0.1	L−PBF	AP	450	640	59		[77]
										L−PBF	AP	590	700	36		
										L−DED	AP	470	675	52.5		
										L−DED	AP	535	665	35		[78]
										L−DED	AP	405	655	57		
										L−DED	AP	505	670	41.58		
										L−PBF	AP	602	664	30		[79]
										L−PBF	AP	557	591	42		
										L−PBF	AP	534	653	16.2		[80]
										L−PBF	AP	444	567	8		
										L−DED	AP	490	685	51		[42]
										L−DED	AP	280	580	62		
17-4 PH	<0.07	15–17.5	3–5		<1.0	<1.0	/	/	Nb 0.15–0.45	L−PBF	L−PBF	452	1119	15.2		[81]
										L−PBF	AP	798	1101	15.8	346.3 HV	
										L−PBF	AP	824	916	4.2	356.1 HV	
										L−PBF	AP	810	948	4.8	350.2 HV	
										L−PBF	AP	773	1043	17.6	355.3 HV	[82]
										L−PBF	AP	873	951	5.3	346.7 HV	
										L−PBF	AP	866	935	3.3	350.3 HV	
										L−PBF	AP	1190	1370	8.3	380 HV	[83]
										L−PBF	AP	570	944	50		[84]

Table 2. *Cont.*

Steel Type	Elements (wt%)									3D Printing Techniques	Heat Treatment Process	Mechanical Properties				Ref.
	C	Cr	Ni	Mo	Mn	Si	Ti	Al	Others			YS (MPa)	UTS (MPa)	Elongation (%)	Hardness (HV or HRC)	
18Ni−300	<0.03	<0.5	17–19	4.5–5.2	<0.1	<0.1	0.6–0.8	0.05–0.15	Co 8.5–9.5	L−PBF	AP	815–1080	1010–1205	8.3–12	420 HV	
										L−PBF	SA	800	950	13.5	320 HV	[85]
										L−PBF	AH	1750	1850	5.1	600 HV	
										L−PBF	AP		1085–1192	5–8	33 HRC	[86]
										L−PBF	AP	985	1152	7.6	34 HRC	[87]
										L−PBF	AP	915	1188	6.1	371 HV	[88]
										L−PBF	AH	1957	2017	1.5	600 HV	
										L−PBF	AP		1290	13.3	40 HRC	[89]
										L−PBF	AH		2217	1.6	58 HRC	
										L−PBF	AP	915	1165	12.4	35 HRC	
										L−PBF	AH	1967	2014	3.3	54 HRC	[90]
										L−PBF	SA	962	1025	14.4	28 HRC	
										L−PBF	SA+AH	1882	1943	5.6	53 HRC	
										L−PBF	AP	1003	1370	1.7	59 HRC	[91]
										L−PBF	AH	1580	1860	2.2	51 HRC	
										DED	AP	1288–1564	2033–2064	5–6	660 HV	[92]
H13	0.32–0.45	4.75–5.5		1.1–1.75	0.2–0.6	0.8–1.2	/	/	V 0.8–1.2	L−PBF	AP(BP 240 °C)	892	1440	1.5	575 HV	[93]
										L−PBF	AP	1236	1712	4.1		
										L−PBF	AP(BP 200 °C)	835	1620	4.1		[94]
										L−PBF	AP(BP 400 °C)	1073	1965	3.7		
										L−PBF	AP 100 °C	1150–1275	1550–1650	1.5–2.25		[95]

Table 2. *Cont.*

Steel Type	Elements (wt%)									3D Printing Techniques	Heat Treatment Process	Mechanical Properties				Ref.
	C	Cr	Ni	Mo	Mn	Si	Ti	Al	Others			YS (MPa)	UTS (MPa)	Elongation (%)	Hardness (HV or HRC)	
Ferritic SS 441	<0.03	18	<1.0		<1	<1	/	/	Nb < 0.9, Ti0.1–0.5	L–PBF	AP	679	874	30		[96,97]
										L–PBF	AP	741	896	28		[96,98]
Duplex SS 2205	<0.03	21–23	4.5–6.5	2.5–3.5	<2.0	<1.0	/		N0.08–0.2	L–PBF	AP	950	1071.3	16		[99]
										L–PBF	AP		940	12		[100]
Duplex SS 2507	<0.03	24–26	6–8	3–5	<1.2	<0.8			Cu < 0.5, N 0.24–0.32	L–PBF	AP	1214	1321	8	450 HV	[101,102]
Other steels	Also includes duplex stainless steel (SAF2705), ODS steel (PM200), tool steel (M2), etc.															[103–105]

3.1. Porosity in Additively Manufactured Steel

Porosity is an important concern in metal additive manufacturing, as it worsens the apparent strength of the sample and decreases the fatigue life of the products. Among additively manufactured specimens, porosity formation is a hot research topic [106]. The main causes of porosity in additively manufactured specimens are residual gases from metal powder raw materials and powder solidification without fusion, which exerts a notable influence on the corrosion performance of AM samples [107]. Itzhak et al. [108] conducted a study by putting 316L stainless steel into sulfuric acid. The results showed that the porosity in the specimen was the major determinant of anti−corrosion. To study the occurrence of pitting corrosion of stainless steel, Prieto et al. [109] prepared 316L with a direct metal laser sintering process, finding that the residual porosity and microstructural deformation rendered the sample more susceptible to pitting. Laleh et al. [110] observed that the erosion and corrosion of the 316L sample prepared by SLM were poor and it was closely related to the porosity in the sample. To study the sliding wear behavior of 316L, Sun et al. [111] prepared 316L by LSM. It was found that the presence of porosity in the specimen had a more important effect on the wear behavior than the microhardness.

To examine the influence of heat treatment temperatures on the mechanical performance and wear behavior of AM samples, Emre et al. [112] conducted experiments with laser selective melting using three heat treatment temperature methods: HT−1, HT−2, and HT−3 were respectively kept at 600, 850, and 1100 °C for 2 h, respectively, and then air−cooled. Figure 2a presents a schematic drawing of the wear test. The porosity of the four heat−treated specimens is presented in Figure 2b. The porosity of the as−prepared specimen, HT−1, HT−2, and HT−3 was measured by optical microscopy, with values of 0.43%, 0.38%, 0.29%, and 0.08%, respectively. As the heat treatment temperature increased, the sample structure was homogenized and the porosity was further reduced, which was closely related to factors such as process parameter optimization, gas overflow, etc. Figure 2c shows the wear curves of the sample wear test. The figure displays that the various heat treatments exerted a distinct influence on the wear behavior of the samples. The wear depth of the as−prepared sample was approximately 52 µm, whereas the wear depth of the samples with HT−1, HT−2, and HT−3 treatments were 62, 53, and 45 µm, respectively. This study shed light on the fact that the wear resistance of 316L produced by SLM was significantly affected by the porosity, and the wear resistance increased with decreasing porosity.

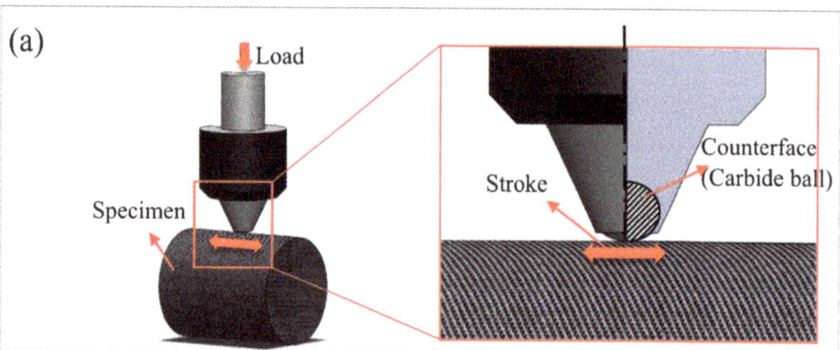

Figure 2. *Cont.*

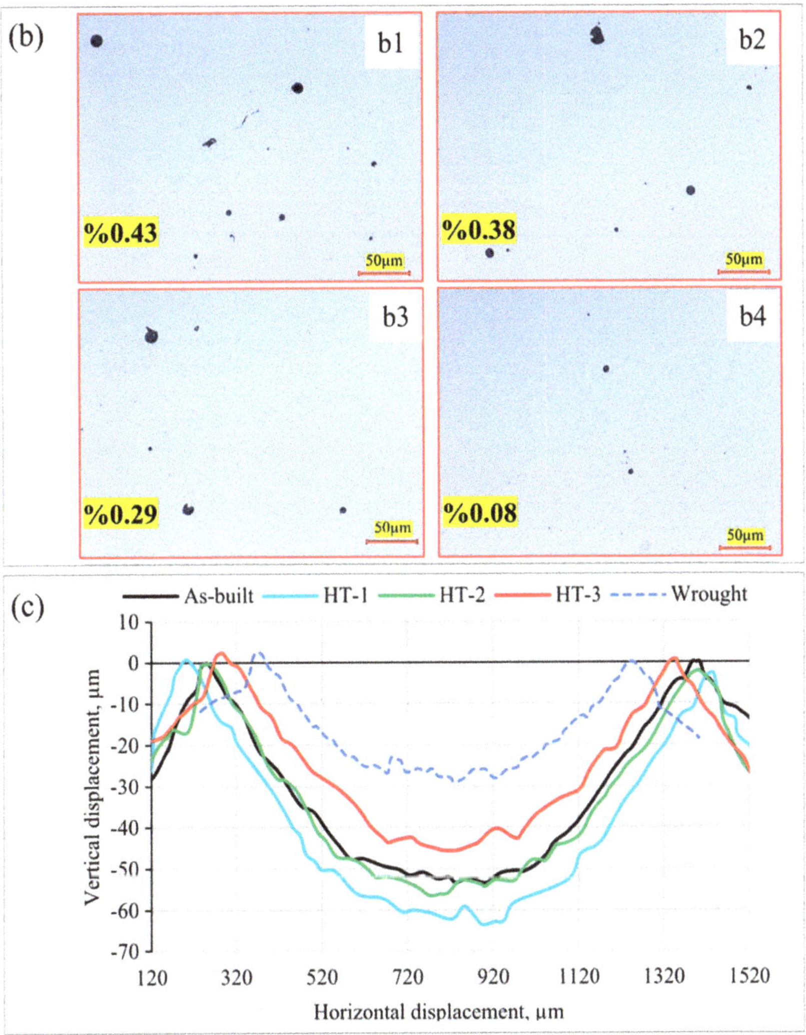

Figure 2. The influence of heat treatment temperature on the porosity of additively manufactured 316L. (**a**) Schematic drawing of the wear test; (**b**) porosity in different samples: (**b1**) as−built sample, (**b2**) HT−1 sample, (**b3**) HT−2 sample, (**b4**) HT−3 sample; (**c**) wear profiles comparing stainless steel 316L of as−built, HT−1, HT−2, and HT−3 treatments. (Reprinted with permission from ref. [112]. Copyright 2020, International Journal of Advanced Manufacturing Technology).

3.2. Dendrite Structures in Additively Manufactured Steel

During the process of printing metal samples via additive manufacturing, dendrite structures are developed. To study the dendritic microstructure and its effect on the mechanical performance of 316L, Chen et al. [113] produced 316L samples using the gas metal arc additive manufacturing (GMA−AM) technique. It was found that austenite dendrites were aligned vertically to the GMA−AM 316L sheet. To investigate the microstructure and its mechanical performance of 308L stainless steel, Le et al. [49] fabricated thin−walled 308L samples by gas−shielded welding additive manufacturing (GMAW−AM). It was found that columnar dendrites mainly existed in the GMAW−AM thin−walled 308L microstructure, and the columnar dendrites grew towards the deposited direction with an increasing number of layers in the printed samples. In addition to the arc additive manufacturing process, cold metal transfer is another potential technique [37].

To investigate the microstructure and the corresponding mechanical performance of additive manufactured 304L steel, Ji et al. [114] conducted tensile tests and metallographic experiments. It was found that as the number of printing layers increased, a slower cooling rate, thicker dendrites, and much more stable dendrite morphology were observed. Figure 3a illustrates a schematic drawing of the developed experimental setup. As shown in Figure 3b, coarse columnar grains appeared on the surface of the chemically etched samples. To better identify the microstructural features, 4 regions were selected within the sample, which were labeled b2, b3, b4, and b5 from top to bottom, respectively. Figure 3b presents the microstructure of dendritic growth within the whole sample, showing that the dendrites became thicker and more stable as the number of layers increased. From top to bottom, the spacing between primary dendrite arms in Figure 3(b2–b5) is 15.44, 13.59, 8.23, and 4.94 um, respectively, and the spacing between the secondary dendrite arms is 7.14, 8.93, 4.98, and 3.52 um, respectively. A total of six tensile specimens (T1, T2, T3, L1, L2, and L3) were fabricated to identify the mechanical performance of the specimen, and the results of the six tensile tests are displayed in Figure 3c. As displayed in Figure 3c, the strength of the T series specimens was higher than that of the L series specimens but the elongation was much smaller than that of the L series specimens. On the whole, the performance of the L series samples was superior. The high strength of the T series samples and high elongation of the L series samples were due to the presence of dendrites, wherein the T series samples were perpendicular to the dendrite and the L series samples were parallel to the dendrite. During the tensile process of the T series samples, the dislocation movement was hindered by the dendrite boundary, which resulted in the increased strength and decreased elongation. Meanwhile, the tensile test of the L series samples displayed an opposite trend.

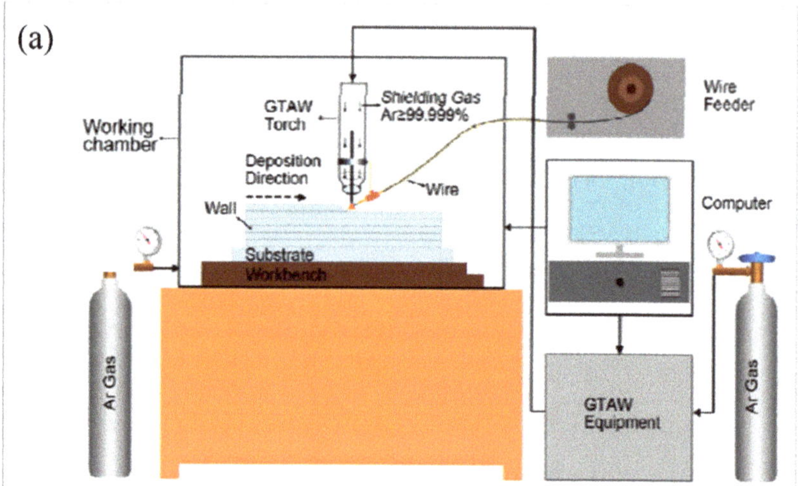

Figure 3. *Cont.*

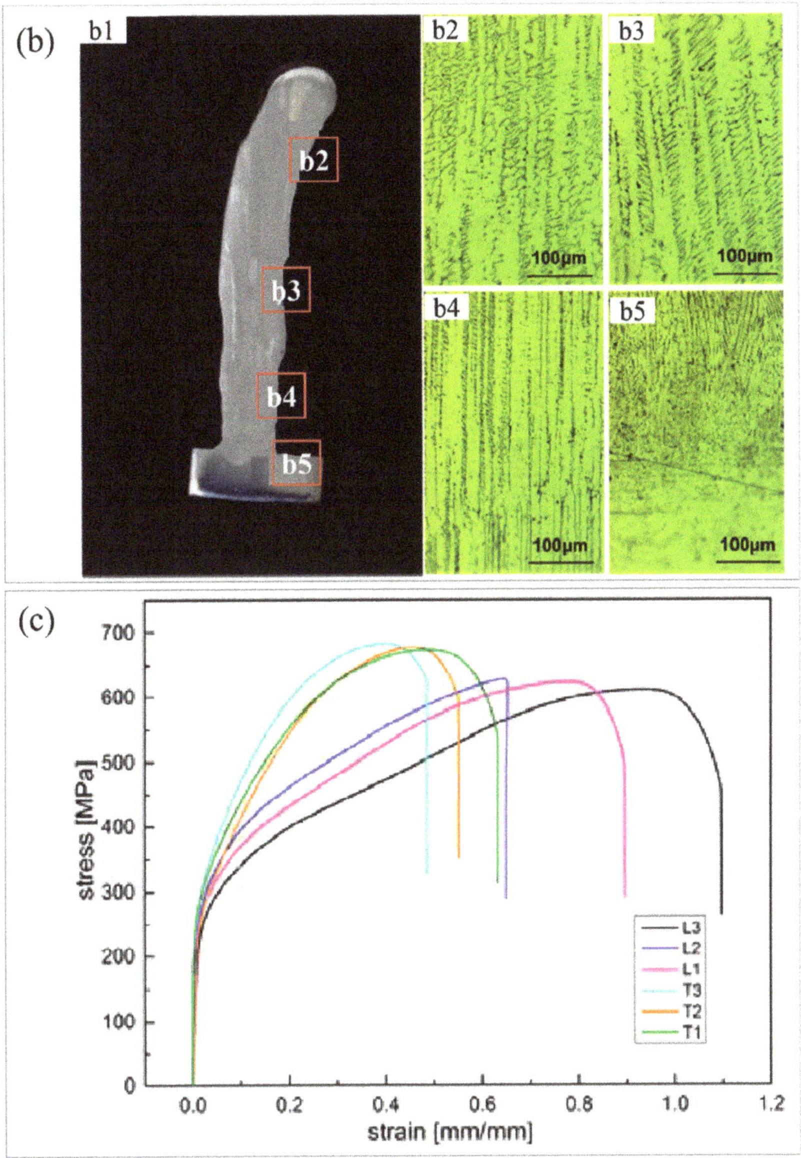

Figure 3. The influence of dendritic structure on the mechanical properties of the additively manufactured 304L stainless steel products. (**a**) Schematic drawing of the experimental device; (**b**) sample structure diagram (**b1**) and metallographic micrographs of four regions (**b2–b5**); (**c**) stress−strain curves of the longitudinal (L1, L2, L3) and transverse (T1, T2, T3) tensile specimens. (Reprinted with permission from ref. [114]. Copyright 2017, Electronic Material).

3.3. Dislocation Cells in Additively Manufactured Steel

Unlike conventional manufacturing processes, cellular structures are usually generated inside the grains of some additively manufactured alloys. Due to the high internal stress inside the sample, a high density of dislocations are formed on the cell wall [76], which are usually closely related to the excellent yield strength of additively manufactured samples [115,116].

To study and examine the cellular boundary of SLM 316L stainless steel, Hong et al. [117] employed different heat treatments to tune the cellular sub−grains of laser−melted 316L. It was found that the dislocation density at cellular boundaries exerted a critical role in interfacial strengthening. As shown in Figure 4(a1), the experimental sample size was 5 mm × 20 mm × 80 mm, which was fabricated by SLM equipment, and the tensile specimen is presented in Figure 4(a3). Four different heat treatment temperatures (500, 900, 950, and 1100 °C) were applied for 1 h and then the AM samples were furnace cooled. Ar gas was filled as a protective gas through the heat treatment operation. According to the temperature of heat treatment, the four samples were designated HT500, HT900, HT950, and HT1100, respectively. The TEM images shown in Figure 4b are the as−prepared specimen and the HT1100 sample. The TEM image shown in Figure 4(b1) displays that the cellular sub−grains include a high density of dislocations. Figure 4(b2) presents many dislocations around the cell boundaries. Thermal shrinkage stress during rapid solidification greatly contributed to as−received high dislocation density [118]. The dislocation density shown in Figure 4(b3) was lower than that shown in Figure 4(b1), which demonstrated that the dislocation density at the unit cell boundary was greatly reduced after the specimen was heat−treated. Figure 4(b4) is an enlarged view of the unit cell boundary in Figure 4(b3), showing that the dislocations were uniformly distributed on the unit cell boundaries. Figure 4c shows a set of tensile curves for the SLM samples with various heat treatments; the yield strength reduced from 578 ± 5 MPa for the as−built sample to 326 ± 5 MPa for the HT1100 sample.

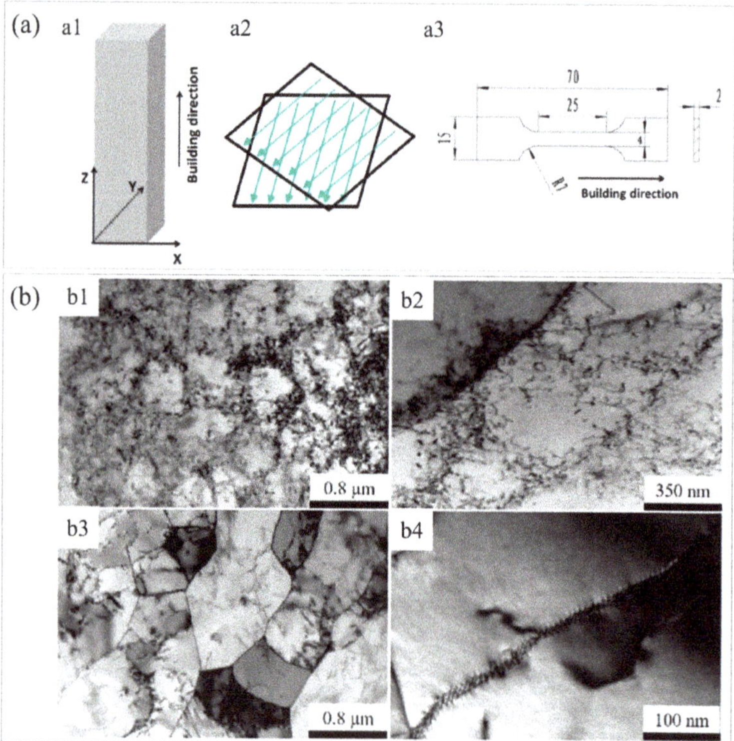

Figure 4. *Cont.*

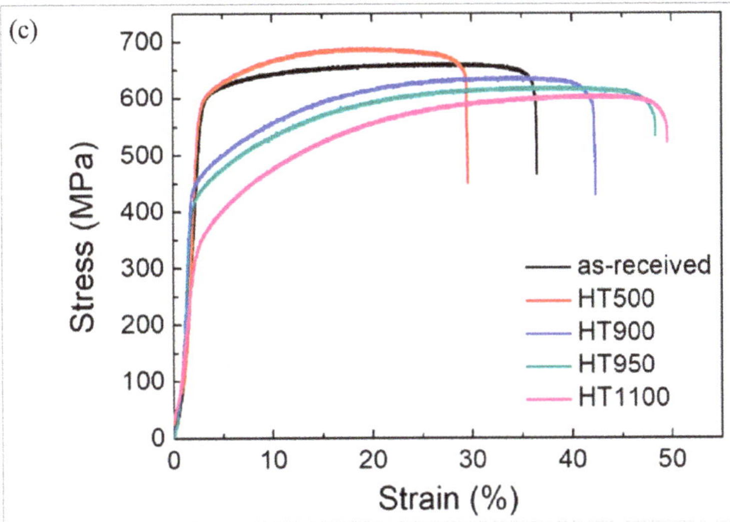

Figure 4. The effect of dislocation density on the performance of additively manufactured 316L. (**a**) Experimental results and process parameters, (**a1**) schematic diagram of the sample fabricated by SLM, the X and Y axes denote directions equal to the powder bed, whereas the Z axis represents the sample build direction, (**a2**) laser scanning strategy of SLM, (**a3**) schematic diagram of the tensile sample size. (**b**) TEM images of as−prepared and HT1100 samples, (**b1**) TEM image of as−prepared sample, (**b2**) TEM image of an enlarged view of (**b1**), (**b3**) TEM image of HT1100 sample, (**b4**) TEM image of an enlarged view of (**b3**). (**c**) Tensile stress−strain curves for the studied samples. (Reprinted with permission from ref. [117]. Copyright 2021, Materials Science Engineering: A).

3.4. Residual Stress in Additively Manufactured Steel

During the additive manufacturing process, the powder layers are melted and solidified layer−by−layer and the expansion and contraction stress accumulates, forming high residual internal stress, which may be released or redistributed during long−term service, resulting in fatigue cracks, brittle fracture, and stress corrosion failure. For example, austenitic stainless steel for nuclear applications is prone to stress corrosion cracking (SCC) in high−temperature water, which is closely related to residual stress. Furthermore, the main factors affecting SCC are temperature, radiation damage, electrochemical potential, water chemistry, sensitization, etc. [119–121].

To investigate SCC growth in high−temperature water with 316L manufactured using the laser powder bed method, Lou et al. [122] evaluated the experimental parameters and their influence, including crack orientation, microstructure, and stress intensity factor. Figure 5a shows the direction of the sample compared to the powder bed, while the X and Y axes are parallel to the powder bed and the Z axis is the orientation in which the specimen was built. Z−X indicated crack growth in the X direction and loading in the Z direction. It was found that stress−relieved 316L stainless steel showed two special cracking characteristics: (1) the cracks grew in the build direction; (2) the hydrogen water chemistry did not affect the cracks in the X−Z direction when forging along the X direction. Figure 5b presents the EBSD plot of the crack formation in the AM 316L stainless steel without cold work. From the crystal boundary diagram and inverse pole diagram, it could be seen that some cracks developed along the high−angle boundary of adjacent grains. High−angle grain boundaries were not the only path for SCC propagation, cracks also propagated along sub−grain structures, such as low−angle grain boundaries and dislocations. According to Figure 5c, the SCC growth rate was dependent on the stress intensity factor (K) in AM 316L stainless steel under HIP + SA conditions (annealed and 20% cold work).

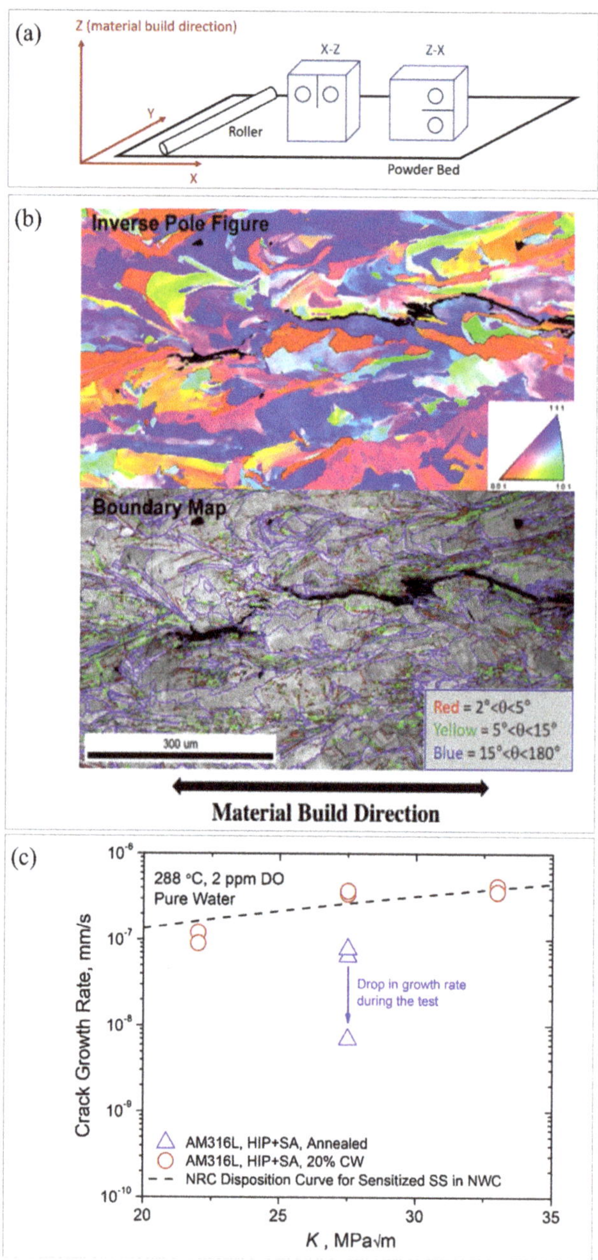

Figure 5. In high−temperature water, SCC develops in additively manufactured stainless steel. (**a**) Schematic drawing of the direction of the tensile specimen compared to the powder bed, the X and Y axes denote the directions parallel to the powder bed, while the Z axis represents the sample build direction. (**b**) EBSD plot of SCC in stress−relieved additively manufactured 316L without additional cold working in the X−Z direction, including inverse pole figure and grain boundary map. (**c**) The influence of stress intensity K on the growth rate of SCC in AM 316L stainless−steel in normal water chemistry. (Reprinted with permission from ref. [122]. Copyright 2017, Corrosion Science).

3.5. Element Segregation in Additively Manufactured Steel

In additive manufacturing, due to the fast cooling rate, the elements in the solidification zone do not have enough time to fully diffuse, thus element aggregation may occur. Elemental segregation between dendrites of additively manufactured samples plays an important influence on corrosion resistance [123,124]. To investigate the effect of linear heat input (LHI) on the microstructure and corrosion behavior of austenitic stainless steel, Wen et al. [123] prepared austenitic stainless steel using wire−arc additive manufacturing, observing that steel with high LHI was strongly related to the segregation of Cr and Mo atoms in Ni−poor δ−ferrite.

To develop a 316L stainless steel (SS) with high yield strength and ductility, Wang et al. [77] employed two different laser powder bed fusion (L−PBF) techniques ('Concept' and 'Fraunhofer'), which showed strength and ductility beyond traditional 316L SS. Figure 6a mainly shows the microstructure of 316L SS fabricated by L−PBF, which includes grain morphology, grain boundary angle, dislocation density, composition segregation, etc. The detailed EBSD analysis is shown in Figure 6(a3), demonstrating a large number of small−angle grain boundaries inside the grains. It can be seen from Figure 6(a4) that the elements Cr and Mo segregated at the cell wall. Elemental segregation and a large number of low−angle grain boundaries enhanced dislocation pinning and promoted twinning, which in turn affected the strength and ductility of the additively manufactured samples. Figure 6b shows the component analysis of the cell wall and cell interior of the PBF 316L SS manufactured by Fraunhofer machines. It can be seen from the figure that there was very little elemental segregation in the Fraunhofer samples. As shown in Figure 6c, both L−PBF−fabricated 316L samples were stronger than the cast and forged samples in terms of strength and ductility. Figure 6(c2) shows a summary of yield stress versus uniform elongation for various 316L SS. The outstanding strength and ductility of 3D−printed steels exceeded that of conventional 316L SS. The excellent performance depended on many factors, including that the compositional segregation at the grain boundary may have pinned the dislocation motion and, meanwhile, a large number of small−angle grain boundaries would also hinder the movement of dislocations, resulting in increased strength.

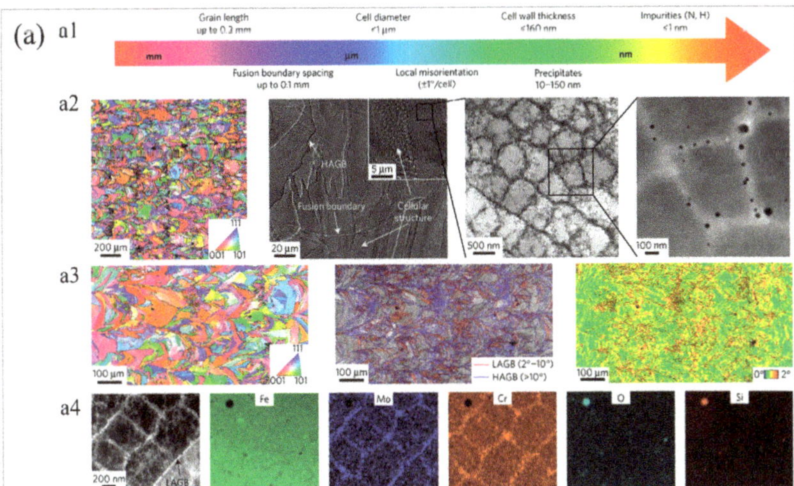

Figure 6. *Cont.*

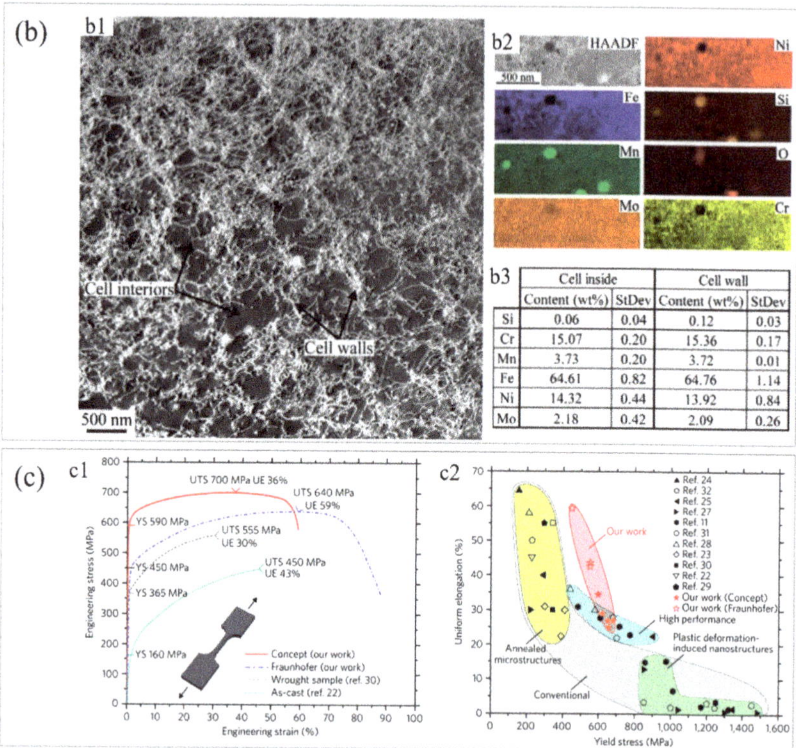

Figure 6. Typical microstructure and mechanical properties of 316L SS prepared by L−PBF. (**a**) Typical microstructures of 316L SS produced by L−PBF, (**a1**) illustration of the multi−scale microstructures in 316L SS, (**a2**) cross−sectional EBSD map, SEM image, bright−field TEM image, and dark−field STEM image of t316L SS, (**a3**) detailed EBSD analysis, (**a4**) HAADF STEM (Z contrast) image with corresponding Cr Fe, and Mo EDS maps showing Cr and Mo segregation at the solidification cellular walls. (**b**) Compositional analysis of cell walls and cell interior for L−PBF 316L SS fabricated using the Fraunhofer machine, (**b1**) bright−field TEM image of cells, (**b2**) high−angle annular dark field image with corresponding elemental maps, (**b3**) detailed composition of the selected regions. (**c**) Mechanical properties of L−PBF 316L SS, (**c1**) tensile engineering stress−strain curves, (**c2**) summary of yield stress vs. uniform elongation for various 316L SS, wherein the references in this figure should be referred to the original article [77]. (Reprinted with permission from ref. [77]. Copyright 2018, Nature Materials).

3.6. Other Structural Characteristics in Additively Manufactured Steel

In addition to the above five characteristic microstructures in additively manufactured products, some other characteristic structures also exert an important effect on the service properties of additively manufactured samples. Due to local melting and non−uniform heating during additive manufacturing, there may be defects such as poor powder fusion that could affect the surface roughness of the printed sample. Melting and directional solidification during printing can lead to periodic cracks, which is one of the main limitations of the wide application of additive manufacturing in metallic materials.

The DED technique provides an opportunity to print graded [125] or layered [126] functionally graded materials by changing the powder supply [125]. For instance, changing the percentage of stainless steel to Inconel during printing can create hardness gradients in the specimen [127,128]. Hofmann et al. [129] designed a path from one alloy to another based on a multi−component phase diagram to evade harmful phase formation between the two components. On this basis, using a multi−hopper LMD system, linear and radial gradient alloys could be designed and manufactured.

To study the microstructure and performance of 316L−Inconel 718 composition gradient stainless steel alloy, Wen et al. [130] used a laser powder bed to prepare composition gradient alloy (CGA) 316L and Inconel 718 alloys. It was found that the mechanical properties of the specimens varied with the composition gradient. Figure 7(a1) presents a schematic diagram of the powder bed. Figure 7(a2) shows the shape of the printed sample with the manufacturing direction and compositional gradient direction (GD). The authors mainly studied the composition changes of five main elements, Cr, Mo, Fe, Nb, and Ni, in 316L/IN718 stainless steel constructed along GD. Figure 7b shows the results of the weight percent of elements measured by EDS. Figure 7c displays a cross−section of the sample showing the XRF compositional patterns of the five main components. The gradual change in element color along the GD direction confirmed the existence of a compositional gradient in the CGA. At different positions along the BD orientation of the sample, the consistent color of the elements indicated that the powder deposition process was stable. Various cross−sectional positions of CGA slices were tested for uniaxial tensile performance in two conditions: as−prepared and heat−treated samples. Figure 7(d1) and Figure 7(d2) show the relationship between engineering stress and engineering strain for IN718 at 0, 8, 25, 48, 65, 82, and 100 wt% cross−sections, respectively. In the slices with 0, 8, and 25 wt% IN718, σ_y gradually decreased. The results indicated that σ_y was reduced with increased IN718 content at low IN718 content ($\leq$25 wt%). As the proportion of IN718 increased to 48 wt%, σ_y did not decrease but increased to 647 $\pm$ 28 MPa. Then, σ_y and UTS significantly increased with further increase in IN718 content. When the IN718 proportion further increased to $\geq$82 wt%, YS and UTS were higher while ductility was lower in the CGA sample relative to the as−prepared sample.

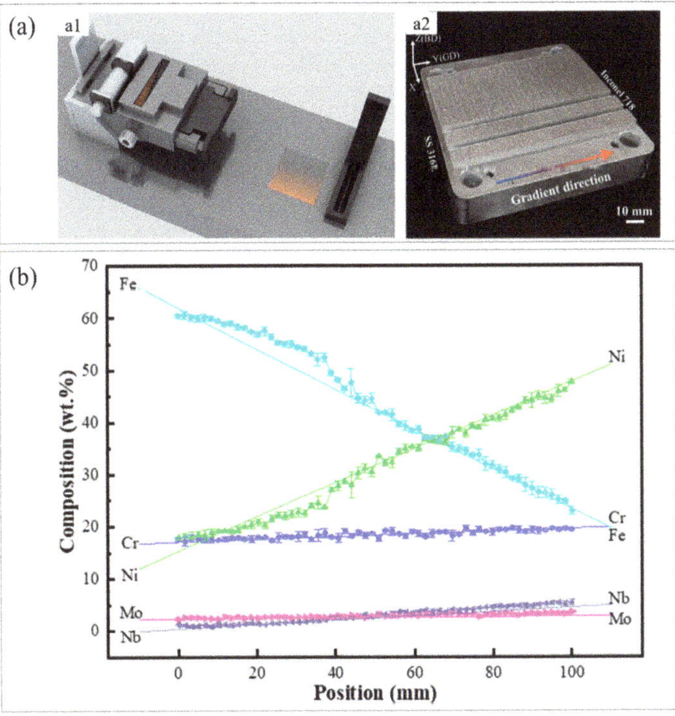

Figure 7. *Cont.*

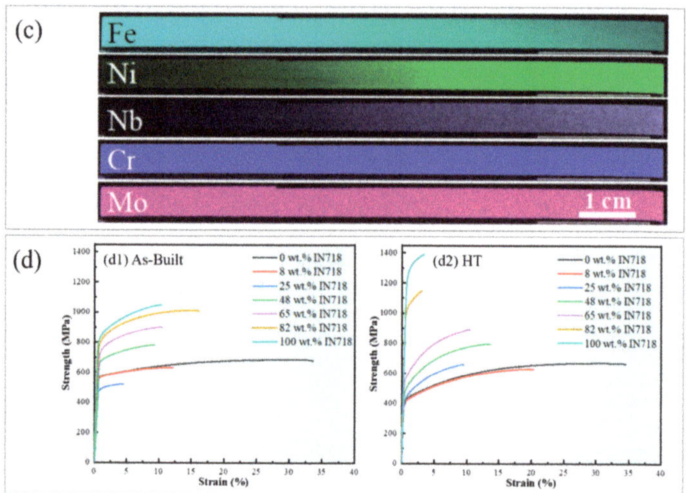

Figure 7. Composition variation of gradient specimens prepared by powder bed. (**a**) Schematic diagram and sample image, (**a1**) schematic illustration of the CGA LPBF system, (**a2**) image of fabricated CGA samples; (**b**) component distribution map measured by EDS technology along the gradient orientation; (**c**) composition gradient distribution map of CGA measured by XRF on the plane; (**d**) engineering stress−strain curves of (**d1**) as−prepared and (**d2**) heat−treated CGA specimens. (Reprinted with permission from ref. [130]. Copyright 2022, Materials Science Engineering: A).

4. Conclusions and Perspectives

In summary, AM has been widely used in the design and manufacture of high−performance steels, effectively saving processing time and improving material utilization; however, the physical metallurgical process of additively manufactured steel is very complex. This work summarizes some typical microstructures of additively manufactured steel−based alloys. Specifically, the high−temperature gradients in additive manufacturing processes can form dislocation cells and alloying elements may segregate and aggregate at defects; the repeated melting and solidification of the powder layer can cause large residual stress; improper processing parameters of the laser can generate pores in the specimen, etc. These characteristic microstructures exert a significant influence on the properties of additively manufactured products. The characteristic microstructures summarized in this work will be helpful for follow−up research, and this work may promote the application of additive manufacturing technology in the field of steel−based alloys. In recent years, the development of AM has presented a diversified scene, considering the multi−scale and complex phase transformation characteristics of the steel itself. Nevertheless, there are some trends in developing high−performance steel−based materials via additive manufacturing.

1. The additive manufacturing technique is a non−equilibrium solidification process, and the microstructure structure exhibits multi−level and cross−scale characteristics. It is difficult to quantitatively characterize the microscopic mechanism of additively manufactured products in experiments. Therefore, it is highly desirable to develop advanced multi−scale computing techniques to shed light on the complex mechanism of microstructure evolution and thus improve the macro−performance.
2. As a potential high−throughput experimental method, the additive manufacturing technique can effectively accelerate the composition and process optimization design of high−performance steel−based materials by gradient printing.
3. Steel is born with complex solid−state phase transition; therefore, learning from the abundant traditional heat treatment experience and developing a heat treatment scheme suitable for additive manufacturing is one of the future research directions.

4. At present, all grades of steel are proposed for traditional steel preparation processes, but it is urgent to establish a set of steel grades suitable for additively manufactured steel.
5. Data−driven additive manufacturing technology is another future direction. Unlike traditional steel preparation, 3D printing of metal specimens lacks a large amount of high−quality data at present, thus it is also urgent to develop a database and data−driven strategies for additive manufacturing.

Funding: The present work is supported by the National Natural Science Foundation of China (Grant Nos. 52122408, 51901013, 52071023).

Institutional Review Board Statement: Not applicable.

Informed Consent Statement: Not applicable.

Data Availability Statement: Not applicable.

Acknowledgments: H.H. Wu also acknowledges financial support from the Fundamental Research Funds for the Central Universities (University of Science and Technology Beijing, No. FRF−TP−2021−04C1, and 06500135). We appreciate Jiaming Zhu from Shandong University for helping proofread this revised manuscript. The computing work was supported by USTB MatCom of the Beijing Advanced Innovation Center for Materials Genome Engineering.

Conflicts of Interest: The authors declare no conflict of interest.

References

1. Wu, B.; Pan, Z.; Ding, D.; Cuiuri, D.; Li, H.; Xu, J.; Norrish, J. A review of the wire arc additive manufacturing of metals: Properties, defects and quality improvement. *J. Manuf. Process.* **2018**, *35*, 127–139. [CrossRef]
2. He, B.; Wu, W.; Zhang, L.; Lu, L.; Yang, Q.; Long, Q.; Chang, K. Microstructural characteristic and mechanical property of Ti6Al4V alloy fabricated by selective laser melting. *Vacuum* **2018**, *150*, 79–83. [CrossRef]
3. Cui, L.; Deng, D.; Jiang, F.; Peng, R.; Xin, T.; Mousavian, R.; Yang, Z. Superior low cycle fatigue property from cell structures in additively manufactured 316L stainless steel. *J. Mater. Sci. Technol.* **2022**, *111*, 268–278. [CrossRef]
4. Ma, M.; Wang, Z.; Gao, M.; Zeng, X. Layer thickness dependence of performance in high-power selective laser melting of 1Cr18Ni9Ti stainless steel. *J. Mater. Process. Technol.* **2015**, *215*, 142–150. [CrossRef]
5. Luo, Y.; Wang, M.; Tu, J.; Jiang, Y.; Jiao, S.-Q. Reduction of residual stress in porous Ti6Al4V by in situ double scanning during laser additive manufacturing. *Int. J. Miner. Metall. Mater.* **2021**, *28*, 1844–1853. [CrossRef]
6. Martin, J.H.; Yahata, B.D.; Hundley, J.M.; Mayer, J.A.; Schaedler, T.A.; Pollock, T.M. 3D printing of high-strength aluminium alloys. *Nature* **2017**, *549*, 365–369. [CrossRef]
7. Tang, W.; Yang, X.; Tian, C.; Xu, Y. Microstructural heterogeneity and bonding strength of planar interface formed in additive manufacturing of Al−Mg−Si alloy based on friction and extrusion. *Int. J. Miner. Metall. Mater.* **2022**, *29*, 1755–1769. [CrossRef]
8. Wei, K.; Gao, M.; Wang, Z.; Zeng, X. Effect of energy input on formability, microstructure and mechanical properties of selective laser melted AZ91D magnesium alloy. *Mater. Sci. Eng. A* **2014**, *611*, 212–222. [CrossRef]
9. Yue, H.Y.; Peng, H.; Su, Y.J.; Wang, X.-P.; Chen, Y.-Y. Microstructure and high-temperature tensile property of TiAl alloy produced by selective electron beam melting. *Rare Metals* **2021**, *40*, 3635–3644. [CrossRef]
10. Li, H.G.; Che, P.C.; Yang, X.K.; Huang, Y.-J.; Ning, Z.-L.; Sun, J.-F.; Fan, H.-B. Enhanced tensile properties and wear resistance of additively manufactured CoCrFeMnNi high-entropy alloy at cryogenic temperature. *Rare Metals* **2022**, *41*, 1210–1216. [CrossRef]
11. Liu, Z.Y.; Zhao, X.Y.; Wu, Y.W.; Chen, Q.; Yang, B.-H.; Wang, P.; Chen, Z.-W.; Yang, C. Homogenization heat treatment for an additively manufactured precipitation-hardening high-entropy alloy. *Rare Metals* **2022**, *41*, 2853–2863. [CrossRef]
12. Hussein, A.; Hao, L.; Yan, C.; Everson, R. Design, Finite element simulation of the temperature and stress fields in single layers built without-support in selective laser melting. *Materials* **2013**, *52*, 638–647.
13. Wang, H. Materials' fundamental issues of laser additive manufacturing for high-performance large metallic components. *Acta Aeronaut. Astronaut. Sin.* **2014**, *35*, 2690–2698.
14. Yan, C.; Hao, L.; Hussein, A.; Young, P. Ti–6Al–4V triply periodic minimal surface structures for bone implants fabricated via selective laser melting. *J. Mech. Behav. Biomed. Mater.* **2015**, *51*, 61–73. [CrossRef]
15. Caiazzo, F.; Cardaropoli, F.; Alfieri, V.; Sergi, V.; Cuccaro, L. In Experimental analysis of selective laser melting process for Ti-6Al-4V turbine blade manufacturing. In *XIX International Symposium on High-Power Laser Systems and Applications*; SPIE: Bellingham, WA, USA, 2012; pp. 381–390.
16. Andersson, J.; Sjöberg, G.P. Repair welding of wrought superalloys: Alloy 718, Allvac 718plus and Waspaloy. *Sci. Technol. Weld. Join.* **2012**, *17*, 49–59. [CrossRef]

17. Guo, C.; Zhou, Y.; Li, X.; Hu, X.; Xu, Z.; Dong, E.; Zhu, Q.; Ward, R.M. A comparing study of defect generation in IN738LC superalloy fabricated by laser powder bed fusion: Continuous-wave mode versus pulsed-wave mode. *J. Mater. Sci. Technol.* **2021**, *90*, 45–57. [CrossRef]

18. Liu, Z.; Zhao, D.; Wang, P.; Yan, M.; Yang, C.; Chen, Z.; Lu, J.; Lu, Z. Additive manufacturing of metals: Microstructure evolution and multistage control. *J. Mater. Sci. Technol.* **2022**, *100*, 224–236. [CrossRef]

19. Ribeiro, K.S.; Mariani, F.E.; Coelho, R.T. A study of different deposition strategies in direct energy deposition (DED) processes. *Procedia Manuf.* **2020**, *48*, 663–670. [CrossRef]

20. Chen, H.; Gu, D.; Ge, Q.; Shi, X.; Zhang, H.; Wang, R.; Zhang, H.; Kosiba, K. Role of laser scan strategies in defect control, microstructural evolution and mechanical properties of steel matrix composites prepared by laser additive manufacturing. *Int. J. Miner. Metall. Mater.* **2021**, *28*, 462–474. [CrossRef]

21. Veiga, F.; Arizmendi, M.; Suarez, A.; Bilbao, J.; Uralde, V. Different path strategies for directed energy deposition of crossing intersections from stainless steel SS316L-Si. *J. Manuf. Process.* **2022**, *84*, 953–964. [CrossRef]

22. Barrionuevo, G.O.; Ramos-Grez, J.A.; Walczak, M.; Sánchez-Sánchez, X.; Guerra, C.; Debut, A.; Haro, E. Microstructure simulation and experimental evaluation of the anisotropy of 316 L stainless steel manufactured by laser powder bed fusion. *Rapid Prototyp. J.* **2022**; *ahead-of-print*.

23. Basak, A.; Lee, A.; Pramanik, A.; Neubauer, K.; Prakash, C.; Shankar, S. Material extrusion additive manufacturing of 17–4 PH stainless steel: Effect of process parameters on mechanical properties. *Rapid Prototyp. J.* **2023**; *ahead-of-print*.

24. Wang, Z.; Palmer, T.A.; Beese, A.M. Effect of processing parameters on microstructure and tensile properties of austenitic stainless steel 304L made by directed energy deposition additive manufacturing. *Acta Mater.* **2016**, *110*, 226–235. [CrossRef]

25. Helmer, H.; Bauereiß, A.; Singer, R.; Körner, C. Grain structure evolution in Inconel 718 during selective electron beam melting. *Mater. Sci. Eng. A* **2016**, *668*, 180–187. [CrossRef]

26. Zhao, C.; Bai, Y.; Zhang, Y.; Wang, X.; Xue, J.M.; Wang, H. Influence of scanning strategy and building direction on microstructure and corrosion behaviour of selective laser melted 316L stainless steel. *Mater. Des.* **2021**, *209*, 109999. [CrossRef]

27. Yang, X.; Ren, Y.; Liu, S.; Wang, Q.; Shi, M. Microstructure and tensile property of SLM 316L stainless steel manufactured with fine and coarse powder mixtures. *J. Cent. South Univ.* **2020**, *27*, 334–343. [CrossRef]

28. Khorasani, M.; Gibson, I.; Ghasemi, A.H.; Hadavi, E.; Rolfe, B. Laser subtractive and laser powder bed fusion of metals: Review of process and production features. *Rapid Prototyp. J.* **2023**; *ahead-of-print*.

29. Luecke, W.E.; Slotwinski, J.A. Mechanical properties of austenitic stainless steel made by additive manufacturing. *J. Res. Natl. Inst. Stand. Technol.* **2014**, *119*, 398. [CrossRef]

30. Gong, G.; Ye, J.; Chi, Y.; Zhao, Z.; Wang, Z.; Xia, G.; Du, X.; Tian, H.; Yu, H.; Chen, C. Research status of laser additive manufacturing for metal: A review. *J. Mater. Res. Technol.* **2021**, *15*, 855–884. [CrossRef]

31. Sachs, E.M.; Haggerty, J.S.; Cima, M.J.; Williams, P.A. Three-Dimensional Printing Techniques. U.S. Patent 5340656A, 23 August 1994.

32. Ladd, C.; So, J.H.; Muth, J.; Dickey, M.D. 3D printing of free standing liquid metal microstructures. *Adv. Mater.* **2013**, *25*, 5081–5085. [CrossRef]

33. Henry, T.C.; Morales, M.A.; Cole, D.P.; Shumeyko, C.M.; Riddick, J.C. Mechanical behavior of 17-4 PH stainless steel processed by atomic diffusion additive manufacturing. *Int. J. Adv. Manuf. Technol.* **2021**, *114*, 2103–2114. [CrossRef]

34. White, D. Ultrasonic Object Consolidation. U.S. Patent US6519500B1, 11 February 2003.

35. White, D.R. Ultrasonic consolidation of aluminum tooling. *Adv. Mater. Process.* **2003**, *161*, 64–66.

36. Zenou, M.; Sa'Ar, A.; Kotler, Z. Laser jetting of femto-liter metal droplets for high resolution 3D printed structures. *Sci. Rep.* **2015**, *5*, 17265. [CrossRef]

37. Cong, B.; Ding, J.; Williams, S. Effect of arc mode in cold metal transfer process on porosity of additively manufactured Al-6.3% Cu alloy. *Int. J. Adv. Manuf. Technol.* **2015**, *76*, 1593–1606. [CrossRef]

38. Liu, S.; Li, Y.; Wang, Y.; Wei, Y.; Zhang, L.; Wang, J.; Yang, X. Selective laser melting of WC-Co reinforced AISI 1045 steel composites: Microstructure characterization and mechanical properties. *J. Mater. Res. Technol.* **2022**, *19*, 1821–1835. [CrossRef]

39. Lehmann, T.; Rose, D.; Ranjbar, E.; Ghasri-Khouzani, M.; Tavakoli, M.; Henein, H.; Wolfe, T.; Jawad Qureshi, A. Large-scale metal additive manufacturing: A holistic review of the state of the art and challenges. *Int. Mater. Rev.* **2022**, *67*, 410–459. [CrossRef]

40. Majumdar, J.D.; Pinkerton, A.; Liu, Z.; Manna, I.; Li, L. Microstructure characterisation and process optimization of laser assisted rapid fabrication of 316L stainless steel. *Appl. Surf. Sci.* **2005**, *247*, 320–327. [CrossRef]

41. Griffith, M.L.; Ensz, M.T.; Puskar, J.D.; Robino, C.V.; Brooks, J.A.; Philliber, J.A.; Smugeresky, J.E.; Hofmeister, W. Understanding the microstructure and properties of components fabricated by laser engineered net shaping (LENS). *MRS Online Proc. Libr.* **2000**, *9*, 625. [CrossRef]

42. Yu, J.; Rombouts, M.; Maes, G. Cracking behavior and mechanical properties of austenitic stainless steel parts produced by laser metal deposition. *Mater. Des.* **2013**, *45*, 228–235. [CrossRef]

43. Campanelli, S.L.; Angelastro, A.; Signorile, C.G.; Casalino, G. Investigation on direct laser powder deposition of 18 Ni (300) maraging steel using mathematical model and experimental characterisation. *Int. J. Adv. Manuf. Technol.* **2017**, *89*, 885–895. [CrossRef]

44. Yang, N.; Yee, J.; Zheng, B.; Gaiser, K.; Reynolds, T.; Clemon, L.; Lu, W.; Schoenung, J.; Lavernia, E. Process-structure-property relationships for 316L stainless steel fabricated by additive manufacturing and its implication for component engineering. *J. Therm. Spray Technol.* **2017**, *26*, 610–626. [CrossRef]

45. Zhang, K.; Wang, S.; Liu, W.; Shang, X. Characterization of stainless steel parts by laser metal deposition shaping. *Mater. Des.* **2014**, *55*, 104–119. [CrossRef]

46. El Moghazi, S.N.; Wolfe, T.; Ivey, D.G.; Henein, H. Plasma transfer arc additive manufacturing of 17-4 PH: Assessment of defects. *Int. J. Adv. Manuf. Technol.* **2020**, *108*, 2301–2313. [CrossRef]

47. Liberini, M.; Astarita, A.; Campatelli, G.; Scippa, A.; Montevecchi, F.; Venturini, G.; Durante, M.; Boccarusso, L.; Minutolo, F.M.C.; Squillace, A. Selection of optimal process parameters for wire arc additive manufacturing. *Procedia Cirp* **2017**, *62*, 470–474. [CrossRef]

48. Rafieazad, M.; Ghaffari, M.; Vahedi Nemani, A.; Nasiri, A. Microstructural evolution and mechanical properties of a low-carbon low-alloy steel produced by wire arc additive manufacturing. *Int. J. Adv. Manuf. Technol.* **2019**, *105*, 2121–2134. [CrossRef]

49. Mai, D.S. Microstructural and mechanical characteristics of 308L stainless steel manufactured by gas metal arc welding-based additive manufacturing. *Mater. Lett.* **2020**, *271*, 127791.

50. Elmer, J.W.; Gibbs, G. The effect of atmosphere on the composition of wire arc additive manufactured metal components. *Sci. Technol. Weld. Join.* **2019**, *24*, 367–374. [CrossRef]

51. Lunde, J.; Kazemipour, M.; Salahi, S.; Nasiri, A. In Microstructure and mechanical properties of AISI 420 stainless steel produced by wire arc additive manufacturing. In *TMS 2020 149th Annual Meeting & Exhibition Supplemental Proceedings*; Springer: Berlin/Heidelberg, Germany, 2020; pp. 413–424.

52. Wang, L.; Xue, J.; Wang, Q. Correlation between arc mode, microstructure, and mechanical properties during wire arc additive manufacturing of 316L stainless steel. *Mater. Sci. Eng. A* **2019**, *751*, 183–190. [CrossRef]

53. Caballero, A.; Ding, J.; Ganguly, S.; Williams, S. Wire+ Arc Additive Manufacture of 17-4 PH stainless steel: Effect of different processing conditions on microstructure, hardness, and tensile strength. *J. Mater. Process. Technol.* **2019**, *268*, 54–62. [CrossRef]

54. Wu, W.; Xue, J.; Wang, L.; Zhang, Z.; Hu, Y.; Dong, C. Forming process, microstructure, and mechanical properties of thin-walled 316L stainless steel using speed-cold-welding additive manufacturing. *Metals* **2019**, *9*, 109. [CrossRef]

55. Silwal, B.; Santangelo, M. Effect of vibration and hot-wire gas tungsten arc (GTA) on the geometric shape. *J. Mater. Process. Technol.* **2018**, *251*, 138–145. [CrossRef]

56. Zhang, Y.; Wu, L.; Guo, X.; Kane, S.; Deng, Y.; Jung, Y.-G.; Lee, J.-H.; Zhang, J. Additive manufacturing of metallic materials: A review. *J. Mater. Eng. Perform.* **2018**, *27*, 1–13. [CrossRef]

57. Herzog, D.; Seyda, V.; Wycisk, E.; Emmelmann, C. Additive manufacturing of metals. *Acta Mater.* **2016**, *117*, 371–392. [CrossRef]

58. Yap, C.Y.; Chua, C.K.; Dong, Z.L.; Liu, Z.H.; Zhang, D.Q.; Loh, L.E.; Sing, S.L. Review of selective laser melting: Materials and applications. *Appl. Phys. Rev.* **2015**, *2*, 041101. [CrossRef]

59. Gibson, I.; Rosen, D.W.; Stucker, B.; Khorasani, M.; Rosen, D.; Stucker, B.; Khorasani, M. *Additive Manufacturing Technologies*; Springer: Berlin/Heidelberg, Germany, 2021; Volume 17.

60. Das, M.; Balla, V.K.; Basu, D.; Bose, S.; Bandyopadhyay, A. Laser processing of SiC-particle-reinforced coating on titanium. *Scr. Mater.* **2010**, *63*, 438–441. [CrossRef]

61. Gu, D.; Wang, H.; Zhang, G. Selective laser melting additive manufacturing of Ti-based nanocomposites: The role of nanopowder. *Metall. Mater. Trans. A* **2014**, *45*, 464–476. [CrossRef]

62. Kelly, S.; Kampe, S. Microstructural evolution in laser-deposited multilayer Ti-6Al-4V builds: Part I. Microstructural characterization. *Metall. Mater. Trans. A* **2004**, *35*, 1861–1867. [CrossRef]

63. Jägle, E.A.; Choi, P.-P.; Van Humbeeck, J.; Raabe, D. Precipitation and austenite reversion behavior of a maraging steel produced by selective laser melting. *J. Mater. Res.* **2014**, *29*, 2072–2079. [CrossRef]

64. Pham, M.; Dovgyy, B.; Hooper, P. Twinning induced plasticity in austenitic stainless steel 316L made by additive manufacturing. *Mater. Sci. Eng. A* **2017**, *704*, 102–111. [CrossRef]

65. Pokharel, R.; Balogh, L.; Brown, D.; Clausen, B.; Gray III, G.; Livescu, V.; Vogel, S.; Takajo, S. Signatures of the unique microstructure of additively manufactured steel observed via diffraction. *Scr. Mater.* **2018**, *155*, 16–20. [CrossRef]

66. Wu, W.; Lai, K. Process analysis of rapid prototyping with selective laser melting. *J. South China Univ. Technol.* **2007**, *3*, 22–27.

67. Xu, W.; Sun, S.; Elambasseril, J.; Liu, Q.; Brandt, M.; Qian, M. Ti-6Al-4V additively manufactured by selective laser melting with superior mechanical properties. *Jom* **2015**, *67*, 668–673. [CrossRef]

68. Zhang, B.; Li, Y.; Bai, Q. Defect formation mechanisms in selective laser melting: A review. *Chin. J. Mech. Eng.* **2017**, *30*, 515–527. [CrossRef]

69. Li, Y.; Lin, X.; Hu, Y.; Yu, J.; Zhao, J.; Dong, H.; Huang, W. Seynergistic effect of Mo and Zr additions on microstructure and mechanical properties of Nb-Ti-Si-based alloys additively manufactured by laser directed energy deposition. *J. Mater. Sci. Technol.* **2022**, *103*, 84–97. [CrossRef]

70. Zhang, X.; Dang, X.; Yang, L. Study on balling phenomena in selective laser melting. *Laser Optoelectron. Prog.* **2014**, *6*, 131–136.

71. Gong, X.; Anderson, T.; Chou, K. Review on powder-based electron beam additive manufacturing technology. *Manuf. Rev.* **2014**, *1*, 2. [CrossRef]

72. Murr, L.E.; Martinez, E.; Amato, K.N.; Gaytan, S.M.; Hernandez, J.; Ramirez, D.A.; Shindo, P.W.; Medina, F.; Wicker, R.B. Fabrication of metal and alloy components by additive manufacturing: Examples of 3D materials science. *J. Mater. Res. Technol.* **2012**, *1*, 42–54. [CrossRef]

73. Mudge, R.P.; Wald, N.R. Laser engineered net shaping advances additive manufacturing and repair. *Weld. J.* **2007**, *86*, 44.

74. Yong, L.; Xianghui, R.; Yunlong, C. Research status of metal additive manufacturing technology. *Hot Work. Technol.* **2018**, *47*, 15–19.

75. Liu, J.; Zheng, Y.; Ma, Y.; Qureshi, A.; Ahmad, R. A topology optimization method for hybrid subtractive–additive remanufacturing. *Int. J. Precis. Eng. Manuf. Green Technol.* **2020**, *7*, 939–953. [CrossRef]

76. Yang, X.; Ma, W.; Zhang, Z.; Liu, S.; Tang, H. Ultra-high specific strength Ti6Al4V alloy lattice material manufactured via selective laser melting. *Mater. Sci. Eng. A* **2022**, *840*, 142956. [CrossRef]

77. Wang, Y.M.; Voisin, T.; McKeown, J.T.; Ye, J.; Calta, N.P.; Li, Z.; Zeng, Z.; Zhang, Y.; Chen, W.; Roehling, T.T. Additively manufactured hierarchical stainless steels with high strength and ductility. *Nat. Mater.* **2018**, *17*, 63–71. [CrossRef]

78. Yu, J.; Lin, X.; Ma, L.; Wang, J.; Fu, X.; Chen, J.; Huang, W. Influence of laser deposition patterns on part distortion, interior quality and mechanical properties by laser solid forming (LSF). *Mater. Sci. Eng. A* **2011**, *528*, 1094–1104. [CrossRef]

79. Tolosa, I.; Garciandía, F.; Zubiri, F.; Zapirain, F.; Esnaola, A. Study of mechanical properties of AISI 316 stainless steel processed by "selective laser melting", following different manufacturing strategies. *Int. J. Adv. Manuf. Technol.* **2010**, *51*, 639–647. [CrossRef]

80. Mertens, A.; Reginster, S.; Contrepois, Q.; Dormal, T.; Lemaire, O.; Lecomte-Beckers, J. Microstructures and Mechanical Properties of Stainless Steel AISI 316L Processed by Selective Laser Melting. *Mater. Sci. Forum* **2014**, *783–786*, 898–903. [CrossRef]

81. Lass, E.A.; Stoudt, M.R.; Williams, M.E. Additively manufactured nitrogen-atomized 17-4 PH stainless steel with mechanical properties comparable to wrought. *Metall. Mater. Trans. A* **2019**, *50*, 1619–1624. [CrossRef]

82. Kudzal, A.; McWilliams, B.; Hofmeister, C.; Kellogg, F.; Yu, J.; Taggart-Scarff, J.; Liang, J.J.M. Design, Effect of scan pattern on the microstructure and mechanical properties of Powder Bed Fusion additive manufactured 17-4 stainless steel. *Mater. Des.* **2017**, *133*, 205–215. [CrossRef]

83. Murr, L.E.; Martinez, E.; Hernandez, J.; Collins, S.; Amato, K.N.; Gaytan, S.M.; Shindo, P.W. Microstructures and properties of 17-4 PH stainless steel fabricated by selective laser melting. *J. Mater. Res. Technol.* **2012**, *1*, 167–177. [CrossRef]

84. Rafi, H.K.; Pal, D.; Patil, N.; Starr, T.L.; Stucker, B.E. Microstructure and mechanical behavior of 17-4 precipitation hardenable steel processed by selective laser melting. *J. Mater. Eng. Perform.* **2014**, *23*, 4421–4428. [CrossRef]

85. Hermann Becker, T.; Dimitrov, D. The achievable mechanical properties of SLM produced Maraging Steel 300 components. *Rapid Prototyp. J.* **2016**, *22*, 487–494. [CrossRef]

86. Casalino, G.; Campanelli, S.; Contuzzi, N.; Ludovico, A. Experimental investigation and statistical optimisation of the selective laser melting process of a maraging steel. *Opt. Laser Technol.* **2015**, *65*, 151–158. [CrossRef]

87. Casavola, C.; Campanelli, S.; Pappalettere, C. Preliminary investigation on distribution of residual stress generated by the selective laser melting process. *J. Strain Anal. Eng. Des.* **2009**, *44*, 93–104. [CrossRef]

88. Casati, R.; Lemke, J.N.; Tuissi, A.; Vedani, M. Aging behaviour and mechanical performance of 18-Ni 300 steel processed by selective laser melting. *Metals* **2016**, *6*, 218. [CrossRef]

89. Yasa, E.; Kempen, K.; Kruth, J.-P. Microstructure and Mechanical Properties of Maraging Steel 300 after Selective Laser Melting. In Proceedings of the 2010 International Solid Freeform Fabrication Symposium, Austin, TX, USA, 9–11 August 2010.

90. Yu, S.; Liu, J.; Wei, M.; Luo, Y.; Zhu, X.; Liu, Y. Compressive property and energy absorption characteristic of open-cell ZA22 foams. *Mater. Des.* **2009**, *30*, 87–90. [CrossRef]

91. Mazur, M.; Brincat, P.; Leary, M.; Brandt, M. Numerical and experimental evaluation of a conformally cooled H13 steel injection mould manufactured with selective laser melting. *Int. J. Adv. Manuf. Technol.* **2017**, *93*, 881–900. [CrossRef]

92. Xue, L.; Chen, J.; Wang, S.-H. Freeform laser consolidated H13 and CPM 9V tool steels. *Metallogr. Microstruct. Anal.* **2013**, *2*, 67–78. [CrossRef]

93. Dörfert, R.; Zhang, J.; Clausen, B.; Freiße, H.; Schumacher, J.; Vollertsen, F. Comparison of the fatigue strength between additively and conventionally fabricated tool steel 1.2344. *Addit. Manuf.* **2019**, *27*, 217–223. [CrossRef]

94. Mertens, R.; Vrancken, B.; Holmstock, N.; Kinds, Y.; Kruth, J.-P.; Van Humbeeck, J. Influence of powder bed preheating on microstructure and mechanical properties of H13 tool steel SLM parts. *Phys. Procedia* **2016**, *83*, 882–890. [CrossRef]

95. Holzweissig, M.J.; Taube, A.; Brenne, F.; Schaper, M.; Niendorf, T. Microstructural characterization and mechanical performance of hot work tool steel processed by selective laser melting. *Metall. Mater. Trans. B* **2015**, *46*, 545–549. [CrossRef]

96. Karlsson, D.; Chou, C.-Y.; Pettersson, N.H.; Helander, T.; Harlin, P.; Sahlberg, M.; Lindwall, G.; Odqvist, J.; Jansson, U. Additive manufacturing of the ferritic stainless steel SS441. *Addit. Manuf.* **2020**, *36*, 101580. [CrossRef]

97. Liu, Z.; Chua, C.; Leong, K.; Kempen, K.; Thijs, L.; Yasa, E.; Van-Humbeeck, J.; Kruth, J. A preliminary investigation on selective laser melting of M2 high speed steel. In Proceedings of the 5th International Conference on Advanced Research and Rapid Prototyping, Leiria, Portugal, 16 September 2011; pp. 339–346.

98. Kempen, K.; Vrancken, B.; Buls, S.; Thijs, L.; Van Humbeeck, J.; Kruth, J.-P. Selective laser melting of crack-free high density M2 high speed steel parts by baseplate preheating. *J. Manuf. Sci. Eng.* **2014**, *136*, 061026. [CrossRef]

99. Papula, S.; Song, M.; Pateras, A.; Chen, X.-B.; Brandt, M.; Easton, M.; Yagodzinskyy, Y.; Virkkunen, I.; Hänninen, H. Selective laser melting of duplex stainless steel 2205: Effect of post-processing heat treatment on microstructure, mechanical properties, and corrosion resistance. *Materials* **2019**, *12*, 2468. [CrossRef]

100. Hengsbach, F.; Koppa, P.; Duschik, K.; Holzweissig, M.J.; Burns, M.; Nellesen, J.; Tillmann, W.; Tröster, T.; Hoyer, K.-P.; Schaper, M. Duplex stainless steel fabricated by selective laser melting-Microstructural and mechanical properties. *Mater. Des.* **2017**, *133*, 136–142. [CrossRef]

101. Saeidi, K.; Alvi, S.; Lofaj, F.; Petkov, V.I.; Akhtar, F. Advanced mechanical strength in post heat treated SLM 2507 at room and high temperature promoted by hard/ductile sigma precipitates. *Metals* **2019**, *9*, 199. [CrossRef]

102. Saeidi, K.; Kevetkova, L.; Lofaj, F.; Shen, Z. Novel ferritic stainless steel formed by laser melting from duplex stainless steel powder with advanced mechanical properties and high ductility. *Mater. Sci. Eng. A* **2016**, *665*, 59–65. [CrossRef]

103. Bajaj, P.; Hariharan, A.; Kini, A.; Kürnsteiner, P.; Raabe, D.; Jägle, E.A. Steels in additive manufacturing: A review of their microstructure and properties. *Mater. Sci. Eng. A* **2020**, *772*, 138633. [CrossRef]

104. Yin, Y.; Tan, Q.; Bermingham, M.; Mo, N.; Zhang, J.; Zhang, M. Laser additive manufacturing of steels. *Int. Mater. Rev.* **2022**, *67*, 487–573. [CrossRef]

105. Zhao, D.; Guo, Y.; Lai, R.; Wen, Y.; Wang, P.; Liu, C.; Chen, Z.; Yang, C.; Li, S.; Chen, W. Abnormal three-stage plastic deformation in a 17-4 PH stainless steel fabricated by laser powder bed fusion. *Mater. Sci. Eng. A* **2022**, *858*, 144160. [CrossRef]

106. Sames, W.J.; List, F.; Pannala, S.; Dehoff, R.R.; Babu, S.S. The metallurgy and processing science of metal additive manufacturing. *Int. Mater. Rev.* **2016**, *61*, 315–360. [CrossRef]

107. Sander, G.; Tan, J.; Balan, P.; Gharbi, O.; Feenstra, D.; Singer, L.; Thomas, S.; Kelly, R.; Scully, J.R.; Birbilis, N. Corrosion of additively manufactured alloys: A review. *Corrosion* **2018**, *74*, 1318–1350. [CrossRef]

108. Itzhak, D.; Aghion, E. Corrosion behaviour of hot-pressed austenitic stainless steel in H2SO4 solutions at room temperature. *Corros. Sci.* **1983**, *23*, 1085–1094. [CrossRef]

109. Prieto, C.; Singer, M.; Cyders, T.; Young, D. Investigation of pitting corrosion initiation and propagation of a type 316L stainless steel manufactured by the direct metal laser sintering process. *Corrosion* **2019**, *75*, 140–143. [CrossRef]

110. Laleh, M.; Hughes, A.E.; Xu, W.; Gibson, I.; Tan, M.Y. Unexpected erosion-corrosion behaviour of 316L stainless steel produced by selective laser melting. *Corros. Sci.* **2019**, *155*, 67–74. [CrossRef]

111. Sun, Y.; Moroz, A.; Alrbaey, K. Sliding wear characteristics and corrosion behaviour of selective laser melted 316L stainless steel. *J. Mater. Eng. Perform.* **2014**, *23*, 518–526. [CrossRef]

112. Tascioglu, E.; Karabulut, Y.; Kaynak, Y. Influence of heat treatment temperature on the microstructural, mechanical, and wear behavior of 316L stainless steel fabricated by laser powder bed additive manufacturing. *Int. J. Adv. Manuf. Technol.* **2020**, *107*, 1947–1956. [CrossRef]

113. Chen, X.; Li, J.; Cheng, X.; He, B.; Wang, H.; Huang, Z. Microstructure and mechanical properties of the austenitic stainless steel 316L fabricated by gas metal arc additive manufacturing. *Mater. Sci. Eng. A* **2017**, *703*, 567–577. [CrossRef]

114. Ji, L.; Lu, J.; Liu, C.; Jing, C.; Fan, H.; Ma, S. Microstructure and mechanical properties of 304L steel fabricated by arc additive manufacturing. *MATEC Web Conf.* **2017**, *128*, 03006. [CrossRef]

115. Saeidi, K.; Gao, X.; Zhong, Y.; Shen, Z.J. Hardened austenite steel with columnar sub-grain structure formed by laser melting. *Mater. Sci. Eng. A* **2015**, *625*, 221–229. [CrossRef]

116. Ziętala, M.; Durejko, T.; Polański, M.; Kunce, I.; Płociński, T.; Zieliński, W.; Łazińska, M.; Stępniowski, W.; Czujko, T.; Kurzydłowski, K.J. The microstructure, mechanical properties and corrosion resistance of 316 L stainless steel fabricated using laser engineered net shaping. *Mater. Sci. Eng. A* **2016**, *677*, 1–10. [CrossRef]

117. Hong, Y.; Zhou, C.; Zheng, Y.; Zhang, L.; Zheng, J. The cellular boundary with high density of dislocations governed the strengthening mechanism in selective laser melted 316L stainless steel. *Mater. Sci. Eng. A* **2021**, *799*, 140279. [CrossRef]

118. Gorsse, S.; Hutchinson, C.; Gouné, M.; Banerjee, R. Additive manufacturing of metals: A brief review of the characteristic microstructures and properties of steels, Ti-6Al-4V and high-entropy alloys. *Sci. Technol. Adv. MaTerialS* **2017**, *18*, 584–610. [CrossRef]

119. Andresen, P.; Morra, M. Stress corrosion cracking of stainless steels and nickel alloys in high-temperature water. *Corrosion* **2008**, *64*, 15–29. [CrossRef]

120. Andresen, P. Environmentally assisted growth rate response of nonsensitized AISI 316 grade stainless steels in high temperature water. *Corrosion* **1988**, *44*, 450–460. [CrossRef]

121. Andresen, P.; Briant, C. Environmentally assisted cracking of types 304L/316L/316NG stainless steel in 288 C water. *Corrosion* **1989**, *45*, 448–463. [CrossRef]

122. Lou, X.; Song, M.; Emigh, P.W.; Othon, M.A.; Andresen, P.L. On the stress corrosion crack growth behaviour in high temperature water of 316L stainless steel made by laser powder bed fusion additive manufacturing. *Corros. Sci.* **2017**, *128*, 140–153. [CrossRef]

123. Wen, D.; Long, P.; Li, J.; Huang, L.; Zheng, Z. Effects of linear heat input on microstructure and corrosion behavior of an austenitic stainless steel processed by wire arc additive manufacturing. *Vacuum* **2020**, *173*, 109131. [CrossRef]

124. Tokita, S.; Kadoi, K.; Aoki, S.; Inoue, H. Relationship between the microstructure and local corrosion properties of weld metal in austenitic stainless steels. *Corros. Sci.* **2020**, *175*, 108867. [CrossRef]

125. Carroll, B.E.; Otis, R.A.; Borgonia, J.P.; Suh, J.; Dillon, R.P.; Shapiro, A.A.; Hofmann, D.C.; Liu, Z.-K.; Beese, A.M. Functionally graded material of 304L stainless steel and inconel 625 fabricated by directed energy deposition: Characterization and thermodynamic modeling. *Acta Mater.* **2016**, *108*, 46–54. [CrossRef]

126. Lewis, G.K.; Schlienger, E. Practical considerations and capabilities for laser assisted direct metal deposition. *Mater. Des.* **2000**, *21*, 417–423. [CrossRef]

127. Udupa, G.; Rao, S.S.; Gangadharan, K. Functionally graded composite materials: An overview. *Procedia Mater. Sci.* **2014**, *5*, 1291–1299. [CrossRef]
128. Shakil, M.; Ahmad, M.; Tariq, N.; Hasan, B.; Akhter, J.; Ahmed, E.; Mehmood, M.; Choudhry, M.; Iqbal, M. Microstructure and hardness studies of electron beam welded Inconel 625 and stainless steel 304L. *Vacuum* **2014**, *110*, 121–126. [CrossRef]
129. Hofmann, D.C.; Roberts, S.; Otis, R.; Kolodziejska, J.; Dillon, R.P.; Suh, J.; Shapiro, A.A.; Liu, Z.; Borgonia, J. Developing gradient metal alloys through radial deposition additive manufacturing. *Sci. Rep.* **2014**, *4*, 1–8. [CrossRef] [PubMed]
130. Wen, Y.; Gao, J.; Narayan, R.L.; Wang, P.; Zhang, L.; Zhang, B.; Ramamurty, U.; Qu, X. Microstructure-property correlations in as-built and heat-treated compositionally graded stainless steel 316L-Inconel 718 alloy fabricated by laser powder bed fusion. *Mater. Sci. Eng. A* **2022**, *862*, 144515. [CrossRef]

 *materials*

Article

Sustainability Study of Concrete Blocks with Wood Chips Used in Structural Walls in Seismic Areas

Simon Pescari *, Laurentiu Budau, Razvan Ciubotaru and Valeriu Stoian

Department of Civil Engineering and Building Services, Faculty of Construction, Polytechnic University Timișoara, Victoriei Square, No. 2, 300006 Timișoara, Romania
* Correspondence: simon.pescari@upt.ro

Abstract: The concept of sustainability has become a priority in the construction field, in a context where there is an increasing discussion about reducing carbon dioxide emissions, as the construction industry is one of the most polluting industries with a focus on the production of building materials. At present, the classic solution used for structural masonry walls worldwide is the ceramic block. Given that the production of ceramic blocks represents an environmentally polluting process, the alternative solution of using concrete blocks with wood chips is proposed. The proposed solution is more environmentally friendly, both in terms of production technology and materials used, as it is made of wood chips, wood being a sustainable material. These types of blocks are currently used in non-seismic areas due to their poor structural performance. This paper deals with a study on the use of recyclable materials, such as wood chips, from waste materials and aims to propose viable solutions for the use of this type of blocks for structural walls in seismic areas. Two solutions, including concrete blocks with wood chips, have been proposed and numerical analyses have been carried out. Numerical analyses were also carried out for the classical solutions, so that, finally, a comparison could be made between them from a structural point of view. Following the numerical analysis of four types of walls, the two proposed solutions of concrete blocks with wood chips had the best results in terms of force–displacement relationship. Moreover, the quantitative results are presented in a force–displacement graph for the four wall types. This stage represents the first phase of the research, while phase II will continue with experimental tests of the proposed solutions.

Keywords: sustainable materials; wood chips; theoretical study; concrete blocks; structural elements

check for
updates

Citation: Pescari, S.; Budau, L.;
Ciubotaru, R.; Stoian, V.
Sustainability Study of Concrete
Blocks with Wood Chips Used in
Structural Walls in Seismic Areas.
Materials **2022**, *15*, 6659. https://
doi.org/10.3390/ma15196659

Academic Editor: Raul D.S.G.
Campilho

Received: 2 September 2022
Accepted: 21 September 2022
Published: 26 September 2022

Publisher's Note: MDPI stays neutral
with regard to jurisdictional claims in
published maps and institutional affiliations.

1. Introduction

A lot of studies on the use of structural walls made of ceramic blocks in seismic areas are found in the literature. Studies exist about concrete blocks with different materials such as ceramic aggregates, plastic fiber, ornamental stone waste, and alkali-activated slag [1–3]. The topic of numerical modeling of walls made of ceramic blocks is extensively researched and debated in specialized works [4–10]. The weak mechanical characteristics of the walls of brick masonry that can even lead to degradation in the structure due to moderate seismic actions [11–16], makes finding a solution regarding the use of walls to be a research topic of great interest. On the other hand, the European Commission has made an intense campaign on environmental issues lately which also consisted in the elaboration of scientific reports on this subject [17–19]. As a result, several directives have been adopted on the reduction of the environmental impact of the construction sector. The production of building materials, the production of which significantly pollutes the air quality by releasing dust, and then the entire construction process, which is also highly polluting and unfriendly to the environment [20,21].

Sustainability is the key word around which most of the European directives adopted in the field of construction have focused lately [22–24]. The desire to have a healthier environment, breathe cleaner air, and mitigate the effects of global warming led to the adoption of these directives. In addition to reducing the number of pollutants emitted by the production of various building materials, the aim is, also, to reduce greenhouse gas emissions caused by the burning of fossil fuels used to heat buildings [25,26]. The manufacture of ceramic blocks used in construction is a process that generates pollution by releasing dust into the atmosphere. The environmental factors affected by the pollution generated are water, air, soil, and vegetation [4,27]. Thus, nowadays, alternative solutions are being sought for the use of building materials that meet environmental requirements.

The novelty of this study is the use of concrete blocks with wood chips in seismic areas. Finding a solution such that these sorts of blocks are practical in seismic areas can have a considerable benefit, bringing innovation in terms of building structural walls in previously mentioned areas.

Consequently, this work is to analyze concrete blocks with wood chips, a solution which is known to use wood as a raw material, as wood is considered a sustainable material, and, furthermore, to find a viable structural solution so that these blocks can be used in seismic areas. Due to the fact that these blocks have poor mechanical characteristics in terms of tensile strength, compressive strength, and modulus of elasticity, solutions must be proposed to improve the configuration of the blocks so that they become viable solutions for use in seismic zones. In addition, the paper presents a comparison between ordinary ceramic blocks and the proposed solutions for concrete blocks with wood chips.

Therefore, due to the aspects mentioned above, and due to the European directives which stipulate the need for construction with a low impact on the environment, as well as starting from the topic of evaluation of structural walls of brick masonry [28,29]. This paper addresses the topic of walls made of concrete blocks with wood chips in seismic areas that can prove to be a long-term solution for reaching structural safety, the environmental, and sustainability requirements [30]. Wood can provide added value in terms of energy performance of the building due to its thermal conductivity [31]. The first phase of the work aims to carry out numerical analyses, after which experimental tests are planned for the proposed solutions.

2. Concrete Blocks with Wood Chips Solutions

The high wood content masonry blocks have been known worldwide for more than 50 years. The first element of this type appeared in 1934, which is nowadays a quite widespread and frequently used product, being produced and put into use all over the world. The conformity and quality of these products are certified by procedures, which strictly comply with the legislation in force, and all items have a declaration of performance. These masonry blocks are a prefabricated building element made of a mixture of cement, water, and mineralized wood chips. Renewable resources, such as wood chips, are needed to produce the base material and can be in different forms. The wood used to make these elements is waste wood made of old pallets and cuttings that would have been stored as waste, thus releasing carbon dioxide into the atmosphere. Composition of concrete block with wood chips is detailed in Table 1.

Table 1. Composition of concrete block with wood chips.

Type of Materials	Ratio
Wood chips	85%
Cement	10%
Polymer	5%

Because the wood is mineralized and enclosed in cement it will not burn or rot, so the embedded carbon is captured and permanently locked in. Concrete blocks with wood chips are fully recyclable, they contain no toxic elements and are safe for the environment as they are truly green and healthy products. These types of blocks are highlighted in the following images, thus, in Figure 1, a concrete block with wood chips of greater thickness partially filled with concrete is illustrated. The image also shows the texture of the material resulting from the mix of concrete and wood chips.

Figure 1. Concrete block with wood chips.

The configuration of the concrete block with wood chips consists of longitudinal walls and transverse ribs delimiting the interior voids, the blocks also act as lost formwork. These types of elements are used in the construction of all types of walls, either interior walls, end walls, partition walls, or load-bearing walls. The thickness of concrete blocks with wood chips ranges from 15 to 37.5 cm, depending on the type and purpose of the construction: houses, blocks of flats, public buildings, or industrial buildings. The assembly of the elements is performed by hand, with the blocks being connected using concrete poured into the elements.

This type of system combines all the advantages of wood and concrete, being thermally and acoustically insulating and has a very good heat storage capacity. This system gives an advantage in terms of water vapor permeability, ensuring a balance between the air temperature in the room and its relative humidity, thus ensuring a healthy and pleasant environment. The wood chip blanket gives an advantage in terms of thermal insulation properties as well as the heat storage capacity of the concrete core, all of which, in turn, give cost advantages for heating during the cold season or for cooling the air in periods of high temperatures. For a standard concrete block with wood chips, with a thickness of 375 mm and a length of 500 mm filled with thermal insulation of 175 mm and a height of 250 mm, the heat transfer resistance for an unglazed wall is R = 5.26 m^2 K/W and the heat transfer coefficient for an exterior wall with plaster interior plaster + lime-cement exterior plaster is U = 0.18 W/m^2 K [5].

Considering the fire resistance of these types of blocks, although they have a wood chip content of 85% and some concrete blocks with wood chips also include a polystyrene thermal insulation layer, which is due to the manufacturing process they are 100% fire resistant REI 180; a resistance higher than 180 min for a plastered wall. Regarding the load-bearing characteristics of these wood chip blocks, there is research, currently, which shows that the characteristic compressive strength is comparable to that of a low-quality ceramic block Rc, average = 2.5 MPa, modulus of elasticity E = 300 MPa, and specific gravity g = 700 kg/m^3 [32]. Like commonly used ceramic blocks concrete blocks with wood chips have a low tensile strength. The ability to take up tensile stresses can be increased by introducing appropriate amounts of reinforcement by effective anchoring. Referring to the load bearing system, for the concrete block system with wood chips, the loads acting on them are taken up by the concrete core. The concrete core can be plain or reinforced and as a method of placing, it must be cast in place in the voids of the elements. Mix ratio of concrete use is detailed in Table 2.

Table 2. Mix ratio of concrete use.

Type of Materials	Ratio
Cement	10%
Water	18%
Sand	25%
Gravel	41%
Air	6%

3. Proposals for Evaluation against Existing Walls

3.1. Solid Brick Wall

The evaluation of solid brick masonry walls has been the subject of recent research carried out in the Department of Civil Construction and Installations of the University "Politechnica" of Timisoara. The aim of the research was to evaluate the behavior of solid brick masonry walls to lateral loading both in their initial and reinforced state. The research was carried out for a wall with the following dimensions: length = 150 cm; height = 150 cm; thickness = 24 cm, with the ceramic blocks having an average compressive strength of 10 N/mm^2 and a cement-based mortar with a compressive strength of 13–16 N/mm^2. The tested wall was subjected, simultaneously, to a vertical force applied constantly and a horizontal force applied cyclically by means of an actuator. The results of the experimental test revealed the shortcomings of solid brick masonry walls in terms of seismic energy dissipation induced by horizontal forces. The wall failure occurred through a diagonal crack along the wall at a horizontal load of 173 kN, while the maximum horizontal displacement reached 10.488 mm. The consolidation solution achieved was the consolidation with metal fiber reinforced mortar. A high-strength, metal fiber-reinforced mortar was used for the reinforcement, which has superior characteristics in terms of tensile strength. The test results are shown in the graph in Figure 2.

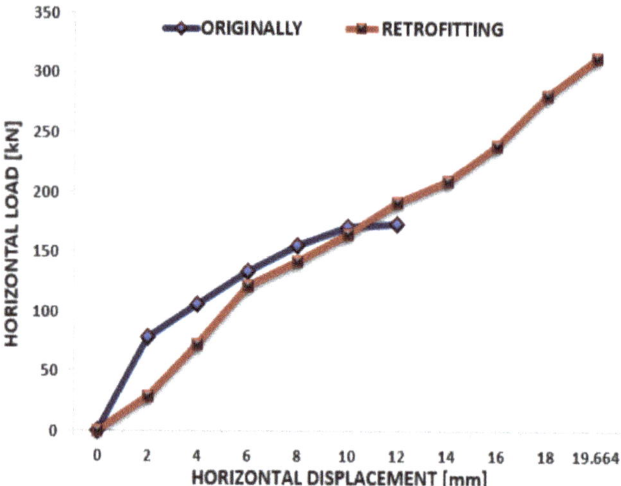

Figure 2. Graph load—displacement solid brick wall originally and retrofitting.

The conclusions of the study show that after the wall reinforcement, the bearing capacity increases by 80% compared to the initial one, and the wall displacements also increase two-fold, thus demonstrating the increase in wall ductility and seismic energy absorption capacity. [33]

3.2. Masonry Wall of Ceramic Blocks with Vertical Voids

Moreover, within the Department of Civil Construction and Installations of the University "Politechnica" of Timisoara, research was carried out on masonry walls made of ceramic blocks with vertical voids, being tested a group of nine walls, consisting of both simple masonry and masonry confined with column. The elements are made of Porotherm 25 vertically hollow ceramic blocks, with dimensions 375 mm $\times$ 250 mm $\times$ 238 mm with an average compressive strength of 10 N/mm^2 and a cement-based mortar with a compressive strength between 5–6 N/mm^2. The failure of the elements made of simple masonry occurred at horizontal loads between 105–140 kN with horizontal displacements between 5.4–11 mm. For elements made of masonry confined with studs, an increased load-bearing capacity was observed at horizontal actions, thus failure occurred at horizontal loads between 200–230 kN with maximum horizontal displacements of 6.0 mm.

The reinforcement solution implemented was polymer composite reinforcement. The tests carried out for the reinforcement of the walls with composite materials showed a recovery of the maximum horizontal force with values between 80% and 115%, and for the maximum lateral displacements, a recovery of between 50% and 80%. The walls reinforced with polymeric materials had a lower ductility than the walls in the original condition. [34]

3.3. Concrete Block Wall with Wood Chips

The study of masonry walls made of concrete blocks with wood chips is proposed to be the next research topic carried out in the Department of Civil Construction and Installations of the University "Politechnica" of Timisoara. The research will focus on walls made of concrete blocks with wood chips with dimensions 150 cm $\times$ 150 cm $\times$ 25 cm. The test procedure adopted will be similar to the other two cases mentioned above, namely the application of a constant vertical force alongside a cyclic horizontal force. The purpose of the tests is to evaluate the behavior of the walls to lateral loading. It is proposed to evaluate a group of three elements for a better accuracy of the results. Compared to the classical ceramic blocks, the advantage of these blocks is the use of wood chips which are considered as sustainable materials.

4. Numerical Analysis

In the research, a series of four numerical models have been built to simulate the behavior of each proposed masonry type under lateral loading. The analyzed elements were modelled as shell elements. Regarding the boundary conditions of the FEM models in the numerical analysis it was considered that at the lower part of the element, the displacements and rotations of the element were blocked, simulating the embedding of the element, while at the upper part of the element the rotations were blocked as well as the vertical displacement at the same time being allowed the displacement of the element in the horizontal plane. Moreover, with reference to the loading conditions, the numerical models were subjected to general static vertical loads with values between 150–200 kN, and, also, to a dynamic load of the Static Riks type intended to simulate the effect of a horizontal movement induced by the earthquake.

The discretization of the model was carried out at the whole element level, so that the block-mortar assembly is idealized as a homogeneous medium with equivalent properties.

4.1. Solid Brick Masonry Wall URM

The numerical model is designed to demonstrate that the assumed experimental model is properly accepted. The numerical model is designed to demonstrate that the assumed experimental model is properly accepted in the calculation, therefore, we used an average compressive strength of 10 [N/mm^2] for the solid brick masonry. Moreover, in the analysis we used a plastic strain of 0.0025, a failure ratio R1 = 1.16; R2 = 1.08, and tension stiffening with displacement of 1.5. The modulus of elasticity of the ceramic block was of E = 13,000 N/mm^2 with Poisson ratio of 0.25. As such, a model has been constructed which fully reflects the study carried out in the laboratory. Thus, a vertical force V = 200 kN and a horizontal increase force was applied in order to observe the mode and moment of failure of the element. The analysis was performed using the ABAQUS program using the finite element method. Following the numerical simulations performed, the maximum horizontal force recorded was H = 144 Kn with a maximum displacement at the top of the element of 10.64 mm.

In the following pictures are illustrated the deformed element with the displacement at the top of the element (Figure 3a), the deformation mode as well as the area where the failure of the element occurred (Figure 3b), the force–displacement graph for the numerical and experimental analysis performed (Figure 3c), and the failure mode of wall of solid brick masonry from the experimental program (Figure 3d).

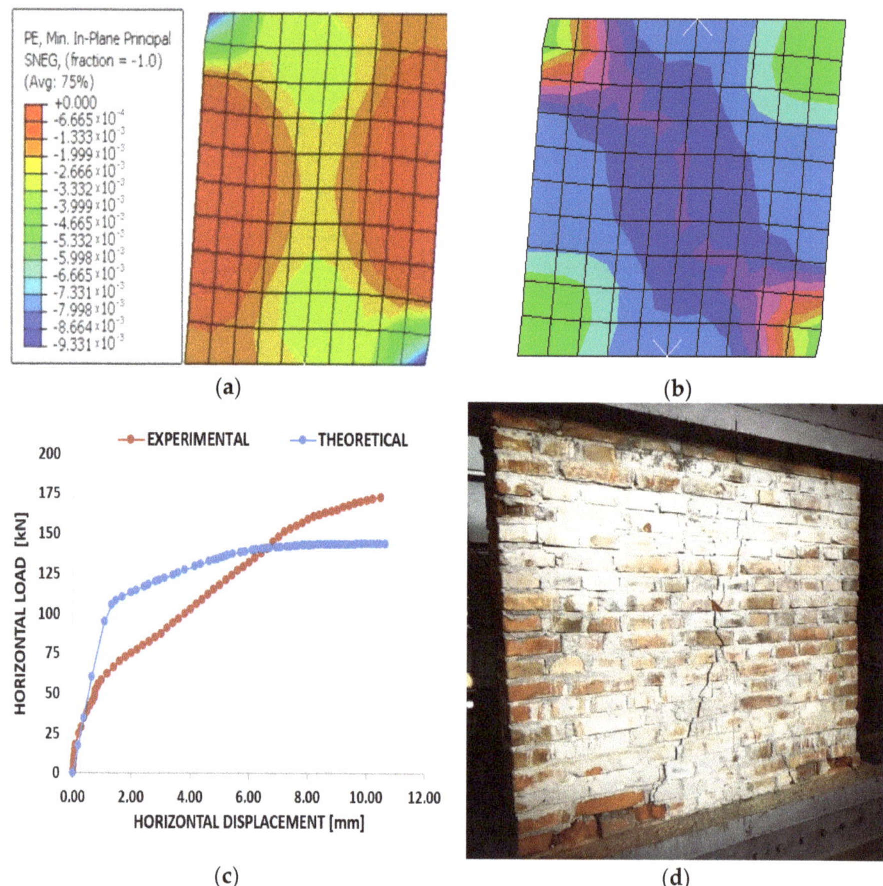

(**a**)

(**b**)

(**c**)

(**d**)

Figure 3. Compression plastic strain (**a**); strain components at integration point (**b**); graph of force–displacement to URM experimental and theoretical (**c**); failure mode of wall of solid brick masonry from the experimental program (**d**).

4.2. Masonry Wall from Ceramic Blocks with Vertical Holes VHB

For the wall realized of masonry of ceramic blocks with vertical holes of the Porotherm type, a numerical model was realized that should validate the experimental test carried out in the Civil Engineering and Installation Department from Timisoara. In the calculation, we used an average compressive strength of 10 [N/mm^2] for the masonry wall from ceramic blocks with vertical holes. Moreover, in the analysis we used a plastic strain of 0.0025, a failure ratio R1 = 1.16; R2 = 1.08, and tension stiffening with displacement of 1.5. The modulus of elasticity of the ceramic block was of E = 16000N/mm^2 with a Poisson ratio of 0.25. The numerical simulation was performed by applying a constant vertical force V = 150 kN and an increasing horizontal force until the failure. The results of the numerical analysis performed show that at the moment of wall failure the displacement at the top of the element was 12.17 mm. Making a comparison with the displacement of the element obtained from the experimental test, the ultimate displacement of which was 11 mm, it can be said that the results are almost identical which confirms veracity of the experimental test. At the same time the mode of failure is a classic one, with failure at shear force by generating a main stress level on the diagonal of the tested element. The results of the numerical analysis they are illustrated in the following images (Figure 4a,b) as well as in the graph in (Figure 4c).

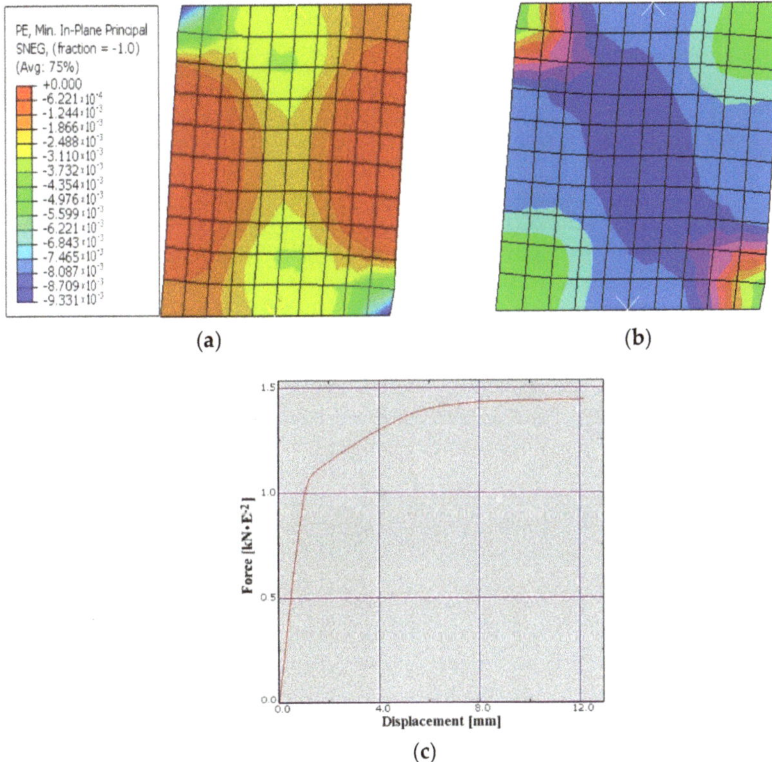

Figure 4. Compression plastic strain (**a**); strain components at integration point (**b**); graph of force–displacement to VHB (**c**).

4.3. Masonry Wall from Concrete Blocks with Wood Chips CBWC1

In order to numerically evaluate the walls made of concrete blocks with wood chips, a numerical model was made than can be replicate the mechanical characteristics of concrete blocks with wood chips. Thus, the model realized has the following dimensions: length = 150 [cm]; height = 150 [cm]; thickness = 25 [cm], being made from blocks with dimension of 250 mm × 250 mm × 500 mm. After a few of simulations performed on the configuration of the concrete blocks with wood chips it was decided to adopt the solution from Figure 5, the solution in which the wood chips block is full with concrete class C20/25 and reinforced with two steel bars of diameter Φ8.

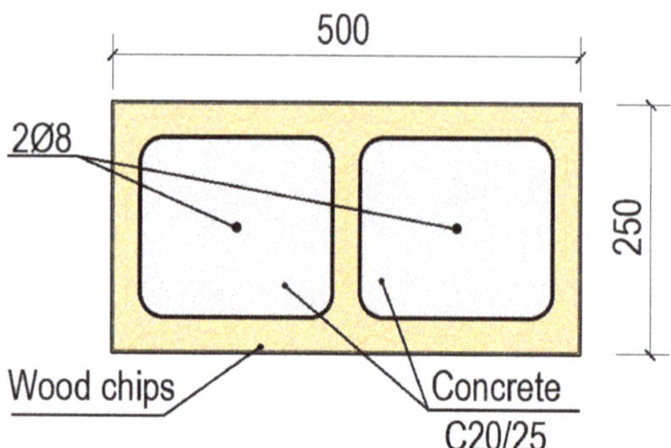

Figure 5. Configuration of block CBWC1.

The concrete block with wood chips consists of a percentage of 41.80% wood chips and 58.20% of concrete. The block used without the addition of concrete has a compressive strength of 2.5 N/mm^2, while the compressive strength of concrete is 20 N/mm^2, and in the calculation using an average strength of 12.69 [N/mm^2]. Moreover, in the analysis we used a plastic strain of 0.0035, a failure ratio R1 = 1.21; R2 = 1.08, and tension stiffening with displacement of 1.5. The modulus of elasticity obtained from the equivalence between the two materials was of E = 21,640 [N/mm^2] with a Poisson ratio of 0.25. The numerical simulation was performed applying a constant vertical force V = 200 kN and an increasing horizontal force until failure.

The results of the numerical analysis show that in the failure moment of the wall, the displacement at the top of the element was of 10.40 mm and the maximum horizontal force recorded was of H = 177.8 kN. In the following images there are illustrated the deformed element with the displacement at the top (Figure 6), the deformation mode as well as the area where the failure of the element is to occur (Figure 6b), as well as the force–displacement graph related to the numerical analysis performed (Figure 6c).

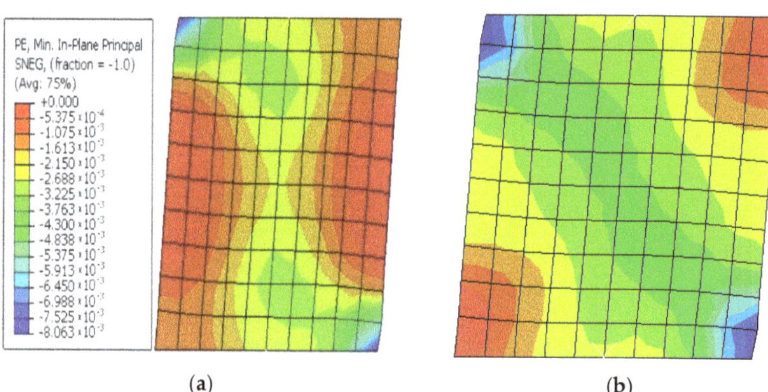

Figure 6. *Cont.*

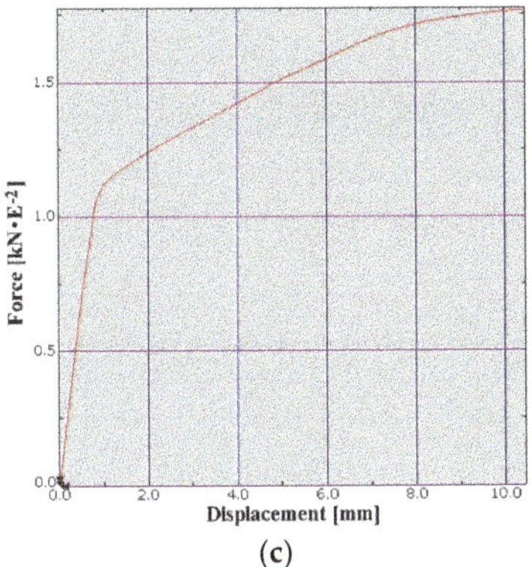

(c)

Figure 6. Compression plastic strain (**a**); strain components at integration point (**b**); graph of force–displacement to CBWC1 (**c**).

4.4. Masonry Wall from Concrete Blocks with Wood Chips CBWC2

The CBWC2 wall it consists of a block with wood chips with a different geometric configuration than the CBWC1 wall. Thus, the realized model has the following dimensions: length = height = 150 [cm]; thickness = 37.5 [cm], being made from blocks with dimension of 375 mm × 250 mm × 498 mm. After a few simulations performed on the configuration of the concrete blocks with wood chips it was decided to adopt the solution from Figure 7, the solution in which the block with wood chips it is partially filled with concrete, class C20/25, and reinforced with two steel bars of diameter Φ8 and thermal insulation.

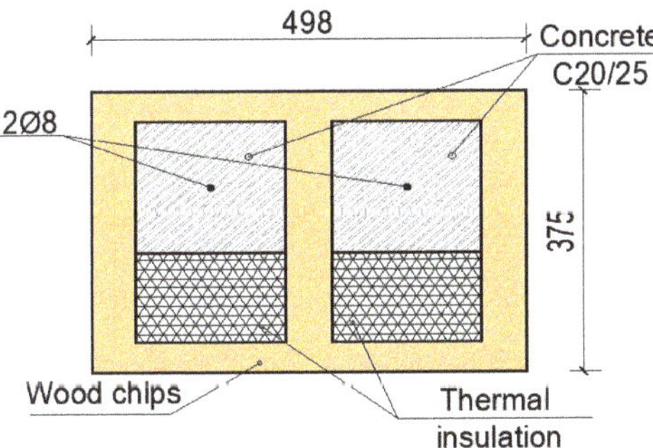

Figure 7. Configuration of block CBWC2.

The concrete block with wood chips consists of a percentage of 45.98% wood chips, 32.05% of concrete, and 21.98% thermal insulation. In the calculation using an average compressive strength of 7.63 [N/mm^2]. Moreover, in the analysis we used a plastic strain of 0.0035, a failure ratio R1 = 1.16; R2 = 1.08, and tension stiffening with displacement of 1.5. The modulus of elasticity obtained from the equivalence between the materials was of E = 14,215 N/mm^2 with a Poisson ratio of 0.25. The test procedure was the same as that of the CBWC1 wall. The results of the numerical analysis show that in the failure moment of the wall, the displacement at the top of the element was of 9.99 mm, and the maximum horizontal force recorded was of H = 165.2 kN. In the following images there are illustrated the deformed element with the displacement at the top (Figure 8), the deformation mode as well as the area where the failure of the element is to occur (Figure 8b), as well as the force–displacement graph related to the numerical analysis performed (Figure 8c).

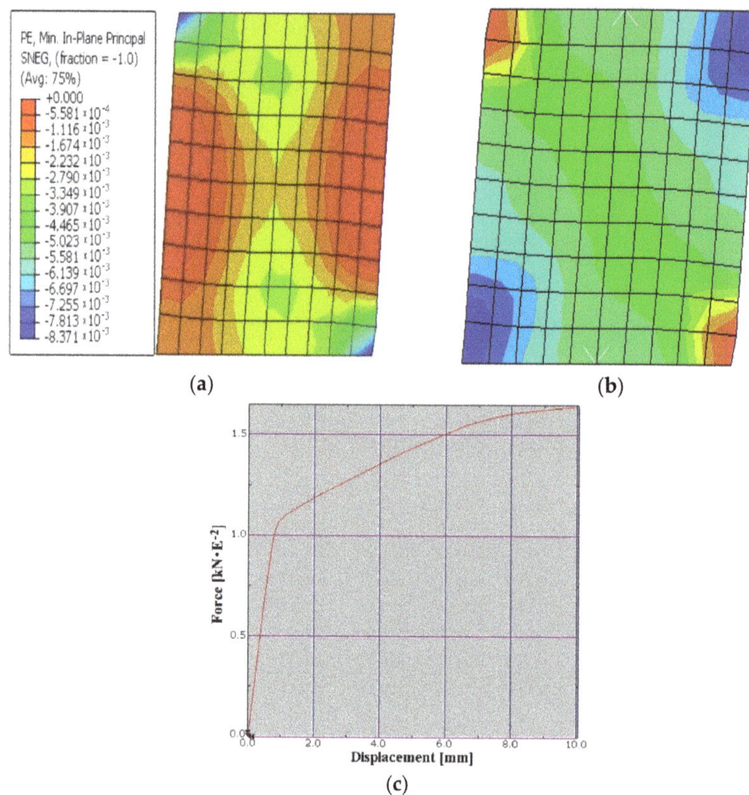

Figure 8. Compression plastic strain (**a**); strain components at integration point (**b**); graph of force–displacement to CBWC2 (**c**).

After obtaining the numerical results for the four types of blocks, a graph such as the one in Figure 9, was made to be able to make a comparison between solutions to highlight their performance from a structural point of view.

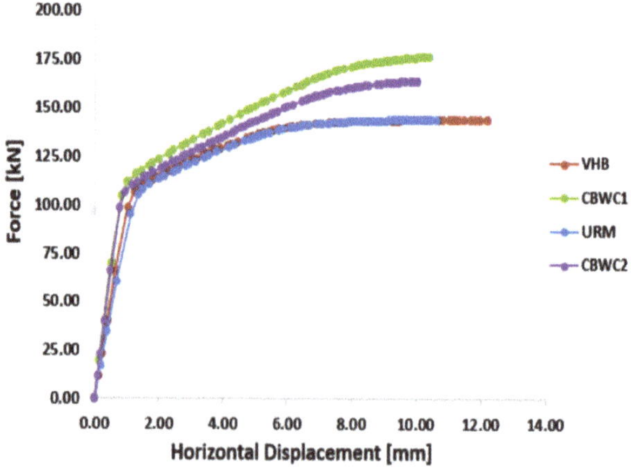

Figure 9. Graph of force–displacement for all four types of masonry blocks.

5. Conclusions

Following the theoretical study carried out on the walls made of concrete blocks with wood chips a series of aspects were observed that will be detailed in the following. Thus, performing the numerical analysis on the walls made up of the four types of masonry blocks the conclusions are as follows:

- The wall made of solid brick masonry and the wall made of bricks with vertical holes have a similar behavior, practically fail at almost the same horizontal forces, with the ultimate horizontal displacement being higher in the case of wall made of masonry brick with vertical holes.
- The solution of integral filling with concrete class C20/25, blocks with wood chips, obtained the best behavior between those four types of blocks. According to Figure 9, the wall made of blocks with wood chips integral filling with concrete, failed at a horizontal load of 177.8 kN, while the ultimate horizontal force recorded for the other three types of walls was between 144 kN and 165.2 kN.
- The force at which elements made of concrete blocks with wood chips failed was 23.5% higher than the force at which elements made of traditional brick failed; moreover, the ultimate horizontal displacement for elements made of concrete blocks with wood chips was 6.5% lower than for elements made of traditional brick.
- The solution of CBWC2 is previewed to be a beneficial solution, both mechanically and energetically, with this solution being in trend with the latest European directives on sustainability and energy building efficiency.

In conclusion, it is proposed to carry out an experimental program and to perform experimental tests on walls such as CBWC1 and CBWC2, made of blocks with wood chips filling with concrete class C20/25 and thermal insulation, for the validation of the results from the numerical analysis. We plan to perform additional experimental test on walls completely built with concrete blocks with wood chips in order to gain a better understanding of their physical characteristics under a seismic load. This will allow us to draw a wider range of conclusions on the behavior of concrete blocks with woodchips used in seismic areas.

Author Contributions: Formal analysis, S.P. and L.B.; Funding acquisition, R.C.; Investigation, L.B.; Methodology, S.P.; Project administration, S.P. and V.S.; Resources, R.C.; Validation, L.B. and V.S.; Visualization, V.S.; Writing—original draft, S.P.; Writing—review & editing, L.B. All authors have read and agreed to the published version of the manuscript.

Funding: The Polytechnic University of Timișoara in the frame of the Ph.D. program of studies.

Institutional Review Board Statement: Not applicable.

Informed Consent Statement: Not applicable.

Data Availability Statement: The data that support the findings of this study are available from the corresponding author upon reasonable request.

Acknowledgments: The paper was made in the frame of the Ph.D. program of studies.

Conflicts of Interest: The authors declare no conflict of interest.

References

1. Carvalho, A.; de Castro Xavier, G.; Alexandre, J.; Pedroti, L.G.; de Azevedo, A.R.G.; Vieira, C.M.F.; Monteiro, S.N. Environmental Durability of Soil-Cement Block Incorporated with Ornamental Stone Waste. *MSF* **2014**, *798–799*, 548–553. [CrossRef]
2. Awoyera, P.O.; Olalusi, O.B.; Ibia, S.; Prakash A., K. Water absorption, strength and microscale properties of interlocking concrete blocks made with plastic fibre and ceramic aggregates. *Case Stud. Constr. Mater.* **2021**, *15*, e00677. [CrossRef]
3. Luo, W.; Liu, S.; Jiang, Y.; Guan, X.; Hu, Y.; Hu, D.; Li, B. Utilisation of dewatered extracted soil in concrete blocks produced with Portland cement or alkali-activated slag: Engineering properties and sustainability. *Case Stud. Constr. Mater.* **2021**, *15*, e00760. [CrossRef]
4. Citto, C. *Two-Dimensional Interface Model Applied to Masonry Structures*; University of Colorado at Boulder: Boulder, CO, USA, 2008; pp. 23–34.

5. Kumar, N.; Amirtham, R.; Pandey, M. Plasticity based approach for failure modelling of unreinforced masonry. *Eng. Struct.* **2014**, *80*, 40–52. [CrossRef]

6. Kumar, N.; Barbato, M. New constitutive model for interface elements in finite element modeling of masonry. *J. Eng. Mech.* **2019**, *145*, 04019022. [CrossRef]

7. Giambanco, G.; Rizzo, S.; Spallino, R. Numerical analysis of masonry structures via interface models. *Comput. Methods Appl. Mech. Eng.* **2001**, *190*, 6493–6511. [CrossRef]

8. Macorini, L.; Izzuddin, B. A non-linear interface element for 3D mesoscale analysis of brick-masonry structures. *Int. J. Numer. Meth. Eng.* **2011**, *85*, 1584–1608. [CrossRef]

9. Milani, G. Simple homogenization model for the non-linear analysis of in-plane loaded masonry walls. *Comput. Struct.* **2011**, *89*, 1586–1601. [CrossRef]

10. Nazir, S.; Dhanasekar, M. A non-linear interface element model for thin layer high adhesive mortared masonry. *Comput. Struct.* **2014**, *144*, 23–39. [CrossRef]

11. Lagomarsino, S. Seismic assessment of rocking masonry structures. *Bull. Earthq. Eng.* **2015**, *13*, 97–128. [CrossRef]

12. Lourenço, P.B.; Mendes, N.; Ramos, L.F.; Oliveira, D.V. Analysis of masonry structures without box behavior. *Int. J. Archit. Herit.* **2011**, *5*, 369–382. [CrossRef]

13. Sorrentino, L.; Cattari, S.; Da Porto, F.; Magenes, G.; Penna, A. Seismic behaviour of ordinary masonry buildings during the 2016 central Italy earthquakes. *Bull. Earthq. Eng.* **2019**, *17*, 5583–5607. [CrossRef]

14. Clementi, F.; Gazzani, V.; Poiani, M.; Mezzapelle, P.A.; Lenci, S. Seismic assessment of a monumental building through nonlinear analyses of a 3D solid model. *J. Earthq. Eng.* **2018**, *22*, 35–61. [CrossRef]

15. Fonti, R.; Borri, A.; Barthel, R.; Candela, M.; Formisano, A. Rubble masonry response under cyclic actions: Experimental tests and theoretical models. *Int. J. Mason. Res.* **2017**, *2*, 30–60. [CrossRef]

16. Valente, M.; Milani, G. Effects of Geometrical Features on the Seismic Response of Historical Masonry Towers. *J. Earthq. Eng.* **2017**, *22*, 1–33. [CrossRef]

17. Pelletier, N.; Allacker, K.; Pant, R.; Manfredi, S. The European Commission Organisation Environmental Footprint method: Comparison with other methods, and rationales for key requirements. *Int. J. Life Cycle Assess.* **2014**, *19*, 387–404. [CrossRef]

18. Eckert, E.; Kovalevska, O. Sustainability in the European Union: Analyzing the Discourse of the European Green Deal. *J. Risk Financial Manag.* **2021**, *14*, 80. [CrossRef]

19. Dell'Anna, F. Green jobs and energy efficiency as strategies for economic growth and the reduction of environmental impacts. *Energy Policy* **2021**, *149*, 112031. [CrossRef]

20. Cheriyan, D.; Choi, J. A review of research on particulate matter pollution in the construction industry. *J. Clean. Prod.* **2020**, *254*, 120077. [CrossRef]

21. Khamraev, K.; Cheriyan, D.; Choi, J. A review on health risk assessment of PM in the construction industry—Current situation and future directions. *Sci. Total Environ.* **2021**, *758*, 143716. [CrossRef]

22. Sfakianaki, E. Resource-efficient construction: Rethinking construction towards sustainability. *World J. Sci. Technol. Sustain. Dev.* **2015**, *12*, 233–242. [CrossRef]

23. Bonoli, A.; Zanni, S.; Serrano-Bernardo, F. Sustainability in Building and Construction within the Framework of Circular Cities and European New Green Deal. *Contrib. Concr. Recycling. Sustain.* **2021**, *13*, 2139. [CrossRef]

24. Larsen, V.G.; Tollin, N.; Sattrup, P.A.; Birkved, M.; Holmboe, T. What are the challenges in assessing circular economy for the built environment? A literature review on integrating LCA, LCC and S-LCA in life cycle sustainability assessment, LCSA. *J. Build. Eng.* **2022**, *50*, 104203. [CrossRef]

25. Billimoria, S.; Guccione, L.; Henchen, M.; Prescott, L.L. The economics of electrifying buildings: How electric space and water heating supports decarbonization of residential buildings. In *World Scientific Encyclopedia of Climate Change, Chapter 33*; World Scientific: Singapore, 2021; pp. 297–304. [CrossRef]

26. Kern, C.; Jess, A. Reducing Global Greenhouse Gas Emissions to Meet Climate Targets—A Comprehensive Quantification and Reasonable Options. *Energies* **2021**, *14*, 5260. [CrossRef]

27. Plopeanu, E.F. Experimental Research on the Capture and Recovery of Fine-Grained Oxide Waste Generated in the Ceramic and Refractory Materials Industry. Ph.D. Thesis, University Politehnica of Bucharest, Bucharest, Romania, 2020.

28. Oliveira, R.G.; Rodrigues, J.P.C.; Pereira, J.M.; Lourenço, P.B.; Lopes, R.F.R. Experimental and numerical analysis on the structural fire behaviour of three-cell hollowed concrete masonry walls. *Eng. Struct.* **2021**, *228*, 111439. [CrossRef]

29. Corradi, M.; Schino, A.D.; Borri, A.; Rufini, R. A review of the use of stainless steel for masonry repair and reinforcement. *Constr. Build. Mater.* **2018**, *181*, 335–346. [CrossRef]

30. Lima, L.; Trindade, E.; Alencar, L.; Alencar, M.; Silva, L. Sustainability in the construction industry: A systematic review of the literature. *J. Clean. Prod.* **2021**, *289*. [CrossRef]

31. Anh, L.D.H.; Pásztory, Z. An overview of factors influencing thermal conductivity of building insulation materials. *J. Build. Eng.* **2021**, *44*, 125730. [CrossRef]

32. Scotta, R.; Vitaliani, R.B. *Concrete Walls Poured into Mineralized Wood Formwork Blocks*, 1st ed.; University of Padova: Padova, Italy, 2009; pp. 41–131.

33. Budău, L.; Pescari, S.; Ciubotaru, R.; Stoian, V. Research on consolidation of masonry structural walls, experimental tests. In Proceedings of the National Technical-Scientific Conference—The 20th Edition—Modern Technologies for the 3rd Millenium, Oradea, Romania, 9 December 2021.
34. Partene, E.E. Study of the Behaviour of Masonry Walls at Seismic Loads Strengthening Masonry Walls with Polymer Materials. Ph.D. Thesis, Polytechnic University of Timisoara, Timisoara, Romania, 2018.

MDPI

St. Alban-Anlage 66

4052 Basel

Switzerland

www.mdpi.com

Materials Editorial Office

E-mail: materials@mdpi.com

www.mdpi.com/journal/materials